Soil Physics

Designed for undergraduate and graduate students interested in learning basic soil physics and its application to the environment, soil health, water quality, and productivity, this book provides readers with a clear coverage of the basic principles of water and solute transport through the vadose zone, the theory behind the transport, and step-by-step guidance on how to use current computer models in the public domain along with soil erosion and contaminant remediation. Students will develop a deeper understanding of the fundamental processes within the soil profile that control water infiltration, redistribution, evapotranspiration, drainage, and erosion. The updated second edition features one new chapter, highlighting new problems, new computer models, and remediation.

Features

- Serves as the most up-to-date textbook on soil physics available.
- Includes one new chapter with three new topics and many new numerical examples.
- Offers mathematical descriptions supported by simplified explanations.
- Provides case studies and step-by-step guidance on how to use public-domain computer models.
- Covers all principles and processes in an easy-to-understand format with numerous illustrations and sample problems.

Students studying in the fields of soil science, environment science, natural resources, agriculture engineering, civil engineering, environmental engineering, range sciences, horticulture, crop sciences, and forestry will find this book provides a solid foundation for their studies. Professionals, researchers, academicians, and companies working in fields related to environmental science, soil physics, hydrology, and irrigation will find this book a great reference tool as it is the most up-to-date in its field.

Soil Physics
An Introduction

Second Edition

Manoj K. Shukla

CRC Press
Taylor & Francis Group
Boca Raton London New York

CRC Press is an imprint of the
Taylor & Francis Group, an **informa** business

Second edition published 2023
by CRC Press
6000 Broken Sound Parkway NW, Suite 300, Boca Raton, FL 33487-2742

and by CRC Press
4 Park Square, Milton Park, Abingdon, Oxon, OX14 4RN

CRC Press is an imprint of Taylor & Francis Group, LLC

First edition published by CRC Press 2013

Library of Congress Cataloging-in-Publication Data

Names: Shukla, Manoj, author.
Title: Soil physics : an introduction / Manoj K. Shukla.
Description: Second edition. | Boca Raton, FL : CRC Press, 2023. | Includes bibliographical references and index. | Summary: "For undergraduate and graduate students interested in learning basic soil physics and its application to environment, soil health, water quality and productivity, this book provides readers with the basic principles of water and solute transport through vadose zone, the theory behind transport and step-by-step guidance on how to use current computer models in the public domain. Students will develop a deeper understanding of the fundamental processes within the soil profile that control water infiltration, redistribution, evapotranspiration, drainage, and erosion. The second edition features one new chapter, highlights new problems, new computer models, and remediation"— Provided by publisher.
Identifiers: LCCN 2022057772 (print) | LCCN 2022057773 (ebook) | ISBN 9780367210168 (hbk) | ISBN 9781032501215 (pbk) | ISBN 9780429264849 (ebk)
Subjects: LCSH: Soil physics. | Soil moisture.
Classification: LCC S592.3 .S58 2023 (print) | LCC S592.3 (ebook) | DDC 631.4/3—dc23/eng/20221229
LC record available at https://lccn.loc.gov/2022057772
LC ebook record available at https://lccn.loc.gov/2022057773

ISBN: 978-0-367-21016-8 (hbk)
ISBN: 978-1-032-50121-5 (pbk)
ISBN: 978-0-429-26484-9 (ebk)

DOI: 10.1201/9780429264849

Typeset in Times
by Deanta Global Publishing Services, Chennai, India

Contents

Preface

Soil physics combines the disciplines of soil and physics. Soil is the interface between the lithosphere and the atmosphere, and is the most basic resource that interacts with the biosphere and the hydrosphere to support life on earth. Physics is the science dealing with everything in physical existence and their interactions. The understanding of the complex interactions between soil, water, agriculture, and land use and management is critically important for sustaining soil and water resources and increasing agriculture production. However, to understand these complex processes, knowledge of fundamental soil properties and processes is a prerequisite. Soil physics deals with the study of physical properties and the processes of soil occurring within the vadose zone, which is the portion of the soil profile located between the atmosphere–soil interface and soil–groundwater table interface. Soil physics studies focus on the measurement and modeling of complex interactions among various physical processes occurring due to natural and anthropogenic perturbations.

This book, *Soil Physics: An Introduction*, Second Edition, is designed for undergraduate and graduate students interested in learning basic soil physics principles and their application to soil, water, and environment quality. The content of the book meets the curriculum needs of students of soil science, environmental science, natural resources, agricultural engineering, civil engineering, environmental engineering, range science, horticulture and crop sciences, and forestry. This book will help students gain an understanding of important physical properties of soil, critical physical processes involving energy and mass transport, movement and retention of water and solute through the soil profile, soil erosion, contaminant remediation, soil temperature regimes and aeration, and plant–water relations.

Since the focus of the discipline of soil physics has changed significantly over the past several decades and mathematical descriptions of water and solute transport have become an integral part of the soil physics curriculum, every attempt is made to present concepts mathematically, provide a simple explanation, and use numerical examples for a better understanding of the fundamental principles of soil physics. This book aims to provide undergraduate and graduate students with a clear coverage of the basic principles of water and solute retention and transport through the vadose zone, the theory behind the transport, and introductory guidance on the use of some of the existing public-domain computer models. Students are expected to develop an improved understanding of the fundamental processes within the soil profile that control water infiltration, redistribution, evapotranspiration, and drainage; understand impacts of anthropogenic contaminants on soil and groundwater quality, soil erosion and contamination, and remediation techniques; and know the effects of land use on flow and transport.

This book consists of 18 chapters, starting with an introduction to soil physics (Chapter 1). The chapter describes the importance of soil physics for a graduate curriculum, how soil physics principles and processes are applied to understand soil and environmental quality, and interactions of soil physics with other disciplines. The chapter also touches upon the projected drastic changes in soil moisture, temperature regimes, and the growing season, provided the forecast models on climate-change prediction are accurate.

Chapter 2 discusses units and dimensions, nondimensional quantities, systems of units, use of units and dimensions, and unit conversion. It is important for students to learn about different systems of units so that it is easier for them to convert quantities from one unit system to another. Chapter 3 focuses on the characteristics of soils of the vadose zone and describes soil formation, soil classification, soil profile, and soil properties, including structure, texture, particle shape, color, and density. As physical properties of soil are usually determined by performing soil sampling, Chapter 4 addresses the practical aspects of sampling and the need for collecting background information prior to sampling. It also provides guidelines for collecting good datasets, sample size, and sampling designs. Soil properties are known to be spatially and temporally variable for many years and display both short- and long-range variability, and Chapter 5 addresses the sources and scales of variability and statistical and geostatistical analysis of variability.

The fundamentals of hydrology are presented in Chapter 6 and the properties of water in Chapter 7. Soil water content, soil water potential, and their relationships are covered in Chapter 8, which lays the foundation for understanding soil water movement through saturated and unsaturated porous media, which is discussed in Chapter 9. Infiltration processes, measurement techniques, management of infiltration, and existing infiltration models are detailed in Chapter 10. Chapter 11 discusses water and energy transport and its influence on atmospheric energy balance, soil water dynamics, and soil temperature; while Chapter 12 presents soil water redistribution due to evaporation, water uptake by roots, and evapotranspiration.

The microscopic and macroscopic approaches associated with root water uptake are presented in Chapter 13. Chapters 14 and 15 cover the theory of airflow and solute transport through the vadose zone, respectively. Chapters 16 and 17 are for students to apply the soil physics principles and processes using public domain programs. Chapter 16 presents a step-by-step guide on the use of HYDRUS-1D model, and Chapter 17 on the Root Zone Water Quality Model (RZWQM), for simulating soil water dynamics through the unsaturated soil profile. Concepts related to solute transport, hydrology, temperature, and energy fluxes through soil are also presented in these chapters, encouraging students to apply these models to their studies related to water and energy transport through unsaturated soil with relative ease. Soil erosion, contaminants and their remediation, and soil salinity and sodicity and their influence on plants are included in Chapter 18 as special topics.

Every effort is made to present concepts mathematically and in a simplified manner. Solved numerical problems using practical examples have also been included with many mathematical formulations to provide a feel for the properties and processes through the vadose zone. These examples should enhance student learning and understanding of the concepts. I have extensively borrowed material and data from various authors, books, journals, a graduate curriculum, and other data sources, and have made sincere attempts to acknowledge the sources of the material. I am sure there are several classical papers that may not have been acknowledged in the book. Chapters 13, 16, and 17 include information provided by Dr. Sanjit Deb. The examples in Chapters 16 and 17 are from the research work carried out by the Soil Physics Research Group and the papers published by the group, with Dr. Deb as the lead author. Material from the lecture notes provided by Dr. Pete Wierenga, emeritus professor of soil physics at New Mexico State University (NMSU), is also incorporated in some chapters of this book, including Chapter 12. Looking at the course requirements of an environmental science program, Chapter 18 discusses soil erosion, soil salinity, contaminants found in soil naturally and those added due to anthropogenic activities, and their remediation techniques.

I would like to thank CRC Press for its efforts in publishing this book. Special thanks are due to Irma Britton for all her help in a very professional manner. I would also like to thank Chelsea Reeves, Vijay Bose at Deanta, Christine Selvan at Deanta, and Richard Tressider at CRC Press for language editing of all the chapters. Thanks to Dr. Laj Ahuja and Dr. Jiri Simunek for the permission to use screenshots of RZWQM and HYDRUS, respectively, and to Sharon Debach for useful suggestions on Chapter 13. I would like to extend my sincere thanks to the NMSU Agricultural Experiment Station, and graduate students and postdocs of the Soil Physics Group at NMSU. Last but not least, I would like to thank my parents Ravi and Shobha Shukla; my wife, Neeta; my daughter-in-law Beverly; and my sons, Utkarsh and Amogh, for their constant support and encouragement.

Author

Manoj K. Shukla, PhD, is a professor of environmental soil physics at New Mexico State University, Las Cruces, New Mexico. He holds the Nakayama Research Excellence Professorship. At NMSU, he also serves as director of Global Programs of the College of Agricultural, Consumer and Environmental Sciences. He has held faculty positions at various universities in India and has served as a research scientist at the University of Agricultural Sciences, Vienna; as a postdoctoral research associate at the University of Illinois; and as a research scientist at The Ohio State University. He earned his PhD in soil physics from the University of Agricultural Sciences, Vienna, Austria, in a collaborative program with the University of California, Davis; his Master's degree in soil and water engineering from the Water and Land Management Institute, India; and his Bachelor's degree in agricultural engineering from the National Institute of Hydrology, India. Dr. Shukla is a recipient of the Patricia Christmore Faculty Teaching Award, the North American Colleges and Teachers of Agriculture Teaching Award of Merit, RGBI Integrated Program Team Award, and an Award of Excellence for outstanding contribution to Western Regional Multistate Research. He also served as an executive board member for the International Arid Lands Consortium and US-Mexico Border Health Commission expert panel member. He is a member of the Association of Public Land-Grant Universities – Commission on International Initiatives and a visiting professor to China Agricultural University. Dr. Shukla is a Fellow of Soil Science Society of America.

Dr. Shukla serves as the technical editor of the Soil Physics and Hydrology section of *Soil Science Society of America Journal*. He also serves as a consulting editor of *Vadose Zone Journal*. He has served as the book review editor for *Vadose Zone Journal* and editorial board member of the *NACTA Journal* (North American Colleges and Teachers of Agriculture). He was the chair of the mentoring committee of Soil Physics and Hydrology division in 2022, and a member of L.R. Ahuja systems modeling award committee of the Soil Science Society of America and member of the unsaturated zone technical committee of the American Geophysical Union. He has previously served as a secretary of the western regional coordination committee and as a presiding officer and convener at several international meetings. He is a member of many professional societies, including the American Geophysical Union (AGU), Soil Science Society of America (SSSA), and American Society of Agricultural and Biological Engineers (ASABE). Dr. Shukla regularly teaches environmental soil physics and advanced soil physics, as well as on occasion, contaminant transport modeling, spatial variability analysis, and introductory soils. He has helped more than 35 students graduate with a Master's or a PhD as chair of their committees, and many more as a committee member. His research interests are related to soil physics, spatial and temporal variability of soil properties, water use efficiency, and water and solute transport under different land use and management systems at multiple scales. Recently, he worked on reuse of brackish and produced waters for ensuring food security and human sustainability. He has received more than $25 million in research funding from various agencies. He coauthored the textbook *Principles of Soil Physics*; edited the book *Soil Hydrology, Land Use and Management: Measurement and Modeling*; and has coauthored 156 peer-reviewed journal articles and several research reports, book chapters, and proceedings papers.

1 Introduction to Soil Physics

Soil physics is the study of the physical properties and processes occurring within the vadose zone at various spatial scales. The vadose zone is the portion of the soil profile located between the soil–groundwater table interface and the atmosphere–soil interface. Soil physics focuses on complex interactions among various physical processes occurring due to natural and anthropogenic perturbations (Shukla, 2011). Soil physics is also known as soil hydrology or vadose zone hydrology. The term *hydrology* is derived from the Greek words *hydor* and *logos*, meaning "water" and "study," respectively. Hence, soil hydrology can further be explained as a branch of science pertaining to the study of water including the study of occurrence, movement, and changes in quantity, quality, and state of atmospheric, surface, vadose zone, plant-borne, and subsurface water on earth.

Water occupies nearly 71% of the earth's total surface and is mostly found in oceans (~97%). About 3% of the total water is considered freshwater, out of which almost 68% is found in glaciers and icecaps and about 30% as groundwater. Water has some unique properties: it is highly dynamic, is available in all three forms (solid, liquid, and gas) at room temperature, and has a positive as well as a negative charge (i.e., dipole). The interaction of soil with various components of hydrology is important for the hydrologic cycle. The hydrologic cycle is the movement of water through the environment. It is the circulation of water between the atmosphere, water bodies, and soil. Water moves into the atmosphere as evaporation from water bodies, plants, and soil; condenses and falls back on earth; moves through plants, soils, and rivers and drains into lakes and oceans; and goes back again into the atmosphere due to evaporation, completing the cycle. The movement of water takes place between the lithosphere (solid inorganic portion of earth), hydrosphere (water in various storages on earth), biosphere (portion of earth where all living things, including plants, are found), and atmosphere (gaseous envelope surrounding the earth). The sun, or solar energy, is the driving force for the hydrologic cycle that describes the continuous movement of liquid, vapor, and/or solid (ice) water in the atmosphere, on soil and plant surfaces, and through soil and plant systems.

Despite the dynamic nature of various components of water, the total mass of water on earth stays fairly constant over time. However, the resident time (time for which water stays in each storage) for water in different states varies from a few days (water vapor) to several thousand years (deep groundwater and ice on poles) (Table 1.1).

Soil, the interface between the lithosphere and the atmosphere, is the most basic resource that interacts with the biosphere and the hydrosphere to support life on earth. Soil hydrology has a strong influence on the water uptake and release by plants during photosynthesis. Thus, soil hydrology can also be defined as the study pertaining to agricultural water management. Soil hydrology considers all the components

DOI: 10.1201/9780429264849-1

TABLE 1.1
Average Residence Time of Water in Different Stores

Storage	Residence Time (Years)
Atmosphere	9 days
Soil moisture	1–2 months
Snow cover (mostly seasonal)	2–6 months
Rivers	2–6 months
Glaciers	20–100
Lakes	50–100
Shallow aquifers (unconfined)	100–200
Oceans	3,200
Deep aquifers (may be confined)	10,000
Antarctica	20,000

Sources: PhysicalGeography.net, Chapter 8: Introduction to the hydrosphere, retrieved on October 6, 2011; Shukla, M. K., Introduction to soil hydrology, in Shukla, M. K., ed., *Soil Hydrology, Land Use and Agriculture: Measurement and Modeling*, CAB International, Wallingford, UK, pp. 1–23, 2011.

of water related to irrigation and drainage, percolation and recharge to groundwater, capillary rise, root and plant water uptake and release, evaporation from soil and plants, and transpiration (Deb and Shukla, 2011). The fate of the total amount of water applied to the soil as irrigation or natural rainfall is usually determined by studying these components and conducting water balance.

1.1 IMPORTANCE OF SOIL PHYSICS

Soil physics or environmental soil physics can be defined as a systematic study of the physical properties taking place within the vadose zone or critical zone. It includes measurement, assessment, prediction, as well as modeling of vadose zone properties and processes for a variety of ecosystems and for land use and management systems in relation to the valuable functions essential for the sustenance of life on earth.

Important soil processes occurring at the pore scale in the vadose zone include infiltration, drainage, water redistribution, evaporation, transpiration through plants, and deep percolation (Figure 1.1). These soil processes are governed by the amount, orientation, size, distribution, and connectivity of the complex network of irregular, interconnected, tortuous flow channels. It is important to note here that for water movement within the vadose zone, pore connectivity is more important than the total number of micro- or macropores. The geometry of pores also has a profound influence on water retention in the vadose zone.

Soil pores are formed due to the natural arrangement of mineral particles into different shapes; occurrence of swelling and shrinking of soils; development of meniscus pressures; onset of physical and chemical reactions; biological activities, including root growth, migration, and decay; earthworm and other soilborne animal

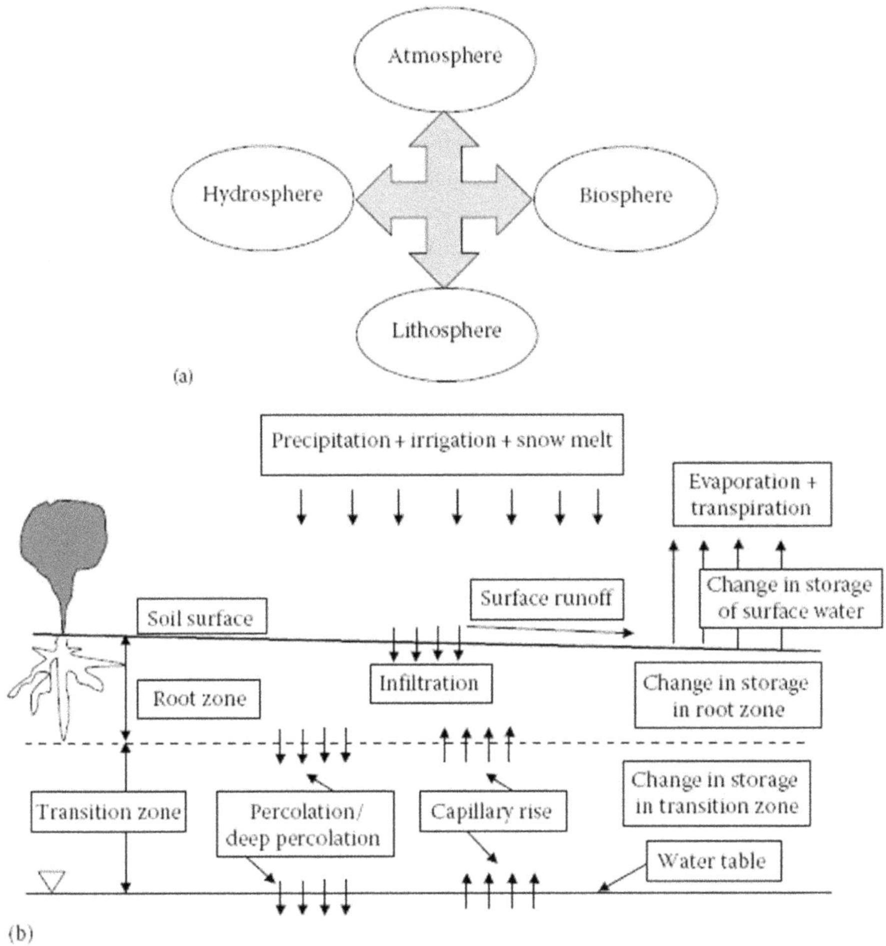

FIGURE 1.1 (a) Hydrologic cycle and (b) components of water balance on soil surface, root zone, and transition zone.

activities; and formation, breakdown, and reformation of micro- or macroaggregates due to land use and management. Soil water storage and transport can also be influenced by the rate and amount of rainfall, soil types, types of vegetation, and the geology and chemistry of the vadose zone. Soil processes are important components of a water budget and directly influence plant growth and sustenance.

Soil properties manifest variability in time and space at multiple scales (short and long range) (Nielsen et al., 1973; Scott et al., 1994; Nielsen and Wendroth, 2003) and are strongly influenced by intrinsic factors of soil formation as well as extrinsic factors associated with land use and management (van Es, 2002). Understanding the variability of soil properties and processes, their complex interactions, and interactions with plants and the environment is important for designing strategies for efficient resource management.

Soil physics addresses practical problems encountered by growers and researchers in real-life situations and can play a pivotal role in man's endeavor to sustain agricultural productivity while maintaining soil, water, and environmental quality. With growing world population and demand for food and fiber, the role of soil physics, together with other disciplines, is becoming critical for sustaining and increasing total grain production and maintaining soil, water, and environmental quality.

1.2 INTERACTIONS OF SOIL PHYSICS WITH OTHER DISCIPLINES

Soil physics is an interdisciplinary as well as transdisciplinary field that interacts closely with several other disciplines (Figure 1.2; Lal and Shukla, 2004; Shukla, 2011). The entry, retention, and movement of water within the vadose zone and its loss via evaporation and transpiration encourage closer cooperation of soil physics with the disciplines of hydrology, soil science, agricultural engineering, horticulture, and plant sciences. Horticulture is defined as the science that deals with plant cultivation. Energy transport during the evapotranspiration process involves fundamentals of physics and climatology. Physics is defined as the science that deals with matter and its spatial and temporal motion. Climatology is the branch of science that deals with the study of climate. The physical properties of soil

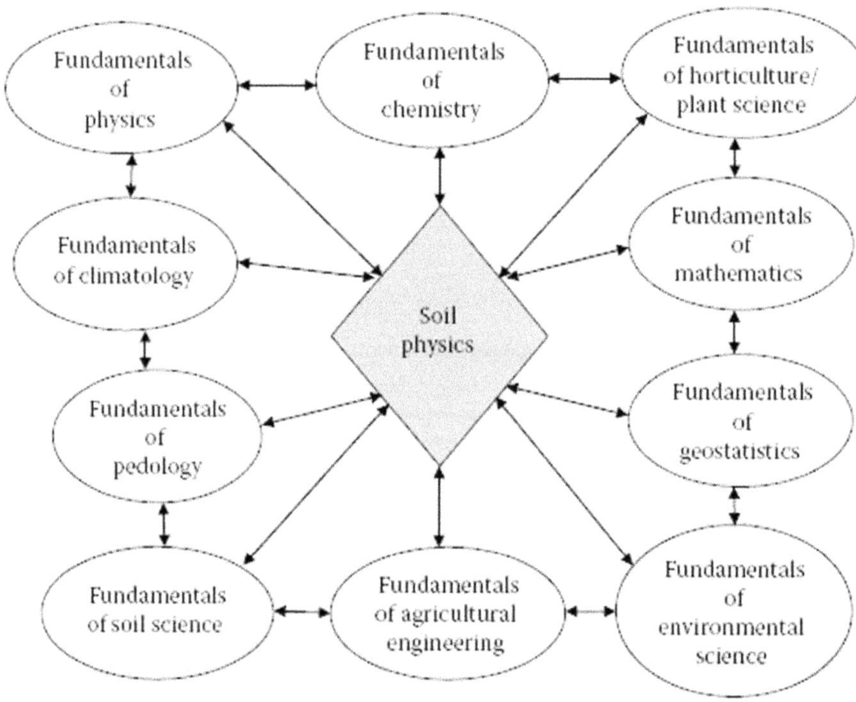

FIGURE 1.2 Interactions of soil physics with other disciplines. *Note:* Several disciplines included in this figure can also interact directly with each other.

are strongly influenced by the natural process of soil development or pedogenesis that takes place at a large timescale, necessitating interactions with the discipline of pedology. Pedology is defined as the study of soils in nature and includes the study of morphology, genesis, and classification of soils. Morphology deals with the observable attributes in soil horizons, for example, soil structure, color, root, and pore distributions. Pedogenesis is the process by which soil is created, and soil classification is the categorization of soils into different groups based on clearly distinguishable criteria. The principles and theories of soil physics involve complex mathematical modeling derived from the field of mathematics. The modeling efforts cover water as well as solute transport or migration through the vadose zone toward the groundwater, and this interacts with the fields of environmental science and groundwater hydrology. The knowledge of solute transport within the vadose zone also requires an in-depth understanding of the complex interactions between soil solids and solutes in water. Thus, soil physics closely interacts with soil chemistry as well. Soil properties and processes show a large degree of spatial and temporal variability and thus utilize the principles of geostatistics for their assessments at scales greater than point scale. As an example, the temporal variability of unsaturated hydraulic conductivity and soil moisture content is presented in Figure 1.3.

1.3 SOIL PHYSICS, SOIL AND ENVIRONMENTAL QUALITY, AND QUALITY OF LIFE

Understanding the principles and processes of soil physics is important for understanding the migration of substances, dust particles, gases, solutes, nutrients, and pollutants through the soil, water, and atmosphere (Figure 1.4). The migration of these substances can influence soil, water, and air quality. Soil quality or soil health, in a simplistic way, can be defined as the capacity of soil to perform specific functions (SSSA, 1987). It cannot be directly measured in a field but can be expressed indirectly based on its physical, chemical, or biological properties, either individually or by grouping some soil types in different categories (Lal, 1998; Shukla et al., 2004). The goal of sustainable agriculture is to maintain yields under a given amount of inputs or increase yields without increasing the costs associated with production. Good soil quality, therefore, could be a prerequisite for maintaining productivity or for pursuing sustainable agriculture.

There are several approaches to evaluate soil quality and the sustainability of an agricultural production system. Quantitative assessment of the sustainability of a production management system was evaluated using several minimum datasets (Larson and Pierce, 1991; Doran and Parkins, 1994). Other tools used to evaluate soil quality and sustainability include linear (single or multiple) regression analysis (Li and Lindstorm, 2001; Ikemura and Shukla, 2009), pedotransfer functions (Bouma, 1989), and principal component analysis (Brejda et al., 2000; Shukla et al., 2006). However, more emphasis is needed on ecosystem services provided by soils, identifying suites of indicators to determine soil functionality, integration of various

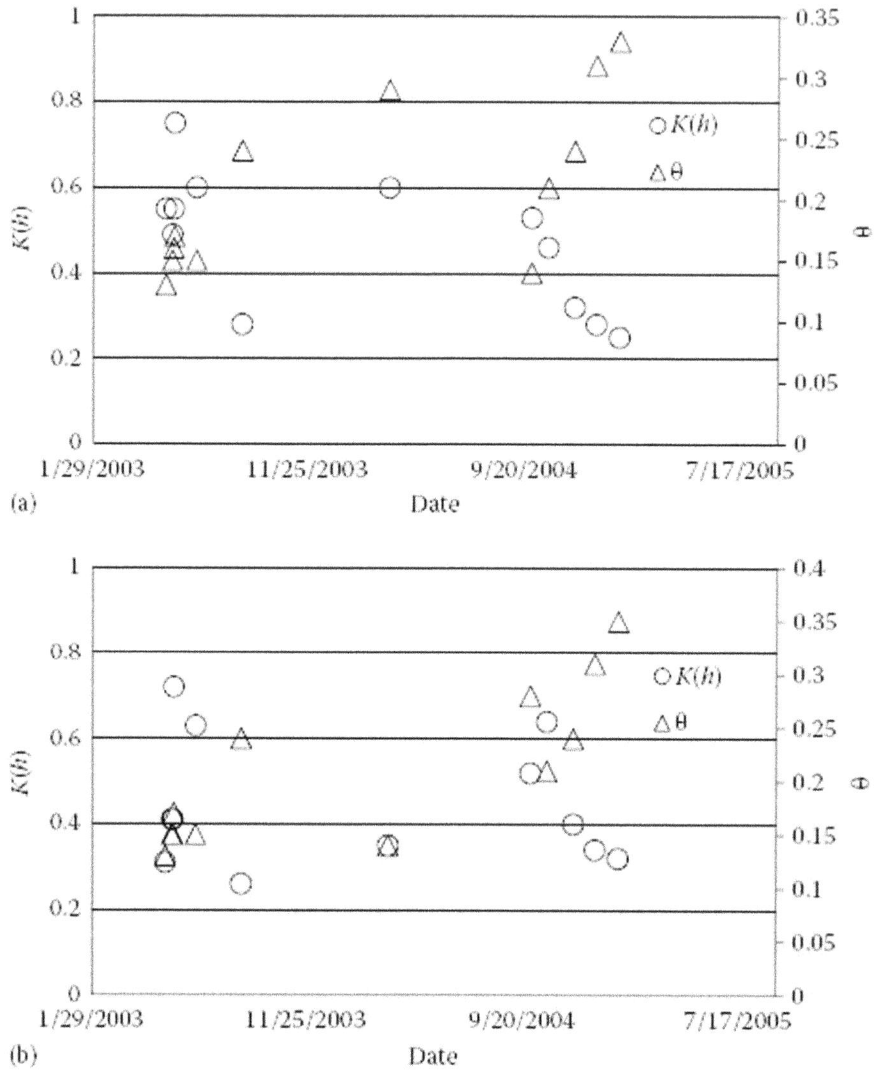

FIGURE 1.3 The temporal variability of mean hydraulic conductivity [$K(h)$] and antecedent soil moisture content (θ) at (a) −0.20 m and (b) −0.15 m matric potential for a field located near College Station, Texas. (Redrawn from Dasgupta, S., et al., *Soil Sci. Soc. Am. J.*, 70, 1872, 2006.)

measurable properties into informative soil-health indices, and appreciation of spatial and timescale variability of the indicators.

Soil properties influencing agricultural productivity can be static (not changing in small timescales, e.g., texture) or dynamic (changing at short timescales, e.g., soil organic carbon). Physical, chemical, and biological properties of soil are important

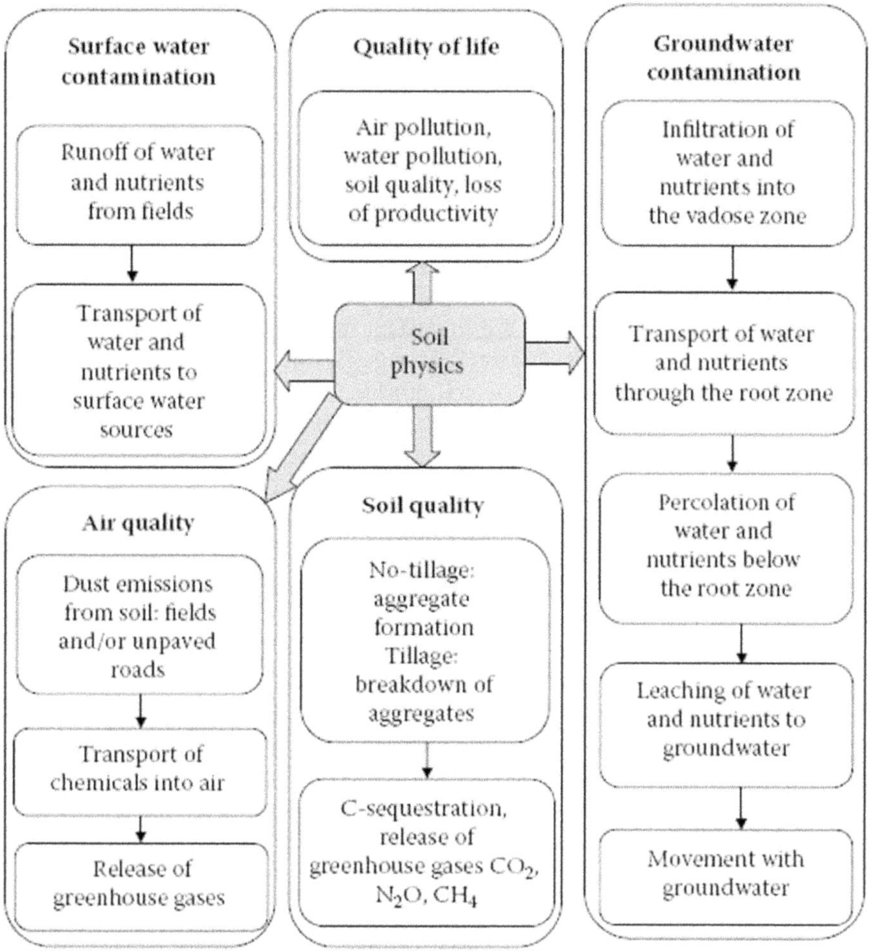

FIGURE 1.4 Influence of soil hydrology on groundwater, surface water, air and soil quality, and overall quality of life on earth.

components of the local land use and management strategies applied to agricultural fields. For example, the amount, intensity, and frequency of tillage are important aspects in production management, as they cause significant changes in soil quality. Frequent and deep tillage can create drastic disturbances in soil, resulting in the breakdown of aggregates, disruption of aggregate formation processes, and atmospheric exposure of the binding agents, such as biomass, which facilitates the attendant release of greenhouse gases into the atmosphere. Frequent tillage can also cause the creation of a plow layer, an increase in soil bulk density with attendant lowering of porosity, and reductions in soil water storage and transport through pores. All these perturbations can cause an overall decline in soil quality and agricultural productivity. On the other hand, organic residues are left undisturbed in the field in no-tillage and conservation-tillage systems. In these systems, improvements in soil quality are observed because of aggrading soil aggregation, increasing concentrations of binding agents, formation of new aggregates, and increasing stability of aggregates. Thus, with decreasing tillage intensity, there is a strong chance of improvement in soil physical condition, soil quality, and agricultural productivity. Improvements in soil physical condition facilitate the increase in porosity and water retention of the soil, decrease in compaction and soil strength, and improvement in the nutrient and organic carbon content of the soil. Thus, the physical properties of soil play an important role in determining soil quality.

The physical properties of soil at the surface (soil–air interface) control the infiltration of water into the soil and are important in dividing the rainfall or applied water into runoff and infiltrated water. The runoff water, depending on its volume and the slope of the field, may cause soil erosion. It may contain dissolved or suspended sediments, chemicals, and nutrients. These pollutants can get transferred to surface water resources, such as lakes, ponds, or rivers. Thus, physical properties of soil are important in maintaining the quality of surface water.

Physical properties of vadose zone soil influence the storage of water and the chemicals dissolved in it, including nutrients and contaminants through the profile, and control the migration of these solutes to the groundwater table. Some of these solutes interact with the soil solids and are retained in the vadose zone, the plant roots take up some others, and the remaining solutes percolate below the root zone and eventually join the groundwater. Thus, knowledge of soil physics is important in understanding the migration of solutes through the vadose zone (partially wet aquifer above the water table) for designing strategies to prevent or decrease groundwater contamination.

The physical properties of soil and the surface soil moisture content can control the emission of dust particles into the atmosphere. Dust emissions happen in agricultural fields due to high natural winds or during agricultural operations, such as plowing, disking, harrowing, leveling, or harvesting. They can also happen on unpaved roads due to high natural winds or vehicles using those roads (Williams et al., 2008). The larger dust particles usually become airborne for a small time due to the aforementioned actions and then fall back; their motion is known as saltation. The finer dust particles can travel long distances and carry chemicals sorbed on their surfaces due to their high specific surface area. Thus, soil physics plays an important role in controlling airborne particulate matter emission, and the physical properties of soil are important in determining air quality.

Land-management practices influence greenhouse gas emissions from agricultural fields. Application of nitrogenous fertilizer increases N_2O emissions from soil. An increase in N_2O emissions is observed immediately after a fertigation event. N_2O emissions are also higher immediately after water is applied to a field because of the denitrification (reduction of nitrate to N_2O and nitrogen gas) of fertilizers in the soil. Similarly, depending on the quality of biomass and management practices, CO_2 and CH_4 emissions can happen at varied intensities. The release of these greenhouse gases can have a profound influence on the degradation of environmental quality. The preservation of resource base and environmental quality is immensely important for sustaining life on earth. The degradations or aggradations of groundwater quality, surface water quality, air or environmental quality, and soil quality directly affect the productivity and sustainability of agriculture and human and animal health, which in turn affect the overall quality of life on earth (Shukla, 2011). Thus, understanding the properties and processes associated with the discipline of soil physics plays a pivotal role in improving the quality of life on earth by managing the soil, water, and air quality efficiently.

1.4 SOIL PHYSICS AND CLIMATE CHANGE

World population is on the rise and several agencies have projected that it could reach between 7.5 and 10.5 billion by the year 2025 (www.worldometers.info/population/; Population Division of the Department of Economic and Social Affairs of the United Nations Secretariat, 2009). Increasing global population has put immense pressure on land use systems and the pressure will continue to grow. The quality of land, as mentioned earlier, is controlled by the physical properties of soil. Drastic changes in land use can cause drastic changes in soil properties. The necessity of increasing per capita grain production to feed an increasing population led to green revolutions in several parts of the world a few decades ago. Vertical and container farming are expensive solutions to increase production. Growing more crops in a year can put enormous stress on the resistance and resilience of soil. Intensive cultivation and cropping can cause loss of soil structure and organic matter and salinity buildup. It also encourages excessive use of fertilizers and other chemicals to control weeds and pests. Barring a few industrialized nations, in general, an increase in water usage is reported worldwide, primarily due to population growth and economic development (Bates et al., 2008). It has also been reported that globally, agricultural irrigated area has increased at a rate of 2% per year since 1960 (Bruinsma, 2003; Bates et al., 2008); however, the quality of ground and surface water has generally declined (UN, 2006).

Limited information is available on the change in land use due to climate change. If future climate predictions using existing forecast models are assumed to be fairly accurate, then climate change can cause substantial changes in the annual precipitation amount and frequency, resulting in drastic changes in soil moisture regimes, soil and air temperature, and the length of growing seasons. These changes will differ in magnitude and rate of change in different parts of the world and will be confounded by the fundamental complex agricultural systems and crop response to the changed frequency of precipitation, moisture, temperature, and carbon dioxide regimes (https://scied.ucar.edu/learning-zone/climate-change-impacts/predictions-future-global-climate). Predicted changes in water availability (median from 12 climate models) at the end of the 21st century (2090–2099 versus 1980–1999) are shown in Figure 1.5 (Bates et al., 2008). Figure 1.5 indicates that there is a likelihood of increased water stress in the Mediterranean basin, Central America, and the subtropical regions of Africa and Australia (Bates et al., 2008). This can cause substantial changes in land use around the world. For a more elaborate discussion of the climate effect on land use, readers are referred to the Intergovernmental Panel on Climate Change (IPCC) report on climate change and water (Bates et al., 2008). Challenges posed by changes

FIGURE 1.5 Predicted changes in water availability at the end of the 21st century with white and hatched area indicating disagreement/agreement among different models. (Adapted from Milly, P. C. D., et al., *Nature*, 347, 2005; Bates, B. C., et al., eds., *Climate Change and Water*, Technical Paper of the Intergovernmental Panel on Climate Change, IPCC Secretariat, Geneva, Switzerland, 210pp, 2008.)

in climate would require the soil physics discipline to increase interactions with other disciplines. Changes in water availability and water stress will force soil physicists, soil hydrologists, and climatologists to develop new technologies for managing soil and water resources more efficiently.

1.5 SOIL PHYSICS CURRICULUM

Soil physics is taught in most land grant universities in the United States and at several agricultural universities worldwide. Although water storage and transport within the vadose zone remains central to the soil physics curriculum, increasing attention is paid to processes leading to environmental pollution. Solute transport has gained a lot more attention and importance in the soil physics curriculum than it had in the past (Jury et al., 1991). Many new mathematical models have enhanced our understanding of the vadose zone theory and processes. The soil physics course is primarily designed for undergraduate and graduate students interested in learning the basic principles and processes of soil physics, water transport, and solute transport through soils (Lal and Shukla, 2004).

The theory of soil physics is applicable not just to soil but to any porous media for increasing our understanding of soil, water, and environmental quality. Some of the objectives of the soil physics curriculum include (1) the study of the methods of measurement and assessment of the physical properties of soil, (2) the identification of those physical properties and processes and their influence on water retention and transport through soil profile, (3) understanding the nature and properties of groundwater, (4) understanding the fundamental principles of water flow and contaminant transport, (5) understanding the fundamental processes within the soil profile that control water balance, erosion, and soil contamination, (6) understanding the impacts of anthropogenic contaminants on soil and groundwater quality, and (7) understanding the effects of land use on flow and transport. These processes are increasingly studied at the soil–plant–atmosphere continuum. The syllabus meets the curriculum needs of students pursuing soil science, water science, environmental science, natural resources, agricultural engineering, civil engineering, chemical engineering, range sciences, horticulture and crop sciences, and forestry. Students gain insight into the important physical properties of soil; critical physical processes involving mass transport, movement, and retention of water and solutes (nutrients and contaminants) through soil profiles; soil temperature regime; aeration; and plant–water relations.

PROBLEMS

1.1 Climate change scenarios have large uncertainty. Do you agree with this statement? Explain your answer with the support of published literature on this issue.

1.2 List some of the physical properties of soil that you come across in everyday life. Explain how these have an impact on your day-to-day life.

1.3 Collect some published refereed journal articles on soil hydraulic conductivity and write a summary on the spatial and temporal variability of soil hydraulic conductivity.
1.4 Write a note on climate change and explain if and how it can affect life on earth.

REFERENCES

Bates, B. C., Z. W. Kundzewicz, S. Wu, and J. P. Palutikof (eds.). 2008. *Climate Change and Water.* Technical Paper of the Intergovernmental Panel on Climate Change, IPCC Secretariat, Geneva, Switzerland .

Bouma, J. 1989. Using soil survey data for quantitative land evaluation. *Adv. Soil Sci.* 9: 177–213.

Brejda, J. J., T. B. Moorman, D. L. Karlen, and T. H. Dao. 2000. Identification of regional soil quality factors and indicators: I. Central and southern high plains. *Soil Sci. Soc. Am. J.* 64: 2115–2124.

Bruinsma, J. 2003. *World Agriculture: Towards 2015/2030. An FAO Perspective.* Earthscan, London, UK.

Dasgupta, S., B. Mohanty, and J. M. Kohne. 2006. Soil hydraulic conductivities and their spatial and temporal variations in a Vertisol. *Soil Sci. Soc. Am. J.* 70: 1872–1881.

Deb, S. and M. K. Shukla. 2011. An overview of some hydrological watershed models. In: M. K. Shukla (ed.), *Soil Hydrology, Land Use and Agriculture: Measurement and Modeling.* CAB International, Wallingford, UK, pp. 75–116.

Doran, J. W. and T. B. Parkin. 1994. Defining and assessing soil quality. In: J. W. Doran D. C. Coleman, B. F. Bezdicek, and B. A. Stewart (eds.), *Defining Soil Quality for a Sustainable Environment.* Soil Science Society of America, Madison, WI, pp. 3–21.

Ikemura, Y. and M. K. Shukla. 2009. Soil quality in organic and conventional farms for an arid ecosystem of New Mexico. *J. Org. Syst.* 4(1): 34–47.

Jury, W. A., W. R. Gardner, and W. H. Gardner. 1991. *Soil Physics.* John Wiley & Sons, Inc., New York, p. 328.

Lal, R. 1998. Soil quality and agricultural sustainability. In: R. Lal (ed.), *Soil Quality and Agricultural Sustainability.* Ann Arbor Press, Chelsea, MI, pp. 3–12.

Lal, R. and M. K. Shukla. 2004. *Principles of Soil Physics.* Marcel Dekker Inc., New York, p. 716.

Larson, W. E. and F. J. Pierce. 1991. Conservation and enhancement of soil quality. In: *Evaluation for Sustainable Land Management in the Developing World. Vol. 2, IBSRAM Proc. 12(2).* International Board for Soil Research and Management, Bangkok, Thailand.

Li, Y. and M. J. Lindstorm. 2001. Evaluating soil quality-soil redistribution relationship on terraces and steep hillslope. *Soil Sci. Soc. Am.* 65: 1500–1508.

Milly, P. C. D., K. A. Dunne, and A. V. Vechhia. 2005. Global pattern of trends in streamflow and water availability in a changing climate. *Nature* 438: 347–350. doi:10.1038/nature04312

Nielsen, D. R. and O. Wendroth. 2003. *Spatial and Temporal Statistics-Sampling Field Soils and Their Vegetation.* Catena Verlag, Reiskirchen, Germany.

Nielsen, D. R., J. W. Biggar, and K. T. Erh, 1973. Spatial variability of field measured soil water properties. *Hilgardia* 42: 215–259.

Population Division of the Department of Economic and Social Affairs of the United Nations Secretariat. 2009. *World Population Prospects: The 2008 Revision.* United Nations, New York.

Scott, H. D., A. Mauromoustakos, I. P. Handayani, and D. M. Miller. 1994. Temporal variability of selected properties of loessial soil as affected by cropping. *Soil Sci. Soc. Am. J.* 58: 1531–1538.

Shukla, M. K. 2011. Introduction to soil hydrology. In: M. K. Shukla (ed.) *Soil Hydrology, Land Use and Agriculture: Measurement and Modeling*. CAB International, Wallingford, UK, pp. 1–23.

Shukla, M. K., R. Lal, and M. Ebinger. 2004. Soil quality indicators for the Northern Appalachian experimental watersheds in Coshocton Ohio. *Soil Sci.* 169(3): 195–205.

Shukla, M. K., R. Lal, and M. Ebinger. 2006. Determining soil quality indicators by factor analysis. *Soil Till. Res.* 87: 194–204.

SSSA (Soil Science Society of America). 1987. *Glossary of Soil Science Terms*. SSSA, Madison, WI.

UN. 2006. *World Water Development Report 2: Water, a Shared Responsibility*. UNESCO, Paris, France.

van Es, H. M. 2002. Sources of soil variability. In: J. Dane and C. Topp (eds.), *Methods of Soil Analysis, Part 4: Physical Properties*. Soil Science Society of America, Madison, WI.

Williams, D. S., M. K. Shukla, and J. Ross. 2008. Mapping airborne particulate matter emitted by a vehicle running on unpaved road. *Atmos. Environ.* 42: 3899–3905.

2 Units and Dimensions

2.1 INTRODUCTION

A quantifiable property is generally known as a quantity. A quantity can be intensive or extensive. An intensive quantity is not dependent on the extent of the system, whereas an extensive quantity is. A quantity that appears in a mathematical application of science or engineering is known as a physical quantity. Thus, a physical quantity can be measured and expressed in numbers, and has a unit (Scott, 2000). A particular physical quantity used to compare other similar physical quantities is known as a unit. In general, the numerical value of a particular quantity depends on its unit (http://physics.nist.gov/cuu/units/index.html). For example, among the 48 states in the continental United States, Mount Whitney in California has the highest peak elevation of 4,418 m (or 14,495 ft) above mean sea level. Mount Everest has a peak elevation of 8,848 m (or 29,029 ft) above mean sea level. Did you notice how numerical values for the same quantity change with its units in these two examples? Did you notice, however, that the conversion factor from one unit to another stays the same and is independent of the actual values in different units? Specification of standard measurement units or dimensions and the conversion among different types of units is essential for almost every field of engineering, soil science, medical science, aviation, agriculture, commerce, industry, etc.

There are several different systems of units, discussed later in the chapter, and societies are more adept at using a particular system of units or mixed systems of units for different physical quantities. For example, if you go to a hardware store in the United States to buy a pipe, you will commonly use inch for specifying the diameter (mostly internal) of the pipe. Often you use a fractional description of the size of the pipe such as 1/4 of an inch. For some other items, you may use more commonly used units that may or may not be part of any of the systems of units. For example, for purchasing fruit juices, we commonly use gallon or quart in the United States. On the other hand, shopping in several European cities may require using the units such as centimeter or meter for internal diameter and liter or cubic meter for volume. Some aircrafts use meters for describing the flight elevation, while others use feet above the mean sea level. In order to be at ease in different parts of the world, knowledge of the different systems of units and conversions from one system of units to another is essential.

2.2 DEFINITIONS OF UNIT AND DIMENSION

There are five commonly used fundamental dimensions representing five fundamental physical quantities. These five fundamental dimensions are mass (M), length (L), time

DOI: 10.1201/9780429264849-2

(T), temperature (θ), and electric charge or current (I). Two other fundamental dimensions are N, for the amount of substance, and J, for intensity of light. The dimensions of some physical quantities are derived from the two or more fundamental dimensions. For example, velocity has the derived dimension of distance (or length) over time L/T.

Units, on the other hand, provide the magnitude of a dimension, and unlike dimensions, where just a few dimensions are needed to express a physical quantity, there are several different units for representing a single physical quantity. Examples of the units of the fundamental quantity mass are gram, kilogram, pounds; example of the unit of the derived quantity velocity are cm/s, m/s, or ft/s (Campbell, 1977).

Units and dimensions are not the same, for example, the dimension of length is unique and is denoted by L; however, units of length could be centimeter, meter, kilometer, foot, nautical mile, Angström, etc. Units also provide a meaning to a physical quantity. For example, the height of a bridge is 100 ft; thus, it provides a magnitude of the height relative to a standard that could be 1 ft. Without units, the height of a bridge cannot be described correctly and sensibly.

2.3 SYSTEMS OF UNITS

There are many systems of units followed in different parts of the world. This chapter lists some of the most commonly used systems of units. They are:

1. Metric system of units
   a. Gravitational system
   b. CGS system or centimeter–gram–second system
   c. MKS system or meter–kilogram–second system
   d. MTS system or meter–tonne–second system (used in USSR)
2. English system of units: FPS system or foot–pound–second system
3. SI system or International System of Units (Système International d'Unités)

In the gravitational system, the magnitude of force is defined as gram force, kilogram force, or kilopond. A gram force is defined as the magnitude of force exerted by a mass of 1 g in a 9.80665 m/s^2 gravitational field. Initially, gram force was not a well-defined quantity until the General Conference on Weights and Measures (CGPM) adopted the acceleration due to gravity at sea level as being equal to 9.80665 m/s^2.

The CGS system of physical units is based on centimeter as the unit for length, gram as the unit for mass, and seconds as the unit for time. It was replaced by the MKS system, where meter replaced centimeter as the unit for length and kilograms replaced grams as the unit for mass; however, the unit for time stayed the same (second). The MKS system was later replaced by the SI system or International System of Units.

The SI or International System of Units is the official system of measurements in several nations around the globe. The United States, Liberia, and Myanmar are probably the only three nations where the SI system is not endorsed formally as the system of measurements. In the United States, the English system is still the main system of measurement, where foot is used as a unit for length; pound, which is a

TABLE 2.1
Base Units in CGS, FPS, and SI Systems

Physical Quantity	Name and Symbol in Various Systems of Units					
	CGS	Symbol	FPS	Symbol	SI	Symbol
Mass	Gram	g	Pound	lb	Kilogram	kg
Length	Centimeter	cm	Foot	ft	Meter	m
Time	Second	s	Second	s	Second	s
Electric current	Biot	Bi			Ampere	A
Temperature	Celsius	°C	Fahrenheit	°F	Kelvin	K
Intensity of light	Lambert	Lb			Candelas	cd
Amount of substance					Mole	mol

unit of weight, as the unit for mass; and second as the unit for time. The use of the metric or SI system is increasing, and a vast majority of scientific journals published in the United States use only the SI system as the system of measurement (American Society of Agronomy et al., 1988).

Table 2.1 presents the fundamental units in CGS, FPS, and SI systems of units; Table 2.2 presents derived units; and Table 2.3 presents derived units with special names and symbols.

The SI unit for mass is kilogram, and unit mass is equal to the mass of the platinum–iridium cylinder stored near Paris, France (Table 2.1). The SI unit for length is meter, and unit length is equal to the distance that light travels in 1/299,793 s. The SI unit for time is second, and unit time is equal to the 1/86,400 of the mean solar day (1 day = 86,400 s = 24 [h] × 60 [min/h] × 60 [s/min]). The SI unit for temperature is kelvin (K = °C + 273.15).

TABLE 2.2
Derived Units in CGS, FPS, and SI Systems

Physical Quantity	Name in Various Systems of Units			
	CGS	FPS	SI	Dimension
Area	**cm^2**	**ft^2**	**m^2**	**L^2**
Volume	cm^3	ft^3	m^3	L^3
Velocity	cm/s	ft/s	m/s	L/T
Acceleration	cm/s^2	ft/s^2	m/s^2	L/T^2
Density	g/cm^3	lb/ft^3	kg/m^3	M/L^3
Force	$g{\cdot}cm/s^2$	$lb{\cdot}ft/s^2$	$kg{\cdot}m/s^2$	ML/T^2
Work	$g{\cdot}cm^2/s^2$	$lb{\cdot}ft^2/s^2$	$kg{\cdot}m^2/s^2$	ML^2/T^2
Pressure	$g{\cdot}cm^{-1}/s^2$	$lb{\cdot}ft^{-1}/s^2$	$kg{\cdot}m^{-1}/s^2$	$M/L/T^2$
Energy	$g{\cdot}cm^2/s^2$	$lb{\cdot}ft^2/s^2$	$kg{\cdot}m^2/s^2$	ML^2/T^2

TABLE 2.3
Derived SI Units with Special Names

Physical Quantity	CGS	SI
Energy or work	Erg	Joule (J)
Force	Dyne	Newton (N)
Frequency	Cycle/s	Hertz (Hz)
Power	Erg/s	Watt (W) (J/s)
Pressure	Barye	Pa
Charge	Franklin	Coulomb (C)
Kinematic viscosity	Stokes	m^2/s
Dynamic viscosity	Poise	Pa·s
Magnetic flux	Maxwell	Weber
Wave number	Kayser	m^{-1}

2.4 NONDIMENSIONAL QUANTITIES

Quantities those do not have dimensions are known as dimensionless or nondimensional quantities. Examples of nondimensional quantities are all trigonometric functions (sine, cosine, tangent, cotangent, secant, cosecant), the exponential function (exp), the logarithmic function (log), angles, quantities that can be counted (number of players in a football team), plain old numbers (2, π, etc.), and the ratio of dimensional quantities (Archimedes number, ratio of gravitational force to viscous force, capillary number, ratio of viscous force to surface tension force).

2.5 DERIVING UNITS OF PHYSICAL QUANTITIES

The derived units of various physical quantities need not be memorized. Instead, the definitions of these terms need to be clearly understood. Here are some of the common physical quantities and an exercise on deriving the units of different quantities.

Example 2.1

Derive the units of physical quantities including area, volume, density, velocity, acceleration, force, work, pressure, and energy using their definitions.

SOLUTION

Area (rectangular shape) = Length × Width = cm × cm = cm^2
Area (circular shape) = π × $Radius^2$ = cm^2 = cm^2
(*Note*: π does not have units; it is dimensionless.)
Volume (rectangular shape) = Length × Width × Depth
= cm × cm × cm = cm^3
Volume (cylindrical shape) = π × $Radius^2$ × Height = cm^2 × cm = cm^3

Density = Mass/Volume = g/cm^3 = g cm^{-3}
Velocity = Distance/Time = Length/Time = cm/s = cm s^{-1}
Acceleration = Velocity/Time = (Length/Time)/Time
= (cm/s)/s = cm/s^2 = cm s^{-2}
Force = Mass × Acceleration = Mass × Velocity/Time
= Mass × (Length/Time)/Time = g × (cm/s)/s = g cm s^{-2}
Work = Force × Distance = g cm s^{-2} × cm = g cm^2s^{-2}
Pressure = Force/Area = g cm s^{-2}/cm^2 = g $cm^{-1}s^{-2}$
Energy = Mass × Acceleration due to gravity × Distance
= g × cm s^{-2} × cm = g cm^2 s^{-2}

2.6 USE OF UNITS AND DIMENSIONS

Units or dimensions are transferred to a mathematical relationship to check if a given mathematical relationship is dimensionally (or unit-wise) balanced or correct according to a given system of units. It is a powerful technique for identifying errors in equations or relationships. However, care must be taken to use units and dimensions consistently in any calculation. If an algebraic relationship has two parts that are added together, then the units or dimensions for both parts must be identical.

Example 2.2

Check whether the following relationships are dimensionally balanced or correct according to the given system of units:

a. $Q = A \times V$, where Q is the volumetric flow rate, A is the cross-sectional area of flow, and V is the flow velocity.

Parameter	Unit	Dimension
Volumetric flow rate (Q)	cm^3/s	L^3/T
Cross-sectional area (A)	cm^2	L^2
Flow velocity (V)	cm/s	L/T
$Q = A \times V$	$cm^3/s = cm^2 \times cm/s = cm^3/s$	$L^3/T = L^2 \times L/T = L^3/T$

Thus, the equation is balanced based on the units and dimensions.

b. $F = M \times g$, where F is the force, M is the mass, and g is the acceleration due to gravity.

Parameter	Unit	Dimension
Force (F)	g cm/s^2	ML/T^2
Mass (M)	G	M
Acceleration due to gravity (g)	cm/s^2	L/T^2
$F = M \times g$	g cm/s^2 = g × cm/s^2 = g cm/s^2	$ML/T^2 = M \times L/T^2 = ML/T^2$

c. $\Delta P = T \times \left(\frac{1}{R_x} + \frac{1}{R_y} \right)$, where P is the pressure difference across a domain, T is the surface tension, R_x is the radius of curvature along the x direction, and R_y is the radius of curvature along the y direction.

Parameter	Unit	Dimension
Pressure difference (ΔP)	g/cm/s^2	M/L/T^2
Surface tension (T)	dyne/cm = g × cm/s^2/cm	M/T^2
Radius of curvature (R_x)	cm	L
Radius of curvature (R_y)	cm	L
$\Delta P = T \times \left(\frac{1}{R_x} + \frac{1}{R_y} \right)$	g/cm/s^2 = g × cm/s^2/cm × (1/ cm + 1/ cm)= g/cm/s^2	M/L/T^2 = M/T^2 × (1/L +1/L) = M/L/T^2

2.7 UNIT CONVERSIONS

Different datasets are reported by different agencies using different systems of units, and it is sometimes necessary to convert the units of some physical quantities or parameters from one system of units to another or from one measurement unit to another. For example, the odometer in the United States displays the speed of a car in mile/h (although km/h is also displayed on the inside), and that must be converted to km/h or m/h for most scientific publications. Similarly, net radiation data in weather stations are generally recorded as Langley per hour (Ly/h), and that has to be converted to watt per square meter (W/m^2) for some model applications in estimating evapotranspiration. Sometimes, a change of unit requires the knowledge of some other physical quantities and their appropriate units. For example, converting concentration from g/kg to g/L requires the knowledge of bulk density, while converting concentration to mol/L requires the knowledge of the molecular weight of the substances present in the solution.

Example 2.3

Convert the units in the following problems:

a. A vehicle is running at a speed of 40 km/h; convert it to m/s.

$$40\ \text{Km/h} = \frac{40\ \text{Km}}{\text{h}} \times \frac{1{,}000\ \text{Km}}{\text{Km}} \times \frac{1\ \text{h}}{60\ \text{min}} \times \frac{1\ \text{min}}{60\ \text{s}} = \frac{40{,}000\ \text{m}}{3{,}600\ \text{s}} = 11.11\ \text{m/s}$$

b. The concentration of nitrate measured in a soil sample was 25 g/kg and bulk density of soil was 1,400 kg/m^3. Convert concentration to g/L.

$$25\ \text{g/Kg} = \frac{25\ \text{g}}{\text{Kg}} \times \frac{1{,}400\ \text{Kg}}{\text{m}^3} \times \frac{1\ \text{m}^3}{10^3\ \text{L}} = \frac{35{,}000\ \text{g}}{1{,}000\ \text{L}} = 35\ \text{g/L}$$

TABLE 2.4
Prefixes Based on the International System of Units

Prefix	Symbol	Factor	Prefix	Symbol	Factor
Deci	d	1.0E–01	Deca	da	1.0E+01
Centi	c	1.0E–02	Hecto	h	1.0E+02
Milli	m	1.0E–03	Kilo	k	1.0E+03
Micro	µ	1.0E–06	Mega	M	1.0E+06
Nano	n	1.0E–09	Giga	G	1.0E+09
Pico	p	1.0E–12	Tera	T	1.0E+12
Femto	f	1.0E–15	Peta	P	1.0E+15
Atto	a	1.0E–18	Exa	E	1.0E+18

c. Organic carbon content of the soil was 2 g/kg. Convert it to kg/ha. Consider that the soil layer is 20 cm thick with a bulk density of 1,400 kg/m³.

$$2\text{ g}/\text{Kg} = \frac{2\text{ g}}{\text{Kg}} \times \frac{1\text{ Kg}}{1{,}000\text{ g}} \times \frac{1{,}400\text{ Kg}}{\text{m}^3} \times 20\text{ cm} \times \frac{1\text{ m}}{100\text{ cm}} \times \frac{10^4\text{ m}^2}{\text{ha}}$$

$$= \frac{56 \times 10^7\text{ Kg}}{10^5\text{ ha}} = 5{,}600\text{ Kg}/\text{ha}$$

The International System of Units (SI) presents a number of prefixes as positive or negative powers of 10 (Table 2.4). Thus, 1 km can be represented as 1.0E + 03 m.

PROBLEMS

2.1 Describe systems of units. Provide examples for each system of units.
2.2 A vehicle is traveling at a speed of 40 miles/h. What will be the speed in m/s?
2.3 Why is the knowledge of units and dimensions important?
2.4 Write down the units and dimensions of at least five quantities that you come across every day in your life.
2.5 Find out the formula to calculate saturated hydraulic conductivity of soil using the constant head method (Chapter 9) and prove the relationship is correct.
2.6 List some of the units used in the United States (or any other country) that are mostly used locally and are not included in this chapter.

REFERENCES

American Society of Agronomy, Crop Science Society of America and Soil Science Society of America. 1988. *Conventions and Style. Chapter 6. Publications Handbook and Style Manual.* American Society of Agronomy, Madison, WI.

Campbell, G. S. 1977. *An Introduction to Environmental Biophysics.* Springer-Verlag, New York.

Scott, D. H. 2000. *Soil Physics: Agriculture and Environmental Applications.* Iowa State University Press, Ames, IA.

3 Characteristics of Soils of the Vadose Zone

3.1 INTRODUCTION

People, in general, are familiar with soil because it is an essential part of their day-to-day life. Soil is recognized by many as most essential for life on earth, and as the most basic nonrenewable resource because several hundreds of years are needed to reproduce it. But, with rapid urbanization, increasing growth of big cities since the industrial revolution, and migration from rural to urban centers, fewer and fewer people are in direct contact with soil. This disconnect also contributes to the reduced understanding of soil as a basic natural resource. The importance of soil for the survival of mankind is less imperative for people living in urbanized areas. With the ever-increasing number of supermarkets and varieties of frozen foods available on the shelves of grocery stores, there is no time to think about soil, which is the paramount resource for producing most of our handy microwavable food.

Soil is defined as a medium for plant growth. More elaborate definitions of soil describe it as a medium to support and grow plants, hold and filter water and other chemicals, and build on. Soil is a natural body that significantly differs from its parent material, including bedrock and other secondary material transported from soil by water or wind, and consists of layers of different materials of variable thickness. According to an engineering definition, soil is a medium composed of loose rock material suitable for building on and through it. Each soil layer consists of inorganic (quartz) and organic materials (biomass, humus). Soils are considered to be ancient, and in terms of the geological period are considered to be more than 65 million years old (Buol et al., 1973).

3.2 SOIL FORMATION

Soil formation takes place very slowly and is known as pedogenesis. Soils are sometimes referred to as "young" or "mature," but these terms only refer to the degree of weathering and not so much the age of soil. During the process of soil formation, the parent material undergoes nonreversible transformation due to physical, chemical, and anthropogenic processes. The parent material disintegrates into successively smaller fractions and is deposited locally or transported to other locations via gravity, water, or wind. The soils developed from local deposition of parent materials are called residual soils; these soils have properties similar to that of the parent material (Figure 3.1). When the parent material is transported by gravity, the soil is deposited at the foot or bottom of the slope and is known as colluvial. The parent material transported by water and deposited in streams is known as alluvial. When water-transported disintegrated parent material gets deposited in lakes, it is known as lacustrine, and that deposited in oceans is marine. Ice-transported parent material

DOI: 10.1201/9780429264849-3

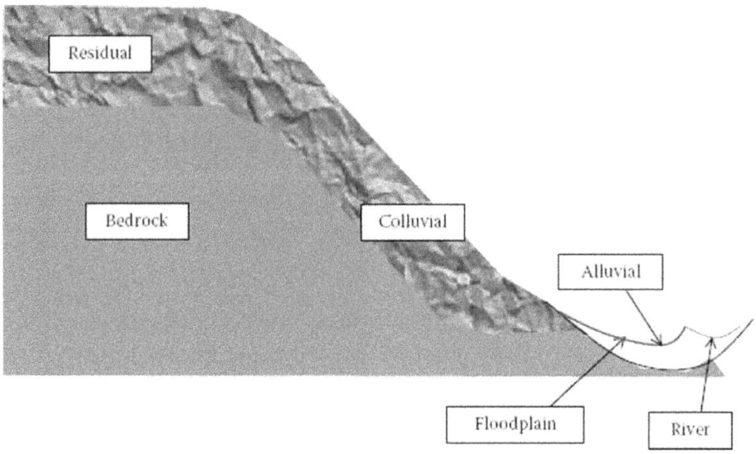

FIGURE 3.1 The topography and deposition of soil at different slope positions. See how different mechanisms involved in the transport of parent material are classified into residual, colluvial, and alluvial deposits.

is known as till, and that transported by wind is known as eolian. The four basic processes of soil formation are transformation or weathering of primary particles, translocation or movement of particles, additions of organic matter or transported particles, and losses due to leaching and erosion.

The disintegration of the parent material also depends on the climate, including local precipitation intensity and amount, atmospheric temperatures; biota or living organisms; and topography of the area. The differences in elevation, slope, and landscape position have a strong influence on the weathering and erosion of parent materials as well as soil and vegetation development. Generally, less soil and vegetation development takes place on steeper slopes. The aspect of a field or slope, measured from true north being zero radians, is also important for soil and vegetation development. In the northern hemisphere, south-facing slopes are more perpendicular to sunrays and are less weathered because of low moisture and less organic matter and vegetation. On the contrary, north-facing slopes in the northern hemisphere are moist and have more vegetation. The areas under steeper slopes can exacerbate the disintegration and transport of parent material or soil development itself.

The transformation of parent material generally takes time. Soil age, a slow transformation of parent material, is the length of time until the landscape became stable and soil formation started. Soil formation is a time-dependent process and is a direct function of the intensity of the activity of other soil-forming factors. The more intense the precipitation, climate, and biological activities, the quicker the development of soil. With time as a soil formation factor, soil development can take from a few hundred to several thousand years. An average soil formation rate has been reported to be about 180 years/cm (Cooper, 2009), but remember that soil development rates are highly variable depending on location. The number of years required for soil development varies by several orders between a continental and a tropical

TABLE 3.1
Soil Orders and a Brief Description for Identification

Order	Description
Entisols	Recently formed soil
Histosols	Organic soil
Gelisols	Permafrost soil (at or below the freezing point of water)
Andisols	Volcanic soil
Inceptisols	Young soil
Aridisols	Desert soil formed under dry conditions
Vertisols	Soil that swells when wet and shrinks when dry
Alfisols	Contain iron and aluminum, found under forests
Mollisols	Soils with thick A-horizon, found on grassland
Ultisols	Heavily leached soils
Spodosols	Soils of cooler climate, found under deciduous or conifer forest
Oxisols	Highly weathered soils

climate. The five major soil-forming factors are parent material, climate, biota, topography, and time. Further discussion of soil formation can be found elsewhere (Brady and Weil, 2004).

The five soil-forming factors are the basis of the detailed soil classification work done by the United States Department of Agriculture (USDA). On the basis of development and the USDA soil taxonomy system, soils are classified into 12 orders. They are Entisols, Histosols, Gelisols, Andisols, Inceptisols, Aridisols, Vertisols, Alfisols, Mollisols, Ultisols, Spodosols, and Oxisols (Table 3.1). A more complete description of the taxonomic classification of soils also involves suborders, great group, subgroup, family, series, and phase. An example of classification of a soil type is presented in Table 3.2. At its lowest level of organization, the US system of soil classification recognizes approximately 19,000 different soil series. Soil classification also identifies five different soil moisture regimes, namely, aquic, udic, ustic, aridic, and xeric. In general, the amount of water in the soil decreases from aquic (which is saturated or poorly aerated) to aridic or xeric (which stays dry for several months during a year). The udic soil moisture regime indicates that soil has sufficient moisture for most of the year and can be found in humid areas. The ustic moisture regime indicates less soil water availability than udic regime coupled with drought for most of the year.

3.3 SOIL PROFILE

A typical soil profile consists of distinctive soil layers of variable thickness, color, and degree of weathering, and each layer is known as a horizon (Figure 3.2a). Horizon formation is a slow process, and a newly formed soil may not have distinct soil horizons. Some examples where only one soil horizon is seen include sand

TABLE 3.2
Classification of a Recently Developed Soil Using the US Taxonomy System

Order	Entisol
Suborder	Fluvent
Great Group	Torrifluvent
Subgroup	Typic Torrifluvent
Family	Fine-loamy, mixed, superactive, calcareous, Typic Torrifluvent
Series	Jocity, Youngston

Note: Torri means arid (dry); fluvent means floodplain soil; calcareous indicates presence of calcium carbonate; mixed indicates presence of swelling as well as nonswelling clays; superactive indicates higher cation exchange capacity.

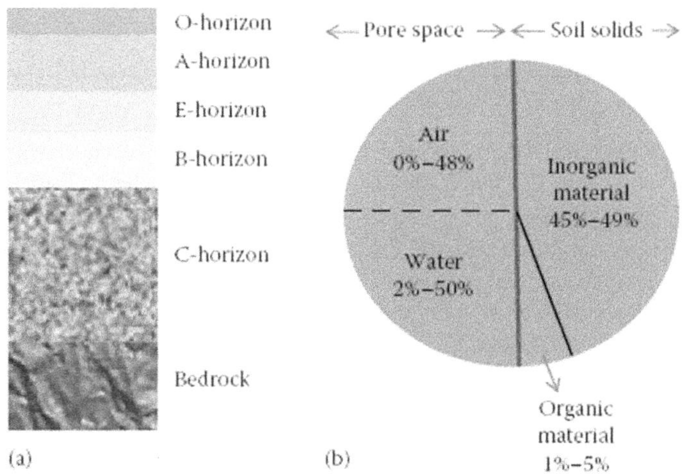

FIGURE 3.2 (a) Soil profile showing distinctive layers known as horizons and (b) a typical volume composition of a soil (remember that the soil profile as well as volume composition vary with soils, locations, and land uses).

dunes, volcanic ash deposits, alluvium, and highly weathered Oxisols. The topmost horizon could consist of plant residues that are partially decomposed and is known as the O-horizon. The subsequent horizon is usually the A-horizon, which is a mixture of inorganic and organic materials. The soil layers below the A-horizon usually have low organic matter content and can be an E- or a B-horizon. If the horizon contains materials, such as organic matter, clay, oxide, and organic compounds, leached (eluviation, opposite of illuviation, or accumulation) from the A-horizon, it is known as the E-horizon. The B-horizon also accumulates leached materials from the A-horizon. The A- and B-horizons together are also known as solum, whereas all horizons up to the parent material are known as regolith. The C-horizon is immediately below the B-horizon and usually consists of material that has not undergone drastic change from the parent material. Additionally, the literature also reports a D- and an R-horizon. The D-horizon indicates material different from the C-horizon and parent material but is not universally recognized. Continuous but only slightly weathered bedrock is called the R-horizon.

It is important to remember that all the horizons shown in Figure 3.2 are not always present in a soil profile. In some arid regions, a distinct O-horizon is seldom seen. Similarly, soil horizons are seldom horizontal to the soil surface, and their thickness is variable in space.

Soil in a three-dimensional horizon can have a volume composition of solid and pore space as indicated in Figure 3.2b. Depending on the climate of the area, type of land use, and parent material, the fraction or percentage of solid and pore spaces can change. However, it is important to remember that the sum of inorganic, organic, air, and water fractions is always equal to one.

3.4 SOIL TEXTURE

Soil consists of primary and secondary particles. Primary particles are discrete units and cannot be subdivided further, whereas secondary particles consist of primary particles and can be subdivided further. These particles are inorganic (quartz, silicates) or organic (humus, organic matter). The inorganic particles, known as soil separates, include sand, silt, and clay particles. Soil texture classification is based on the equivalent diameter of particles present in a given soil volume (USDA Web Soil Survey, http://websoilsurvey.nrcs.usda.gov/app/WebSoilSurvey.aspx). When we look at a three-dimensional volume of soil, it is very hard to imagine that the volume consists of different particle sizes and pores held tightly together. In spite of this, soil texture, primarily determined by the fractions of sand, silt, and clay particles, remains one of the important physical properties of soil. Soil texture is one of the first properties sought after. It is important to know here that soil texture usually changes with increasing soil depth. It also changes spatially and thus can be highly variable across any landscape. There is an important difference between soil texture and particle size distribution. Particle size distribution is the quantitative measure of different particle sizes within a soil domain, whereas, soil texture is more qualitative and uses a range of particle sizes to classify soil into groups.

Soils can have a variety of combinations of gravel, sand, silt, and clay fractions. If the gravel fraction dominates, the soil is classified as gravelly. If smaller particles dominate, the soil is known as clay. Another term based on the small size of particles is colloids, which is defined as inorganic and organic materials of clay size and smaller. Similarly, when the larger particle sizes dominate with very small amounts of finer particles, the soil is classified as sand. According to the USDA soil classification system, between sand and clay there are ten different types of soils, namely, loamy sand, sandy loam, sandy clay loam, sandy clay, loam, silt loam, clay loam, silty clay loam, silty clay, and silt (Figure 3.3).

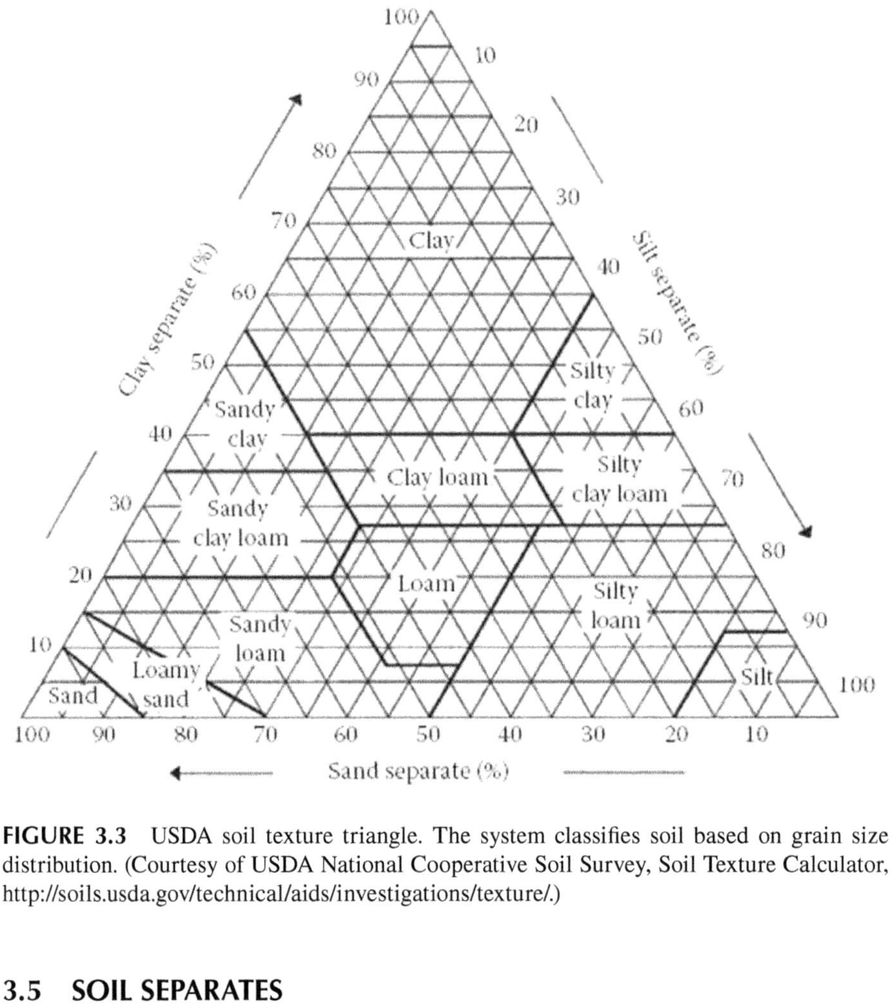

FIGURE 3.3 USDA soil texture triangle. The system classifies soil based on grain size distribution. (Courtesy of USDA National Cooperative Soil Survey, Soil Texture Calculator, http://soils.usda.gov/technical/aids/investigations/texture/.)

3.5 SOIL SEPARATES

Soil separates are essentially divided into four categories: gravel or skeletal material, sand, silt, and clay. Gravel is classified as soil particles with a diameter larger than 2 mm (Table 3.3). Although gravel can have an important influence on water and air movement through the soils as well as on seed germination and plant stand, these particles are not included in the determination of the soil texture. Soil texture is determined using particles smaller than 2 mm or using percentages of sand, silt, and clay.

Sand is a naturally occurring material usually known as silica (SiO_2) or quartz. It consists of highly variable granular particles consisting of rock fragments and mineral matter. According to the USDA classification, particles with a diameter between 2 and 0.05 mm are known as sand particles (Table 3.3). These particles have different shapes ranging from angular to round. The shape depends on the degree of weathering; in general, more weathering makes sand particles more round. Sand particles

TABLE 3.3
Classification to Designate Particles into Sand, Silt, and Clay Based on Their Size

Particle Size Description	USDA	ISSS	USPRA	ASTM
Gravel	>2	>2	>2	>4.75
Very coarse sand	1–2			
Coarse sand	0.5–1.0	0.2–2.0	0.25–2.0	2.0–4.75
Medium sand	0.25–0.50			0.475–2.0
Fine sand	0.10–0.25	0.02–0.2	0.05–0.25	0.075–0.475
Very fine sand	0.05–0.10			
Silt	0.002–0.05	0.002–0.02	0.005–0.05	<0.075
Clay	<0.002	<0.002	<0.005	

Note: Size is expressed as equivalent particle diameter in millimeters (mm); USDA, United States Department of Agriculture; ISSS, International Society of Soil Science; USPRA, United States Public Roads Administration; ASTM, American Society for Testing and Materials.

generally have different colors, which depend on the parent material as well as on the presence of coatings of iron and aluminum oxides. Sand commonly has different shades of colors such as white, yellow, brown, green, red, and black. White-colored sand usually consists of eroded limestone or gypsum (e.g., white sands of New Mexico). Sand containing large amounts of magnetite or basalt is usually darker in color, those with abundance of chlorite and glauconite are greenish, and those with iron are yellowish. Sand particles feel gritty when rubbed between fingers. These particles can be transported by water and winds.

Silt is also a granular material and is smaller than sand. According to the USDA classification, particles between 0.05 and 0.002 mm are designated as silt particles. It is also a by-product of weathering of parent material and breakdown of sand particles into smaller sizes. Similar to sand, silt is also composed of quartz and but can also be made up of feldspar. Unlike sand, silt particles feel smooth when rubbed between fingers. The shape of silt particles is more rounded. However, similar to sand, silt particles do not stick to surfaces when wet and can be transported by water and wind easily.

Clay is a fine-grained material composed of aluminum silicate and is finer than silt-sized particles. Clays are usually formed due to chemical weathering of parent material and other primary particles. According to the USDA classification, clay particles are smaller than 0.002 mm. The shape of clay-sized particles is like a plate or a tiny flake. Since clay particles are very light, they can be transported long distances by water and wind. They also stay suspended in water and air for a long time. These particles are generally very sticky when wet and are very smooth and soothing when rubbed between fingers.

There are several different classifications to designate particles as sand, silt, or clay. These classifications are proposed by the International Society of Soil Science (ISSS), the United States Department of Agriculture (USDA), and the American Association of State Highway and Transportation Officials (AASHTO), among others (Table 3.3). There are other systems of classifications such as the Massachusetts Institute of Technology (MIT), the British Standards Institution (BSI), and the German Standard (DIN). Several countries around the world have a classification system that is similar to one of the previously mentioned systems. The data collected using one system can be converted to another.

In general, particles bigger than 2 mm diameter are classified as gravel by the USDA, the ISSS, and the USPRA systems, and particles finer than 0.002 mm are classified as clay by the USDA and ISSS systems. All three systems define silt particles slightly differently (Table 3.3). The American Society for Testing and Materials (ASTM) and the unified soil classification system (USCS) are quite different from the other three systems, and are used by engineers and geologists to describe particle size and texture of a soil. The ASTM system classifies particles finer than 0.075 mm as fine-grained particles consisting of silt and clay. The particles that exhibit plasticity when wet and strength when dry are clay, while those not exhibiting plasticity when wet are classified as silt.

3.6 METHODS FOR PARTICLE SIZE MEASUREMENT

Soil texture, as presented in Figure 3.3, indicates the relative proportions of particles classified as sand, silt, and clay. The soil texture triangle also tells us that the sum of the percentages of sand, silt, and clay contents in a given volume of soil is 100%. The fractions that are >2 mm are not counted on the texture triangle. These fractions are also known as gravel, pebbles, or skeletal material. The sand, silt, and clay contents in the laboratory can be determined by the hydrometer method or pipette method. The process of determining particle size distribution is also known as mechanical analysis and involves dispersion and fractionation. Dispersion is the removal of cementing materials to break secondary particles into primary particles, and fractionation is the process of physically separating particles into different size fractions. The principle of particle size determination is based on Stokes' law, which states that the resistance to the fall of a spherical particle is a function of the radius and not the surface. During a fall, a particle is subjected to three forces: gravity (F_g), buoyancy (F_b), and friction (F_r). At terminal velocity, when the rate of fall is constant, these three forces can be equated as follows:

$$F_g = F_b + F_r \tag{3.1}$$

$$\frac{4}{3}\pi r^3 \rho_s g = \frac{4}{3}\pi r^3 \rho_l g + 6\pi r \eta \theta \tag{3.2}$$

or

$$\theta = \frac{2}{9\eta} r^2 \left(\rho_s - \rho_l\right) g \tag{3.3}$$

where

θ is the settling (terminal) velocity.
r is the radius of a particle.
ρ_l is the density of liquid.
ρ_s is particle density.
η is the dynamic viscosity of the liquid.

After attaining terminal velocity, if a particle falls to a length of h in time t, then $\theta = h/t$.

In the hydrometer method, about 50 g of air-dried soil sample passing through a 2 mm sieve is taken and mixed with 50 mL of deionized (DI) water and 75 mL of sodium hexametaphosphate (40 g dissolved in 1 L of DI water). The sample is mixed, left to sit for few hours, and stirred to mechanically separate particles of different sizes. The sample is subsequently transferred to a cylinder and filled to the 1 L mark with DI water. The sample is stirred thoroughly with a stirrer or plunger, a hydrometer is lowered into the cylinder, and a reading is taken at 40 s and another at 3 h. The temperature of the suspension is also recorded at these two time intervals. The first reading at 40 s provides the sand fraction and the second reading provides the estimate of clay. For further details on other methods, readers are referred to Dane and Topp (2002).

The soil textural class can be determined in the field by the "feel method." In this method, a soil sample, if dry, is slowly wetted by adding water until it is completely wet. After that, it is rolled in the shape of a ball and then squeezed between the thumb and the side of the forefinger making a ribbon of the soil. The ribbon is made continuously until it breaks. A long and continuous ribbon can be formed for clay soils; however, no ribbon can be formed for sandy soil.

Example 3.1

A hydrometer test was performed in the soil physics laboratory using water with a viscosity of 1×10^{-3} kg/m/s. Calculate the settling velocity for particles of equivalent radius of 0.2 and 0.002 mm. The particle density and density of water are 2.65 and 1 Mg/m³, respectively.

For a particle of 0.2 mm radius, the settling velocity (Equation 3.3) is

$$\theta = \frac{2}{9\eta} r^2 \left(\rho_s - \rho_l\right) g = \frac{2 \times \left(0.2/\left(10 \times 100\right)\right)^2 \left(2.65 \times 10^3 - 1 \times 10^3\right) \times 9.81}{9 \times \left(1/1000\right)}$$

$$= 0.144 \text{m/s}$$

For a particle of 0.002 mm radius, the settling velocity is

$$\theta = \frac{2}{9\eta} r^2 (\rho_s - \rho_1) g = \frac{2 \times (0.002/(10 \times 100))^2 (2.65 \times 10^3 - 1 \times 10^3) \times 9.81}{9 \times (1/1000)}$$

$$= 1.44 \times 10^{-5} \text{m/s}$$

3.7 PARTICLE SHAPES

Soil particles have distinctly different shapes. Larger particles are more regular and closer to being considered spherical in shape; however, finer particles are nowhere near a spherical shape. Particle sizes and shapes reflect the type of weathering (chemical and physical) a mass of mineral has undergone, the mechanisms involved in breakdown, subsequent transport and deposition from the original location, and the mineral composition of the parent material (Santamarina and Cho, 2004). One should remember that with increasing age of a particle, the likelihood of physical or chemical weathering also increases and particles tend to become smoother. There is an inverse relationship between the degree of weathering and angularity (a shape having one or more sharp angles), and with increasing weathering particles become less angular. In general, larger particles are more prone to physical weathering (or abrasion) than smaller particles (<0.5 mm) likely due to the lower mechanical momentum. Larger particles are bigger, denser, and can move faster (less suspended), and thus higher kinetic energy results in higher abrasion (Kuenen, 1960; Santamarina and Cho, 2004).

Particle shape is described by sphericity, roundness, and smoothness. The sphericity of a soil particle indicates its closeness to being a perfect sphere. For example, a soil particle with a sphericity of 0.9 will be closer to being a sphere than a particle with a sphericity of 0.7. A sphere is the most compact solid with minimum surface area. Sphericity is a ratio and therefore it is a dimensionless number. Roundness is a measure of the sharpness of the grain corners. If the corners of a particle are sharp, it has low roundness. Smoothness refers to the condition of the surface of a particle.

Sphericity is calculated as the ratio of the diameters of the largest inscribed (D_i) sphere, with the area of the sphere equal to the particle area, and the diameter of the smallest circumscribing (D_c) sphere (Figure 3.4a). It is also expressed as the ratio of the surface area of a sphere (A_s) having the same volume as that of the particle (V_p) and the surface area of the particle (A_p) itself. Roundness, on the other hand, is calculated using the ratio of the average radius of the corners (r_i) of a particle and the radius of the maximum inscribed (R) sphere in that particle (Figure 3.4b).

$$\text{Sphericity} = \frac{D_i}{D_c} \tag{3.4}$$

$$\text{Sphericity} = \frac{A_s}{A_P} = \frac{\pi^{1/3} (6V_P)^{2/3}}{A_P} \tag{3.5}$$

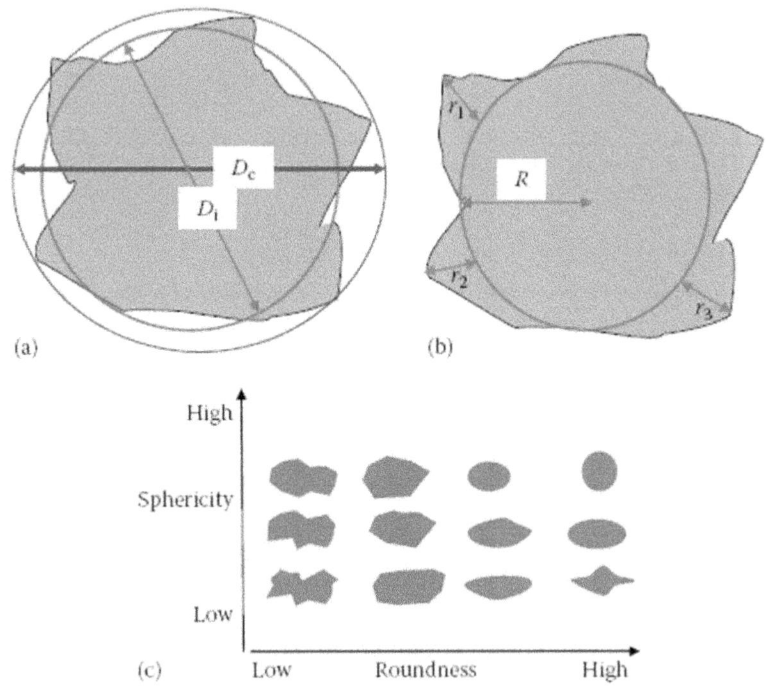

FIGURE 3.4 (a) Sphericity of a particle, (b) roundness of the same particle, and (c) illustration of shapes with respect to the degree of sphericity and roundness.

$$\text{Roundness} = \frac{1}{R}\sum_{i=1}^{n}\frac{1}{n}r_i \tag{3.6}$$

Pi (π) is a mathematical constant and is equal to 22/7 (~3.141593). It is calculated as the ratio of the circumference ($\pi \times d$) and the diameter (d) of a circle, or as the ratio of the area ($\pi \times r^2$) and the square of the radius (r^2) of a circle. All the parameters required to determine sphericity or roundness can also be easily obtained using any image processing program (imageJ, photoshop). There are several public domain as well as advanced/professional programs available for use.

Example 3.2

An equivalent diameter of a sphere is 20 cm (D_i) and has an area equal to the area of the particle. The diameter of the smallest circumscribing sphere is 25 cm (D_c). Calculate the sphericity of the particle.

$$\text{Sphericity} = \frac{D_i}{D_c} = \frac{20\text{cm}}{25\text{cm}} = 0.8$$

Example 3.3

Calculate the sphericity of particles that are cubical, tetrahedral, and spherical in shape.

Assuming the length of a side of the cube or tetrahedron is L and radius of the sphere is r, the volumes and areas of the cube, tetrahedron, and sphere can be calculated as follows:

The volume of a cube = L^3 and the area of a cube = $6L^2$.
The volume of a tetrahedron = $\left(\sqrt{2}/12\right)L^3$ and the area of a tetrahedron = $\sqrt{3}L^2$.
The volume of a sphere = $(4/3)\pi\, r^3$ and the area of a sphere = $4\pi\, r^2$.

$$\text{Sphericity of cube} = \frac{A_s}{A_P} = \frac{\pi^{1/3}\left(6V_P\right)^{2/3}}{A_P} = \frac{3.141593^{0.3333}\times\left(6\times L^3\right)^{0.6667}}{6L^2} = 0.806$$

$$\text{Sphericity of tetrahedron} = \frac{A_s}{A_P} = \frac{3.141593^{0.3333}\times\left(6\times\left(\sqrt{2}/12\right)L^3\right)^{0.6667}}{\sqrt{3}L^2} = 0.671$$

$$\text{Sphericity of sphere} = \frac{A_s}{A_P} = \frac{3.141593^{0.3333}\times\left(6\times(4/3)\pi r^3\right)^{0.6667}}{4\pi r^2} = 1$$

Example 3.4

Calculate the roundness of the particle shown in Figure 3.4b. Assume r_1, r_2, r_3, r_4, and r_5 equal to 5, 4, 5, 6, and 4 cm, respectively. Assume the radius of the inner circle as 14 cm.

$$\text{Roundness} = \frac{1}{R}\sum_{i=1}^{n}\frac{1}{n}r^i = \frac{1}{14}\sum_{1}^{5}\frac{r^i}{n} = \frac{1}{14}\frac{(5+4+5+6+4)}{5} = \frac{24}{70} = 0.343$$

3.8 PROPERTIES OF CLAY PARTICLES

The slow process of chemical weathering of rock by solvents results in the formation of smaller particles known as clay minerals. These solvents are carbonic acids and other acids that leach through the soil profile. In addition, hydrothermal activities involving circulation of hot water can also lead to the formation of clay minerals. Clay deposits can be formed as residual but are usually formed as sedimentary deposits in lakes, ponds, river flood plains, and marine deposits. In addition, they can also be formed by transport by winds.

The clay fraction consists of several elements and ions, including silica (Si^{4+}), aluminum (Al^{3+}), iron (Fe^{3+} and Fe^{2+}), magnesium (Mg^{2+}), zinc (Zn^{2+}), sodium (Na^{+}), calcium (Ca^{2+}), potassium (K^{+}), oxygen (O^{2-}), and hydroxide (OH^{-}). Electron microscope analysis and x-ray diffraction have shown that contrary to earlier belief, clay particles are crystalline in structure (Hendricks, 1942; Marshall, 1967). The clays could be silicates, allophane (poorly crystalline hydrous aluminum silicate), and iron and aluminum oxides. Clay primarily consists of hydrous aluminum silicate ($Al_2(SiO_3)_3$), which is a mixture of aluminum, silica, oxygen, and water. Without water, it is a mineral; when there is water, it is known as clay. Kaolin, a soft white material with the chemical formula $Al_2(SiO_3)_32H_2O$, describes a group of clay minerals. Kaolinite is the main constituent of kaolin. Clays also consist of hydrous magnesium aluminum silicate and synthetic amorphous sodium aluminosilicate.

Although the mineralogical arrangement depends on the type of clay, silicate clays, also known as phyllosilicate (or leaf-like structure), have one silicon atom surrounded by four oxygen atoms in a tetrahedral (four-sided) formation (Figure 3.5a). The other type of formation is known as an octahedral (eight-sided) sheet, in which an aluminum or magnesium atom is surrounded by six oxygen or hydroxyl atoms (Figure 3.5b). The number and arrangement of silica and oxygen or hydroxyl atoms is used to classify silicate clay into a 1:1 or 2:1 type of mineral. When a clay layer is made up of one tetrahedral and one octahedral sheet, it is known as a 1:1 type of clay. In this type of formation, oxygen atoms tightly hold the tetrahedral and octahedral sheets. Subsequent layers stick together due to hydrogen bonding (discussed in Chapter 7), and thus these layers do not swell or expand (no swelling or shrinking) when hydrated. An example of a 1:1 type of clay is kaolinite, along with some other prototypes such as halloysite, nacrite, and dickite. All of these have the same chemical formula, $Al_2Si_2O_5(OH)_4$. Kaolinite crystals are usually hexagonal in shape, larger in size compared to other clay minerals, have low swell–shrink capacity, low surface area, low cation exchange capacity (CEC), and low plasticity. Kaolinite is good for making bricks, building roads and houses, agriculture

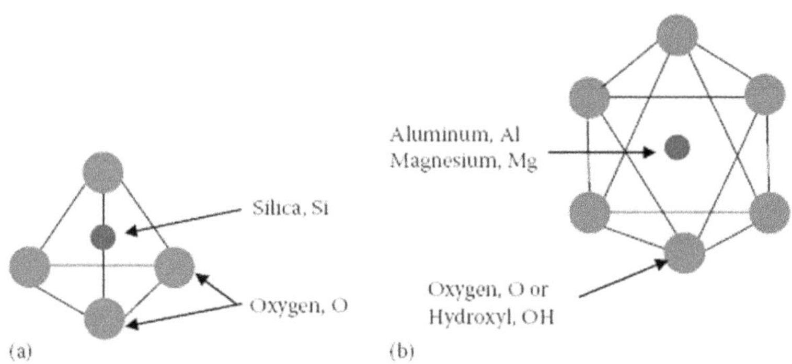

FIGURE 3.5 The structural arrangement of clay particles: (a) a tetrahedral (four-sided) formation and (b) an octahedral (eight-sided) formation.

(although nutrient retention capacity is low), and it is easy to work with. In the rainforests of Brazil, large amounts of kaolinite are found, and the soil is usually well drained and weathered.

The 2:1 type of clays have a structure in which one octahedral sheet is surrounded by two tetrahedral sheets. Examples of the 2:1 type of clays are smectite, illite, vermiculite, and chlorite. Smectite expands when hydrated; in general, the degree of expansion follows the order: smectite < vermiculite < illite < chlorite. The smectite group includes montmorillonite and bentonite. Smectite displays a high swell–shrink capacity, plasticity, CEC, internal surface area, and nutrient retention capacity. Vermiculite has a higher layer charge than smectite, a high CEC, and a medium swell–shrink capacity. Illite and chlorite are also known as 2:1 type nonexpanding clays, and these particles are considerably larger than smectite. In illite, Al^{3+} substitutes Si^{4+} in the tetrahedral sheets, while in chlorite Fe^{2+} or Mg^{2+} occupies most of the octahedral sheets. The expansion in illite and chlorite is very small, and potassium (K^+) acts as a strong bonding agent in the former, while the hydroxide sheet acts as a bonding agent in the latter. The CEC, plasticity, cohesiveness, and stickiness are between smectite and kaolinite. Chlorite is sometimes referred to as a 2:1:1 type of clay or phyllosilicate or sheet silicate. Phyllosilicate minerals include micas, serpentine, talc, and other clay minerals. These soils are low in fertility but are easy to work with.

Oxides of iron and aluminum are noncrystalline and are also known as hydrous oxide clays or sesquioxide clays. These types of clays are dominant in highly weathered soils such as Oxisols and Ultisols. These soils have low CEC and sometimes have an anion exchange capacity. These soils are generally not fertile but can bind phosphorus in the soil, making it unavailable to plants. Allophanic clays are relatively recent, generally have poorly defined structure and low density, and could be from volcanic ash (Andisols).

3.9 PHYSICAL PROPERTIES OF SOIL

Knowledge of physical properties of soil is essential for designing any site-specific management system, including the type of irrigation and water management system, types of crops to be grown in the area, and scheduling of irrigation and fertilizer regimes, among others. Some of the properties will be discussed here, while others will be discussed in other chapters where it is more appropriate.

3.9.1 Soil Color

Color is one of the most obvious and easily determined soil properties. Color can provide limited information about soil properties, soil functions, or soil conditions. For example, soil color can provide information on water table depth, drainage, and chemistry or chemical reactions such as oxidations or reductions that have occurred in the soil. Soil color also provides clues to the mineral content of the soil. For example, yellow could be an indication of goethite mineral being present in the soil; red indicates hematite or lepidocrocite in the soil; black indicates the presence of iron sulfide, pyrite, or humus; and white is an indication of the presence of calcite, dolomite, gypsum, or salts. Glauconite green sands form in shallow ocean water near a coast and become part of the soils that form after the sea level drops. The white color of uncoated calcite, dolomite, and gypsum is common in geologic materials and soils in arid climate. Goethite soil colors occur more frequently in temperate climates, whereas hematite colors are more prevalent in hot deserts and tropical climates.

Color, or lack of it, also tells us about the soil environment. Aerobic soil environments produce brighter colors, primarily due to the oxidation of the minerals. Anaerobic environments occur in areas with a shallow water table and can be distinguished by the dull soil color. Normally, soils are darker when wet. Anaerobic bacteria can reduce ferric iron (Fe^{3+}) in goethite and hematite to colorless, water-soluble ferrous iron (Fe^{2+}), which is returned to the soil. Similarly, Mn^{4+} is reduced to colorless, soluble Mn^{2+}. Loss of pigment leaves a gray color on the underlying mineral. If the water table stays shallow for long periods of time, the entire soil zone turns gray. In coastal areas, tide water can saturate soils twice a day. Anaerobic bacteria can use sulfate as an electron acceptor and release sulfide (S^{2-}), which combines with ferrous iron to precipitate black iron sulfide. When the water table goes down in dry seasons (due to evaporation, drainage, or water uptake by plants), oxygen (or air) reenters the soil and soluble iron oxidizes into orange-colored mottles of lepidocrocite on cracks in the soil.

The colors of the rainbow are red, orange, yellow, green, blue, indigo, and violet. These can be remembered as ROYGBIV. Soil color is easily determined using Munsell color charts (Figure 3.6b). Three indicators of the soil color on the Munsell color chart are hue, value, and chroma and are combined in that order to represent a soil color. Hue is the predominant color and also refers to the dominant spectral wavelength. The symbol for hue is the letter abbreviation for the color of the rainbow (R, red; Y, yellow), preceded by a number from 0 to 10. As an example, the hue for red can be represented as 0R, 2.5R, 5R, 7.5R, or 10R, and yellow red (YR) can be represented as 0YR, 2.5YR, 5YR, 7.5YR, or 10YR. Remember that as the number increases from 0 to 10 in YR, the hue becomes more yellow and less red. The notation for hue is located at the top right corners of the color chart page. Value is the lightness or darkness with a value of 0 indicating pure black, and a value of 10 is absolutely white. Value runs vertical (south–north) with each row having the same value. As value increases, vertically, soil colors become successively lighter. Chroma is the intensity of color and a chroma of 0 indicates natural gray and can go up to 20. Chroma increases to the right (horizontally) across the Munsell chart (Munsell Soil Color Chart, 2000).

A small volume of soil is compared to the color chips on the Munsell color chart. Care should be taken that there is enough light at the time of color determination. Putting a sample in an airtight bag can cause a change in soil color; on the other hand, soil color may also change due to exposure to air or oxygen. If soil is dry, it is recommended to use a squirt bottle to moisten the soil prior to color determination because most hydric indicators and color guidelines refer to moist soil.

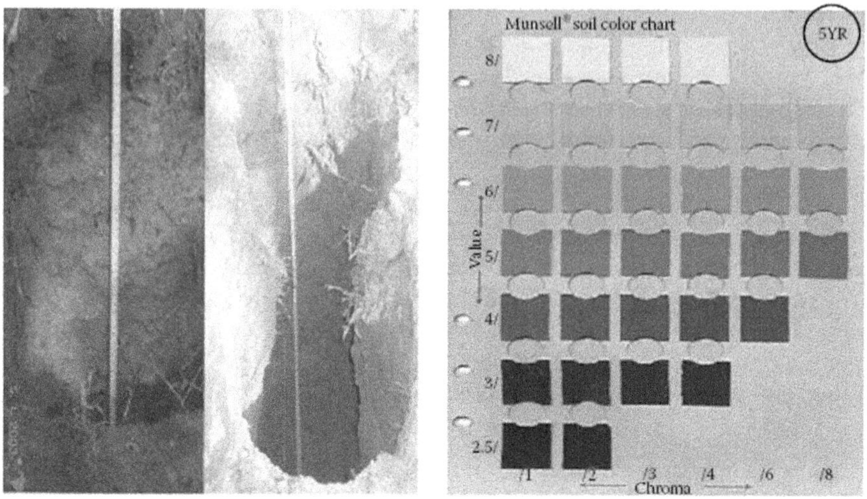

(a) (b)

FIGURE 3.6 (a) Soil profiles. A typical black cover denoting soil organic matter for a profile in South Charleston, Ohio, and a typical light-colored profile denoting no or very low soil organic matter for a profile in Las Cruces, New Mexico. (b) Munsell color chart showing the hue (see circle on the top right—5YR), value (changing south to north from 2.5 to 8), and chroma (changing east to west from 1 to 8).

Example 3.5

What do you understand from 5YR 5/6?

Soil color is always represented by hue, value, and chroma and in that order. Therefore, YR is yellow to red, and color of hue: 5YR; value: 5; and chroma: 6 = 5YR 5/6.

3.9.2 Soil Particle Density

One of the important physical properties of soil is particle density. This is generally denoted by the symbol ρ_p and is also known as density of soil solids or true density. The definition of particle density states that it is the ratio of the mass of the soil solids to its volume:

$$\text{Particle density} = \frac{\text{Mass of soil solids}}{\text{Volume of soil solids}} = \frac{M_S}{V_S} = \frac{M}{L^3} = \frac{\text{g}}{\text{cm}^3} \tag{3.7}$$

For mineral soils, particle density does not vary much because the dominant minerals are silicate minerals (quartz, feldspar, mica). If the particle density of mineral soil is unknown, it is usually taken as 2650 kg/m^3, 2.65 Mg/m^3, or 2.65 g/cm^3. It is reported to vary between 2.60 and 2.75 g/cm^3. It is also reported to follow the order sand < silt < clay. The presence of iron oxides and heavy metals, such as magnetite, limonite, or hematite, can increase the value of particle density. Conversely, the presence of solid organic materials can decrease particle density because organic matter weighs much less than an equal volume of mineral particles.

Particle density is commonly measured in the laboratory by the pycnometer method. For smaller samples but greater degree of accuracy, stoppered bottles are used, whereas volumetric flasks can be used for larger samples with a lesser degree of accuracy (Flint and Flint, 2002). A known mass of dry soil sample, with known moisture content, is placed inside a pycnometer of known weight (W_0) and volume along with accessories (stopper). The pycnometer along with its contents and other accessories are also weighed (W_1). The pycnometer is then filled up to half its volume using a de-aired liquid of known density, usually de-aired water. Entrapped air is carefully removed using a vacuum chamber. The procedure is repeated until the pycnometer is filled up to the capacity and all the entrapped air is removed. At this time, the pycnometer along with its contents and accessories is weighed (W_2) again. The temperature of the water is also measured to obtain the density of the water (ρ_w). The pycnometer is thoroughly cleaned and filled with water and weighed (W_w). The particle density is calculated as follows:

$$\rho_P = \frac{\rho_W (W_1 - W_0)}{(W_1 - W_0) - (W_2 - W_W)} \tag{3.8}$$

Other methods of particle density determination can be found in Dane and Topp (2002). In engineering applications, relative density or specific gravity is commonly used, and it is the ratio of the particle density of the soil and density of water.

Example 3.6

Particle density was obtained at 20°C using a soil pycnometer with a volume of 170 cm^3, weighing 40 g. Dry soil (215 g) was placed in the pycnometer. The mass of the pycnometer, soil, and de-aired water, when the pycnometer was filled to capacity, was 340 g. Calculate the particle density of the soil.

Assuming the density of water at 20°C is 1 g/cm^3.
Mass of water in the pycnometer = 340 − 215 − 40 = 85 g
Volume of water in the pycnometer = 85 cm^3 (mass/density of water)
Volume of soil (solids) (V_s) = Volume of pycnometer − V_w − V_a
= 170 − 85 − 0 = 85 cm^3

$$\text{Particle density} = \frac{\text{Dry mass of solids}}{\text{Volume of solids}} = \frac{215}{85} = 2.53 \text{g}/\text{cm}^3$$

3.9.3 Soil Bulk Density

Soil bulk density is defined as the ratio of the mass of soil to its total volume. The total volume consists of the volume of soil solids (organic and inorganic) and the volume of pores (water filled and air filled). Soil bulk density could be classified as wet soil bulk density and dry soil bulk density. The numerator in wet soil bulk density determination consists of the total in situ mass of soil, including the mass of soil solids as well as the mass of the water in the soil.

$$\text{Wet bulk density} = \frac{\text{Wet mass of soil}}{\text{Total volume of soil}} = \frac{\text{g}}{\text{cm}^3} \tag{3.9}$$

Thus, wet bulk density is not a constant but variable, depending on the water content of the soil. In contrast, the numerator in dry soil bulk density determination consists of only the dry weight of soil solids (organic and inorganic), and thus it is not a function of soil water content.

$$\text{Dry bulk density} = \frac{\text{Dry mass of soil solids}}{\text{Total volume of soil}} = \frac{\text{g}}{\text{cm}^3} \tag{3.10}$$

The SI unit for bulk density is kg/m^3 or Mg/m^3.

Example 3.7

If the dry bulk density of a soil is 1.23 g/cm^3, what will be the dry bulk density in kg/m^3 and Mg/m^3?

$$\text{Dry bulk density} = 1.23\frac{\text{g}}{\text{cm}^3} = 1.23\frac{\text{g}}{\text{cm}^3} \times 100^3\frac{\text{cm}^3}{\text{m}^3} \times \frac{\text{kg}}{1000\text{g}} = 1230\frac{\text{kg}}{\text{m}^3}$$

$$= 1230\frac{\text{kg}}{\text{m}^3} \times \frac{1}{1000}\frac{\text{Mg}}{\text{kg}} = 1.23\frac{\text{Mg}}{\text{m}^3}$$

The dry soil bulk density is also expressed as ρ_b. In general, the numerator or the mass of soil for bulk density determination is the mass after oven-drying the soil at 105°C for 24 h. Dry bulk density is more commonly used for most agricultural applications, and unless otherwise stated, bulk density usually refers to dry soil bulk density. The volume of soil usually consists of particles smaller than 2 mm. If coarser fragments (>2 mm) are found in the soil volume, it is essential to subtract the mass and volume of those coarser particles. The volume of coarser particles can be obtained by dividing the known mass of coarser particles by their densities. In engineering applications, coarser fragments are included in bulk density calculations. Dry bulk density of soil is always smaller than particle density because volumes of air and water are also included along with the volume of solids (in the denominator) in the calculation of dry bulk density.

Soil bulk density can be measured in the field by the core method, clod method, excavation method, or radiation method (Blake and Hartge, 1986). In the core method, a cylindrical metal core of known weight (W_0) and volume (V) is driven inside the soil to a certain depth. The core is removed, and soil at both ends of the core is trimmed so that the volume of soil is equal to the volume of the core. The core with soil is dried in an oven at 105°C for 24 h, cooled to room temperature, and weighed again (W_1). The bulk density of soil is determined as follows:

$$\text{Soil bulk density} = \frac{W_1 - W_0}{V} \tag{3.11}$$

Some of the advantages of the core method are that it is relatively quick and easy, and does not require expensive instrumentation or skilled labor. In this method, the undisturbed soil sample is collected and the original structure of the soil is involved in the bulk density calculation. Some of the disadvantages of this method include compression of soil inside the core, presence of coarser particles (may be skeletal material) inside the core, and small sampling volume. However, some of these drawbacks can be addressed by using large intact cores and making corrections for skeletal fractions.

In the clod method, a large volume of intact soil (or large aggregate or clod) is collected from the field. The clod is coated with a water-repellent substance such as paraffin or wax. It is immersed in water of known volume contained in a graduated jar; the volume of water displaced is recorded and is equal to the volume of the clod. The weight of the clod is recorded after oven-drying at 105°C. This method is useful particularly when collecting an undisturbed core is difficult or not possible. The disadvantage of the method is that some of the interaggregate pore spaces (the pore spaces between aggregates) are excluded, likely causing an overestimation of soil bulk density.

The excavation method involves digging a hole in the soil profile to a required depth. The removed soil is dried in the oven and weighed. The hole in the ground is lined with plastics, and a known volume of water is poured until it is completely full. The ratio of dry weight of the excavated soil and the volume of water that is equal to the volume of the soil excavated provides the bulk density of the soil. This method is very accurate when carefully done and can be used effectively for stony, gravelly, and dry sandy soils, where undisturbed cores are difficult to obtain. The disadvantages are that excavated soil is no longer undisturbed, and there are problems associated with use of plastic, weather conditions, and transport of large amounts of water.

The radiation method involves measurement of radiation transmitted through the soil using a detector mostly in the laboratory. In the field, the backscatter of the rays from the soil can be recorded by a detector. These methods are nondestructive but require knowledge of soil water content. They are expensive and require skilled labor and careful adherence to the nuclear safety protocols.

3.9.4 Soil Porosity

The porosity of soil is the ratio of the volume of the pores filled with water or air to the total volume of soil. It can also be calculated from the known soil bulk density and particle density as follows:

$$\text{Soil porosity}(P) = \frac{\text{Total volume of pores}}{\text{Total volume of soil}} = 1 - \frac{\text{Bulk density}}{\text{Particle density}} \tag{3.12}$$

The percentage solid space is calculated as 100 – Percentage pore space (or porosity). The porosity of soil is inversely related to the particle size and decreases with increasing particle size. Thus, clay soils generally have higher porosities than silt or sand soils. Large pores are usually associated with rapid movement of water through the soil, and smaller pores are associated with the retention of water inside the pore. Thus, sandy soils can transmit water through them rapidly, and clay soils can store large amounts of water in them.

In the engineering field, instead of total porosity, void ratio is commonly used. Void ratio is defined as the ratio of the total volume of the pores and the volume of soil solids (inorganic and organic). Void ratio is more appropriate for soils where the volume of the soil does not stay constant, for example, swelling and shrinking types of soils.

$$\text{Volid ratio}(e) = \frac{\text{Total volume of pores}}{\text{Total volume of soil solids}} \tag{3.13}$$

The porosity and void ratio are related to each other as follows:

$$P = \frac{e}{1+e} \quad \text{or} \quad e = \frac{P}{1-P} \tag{3.14}$$

3.9.5 Soil Water Content

Soil water content is the quantity of water contained in a given soil mass or volume, and denotes the percentage wetness of soil. It can range from extremely dry (or ~0) to the value of the porosity at full saturation. Soil water content is also known as soil moisture content. The water content of soil can be expressed on a mass basis and is known as gravimetric soil water content. It can also be expressed on a volume basis and is known as volumetric water content.

$$\text{Gravimetric watercontent} = \frac{\text{Mass of water}}{\text{Mass of dry soil}} = \frac{\text{g}}{\text{g}} \tag{3.15}$$

$$\text{Volumetric water content} = \frac{\text{Volume of water}}{\text{Total volume of soil}} = \frac{\text{cm}^3}{\text{cm}^3} \tag{3.16}$$

The volumetric water content can be estimated from the known gravimetric water content and soil bulk density as follows:

$$\text{Volumetric water content} = \text{Gravimetric water content} \times \frac{\text{Soil bulk density}}{\text{Density of water}} \tag{3.17}$$

In Equation 3.17, generally the density of water is assumed to be 1 g/cm^3. There are also other methods available in the literature that can be used to indirectly determine the soil water content. These methods are discussed in a separate chapter.

Example 3.8

A soil sample was collected from a chile (also known as chili) field located in southern New Mexico, United States. The undisturbed soil sample was collected in a core of diameter 7.5 cm and length 7.5 cm. Different weights were recorded before and after sample collection:

Weight of empty core = 50 g
Weight of core + Soil = 500 g
Weight of core + Oven-dried soil = 450 g

Calculate: (a) wet and (b) dry bulk density of soil, (c) porosity of soil (assume a particle density of 2.65 g/cm^3), (d) solid fraction, (e) gravimetric soil water content, and (f) volumetric soil water content.

$$\text{Volume of core} = \text{Volume of soil} = \pi \times r^2 \times h = 3.14 \times \left(\frac{7.5}{2}\right)^2 \times 7.5 = 331.34\ \text{cm}^3$$

Weight of the soil (in situ) = 500 − 50 = 450 g
Weight of the dry soil = 450 − 50 = 400 g
Weight of water in the core = 500 − 450 = 50 g

(a) $\text{Wet soil bulk density} = \dfrac{450}{331.34} = 1.36\,\text{g}/\text{cm}^3$

(b) $\text{Dry soil bulk density} = \dfrac{400}{331.34} = 1.21\,\text{g}/\text{cm}^3$

(c) $\text{Porosity} = 1 - \dfrac{1.21}{2.65} = 0.54\,\text{cm}^3/\text{cm}^3$

(d) $\text{Solid space} = 1 - \text{porosity} = 1 - 0.54 = 0.46\,\text{cm}^3/\text{cm}^3$

(e) $\text{Gravimetric water content} = \dfrac{50}{400} = 0.125\,\text{g}/\text{g}$

(f) $\text{Volumetric water content} = \dfrac{1.21}{1}\dfrac{\text{g}}{\text{cm}^3} \times \dfrac{\text{cm}^3}{\text{g}} \times 0.125\dfrac{\text{g}}{\text{g}} = 0.151\,\text{cm}^3/\text{cm}^3$

The dry soil weight can also be determined from the wet soil weight and the gravimetric water content as follows:

$$\text{Dry soil weight} = \frac{\text{Wet weight}}{1+\text{gravimetric water content}} = \frac{450}{1+0.125} = 400\,\text{g}$$

Similarly, dry soil bulk density can be determined from the wet soil bulk density and the gravimetric water content as follows:

$$\text{Dry soil bulk density} = \frac{\text{Wet bulk density}}{1+\text{gravimetric water content}} = \frac{1.36}{1+0.125} = 1.25\,\text{g}/\text{cm}^3$$

3.10 SOIL STRUCTURE

Soil structure can be defined as the arrangement of different constituents of soil. These are pore spaces occupied by water and air surrounded by solid particles consisting of inorganic and organic materials (Marshall and Holmes, 1979). Soil structure cannot be directly measured and is often indirectly inferred by soil physical properties. A soil with good structure has low compaction or bulk density and a large amount of pore space. These soils have high infiltration, quick water movement through the profile, high water retention, high water availability to roots, low crusting on soil surface, high gas exchange, high nutrient availability, easier root penetration, low surface runoff, and reduced erosion. Plants can grow well in good-structured soils, and properly managed agriculture systems in well-structured soils lead to high productivity. On the other hand, a soil with poor structure will have low water retention and transport, low root sustenance, and low growth and that will lead to poor yields from these fields.

Soil structure improves when silt and clay particles bind with inorganic particles to form a larger particle also known as aggregate. Plant material, root exudes, living organisms, and an active microbial population contribute toward stable aggregate formation. Improved and stable aggregate formation enhances the water and nutrient retention in the soil (Tisdall and Oades, 1982; Ochoa et al., 2009). A well-aggregated soil contains large pores, is well aerated, and ultimately leads to increases in crop yields.

Soil structure can be improved or maintained by a large number of practices. Some of the practices include adding biomass to the soil, conservation tillage, no-tillage, reduced tillage, providing ground cover to eliminate raindrop impact, cover crops, crop rotation, and avoiding field operations when the soil is too wet or too dry. Tillage, in general, exposes the organic matter to decomposition, breakdown of the aggregates, and transport of organic matter toward depositional areas, especially in sloping areas (Six et al., 1998; Shukla et al., 2003a, 2003b).

A structureless soil is devoid of aggregation and does not have an orderly arrangement of solids and voids (or pore spaces). Such types of soils can have a single-grain type of structure where soil particles do not cohere to each other. In contrast, a structureless soil can also have a massive structure with the entire soil mass cemented as one big block.

3.10.1 Types of Soil Structure

There are several different types of soil structure, namely, granular or crumb, blocky, columnar or prismatic, platy, and massive. The granular structure is commonly found in the A-horizon of soils (Figure 3.7). The soil particles are joined together in such a way that aggregates have a nearly spherical shape. Soils with a granular type of structure have high porosity, stable aggregates, and high water transport and retention capacity. Soils with granular structure are highly productive soils.

When soil particles adhere to each other in the shape of a square or an angular block, the soil structure is known as blocky or subangular blocky. This type of structure is mostly found in B-horizons where clay accumulation has taken place. These structures can have sharp edges and offer more resistance to entry into the soil and water movement than the granular type of structure. When soil particles are arranged in the shape of vertical columns or pillars, the soil structure is known as columnar or prismatic. This type of structure is also found in B-horizons where clay accumulation has taken place. Soils with prismatic structure are poorly drained and water movement is also fairly slow. When a soil structure is made up of horizontally piled sheets or in the form of thin plates, the soil structure is known as platy structure. This structure can be found in the A-horizon in forest soils as well as in clay pans.

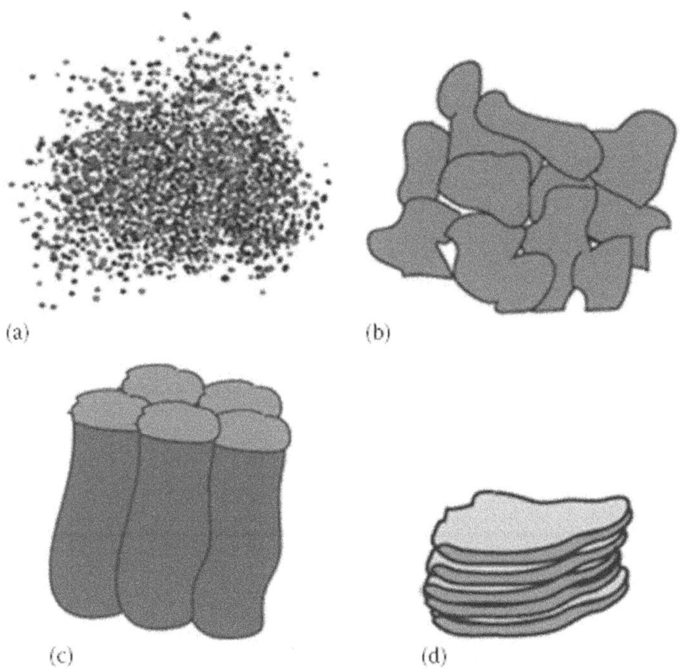

FIGURE 3.7 Types of soil structure: (a) granular, (b) blocky, (c) prismatic, and (d) platy. (From Department of Environment and Primary Industries, Soil structure, Victorian Resources Online, Victoria, Australia, 2013, www.dpi.vic.gov.au/vro. With permission.)

3.10.2 Properties of Soil Aggregates

Soil aggregates are defined as porous peds of various dimensions, consisting of primary particles and binding agents (Figure 3.8). The aggregates inside porous media are usually separated from each other and surrounding soil by interaggregate pore spaces (Gerke and Kohne, 2002). The pore spaces inside an aggregate are known as intra-aggregate pore space, and they are important for water and nutrient retention in the soil. The pore spaces between aggregates are known as interaggregate pore spaces and they are important for rapid transport of water and nutrients through the soil profile (Beven and Germann, 1982). These two flow domains are typically characterized by large differences in porosity, pore connectivity, and pore size, and aggregates in general are known as heterogeneous porous media. There are several different transport models available in the literature that specifically deal with the water movement through these two aggregate pore spaces as well as the exchange between two pore spaces (Gerke and Kohne, 2002). Some of these models are described later.

The transport of organic matter, silicate clay, or hydrous oxides of iron and aluminum can be either mechanical (action by rain or irrigation water) or chemical. The removal of these materials from a soil horizon is known as eluviation, whereas deposition of these materials in a soil layer is termed as illuviation. Cutans are the

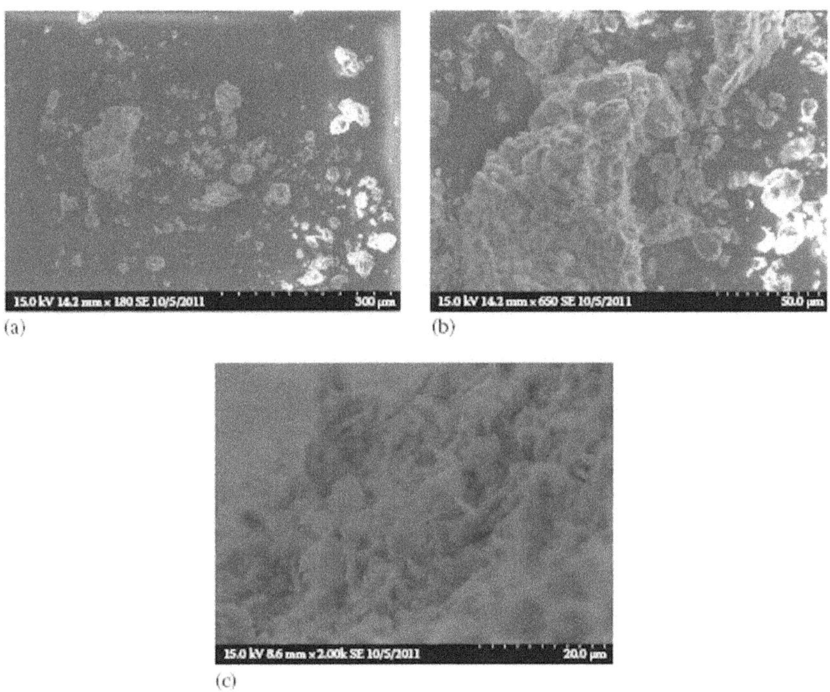

FIGURE 3.8 Electron microscope pictures of a macroaggregate under increasing resolution. *Note:* The resolution increases from 300 to 20 µm.

modifications of soil due to the process of illuviation. Aggregates can be covered with fine-textured clay skins, also called argillans, in Alfisols and Ultisols, especially in the argillic horizons. However, in Vertisols, slickensides, a type of cutan, and pressure faces on surfaces are observed. A slickenside is a surface of a crack in a high swell–shrink soil. Cutans can be homogeneous, they can contain organic matter and hydrous oxides of iron and aluminum, or can have layers of illuviated materials (Cady et al., 1986; Gerke and Kohne, 2002).

Irrigation can also cause aggregate slaking due to the hydration pressure exerted by water. Sodium, a strong deflocculating agent, present in soil or irrigation water, can also cause aggregate breakdown. In addition, chemicals that increase the zeta potential in the soil also act as a deflocculating agent and can cause aggregate breakdown or disruption in aggregate formation. Periodic wetting and drying and freezing and thawing of soils also increase the hydration pressures and can cause aggregate breakdown and loss of soil structure (Chenu et al., 2000; Ghezzehei and Or, 2003). These processes can also change the arrangement of pores, the arrangement of particles, and increasing concentration of clay in the cutans (Horn, 1990). As we move away from the surface of soil aggregates toward the center, the density of aggregates usually decreases.

The aggregates are divided into microaggregates (<250 μm mean diameter) and macroaggregates (>250 μm mean diameter). The bulk density of aggregates can be determined by the clod method (Blake and Hartge, 1986). The intra-aggregate porosity can be determined from the known bulk and particle density. The particle density can be assumed to be equal to 2.65 g/cm^3 for most cases. The available water content (AWC) of an aggregate can be determined by the difference in the volumetric water content of the aggregate at 0.3 bar and at 15 bars using a pressure plate apparatus (Leamer and Shaw, 1941). A discussion on the pressure plate method is provided later in the book.

The aggregate stability is an important property and can be measured as dry or wet aggregate stability. The wet aggregate stability, or the water stability of aggregates (WSA), can be determined by wet-sieving macroaggregates through a nest of sieves to separate water-stable aggregate fractions into different classes. Commonly used aggregate fraction classes are >5 mm (or retained on a 5 mm sieve), 2–5 mm (retained on 2 mm), 1–2 mm (retained on 1 mm), 0.5–1 mm (retained on 0.5 mm), and 0.25–0.5 (retained on 0.25 mm). Aggregate fractions retained on each sieve are usually oven-dried and weighed, and water stability, geometric mean diameter (GMD), and mean weight diameter (MWD) are calculated by the method suggested by Yoder (1936) and Youker and McGuinness (1957).

$$\text{WSA}(\%) = \frac{\sum_{i=1}^{n} w_i}{w_0} \times 100 \tag{3.18}$$

$$\text{GMD} = \frac{\sum_{i=1}^{n} e^{\left(w_i \times LOG(x_i)\right)}}{\sum_{i=1}^{n} w_i} \tag{3.19}$$

$$\text{MWD} = \frac{\sum_{i=1}^{n}(w_i \times x_i)}{w_0} \tag{3.20}$$

where

n is the aggregate fraction class (usually $n = 5$).
w_i is the weight of the aggregate retained on each sieve (g).
w_0 is the total dry weight of macroaggregates placed on the top sieve (g).
x_i is the average size of the sieve diameter in each aggregate fraction class.

So the average sieve size for material retained on a 5 mm sieve will be 6.5 mm [(5 + 8)/2], the average sieve size for material retained on a 2 mm sieve will be 3.5 mm [(2 + 5)/2], and so on.

The conversion from plow tillage to no-tillage is reported to increase WSA as well as diameters of macroaggregates. Table 3.4 shows that compared to a field under a conventional tillage system, WSA in fields under a no-tillage system could be 2–2.6 times higher at 0–10 cm depth (Shukla et al., 2003b). At 0–10 cm depth, GMD and MWD were also higher for fields under no-tillage than the conventional tillage system. Table 3.4 points out that at 10–20 cm depth, WSA was again higher for fields under no-tillage, but no clear-cut trend was evident for GMD and MWD of aggregates. Similarly, Table 3.5 shows that WSA and MWD are higher for fields under no-tillage than fields plowed using a chisel or moldboard.

A positive relationship between WSA and no-tillage duration was also reported by Ochoa et al. (2009) for the macroaggregates collected from a Miamian soil located in South Charleston, Ohio. The AWC associated with macroaggregates was also positively correlated with the no-tillage duration (Figure 3.9). Aggregate skins, generally much less porous, can significantly reduce the rate at which water is absorbed by the macroaggregates (Gerke and Kohne, 2002). Table 3.6 presents the data on hydraulic resistance, hydraulic conductivity, saturated water content of aggregate skins, and interior parts of intact aggregates. The hydraulic resistance at the skin could be 2.4 times greater than that at the interior, and hydraulic conductivities could be 6 times smaller than that for the aggregates without skin (Gerke and Kohne, 2002).

The increase in soil organic content increases aggregation as well as water stability of aggregates. Different soil organic carbon (SOC) fractions (hydrolyzable and nonhydrolyzable OC) also influence different macroaggregate-associated properties. Topographic features such as slope can also influence the SOC in macroaggregates, since considerable amounts of light SOC fractions can be redistributed and concentrated near the soil surface on the toe slope due to water erosion and transport from shoulder or back slope positions (Gregorich et al., 1998). The redistribution of SOC can affect the formation, stability, and hydrological properties of aggregates. Figure 3.10 shows a significant positive correlation between WSA and hydrolyzable SOC of aggregates, and a significant negative relationship between WSA and nonhydrolyzable SOC of aggregates (Ochoa et al., 2009). However, it should be clearly

TABLE 3.4
Analysis of Means for Parameters of Water Stable Aggregation (WSA), Geometric Mean Diameter (GMD), and Mean Weight Diameter (MWD) under Different Land Use Management Systems for Coshocton Watershed of Ohio

	WSA (%)		GMD (mm)		MWD (mm)	
Treatment	**0–10 cm**	**10–20 cm**	**0–10 cm**	**10–20 cm**	**0–10 cm**	**10–20 cm**
NTWM	74.84b	51.72a	1.30b	1.08bc	2.14b	1.08b
NTM	91.09a	60.06a	1.65a	1.19a	3.76a	1.51a
NTCSR	69.74b	51.67a	1.24b	0.95c	1.88b	0.76b
CT	34.91c	39.21b	1.22b	1.25a	1.07c	1.09b
M	85.23a	59.1a	1.55a	1.19a	3.44a	1.50a
LSD (0.05) T × R	9.34	11.85	0.17	0.15	0.65	0.41

Source: Data from Shukla, M. K., et al., *Soil Sci.*, 168, 167, 2003b.

Notes: NTWM, no-tillage without manure; NTM, no-tillage with manure; NTCSR, no-tillage corn–soybean rotation; CT, conventional tillage; M, meadow; LSD, least significant difference; T, land use and soil management practice; R, replicate; alpha = 5%. Different letters indicate significant difference at alpha = 5%.

TABLE 3.5
Effect of Different Tillage Systems on Water Stability of Aggregates and Mean Weight Diameter of Aggregates

	WSA (%)		MWD (mm)	
Treatment	**0–10 cm**	**10–20 cm**	**0–10 cm**	**10–20 cm**
CP	91.56ab	89.755b	3.90	3.34b
MB	89.39b	90.28b	3.60	3.25b
NT	94.32a	94.61a	4.36	4.26a
LSD (0.1)	4.29	21.76	NS	0.51

Source: Data from Shukla, M. K., et al., *Soil Sci.*, 168(11), 802, 2003a.

Notes: CP, chisel plow; MB, moldboard plow; NT, no-tillage; LSD, least significant difference; alpha = 10%. Different letters indicate significant difference at alpha = 5%.

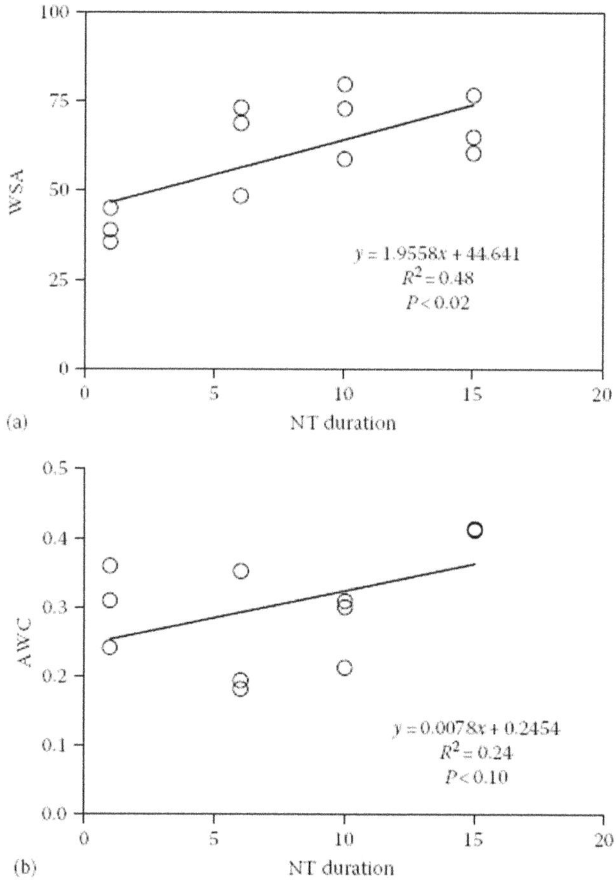

FIGURE 3.9 Relationship between no-tillage (NT) duration and water stability of aggregates (WSA) and macroaggregate-associated available water content (AWC) at 0–10 cm depth of a Miamian soil located in South Charleston, Clark County, Ohio. R^2 is the coefficient of determination; P is the level of significance. (Redrawn from Ochoa, C. G., et al., *Can. J. Soil Sci.*, 89, 319, 2009.)

TABLE 3.6
Hydraulic Parameters Including Hydraulic Resistance (R), Hydraulic Conductivity [K(h)], and Saturated Water Content (θ_s) for Aggregate Skin and Interior Parts of the Intact Aggregates

Sample	R D	K(h) cm/d	θ_s cm³/cm³
Skin	1.42	0.07	0.59
Interior	0.60	0.83	0.37
Intact	2.02	0.30	0.41

Source: Data from Gerke, H. H., and Kohne, J. M., *Soil Sci. Soc. Am. J.*, 66, 26, 2002.

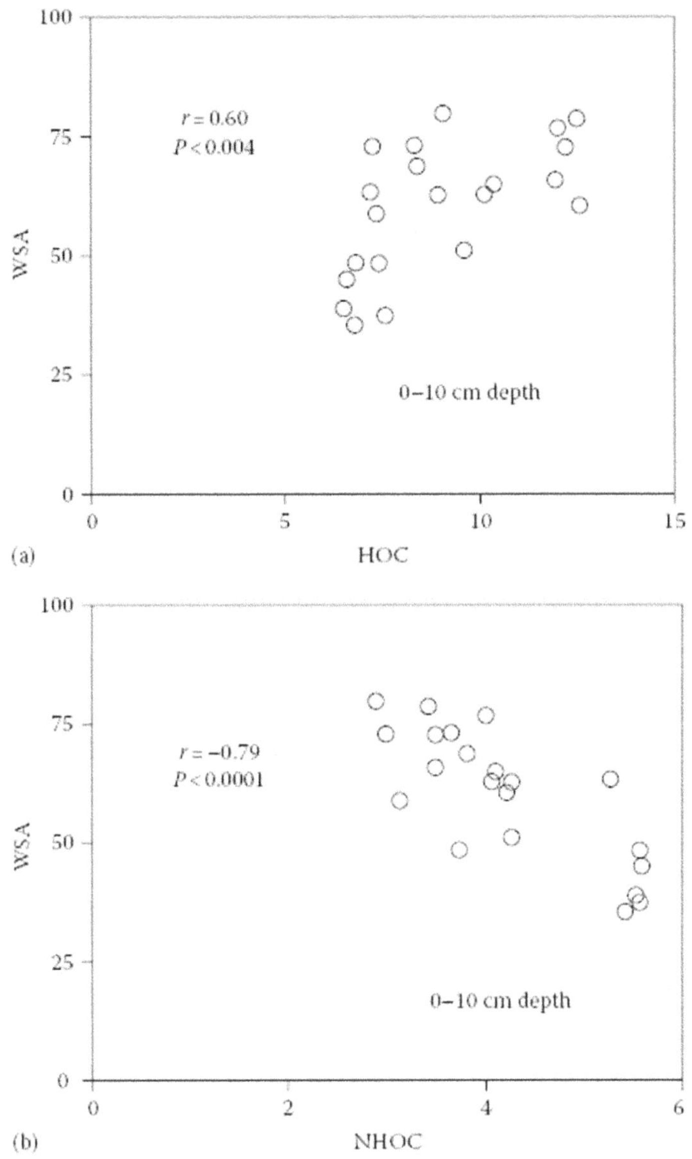

FIGURE 3.10 Relationship between water stable aggregation (WSA,%) and macroaggregate-associated (a) hydrolyzable (HOC) and (b) nonhydrolyzable (NHOC) organic carbon contents (g/kg). The data presented in the figure include macroaggregates from summit as well as the toe slope position of fields under no-tillage for 0–15 years. The macroaggregates were obtained from a Miamian soil in South Charleston, Ohio. *r* is correlation coefficient; *P* is the level of significance. (Redrawn from Ochoa, C. G., et al., *Can. J. Soil Sci.*, 89, 319, 2009.)

understood that mechanisms of aggregate formation and stabilization factors are often site-specific and depend on soil mineralogy, charge characteristics, texture, and land use and management practices prevailing in the area (Tisdall and Oades, 1982; Laird et al., 2001).

A drastic disturbance to soil causes breakdown of aggregates and loss of C (carbon) from the soil; however, restoration of disturbed minesoil improves soil quality and increases the organic C pool (Pederson et al., 1980; Barnhisel and Gray, 2000; Shukla et al., 2004a, 2004b; Shukla and Lal, 2005a, Shukla and Lal, 2005b). C and N (nitrogen) contents vary within the size range of each primary particle, especially clay (Anderson et al., 1981; Tiessen and Stewart, 1983). The C and N concentrations of soil also change with the amount of time since restoration and the aggregate stability, and C and N contents associated with macroaggregates and microaggregates also change with time. Spatial dependence of macroaggregate- and microaggregate-associated C and N also change with the amount of time since restoration (Shukla et al., 2007).

PROBLEMS

3.1 Define processes that are important for soil formation.
3.2 Write a short note on soil textural classification.
3.3 Soil was excavated from a cornfield located in Illinois from an area of 30 × 30 m to a depth of 30 cm. A representative soil sample with a volume of 1000 cm^3 was collected from the excavated soil profile. The weight of the oven-dried soil was 1100 g and the particle density was 2.65 Mg/m^3. Calculate the (a) bulk density of soil, (b) dry weight of soil in the pile, and (c) % solid space and porosity of the soil.
3.4 If the bulk density of the soil in this problem goes down by 15%, what will be the volume of the soil in m^3?
3.5 Write a note on particle shape and from published literature find out how particle shape effects packing density and other properties of soil.
3.6 Describe the role of SOC on soil structure improvement.
3.7 What is the difference between macroaggregates and microaggregates? Which of the two provides more protection to SOC.

REFERENCES

Anderson, D. W., S. Saggar, J. R. Bettany, and J. W. B. Steward. 1981. Particle size fractions and their use in studies of soil organic matter: I. The nature and distribution of forms of carbon, nitrogen, and sulfur. *Soil Sci. Soc. Am. J.* 45: 767–772.

Barnhisel, R. I. and R. B. Gray. 2000. Changes in morphological properties of a prime land soil reclaimed in 1979. In: A New Era of Land Reclamation: Proceedings of the 17th Annual Meeting of the American Society for Surface Mining and Reclamation, Tampa, FL, Vol. 1, June 11–15, 2000. University of Kentucky Extension Service, Lexington, KY, pp. 511–519.

Bevan, K. J. and P. F. Germann. 1982. Macropores and water flow in soils. *Water Resour. Res.* 18: 1311–1325.

Blake, G. R. and K. H. Hartge. 1986. Bulk density. In: A. Klute (ed.), *Methods of Soil Analysis*, Part I, 2nd edn. ASA Monograph No. 9, Madison, WI, pp. 363–376.

Brady, N. and R. Weil. 2004. *Elements of the Nature and Properties of Soils*, 2nd edn. Prentice Hall, Upper Saddle River, NJ, p. 606.

Buol, S. W., F. D. Hole, and R. J. McCracken. 1973. *Soil Genesis and Classification*, 1st edn. Iowa State University Press, Ames, IA.

Cady, J. G., L. Wilding, and L. R. Drees. 1986. Petrographic microscope techniques. In: A. Klute (ed.), *Methods of Soil Analysis. Part 1.* 2nd edn. Agron. Monogr. 9. ASA and SSSA, Madison, WI, pp. 185–218.

Chenu, C., Y. Le Bissonnais, and D. Arrouays. 2000. Organic matter influence on clay wettability and soil aggregate stability. *Soil Sci. Soc. Am. J.* 64: 1479–1486.

Cooper, T. H. 2009. Time as a soil forming factor. Soil 2125 Web Lab, unit 4, chapter 4. University of Minnesota, http://www.swac.umn.edu/classes/soil2125/doc/s4chp4.htm

Dane, J. and Topp (ed.). 2002. *Methods of Soil Analysis Part 4: Physical Methods. Soil Sci. Soc. Am.* Madison, WI, p. 1692.

Flint, A. and L. E. Flint. 2002. Particle density. In: J. H. Dane and G. C. Topp (eds.), *Methods of Soil Analysis Part 4: Physical Methods.* Soil Science Society America, Inc., Madison, WI, pp. 229–240.

Gerke, H. H. and J. M. Kohne. 2002. Estimating hydraulic properties of soil aggregates skins from sorptivity and water retention. *Soil Sci. Soc. Am. J.* 66: 26–36.

Ghezzehei, T. A. and D. Or. 2003. Pore-space dynamics in a soil aggregate bed under a static external load. *Soil Sci. Soc. Am. J.* 67: 12–19.

Gregorich, E. G., K. J. Greer, D. W. Anderson, and B. C. Liang. 1998. Carbon distribution and losses: Erosion and depositional effects. *Soil Tillage Res.* 47: 291–302.

Hendricks, S. G. 1942. Lattice structure of clay minerals and some properties of clay. *J. Geol.* 50: 276–290.

Horn, R. 1990. Aggregate characterization as compared to soil bulk properties. *Soil Tillage Res.* 17: 265–289.

Kuenen, P. H. 1960. Experimental abrasion 4: Eolian action. *J. Geol.* 68: 427–449.

Laird, D. A., D. A. Martens, and W. L. Kingery. 2001. Nature of clay humic complexes in an agricultural soil: I Chemical, biochemical, and spectroscopic analyses *Soil Sci. Soc. Am. J.* 65: 1413–1418.

Leamer, R. W. and B. Shaw. 1941. Simple apparatus for measuring non-capillary porosity on an extensive scale. *J. Am. Soc. Agron.* 33: 1003–1008.

Marshall, C. E. 1967. *The Physical Chemistry and Mineralogy of Soils.* Wiley, New York.

Marshall, T. J. and J. W. Holmes. 1979. *Soil Physics.* Cambridge University Press, Cambridge, UK.

Munsell Soil Color Chart. 2000. *Determination of Soil Color.* GretagMacbeth, Windsor, NY.

Ochoa, C. G., M. K. Shukla, and R. Lal. 2009. Macroaggregate-associated physical and chemical properties of a no-tillage chronosequence in a Miamian soil. *Can. J. Soil Sci.* 89: 319–329.

Pederson, T. A., A. S. Rogowski, and R. Pennock, Jr. 1980. Physical characteristics of some minesoils. *Soil Sci. Soc. Am. J.* 44: 321–328.

Santamarina, J. C. and. G. C. Cho. 2004. Soil behaviour: The role of particle shape. Proceedings of the Skempton Conference, March, London.

Shukla, M. K. and. R. Lal. 2005a. Erosional effects on soil properties in an on-farm study on Alfisol in west central Ohio. *Soil Sci.* 170: 445–456.

Shukla, M. K. and R. Lal. 2005b Temporal changes in soil organic concentration and stocks in reclaimed minesoils of southeastern Ohio. *Soil Sci.* 170: 1013–1021.

Shukla, M. K., R. Lal, and M. Ebinger. 2003a. Tillage effects on physical and hydrological properties of a typic Argiaquolls in central Ohio. *Soil Sci.* 168(11): 802–811.

Shukla, M. K., R. Lal, L. B. Owens, and P. Unkefer. 2003b. Land use and management impacts on structure and infiltration characteristics of soils in the North Appalachian Region of Ohio. *Soil Sci.* 168: 167–177.
Shukla, M. K., R. Lal, and M. Ebinger 2004a. Soil quality indicators for reclaimed minesoils in southeastern Ohio. *Soil Sci.* 169: 133–142.
Shukla, M. K., R. Lal, J. Underwood, and M. Ebinger. 2004b. Physical and hydrological characteristics of reclaimed minesoils in southern Ohio. *Soil Sci. Soc. Am. J.* 68:.1352–1359.
Shukla, M. K., R. Lal, and D. Van Leeuwen. 2007. Variability of soil organic carbon and nitrogen concentrations in macro, meso and micro aggregate fractions in reclaimed minesoils of eastern Ohio. *Soil Sci. Soc. Am. J.* 71: 1748–1757.
Six, J., E. T. Elliott, and K. Paustian. 1998. Aggregate and SOM dynamics under conventional and no-tillage systems. *Soil Sci. Soc. Am. J.* 63: 1350–1358.
Tiessen, H. and J. W. B. Stewart. 1983. Particle size fractions and their use in studies of soil organic matter: II. Cultivation effects on organic matter composition in size fractions. *Soil Sci. Soc. Am. J.* 47: 509–514.
Tisdall, J. M. and J. M. Oades. 1982. Organic matter and water stable aggregates in soils. *J. Soil Sci.* 33: 141–163.
Yoder, R. E. 1936. A direct method of aggregate analysis and a study of physical nature of erosion losses. *J. Am. Soc. Agron.* 28: 337–351.
Youker, R. E. and J. L. McGuinness. 1957. A short method of obtaining mean weight diameter values of aggregate analyses of soils. *Soil Sci.* 83: 291–294.

4 Sampling Concepts and Designs

4.1 INTRODUCTION

In general, people are familiar with sampling, such as food sampling and sampling for cosmetics. Although the purpose of these samplings is to ensure that the quality is acceptable, it is still different from soil sampling. Soil sampling essentially consists of aspects related to site selection, selection of tools, protocol collection, sample collection, sample handling, sample storage, sample preparation, and analysis and interpretation of the data. The quality of sample collection is as important as the quality of analysis or the analytical procedure. When a soil sample of appropriate size and quality, that is representative of the study site, is provided to a laboratory for analysis, the results can be expected to be of good quality. It is also important to identify some clear-cut objectives and criteria for success before collecting soil samples. This in turn may indicate the quality of data required to achieve certain benchmarks. According to the US Environmental Protection Agency (USEPA, 2002), data quality objectives must address precision, accuracy, representativeness, completeness, comparability, and level of detection. Precision is important when the change in a property of the soil is very low. Accuracy, although used as a synonym for precision, can be controlled by the laboratory. Representativeness refers to the degree to which the samples collected from a given area truly represent the property or variability of the property in question. Completeness and comparability are important for making valid comparisons over space and time between the data collected from two different sites. In spite of the guidelines that a researcher should follow to collect good-quality datasets, one must always remember that soil is highly variable, and various processes occurring in the soil or at the soil–atmosphere interface (soil surface) are complex and dynamic in nature and may necessitate employing different sampling strategies for different objectives. In addition to the data quality objectives discussed here, before collecting soil samples from a study site, attempts should be made to gather as much background information as available with different agencies and in the published literature.

4.2 REPRESENTATIVE ELEMENTARY VOLUME

A heterogeneous system of voids and solids is known as a porous medium. A porous medium consists of multiple randomly distributed, interconnected pores or voids of a broad range of sizes, shapes, lengths, and arrangements. Is a given soil volume a porous medium? Yes, it is. Because it has pores, voids, and solids, and air, water, and solute can flow through it. Is a given volume of concrete a porous medium? Yes, it is, provided it is not fully impervious with no interconnected pores. On a macroscopic scale,

DOI: 10.1201/9780429264849-4

a porous medium is a discontinuous mass consisting of solids, voids, cracks, liquids, and gases. In order to determine the properties of a given porous medium, it is usually considered to have a continuous mass or extent rather than a composition of discrete particles. The concept in which a real, discontinuous porous medium is replaced by a hypothetical, (probably) fictitious, but representative contiguous area is known as a continuum concept. A continuum is a mass of media that can be divided into smaller elements with properties of elements being representative of the bulk media. The macroscopic properties (say porosity, or P) of the continuum are measurable and are also differentiable over some specified domain (x) of a given area or volume:

$$\lim_{x \to x_0} (P) = f(x_0) \tag{4.1}$$

This equation has an operator—lim—indicating that the limit of P when the domain of x tends to be x_0 is given by a function (f) of x_0. When macroscopic properties of a porous medium are defined over an area, it is known as the representative elementary area (REA). REA is also defined as the minimum area over which the measurement of a property is unbiased and representative. When macroscopic properties of a porous medium are defined over a volume, it is known as the representative elementary volume (REV). REV is also defined as the minimum or smallest volume of a porous medium over which the measurement of a porous medium property can provide a value that is representative of the field under investigation.

The choice of REV (or REA) can be based on several criteria. In general, the chosen REV must be indifferent to the property, insensitive to the method of determination, and invariant in time and space. Indifferent to property implies that the chosen REV must meet the minimum size for all the properties of interest. The REV choice must be independent of the experimental technique used to measure the property. In addition, the REV must be the same at all locations within a domain as well as in time, although the domain does not have to be necessarily homogeneous.

For any known volume of a porous medium, there is a domain of microscopic effects where pore scale variations are important (Figure 4.1), beyond which the domain of the porous medium starts. The volume at the lower boundary of this domain is known as the REV. The properties of the porous medium beyond the REV can show a distinct change with increasing volume of the medium, especially for heterogeneous porous media. On the other hand, properties could show small to no change for homogeneous porous media (Figure 4.1).

The microscopic and macroscopic variations of porosity as an example are shown in Figure 4.2. There are three volumes—1, 2, and 3—selected in Figure 4.2. Volume 1 shows that the entire volume of the sample is centered in the pore, and therefore the total volume of the sample is equal to the volume of the pore because the volume of solids is zero. Therefore, porosity, the ratio of the volume of pores to total volume of the sample, is at the maximum or equal to one. On the other hand, volume 2 shows that the entire volume of the sample is centered in the solid, and therefore the total volume of the sample is equal to the volume of the solid because the volume of pores is zero. In this case, porosity will be at the minimum or zero. Volume 3 in

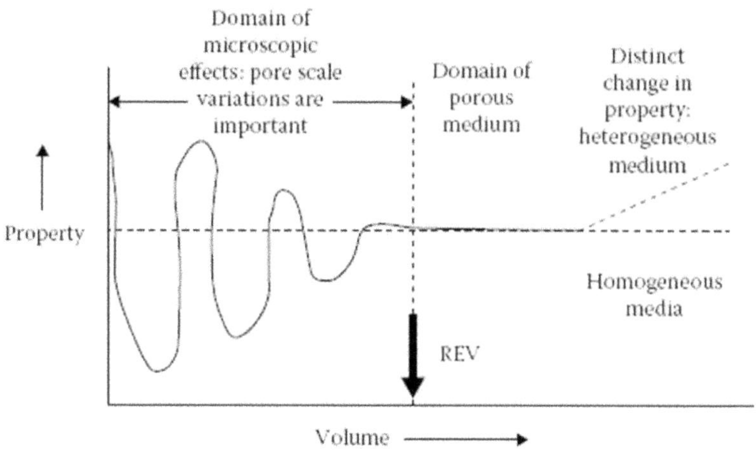

FIGURE 4.1 Domain of microscopic effects, domain of porous media, and domain of homogeneous and heterogeneous porous media.

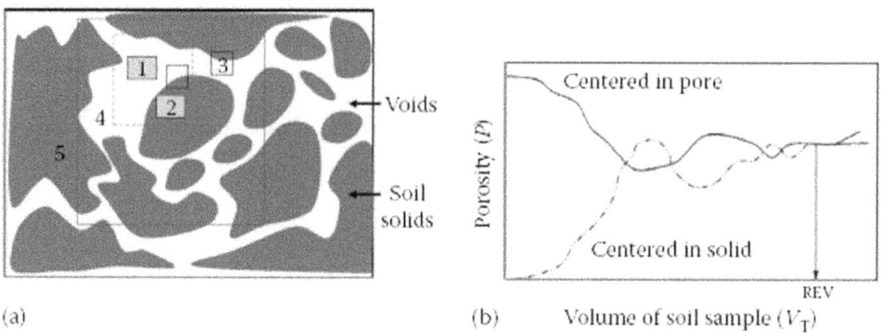

FIGURE 4.2 The representative elementary volume for a soil property such as (a) porosity volume of a soil and (b) change in porosity with increasing volume of soil sample.

Figure 4.2a shows that the volume of the sample is the sum of the volume of the pore and the solid, and therefore the porosity of the sample is between zero and one.

Thus, the estimation of the total volume of the sample (V_T) is crucial. From this example (Table 4.1), it is clear that when

1. V_T is small enough or $V_T = V_P$, $P = 1$ (centered in a pore).
2. V_T is small enough or $V_T = V_S$, $P = 0$ (centered in a solid).
3. V_T increases, we reach a situation when either

$$V_T = \left(V_P + \Delta V_S\right)$$

TABLE 4.1
Calculation of Porosity for the Three Volumes (1, 2, and 3) Presented in Figure 4.2a

Volume Selected	Volume of Pores	Volume of Solids	Total Volume	Porosity
1	V_p	0	$V_p + 0 = V_p$	$V_p/V_p = 1$
2	0	V_s	$0 + V_s = V_s$	$0/V_s = 0$
3	$V_p = 0.7$	$V_s = 0.3$	$V_p + V_s = 1$	$V_p/V_T = 0.7/1 = 0.7$

where Δ represents the change in the quantity, and for the two situations, P is neither one nor zero. As we select larger sample volumes (see volumes 4 and 5 in Figure 4.2a), more and more pore and solid spaces are included in the total volume of the sample. The porosity fluctuates between zero and one, depending on the errors associated with selection of a representative volume until it reaches REV (Figure 4.2b). At REV and later, the changes in porosity values are no longer due to the sample being not a representative sample, but they are due to the natural variability of the porosity of the porous medium.

4.3 SAMPLE SIZE

Knowledge of the number of samples that must be collected is important for adequately representing and characterizing a study area. However, it is one of the most difficult things to determine. When the number of samples collected is less than required, there could be statistical uncertainty. On the other hand, when the number of samples collected is more than required, it could be expensive as well as time consuming. Some considerations prior to sample collection are what is expected of the sample, what population parameters are involved, how much data are already available, what is the variability of the known population, what is the expected precision, what is the cost of the sampling, what is the equipment requirement, and how difficult is it to collect the samples (accessibility to sites, etc.).

There are several equations available in the literature to calculate the number of samples required to estimate a mean of the population (McBratney and Webster, 1983; Gruijter, 2002; Nielsen and Wendroth, 2003). Before we talk about sample size, let us understand "mean." Mean can be estimated and represented as the arithmetic, geometric, or harmonic mean. The arithmetic mean is the central tendency of the collected samples expressed as the ratio of the sum of the samples and the total number of samples collected:

$$\bar{x} = \frac{1}{n}\sum_{i=1}^{n} x_i \tag{4.2}$$

where

$\bar{x}$ is the arithmetic mean.

n is the number of samples.

x_i is the observation of a parameter.

The geometric (G) and harmonic (H) means are estimated as follows:

$$\log G = \frac{1}{n}\sum_{i=1}^{n} \log x_i \tag{4.3}$$

$$H = \frac{n}{\sum_{i=1}^{n} (1/x_i)} \tag{4.4}$$

Variance, the second moment about the mean, is the average of the squares of deviations of a value from the mean (also known as the first moment); standard deviation is the square root of variance:

$$\sigma^2 = \frac{1}{n}\sum_{i=1}^{n} (x_i - \bar{x})^2 \tag{4.5}$$

where

σ^2 is the variance.

σ is the standard deviation about the mean.

The coefficient of variation (CV) is the ratio of the standard deviation and mean. Standard deviation is also known as the measure of absolute variability, whereas the coefficient of variation is a measure of relative variability:

$$CV = \frac{\sigma}{\bar{x}} \tag{4.6}$$

Based on the mean ($\bar{x}$) and the absolute error from the mean ($x_i - \bar{x}$), the number of samples or sample size, for the absolute error to be smaller than the chosen error or chosen accuracy (r), can be calculated as follows:

$$n = \left(\frac{t_\alpha \times \sigma}{r}\right)^2 \tag{4.7}$$

where t_α is the Student's t for a one-sided probability level (confidence level) of $(1 - \alpha)$, and α can also be taken as $(1 - \alpha/2)$ for a two-sided confidence level. The sample size can also be estimated by using the relative error $(x_i - \bar{x})/\bar{x}$ that is smaller than the chosen error (r) in place of the absolute error as follows:

$$n = \left(\frac{t_\alpha \times \sigma}{r\bar{x}}\right)^2 \tag{4.8}$$

Example 4.1

Soil bulk density (BD; g/cm³) was measured at ten random locations in a pecan orchard located in southern New Mexico. The data are provided here. Calculate the number of samples required for an accuracy of 0.05 g/cm³.

No.	1	2	3	4	5	6	7	8	9	10
BD	1.15	1.3	1.45	1.6	1.14	1.48	1.61	1.49	1.31	1.44

Number of samples (n) = 10

The mean BD = (1.15 + 1.3 + 1.45 + 1.6 + 1.14 + 1.48 + 1.61 + 1.49 + 1.31 + 1.44)/10 = 13.97/10 = 1.397 g/cm³

Variance = 0.0251 and standard deviation = 0.1584

Required accuracy = 0.05

Look up the t-table (double tailed) for degree of freedom (df = $n - 1 = 10 - 1 = 9$) and α of 0.025

t-value (from table) = 2.228

$$\text{Achieved accuracy} = \frac{t_\alpha \times \sigma}{\sqrt{n}} = \frac{2.228 \times 0.1584}{10^{0.5}} = 0.1116$$

Thus, the achieved accuracy is more than the required accuracy.

Calculate the number of samples with table value of t for the required accuracy.

$$n = \left(\frac{t \times \sigma}{r}\right)^2 = \left(\frac{2.228 \times 0.1584}{0.05}\right)^2 = 50$$

df = 49, look up t-table for df of 49 and α of 0.025, table value of $t = 2.032$.

$$\text{Achieved accuracy} = \frac{2.032 \times 0.1584}{1400^{0.5}} = 0.0509$$

Thus, achieved accuracy is equivalent to the required accuracy.

$$n = \left(\frac{t \times \sigma}{r}\right)^2 = \left(\frac{2.032 \times 0.1584}{0.05}\right)^2 = 41$$

Thus, 41 samples are required for determining the soil bulk density with an accuracy of ±0.05 g/cm³.

The same problem can also be solved using the relative error in place of the absolute error.

4.4 SAMPLING DESIGNS

In order to estimate the properties from a population, for example, soil physical properties from a large agricultural field, sampling involving much fewer samples than the population is performed. Sampling provides investigators an opportunity to collect as many samples that can be managed within their available resources and still provide a reasonable estimate of the properties of the total population. One of the first questions that is asked is how to collect a sample. In other words, selection of an appropriate technique and a sampling design is a prerequisite to estimating the properties of the population (Carter and Gregorich, 2006).

Before we go any further with the sampling design, which identifies the location, type, and number of sampling units required for measurement, let us discuss some of the terms associated with sampling design concepts. Target population refers to the population about which a decision has to be made. When a part of the target population is selected for making a decision, it is known as the sampled population. A sampling unit is one of the members of the population selected for sampling. When a sampling unit is expressed in terms of a physical quantity, such as size, shape, mass, volume, or area, that physical quantity is known as a sample support. For example, for determining soil bulk density, a core sample was collected from an experimental field. If the volume of soil is equal to the volume of the core, the volume of the core is the sample support (USEPA, 2002).

Sample locations must be selected based on some criteria. It is not recommended to select sampling locations solely based on convenience because it can lead to an inaccurate representation of the area under study. For example, collecting samples from the row nearest to the road or irrigation canal or any other easily accessible location can save money and time but can under- or overestimate the property of interest, such as average water content of soil in the entire field. Sample locations can be selected based on the prior familiarity of the investigator to the study area, and this type of sampling is known as judgmental sampling. For example, in a study on removal and subsequent accumulation of light organic matter due to water erosion, an investigator can look at the study area and identify the different slope areas (summit, back, or toe slopes), including the depositional area. Based on his judgment, he can then collect samples to adequately study the removal–deposition of light organic matter fractions. The accuracy of the method depends on the judgment of the investigator. The other type of sampling approach is known as probability sampling. This approach uses several sampling layouts and calculates probability at each sampling point for collecting random samples (Pennock et al., 2006; Carter and Gregorich, 2006). The probability-based sampling technique is also known as the design-based approach, and it ensures the collection of independent samples required for conducting classical statistical analysis (Brus and de Gruijter, 1997).

Soil sampling could be done in a variety of ways. Some of the sampling designs that will be summarized briefly include simple random sampling, systematic sampling, stratified sampling, and cluster sampling (de Gruijter, 2002; Figure 4.3a–d). More information on sampling designs is available elsewhere (Dane and Topp, 2002).

4.4.1 Simple Random Sampling

In the simple random sampling technique, no partitioning is done, and there is no restriction other than possibly a predetermined sample size. Each sample is randomly

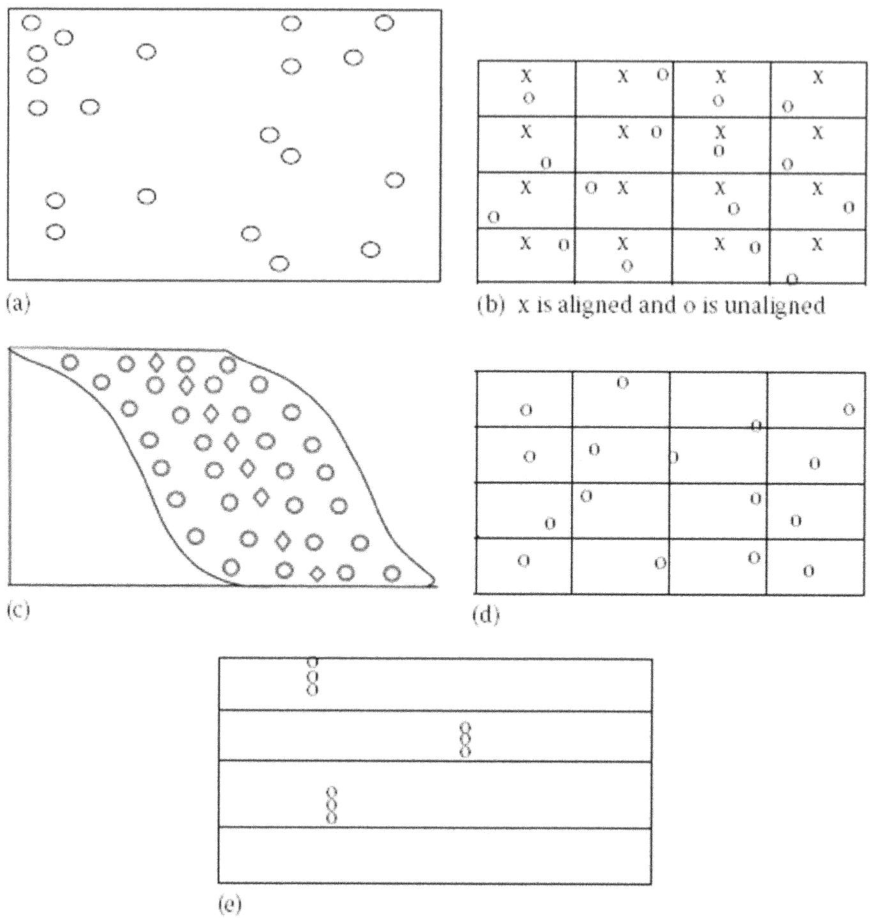

FIGURE 4.3 Sampling designs: (a) simple random sampling, (b) systematic sampling when samples are aligned and unaligned, (c) transect and systematic sampling along a slope, (d) stratified sampling, and (e) cluster sampling. (Modified from de Gruijter, J. J., Soil sampling, in Dane, J. H. and Topp, G. C., *Methods of Soil Analysis—Part 4, Physical Methods*, Soil Science Society America, Inc., Madison, WI, pp. 45–79, 2002.)

chosen and has the same probability of selection during the entire process of sample selection (Figure 4.3a). It is different from collecting a "random sample" because random samples may not truly represent the population and their variation could be simply due to sampling errors. In order for the samples to be truly representative of the population, an unbiased random selection of samples is a prerequisite. The advantages of employing a simple random sampling are that the method is simple, limited knowledge of the population is required, it is cost effective, and it is free from classification errors. It is also very useful for relatively homogeneous areas. The disadvantages of simple random sampling are that the method is time consuming, especially when samples are irregularly distributed and in cases where there are large empty spaces between samples.

4.4.2 Systematic Sampling

Systematic sampling is a type of probability sampling in which samples are selected at a regular interval from a predetermined point following a predetermined scheme (Figure 4.3b). The sampling starts with a random selection but progresses with a predetermined interval. As an example, first a random sample is collected from the list of data, and subsequent samples are collected at a certain fixed interval, say, every 10th or every 15th in the database. As shown in Figure 4.3b, samples can be collected at the intersection of each grid or at the center of each grid. On the other hand, systematic sampling can also be done along a transect. As far as possible, the starting point of a grid or transect should be randomly selected. For level or nearly level fields, grid sampling can be preferred, especially when the objective is to understand the spatial variability of properties using geostatistical tools. Transect sampling can be done for areas having distinct slopes. If the study area has a gradual slope in one direction, from the summit or top to the depositional area only (not usually found), sampling along one transect passing through all different slope positions could be enough (Figure 4.3c). Depending on the number of different distinct slopes identified in an area, a number of transects could be selected. Grid sampling could also be performed in a sloped area by making sure that the grid is large enough to cover all dominant distinct slopes in the study area (Figure 4.3c).

The advantages of the systematic sampling method are that it is simple and systematic, and points can be easily located in the future. It can also reduce the time of travel between samples, especially if the terrain is easily accessible. It can divide the study area into regular grids that could be a square, rectangle, circle, or any other defined grid. This sampling design is useful for determining spatial and temporal trends. Since sampling is not completely random, the unbiased estimates on variation (variance or standard deviation) are not available, and thus quantification of accuracy of the estimate is difficult. This method may not represent the population when the data have some sort of periodicity. When systematic sampling is done in such a way that samples are not collected in a straight line but are unaligned, it is known as systematic unaligned sampling (Figure 4.3b). This type of a design may provide a better representation of the study area. If the study area has small depressions or small elevations, a systematic unaligned sampling can provide a better estimate of the population means than sampling in a straight line.

4.4.3 Stratified Sampling

Stratified sampling involves dividing the sampling area into strata, which are supposed to be more homogeneous, as variations in a soil property are smaller within a stratum than among strata (Figure 4.3d). Strata can be selected based on prior knowledge of the sampling area, and spatial or temporal closeness of strata. Selection of strata in a field could be dividing the area into subgroups by soil texture. A predetermined number of soil samples is collected using simple random sampling in each stratum. Stratified sampling is especially useful for areas where the population contains several discrete groups or classes. This sampling strategy is important because it can maintain the integrity of an individual strata, which could be lost following alternative strategies such as simple random or systematic sampling. It is believed to have higher statistical accuracy, and it provides an opportunity to apply different sampling strategies on different strata. The disadvantages of using the stratified sampling approach include difficulties associated with strata delineation, types of strata present in the study area, inconsistent relationships between different strata, it being less cost effective and more time consuming, and a possible requirement of a large sample size.

4.4.4 Cluster Sampling

Cluster sampling involves collection of a predetermined number of samples in groups or clusters (Figure 4.3e). This sampling technique is different from simple random, systematic, and stratified sampling techniques because it does not select individual samples. Like systematic sampling, a starting sample is collected randomly, and then other samples are identified that can be included in a cluster based on some predetermined classification or definition of the cluster. A cluster can be identified based on some threshold value (hot spot), and more samples are collected around that spot. Once the threshold value is exceeded, it becomes a different cluster, and more samples are collected around that spot for that cluster. Clusters could be later identified as representing hot (e.g., high Na^+ concentration) and cold (low Na^+ concentration) spots in an experimental field. The process is repeated until a certain predetermined number of clusters are selected. This technique has some of the advantages of systematic sampling. Generally, cluster sampling has lower precision than simple random sampling.

4.4.5 Composite Sampling

Composite sampling involves combining samples from several adjacent spots into one homogeneous sample. The advantages of this method include reduction in the number of samples to be analyzed, reduction in the cost of analysis, and greater representation because of the larger volume of the sample. This sampling method is especially helpful if analysis costs are very high as compared to the sampling costs, for example, identification of microbial diversity in an agricultural field.

4.5 PRACTICAL ASPECTS OF SOIL SAMPLING

The goal of soil sampling is to collect one or more samples that are representative of the study area or target population. If samples collected are representative, experimental errors are low. One of the ways to reduce experimental errors is to collect repeated unbiased samples in each class of the population, also known as replicates. For applying a statistical analysis, replicates must be correctly identified and selected, and the population must be clearly defined at the beginning of sample collection. If the population is not clearly defined at the initiation of the research, replicates may not be true replicates, but pseudoreplication can occur (Steel and Torrie, 1980; Pennock et al., 2006). When dealing with a number of treatments in a study, for example, different rates of fertilizers applied on plots, the randomization of plots must be ensured by locating various treatment plots randomly using a computer-generated randomization or random number table or any other scientific method (Pennock et al., 2006).

Sometimes, it is not practically possible to randomize the location of experimental plots, for example, soil sampling performed to quantify soil behavior among different soil textures or among different slope positions. Different soil textural classes in nature usually occur in noncontiguous patches, and thus experimental plots cannot be randomized. Under this situation, care should be taken to ensure that samples are not collected too close to the location of the soil textural change. These types of transitional zones should be avoided as far as possible. Sometimes, a study could have more than one interaction, for example, quantifying light fraction organic carbon in different soil textures and slope positions. Again, slope positions should be clearly identified based on some standard classification, which could be the elevation of points within a certain range, and care should be taken to stay away from transitional areas.

In order to collect samples that accurately represent the population, an appropriate and well-planned sampling design is a prerequisite. One must always remember that a sampling design appropriate for a given objective may not be useful for some other objectives. Therefore, identification of the objective or the purpose of sampling is important before selecting a sampling design. For example, if the objective of the research is to monitor changes in the chemical composition of soil due to leakage in the standpipe of a sprinkler irrigation system, samples can be collected from around the leaking standpipe for chemical analysis (Figure 4.4). However, if the objective is to determine the average chemical composition of soil in the entire irrigated area, samples must be collected from the entire area including the leakage area, because the area around the standpipe is not representative of the entire irrigated area (Figure 4.4). Another point that must be kept in mind is that a faulty sampling design can lead to erroneous conclusions, it could prove costly, and it may be too late to gather the required information. For example, if the objective was to collect the background data from an experimental site to quantify changes in soil quality over time due to wastewater application, a faulty sampling design could result in no good-quality background data being available.

Some other practical aspects of soil sampling are to avoid making sampling design too complicated. It is always good to get a good map of the study area, but it is highly recommended to visit the study site before designing a sampling strategy. Some of the issues related to the site and the terrain of the experimental area can only be fully understood by actually visiting the site. This will also help in making reasonable cost, labor, and time estimates for completing the soil sampling. There are instances when a predetermined sampling design could not be executed on the day of sampling because the experimental site is hilly with variable slopes, with some of the slopes

FIGURE 4.4 The samples collected around the leaking standpipe (see black filled circles) to quantify chemical concentration or contamination due to leakage. More samples near the standpipe, bare areas, and under the canopy (see gray diamonds) are collected to determine the average chemical concentration in the field.

so steep that they are inaccessible. As an example, if the experimental field is large and undulating with no direct vehicular access to certain areas of the field where infiltration tests are to be conducted, a visit to this site will be helpful in deciding how many people will be needed to carry water to different parts of the field and how much time will be required to complete all the tests. Conversely, it can help us decide on the number of tests that can be made within the available resources. Sometimes, there are unexpected delays that are beyond the control of the investigator. I remember an instance when due to heavy rains and winds for a few days, electric poles got uprooted and the site became unreachable. In another instance, a fence was erected, cutting off the approach road to the experimental site. Therefore, suitable allowances must be kept for unexpected delays. In some cases, a test phase can be included before the actual sampling design is executed. Additionally, proper care must be taken with regard to vehicles and weather, as far as possible, and prior permissions must be obtained before going to an experimental site. Appropriate attire must be worn during a visit to the experimental site. A first aid kit should be available either at the experimental site or in the vehicle, and do not forget to carry water with you.

PROBLEMS

4.1 What are REV and REA? Select a physical property of soil and schematically show REV and REA.

4.2 Select a dataset from the literature on a soil property and estimate the number of samples required for an assumed accuracy.

4.3 List various types of soil sampling designs. If your objective is to determine saturated hydraulic conductivity of soil from a 1 ha field consisting of soils with different textures, draw a map of your choice with known locations of the contiguous patches of soils of different textures, design a sampling strategy, and justify your choice.

REFERENCES

Brus, D. J. and J. J. de Gruijter. 1997. Random sampling or geostatistical modelling? Choosing between design-based and model-based sampling strategies for soil (with discussion). *Geoderma* 80: 1–44.

Carter, M. R. and E. G. Gregorich. 2006. *Soil Sampling and Methods of Analysis*, 2nd edn. CRC Press, Boca Raton, FL.

Dane, J. H. and G. C. Topp (eds.). 2002. *Methods of Soil Analysis—Part 4, Physical Methods.* Soil Science Society America, Inc., Madison, WI, p. 1692.

de Gruijter, J. J. 2002. Soil sampling. In: J. H. Dane and G. C. Topp (eds.) *Methods of Soil Analysis—Part 4, Physical Methods.* Soil Science Society America, Inc., Madison, WI, pp. 45–79.

McBratney, A. B. and R. Webster. 1983. How many observations are needed for regional estimation of soil properties? *Soil Sci.* 135: 177–183.

Nielsen, D. R. and O. Wendroth. 2003. *Spatial and Temporal Statistics.* Catena Verlag GMBH, Reiskirchen, Germany.

Pennock, D., T. Yates, and J. Braidek. 2006. Soil sampling designs. In: M. R. Carter and E. G. Gregorich (eds.), *Soil Sampling and Methods of Analysis*, 2nd edn. CRC Press, Boca Raton, FL.

Steel, R.G.D. and Torrie, J. H. 1980. *Principles and Procedures of Statistics. A Biometrical Approach.* McGraw-Hill, New York, p. 633.

USEPA. 2002. *Guidance for Choosing a Sampling Design for Environmental Data Collection (EPA QA/G-5S).* U.S. Environmental Protection Agency Quality Staff (2811R). USEPA, Washington, DC. http://www.epa.gov/quality/qs-docs/g5s-final.pdf, accessed on July 23, 2013.

5 Spatial Variability of Vadose Zone Properties

5.1 INTRODUCTION

Soil properties have been known to be spatially and temporally variable. Soil properties display both short- and long-range variability, are multivariate in nature, and can have systematic or random variation (Nielsen et al., 1973; Burrough, 1983). Systematic variation indicates a gradual change in soil properties across landscapes and could be due to the intrinsic factors of soil formation, geomorphic elements, and landforms. Systematic variation could also be due to extrinsic factors associated with land use and management. Random variations based on intrinsic factors of soil formation are difficult to explain. Intrinsic variability is caused by pedogenesis and usually takes place within large timescales. The variability caused by pedogenetic processes is known as regionalized, with nearby areas considered to be more similar than areas that are farther away (van Es, 2002). The variability caused by extrinsic factors could take effect relatively quickly and could not be treated as regionalized. Two important characteristics of sample populations are measures of central tendency, including mean, median, and mode, and measures of dispersion around the mean, including variance, standard deviation, range, and coefficient of variation (Mulla and McBratney, 2000).

5.2 SOURCES OF VARIABILITY

In addition to the known and unknown sources of variability due to the intrinsic and extrinsic factors, there are several other sources of variability. For example, decisions and choices made by investigators during sampling, sampling designs, availability of resources, number of investigators involved in sampling and analysis, skill level of investigators, type and quality of tools and equipment used to collect samples and analyses, scale of the domain, availability of time, accessibility of sites, criteria of success, and assumptions made for the sampling and analysis have a profound influence on the variability. Methods of measurements also strongly influence variability; for example, infiltration rates measured using a single ring may produce significantly different mean and standard errors than those measured using a double ring. Sample support can also influence the variability; for example, increasing or decreasing the size of the infiltrometer rings can change the mean and variability of the infiltration rates and amounts. Similarly, hydraulic conductivity measured in the field could show a much larger variability than that measured in the laboratory (Shukla, 2011).

DOI: 10.1201/9780429264849-5

5.3 SCALE OF VARIABILITY

Soil maps are one of the most basic tools that are needed before starting any research related to soils. Efforts by the Natural Resources Conservation Service of the US Department of Agriculture have resulted in preparation of soil maps for almost all of the United States. There are five orders of soil survey in the United States (Lin et al., 2005), with order I (1:12,000) being the most detailed and order V (1:250,000) being for general mapping. For areas with less intensive land use, soil surveys are generally conducted at the 1:24,000 scale, and for areas with intensive land use, more detailed surveys are conducted at the scale of 1:12,000 (Soil Survey Division Staff, 1993). For preparing index maps, the scale is usually 1:125,000. However, even at the scale of the most detailed survey, there is not enough information available to assess the variability of soils within a map unit for designing a management-related strategy or quantification of spatial variability at any scale from the available soil survey data.

The variability of soil properties can begin at the molecular scale, where it can be due to the position and number of molecules of ions. The variability can be caused due to the Brownian motion of the molecules inside the filled pore space of a soil. Brownian movement or motion is the irregular or zigzag motion of molecules bumping into each other and thus motion becoming random. The next scale of variability could be at a particle or pore scale. Pore-scale variability could be caused by the thickness of water films inside the pores, the variability of size and shape of pores, the number of cations or anions adsorbed on the pore surface, and the material of the porous media. The next level of variability is the aggregate scale where more pore spaces and solid material are included in the analysis. At the aggregate scale, the variability associated with pores and a solid matrix could show increased variability than that observed at the particle or pore scale. The next scale is the column, soil horizon, or pedon scale. In this case, more types of soils are included and heterogeneity with depth increases due to changes in soil taxonomy, morphology, and types. At this scale, different soil horizons are included, and the variability associated with each of them contributes to the total variability.

From a pedon, the next stage of variability can be quantified at the polypedon scale, or field or watershed scale. More factors associated with extrinsic sources of variability are included at this scale. Management-associated differences, different landowners, crop rotations, land-use differences, and skill level of investigators and instruments used can exacerbate the variability at this scale. The variability is expected to increase further as more and more areas (spatial extent) are included in the investigation (watershed to regional or pedosphere scales). Variability at these scales is probably due to the five factors of soil formation—parent material, climate, biota, relief, and time—as well as space-time factors and the land use and management factors. Some of these factors are difficult to assess because they are the functions of space and time. It is safe to assume that as the spatial and temporal extents increase, the variability of soil also increases.

A large number of studies on spatial variability have been carried out on a single chosen scale (Nielsen et al., 1973; Webster, 1985; Shukla et al., 2004), but only a few attempts have been made on quantifying the spatial variability at multiple scales

(Burrough, 1983; Penmock and de Jong, 1990). It is common knowledge that the modeling of hydrological processes requires an assessment of soil variability at multiple scales. Soil maps, although extremely helpful in making an initial informed decision, usually divide areas into homogeneous (within a map unit), transitional (border area between two map units), and heterogeneous (across different map units). Inclusion of spatial and temporal variability at multiple scales is clearly required in future soil mapping efforts.

5.4 STATISTICAL EVALUATIONS

In classical statistics, sample populations are represented by frequency distributions. Sample class intervals within a range of data (on *x*-axis) and the number of observations occurring within the class intervals (on *y*-axis) are plotted to obtain frequency distributions (Figure 5.1). Frequency distributions are symmetrical when data are normally distributed and asymmetrical when data are non-normal. For normal distributions, values of mean, median, and mode are equal, while for non-normal distributions they are not. The variability of a soil property can be expressed by range, interquartile range, variance, standard deviation, coefficient of variation, skewness, and kurtosis. Range, the difference between the largest and the smallest value, is commonly used to express variability, but it does not truly describe the characteristic of the datasets especially when there are outliers. Interquartile range is the difference between the third quartile and the first quartile for data arranged in ascending order. It is also known as middle fifty and provides a measure of variability or statistical dispersion. It is better than the range because it eliminates the outliers at both ends of the dataset.

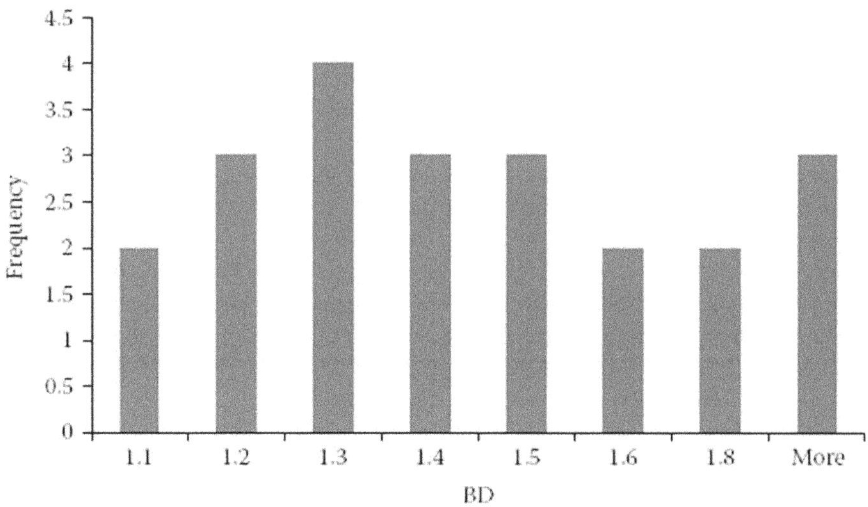

FIGURE 5.1 Frequency distribution for the bulk density (BD) data presented in the table in Example 5.2.

$$\text{Range} = \text{Maximum value} - \text{Minimum value}$$

$$\text{Interquartile range} = Q_3 - Q_1$$

where

Q_3 is the value at third quartile.

Q_1 is the value at first quartile for data arranged in ascending order.

Interquartile range is often preferred over range.

Example 5.1

For the following dataset, where soil bulk density in Mg/m^3 is arranged in ascending order, obtain the range and interquartile range.

1	2	3	4	5	6	7	8	9	10	11
1.0	1.5	1.2	1.25	1.30	1.32	1.36	1.39	1.42	1.45	1.9

Range = 1.9 – 1.0 = 0.9 Mg/m^3

Q_1 is the value at first quartile = 1.20 Mg/m^3

Q_3 is the value at third quartile = 1.42 Mg/m^3

Interquartile range = 1.42 – 1.20 = 0.22 Mg/m^3

Variance is the second moment about the mean and is calculated as the average of the squares of deviations of values from the mean or the first moment. Standard deviation is the square root of variance, whereas coefficient of variation (CV) is the ratio of the standard deviation and the mean. Standard deviation is also known as the measure of absolute variability, whereas coefficient of variation is a measure of the relative variability. Variance, standard deviation, and coefficient of variation of various soil properties are discussed in Chapter 4.

Skewness is calculated as the ratio of the third moment about the mean and the standard deviation, and it is a measure of the asymmetry of the probability distribution of a random variable. The skewness could be positive or negative. A positive skewness indicates that a greater part of the values is to the left of the mean and the tail on the left side is shorter than the tail on the right side (Figure 5.2). On the other hand, a negative skewness indicates that the bulk of the values are to the right of the mean and the tail on the left side is longer than the tail on the right side. However, for a symmetrical distribution, skewness is equal to zero.

$$\text{Skewness} = \frac{1/n\sum_{i=1}^{n}\left(x-\bar{x}\right)^3}{\left(1/n\sum_{i=1}^{n}\left(x-\bar{x}\right)^2\right)^{3/2}} \tag{5.1}$$

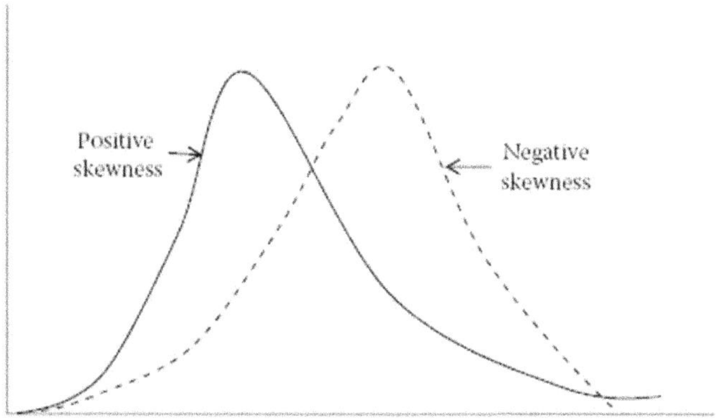

FIGURE 5.2 Positive and negative skewness.

Kurtosis, the ratio of the fourth moment about the mean and the standard deviation, is a measure of the peakedness of the probability distribution of a random variable. Higher peakedness means more variance, probably due to outliers (Isaaks and Srivastava, 1978; Nielsen and Wendroth, 2003).

$$\text{Kurtosis} = \frac{1/n\sum_{i=1}^{n}\left(x-\bar{x}\right)^4}{\left(1/n\sum_{i=1}^{n}\left(x-\bar{x}\right)^2\right)^2} - 3 \tag{5.2}$$

For a normal or symmetric distribution, the kurtosis value is equal to three. The negative three in the formula is used to present the kurtosis of a normal distribution as zero. The positive kurtosis indicates a peaked distribution, while negative kurtosis indicates a flat distribution.

Example 5.2

For the following dataset, where soil bulk density in Mg/m^3 is arranged in ascending order, calculate skewness and kurtosis. *Note*: Soil bulk density values are in the second and fourth row.

1	2	3	4	5	6	7	8	9	10	11
1.0	1.1	1.14	1.16	1.20	1.22	1.25	1.29	1.30	1.32	1.35
12	13	14	15	16	17	18	8	19	20	21
1.40	1.45	1.48	1.50	1.54	1.60	1.36	1.71	1.82	1.85	1.9

The skewness and kurtosis for this dataset can be calculated using the previously listed equations. They can also be calculated using Microsoft Excel's Analysis

ToolPack, which is usually available with the Excel program and can be easily installed. On the data analysis, first select descriptive statistics, then select the input data range and check the box with descriptive statistics. The following values are obtained:

$$\text{Skewness} = 0.406$$

$$\text{Kurtosis} = -0.695$$

The aforementioned results show that the data are slightly skewed and have a slightly negative kurtosis, which indicates that the bulk of the values on the given dataset are to the right of the mean and the tail on the left side is slightly shorter than the tail on the right side.

To explain these indices of soil variability in more detail, let us consider the data collected from an experimental field located near the Ohio State University airport in Columbus, Ohio (Table 5.1). The soil of the experimental site is Kokomo silty clay loam (fine, mixed, superactive, mesic Typic Argiaquoll). A total of 12 plots (each 27 × 15 m) were demarcated in 1993 and were planted with corn (*Zea mays*) in a corn–soybean (*Glycine max*) rotation. The tillage plots were arranged in a randomized complete block design, with three tillage practices replicated four times: moldboard plow (MB), chisel plow (CP), and no-till (NT). The MB treatment consisted of moldboard plowing (20–25 cm deep) in the fall followed by disking (7.5–10 cm deep). The CT plots were disked to a depth of 7.5–10 cm following harvesting of corn in the fall and were cultivated to incorporate the soybean residue prior to planting of corn. There was no soil disturbance in the NT treatment except for planting (Shukla et al., 2003).

The data in Table 5.1 show that the mean and median values are similar for clay content, bulk density (BD), available water content (AWC), and geometric mean weight diameter (GMD) of aggregates. Therefore, outliers, if any, did not affect the measures of central tendency. This is further supported by the low values of skewness and kurtosis. However, data for sand content and water stable aggregates (WSA) showed that mean and median values were slightly different. The dataset showed low dispersion, slight negative skewness, and positive kurtosis. The dataset for the saturated hydraulic conductivity (K_s) showed the largest deviations between mean and median values, with the median value being smaller than the mean. This was supported by high dispersion and high positive skewness values, which indicated that the bulk of the values were to the left of the mean. The data also showed high positive kurtosis or peaked distribution of the K_s.

Variability of the soil is also expressed by the coefficient of variation. Wilding (1985) proposed a ranking of variability in soil properties that occur within landscape units of a few hectares or less. He ranked a CV <0.15 as the least variable, 0.15 < CV < 0.35 as moderate, and a CV >0.35 as the most variable According to these criteria, the variability of clay content, sand content, BD, WSA, and GMD were low (CV <0.15); AWC showed moderate (0.15 < CV < 0.35) variability; and K_s values were highly variable (CV >0.35) (Table 5.1).

TABLE 5.1
Indices for Statistical Variability of Soil Hydrological Properties from a 5 ha Field under Agriculture in Columbus, Ohio

Property	Mean	Median	Var	SD	CV	Min	Max	Range	Skewness	Kurtosis
Clay	42.6	42.7	5.0	2.2	0. 1	38.7	45.4	6.7	−0.2	−0.7
BD	1.3	1.4	0.0	0.1	0.1	1.1	1.5	0.5	−0. 5	0.1
AWC	1.8	1.6	0.2	0.4	0.2	1.1	2.5	1.5	0.3	−0.1
GMD	1.8	1.8	0.0	0.2	0.1	1.4	2.1	0.6	−0 3	−0.5
Sand	13.0	13.3	3. 1	1.8	0.1	8.6	15.3	6.7	−1.3	3.4
WSA	91.8	92.9	13.1	3.6	0.0	82.3	95.4	13.1	−1.8	3.7
K_s	84.0	72.1	7377	85.9	1.0	0.6	327	327	2.3	6.4

Source: Shukla, M. K., et al., *Soil Sci.*, 168(11), 802, 2003.

Notes: Var, variance; SD, standard deviation; CV, coefficient of variation; Min, minimum value; Max, maximum value; BD, bulk density (g/cm^3); K_s, hydraulic conductivity (cm/h); AWC, available water content (cm); WSA, water stability of aggregates (%); GMD, geometric mean weight diameter (mm).

5.5 INFLUENCE OF SAMPLE SUPPORT

It is generally reported that K_s displays a large variability and is influenced by a large number of factors including sample support. The effect of sample support on hydraulic conductivity is presented in Table 5.2 based on infiltration tests that were conducted under four different suctions (Dasgupta et al., 2006).

The experiments used disk infiltrometers with five different disk sizes (0.10, 0.15, 0.17, 0.20, and 0.24 m) at College Station, Texas. The indices of statistical variability showed only minor differences for unsaturated hydraulic conductivity values at any

TABLE 5.2
Influence of Sample Support on Unsaturated Hydraulic Conductivity of Soils Using Tension Infiltrometers with Disk Sizes of 0.10, 0.15, 0.17, 0.20, and 0.24 m Diameter

K(h)	Mean	SD	CV	Max	Min	Range
0.10 m Diameter Disk						
K(−0.10)	2.3×10^{-6}	7.0×10^{-7}	0.3	3.0×10^{-6}	1.6×10^{-6}	1.4×10^{-6}
K(−0.05)	3.0×10^{-6}	2.7×10^{-7}	0.1	3.3×10^{-6}	2.8×10^{-6}	5.3×10^{-7}
K(−0.02)	3.4×10^{-5}	2.0×10^{-7}	0.6	5.4×10^{-5}	1.4×10^{-5}	4.0×10^{-5}
K(0.00)	3.7×10^{-5}	1.9×10^{-5}	0.5	5.5×10^{-5}	1.8×10^{-5}	3.7×10^{-5}
0.15 m Diameter Disk						
K(−0.10)	6.7×10^{-7}	9.1×10^{-7}	1.4	1.8×10^{-6}		
K(−0.05)	3.5×10^{-6}	1.4×10^{-6}	0.4	5.4×10^{-6}	2.0×10^{-6}	3.4×10^{-6}
K(−0.02)	1.9×10^{-5}	8.0×10^{-6}	0.4	2.7×10^{-5}	8.0×10^{-6}	1.9×10^{-5}
K(−0.00)	5.9×10^{-5}	1.6×10^{-5}	0.3	8.2×10^{-5}	4.6×10^{-5}	3.6×10^{-5}
0.17 m Diameter Disk						
K(−0.10)	8.3×10^{-7}	2.4×10^{-7}	0.3	1.1×10^{-6}	5.3×10^{-7}	5.7×10^{-7}
K(−0.05)	4.4×10^{-6}	2.4×10^{-7}	0.5	7.6×10^{-6}	2.1×10^{-6}	5.5×10^{-6}
K(−0.02)	2.9×10^{-5}	1.0×10^{-5}	0.4	4.3×10^{-5}	2×10^{-5}	2.3×10^{-5}
K(−0.00)	6.5×10^{-5}	2.7×10^{-5}	0.4	1.0×10^{-4}	4.3×10^{-5}	6.0×10^{-5}
0.20 m Diameter Disk						
K(−0.10)	9.8×10^{-7}	2.3×10^{-7}	0.23	1.4×10^{-7}	5.4×10^{-7}	0.9×10^{-6}
K(−0.05)	8.2×10^{-6}	6.9×10^{-6}	85	2.1×10^{-5}	1.4×10^{-6}	1.9×10^{-5}
K(−0.02)	8.0×10^{-6}	1.0×10^{-5}	1 28	3.3×10^{-5}	5.6×10^{-7}	3.2×10^{-5}
K(0.00)	2.2×10^{-5}	1.9×10^{-5}	0. 87	5.7×10^{-5}	8.2×10^{-7}	5.6×10^{-5}
0.24 m Diameter Disk						
K(−0.10)	9.2×10^{-7}	5.1×10^{-7}	0 56	1.8×10^{-6}	2.7×10^{-7}	1.5×10^{-6}
K(−0.05)	7.2×10^{-6}	7.6×10^{-6}	1 05	2.5×10^{-5}	6.4×10^{-7}	2.4×10^{-5}
K(−0.02)	7.3×10^{-6}	1.0×10^{-5}	1 36	3.3×10^{-5}	2.0×10^{-7}	3.3×10^{-5}
K(0.00)	2.1×10^{-5}	1.8×10^{-5}	0 86	5.5×10^{-5}	5.7×10^{-7}	5.4×10^{-5}

Source: Data modified from Dasgupta, S., et al., *Soil Sci. Soc. Am. J.*, 70, 1872, 2006.
Notes: SD, standard deviation; CV, coefficient of variation; Min, minimum value; Max, maximum value.

given suction among different disks. The CV for unsaturated hydraulic conductivity seemed to increase with disk size for most suctions, although not consistently. The increase in CV with increasing disk size could be due to the inclusion of larger sample volumes, larger heterogeneity, and possible contribution of macropore channels to the water transport (Shukla, 2011).

5.6 INFLUENCE OF MEASUREMENT DEVICE

The measurement device can have a strong influence on the variability of a measured soil property, for example, K_s. Table 5.3 presents the data where K_s was measured in the laboratory on 7.8 cm long and 7.8 cm diameter cores using the constant head method. The steady-state infiltration rate using double-ring infiltrometers was measured very close to the core sampling locations. Infiltration tests were conducted for 2.5 h using a 27 cm diameter outer ring and 15 cm diameter inner ring. This limited data for Ohio soils under different land use and management practices showed that the mean and dispersion of K_s values were higher for the core method than for the double-ring infiltrometer. Such a trend could be due to several reasons, including smaller support for the laboratory (or core) than the field experiment, water flow

TABLE 5.3
Saturated Hydraulic Conductivity (K_s) Measured in the Laboratory Using Soil Cores and Steady-State Infiltration (i_c) Measured in the Field Using Double-Ring Infiltrometer for Two Locations in Ohio

	Location 1		Location 2	
Index	K_s	i_c	K_s	i_c
Mean	84.0	14.9	23.7	2.5
Median	72.1	11.2	18.2	1.2
Var	7376.8	150.7	446.4	8.7
SD	85.9	12.3	21.1	2.9
CV	1.0	0.8	0.9	1.2
Minimum	0.6	2.0	1.3	0.6
Maximum	327.4	43.7	64. 2	12.0
Range	326.8	41.8	62.9	11.4
Skewness	2.3	1.5	1.0	2.7
Kurtosis	6.4	1.8	0.0	8.3

Source: Shukla, M. K., Introduction to hydrology: Processes and variability of hydrological properties, in Shukla, M. K. (ed.), *Soil Hydrology, Land Use and Agriculture: Measurement and Modeling,* CAB International, Wallingford, UK, pp. 1–23, 2011.

Notes: Var, variance; SD, standard deviation; CV, coefficient of variation; Min, minimum value; Max, maximum value; K_s and i_c are in cm/h.

TABLE 5.4
Saturated Hydraulic Conductivity ($K_s \times 10^{-5}$ m/s) Measured Using Tension Infiltrometer, Pressure Infiltrometer, and Soil Core on Three Different Soils in Native Woodland

Infiltrometer	GM	CV	Max	Min	Range	Soil
Tension	2.1	53.1	6.2	1.0	5.2	Sand
Pressure	21.7	57.8	39.0	7.1	31.8	Sand
Soil core	21.6	95.5	59.3	4.5	54.8	Sand
Tension	4.5	97.4	10.0	0.8	9.2	Loam
Pressure	23.8	63.8	81.6	12.2	69.4	Loam
Soil core	32.4	84.3	88.2	8.6	79.6	Loam
Tension	6.3	83.5	12.3	2.0	10.3	Clay loam
Pressure	7.2	53.3	18.3	3.6	14.7	Clay loam
Soil core	25.2	81.0	78.2	10.8	67.4	Clay loam

Source: Reynolds, W. D., et al., *Soil Sci. Soc. Am. J.*, 64, 478, 2000.
Notes: GM, geometric mean; CV, coefficient of variation; Min, minimum value; Max, maximum value.

between the core wall and the soil, hitting or missing macropores in the soil, and field soil not being completely saturated.

Table 5.4 presents K_s values determined in situ using a tension infiltrometer, pressure infiltrometer, and soil core. K_s values were determined for positive pressures of 1 and 2 cm in the field using a tension infiltrometer and 0.11 ± 0.03 and 0.31 ± 0.10 cm for a pressure infiltrometer. The laboratory tests were conducted by the constant or falling head method (discussed in Chapter 9) using large hydraulic head gradients (up to 10). The tension infiltrometer method consistently produced the lowest geometric mean values for all three soils; however, the CV for K_s for the three methods did not follow a clear trend except that they were always classified as the most variable (Wilding, 1985). The field methods could have a disadvantage of the soil not being fully saturated and measurements being made under quasi-steady-state conditions.

5.7 INFLUENCE OF LAND USE

Land use, management, and cropping systems have an important influence on soil properties in a field. The saturated hydraulic conductivity was measured by the soil core method and a double-ring infiltrometer under three different land uses, i.e., annual tillage by chiseling or moldboard plowing, no-tillage (for 6–15 years), and woodland (Table 5.5). Both the K_s and steady-state infiltration rate (i_c) were higher from woodland than agricultural fields, and i_c values varied in the order woodland > no-tillage > annual tillage. Average values of K_s were higher but standard deviations were smaller for fields under annual tillage than no-tillage. However, based on

TABLE 5.5
Saturated Hydraulic Conductivity (K_s) Measured Using Soil Cores and Steady-State Infiltration (i_c) Measured Using Double-Ring Infiltrometer for Fields under Annual Tillage (Using Chisel, Moldboard), No-Tillage (6–15 years), and Woodland in Ohio

	Annual Tillage		No-Tillage		Woodland	
Index	K_s	i_c	K_s	i_c	K_s	i_c
Mean	48.1	7.9	26. 7	14.6	85.7	24.8
Median	44.3	5.8	10.9	12.9	73.3	22.4
Var	1349.9	37.4	2956.0	117.7	4743.0	204.9
SD	36. 7	6.1	54.4	10.8	68.9	14.3
CV	0.76	0.78	2.03	0.74	0.80	0.58
Minimum	0.6	1.8	0.1	0.6	10.1	6.0
Maximum	101.7	23.2	327.4	43.7	178.0	45.3
Range	101.0	21.4	327.3	43. 1	167.9	39.3
Skewness	0.2	1.3	4.7	0.8	0.3	0.3
Kurtosis	−1.6	1.7	25.0	0.4	−2.1	-0.8

Source: Shukla, M. K., et al., *Soil Sci.*, 168(11), 802, 2003.
Notes: Var, variance; SD, standard deviation; CV, coefficient of variation; Min, minimum value; Max, maximum value; K_s and i_c are in cm/h.

CV (>0.35), the variability of K_s or i_c was classified as the most variable. Higher K_s from annual tilled fields than no-tillage fields could be due to a number of factors, including the larger sample size used for determining the K_s from no-tillage fields than from fields under annual tillage, measurement errors in the field and laboratory while collecting and preparing the core samples, timing of tillage operations, hitting and missing micropores, and errors during sample analyses.

5.8 STATISTICAL ANALYSIS

In the classical statistical analysis, it is assumed that the expected value of a property of soil (e.g., bulk density or BD) at any location (say, x) within the sampling area can be written as the sum of its mean value $\left(\overline{\text{BD}}\right)$ and dispersion [$\varepsilon(x)$]:

$$\text{BD}(x) = \overline{\text{BD}} + \varepsilon(x) \tag{5.3}$$

Equation 5.3 can be used to estimate properties at any unknown location. The analysis assumes that deviations from the mean are normally distributed (Trangmar et al., 1985). However, as shown earlier in this chapter (Tables 5.1, 5.3, and 5.5), soil properties do not always exhibit a normal distribution. Often, distributions are asymmetrical with positive or negative skewness. Occasionally, properties can even show a bimodal distribution, with each mode representing a different population for performing a statistical analysis (Wilding and Drees, 1983). In Chapter 4, the mean is used to predict the number of samples required for acquiring a predefined degree of precision. The number of samples in these approaches is also linearly related to the dispersion (or sample variance). The samples are assumed random, which is debatable, and renders the classical approach inadequate for interpolation of the values at unknown locations within the experimental unit for which the mean and dispersion are known. Within any experimental or sampling unit, soil properties exhibit continuity and a spatial correlation that is likely to vary with increasing sampling distance and changing direction from adjoining samples.

There are several empirical techniques for interpolating soil properties within the sampling distance, such as inverse distance, moving averages, and spline interpolation. According to Burgress and Webster (1980), these methods do not provide optimal estimates of the interpolation errors. These methods are beyond the scope of this book and readers are encouraged to read textbooks on statistical analysis.

5.9 GEOSTATISTICAL ANALYSIS

The frequency distribution and the statistical tools of dispersion do not completely describe the variability because they assume that samples are independent of each other. The independence of samples from each other is possible, but there is more likelihood of samples being spatially correlated. It is commonly known that most soil properties exhibit both short- and long-range variability (Nielsen et al., 1973). It is generally accepted that soil samples collected close to each other are more similar than those collected at greater distances apart, and the similarity decreases as the separation distance between samples increases up to a certain distance beyond which samples become spatially uncorrelated or independent. Geostatistical analysis using autocorrelation, covariance, or semivariogram functions could determine the spatial correlation of soil properties at the scales of measurement.

Figure 5.3 presents contour maps for saturated hydraulic conductivity and clay content for a sandy desert soil located in the Chihuahuan desert near Las Cruces, New Mexico. The area is irrigated with lagoon-treated industrial wastewater. Soil samples were collected at a grid of 50 × 50 m. Thirty additional sampling locations on five grid cells were also randomly selected to obtain lag samples at distances of 2, 7, and 22 m from the grid center in two directions (Adhikari et al., 2012). The kriged contour maps for both properties show that the samples close to each other have more similar values than the samples separated farther apart. This is the important difference between spatially correlated and random samples.

In the regionalized variable theory, differences between pairs of values of a soil property separated by a chosen sampling distance are expressed as their variances (Webster, 1985). If there are a total of n observations of a soil property investigated at an equal spacing from each other, then there could be m (= n −1) pairs of

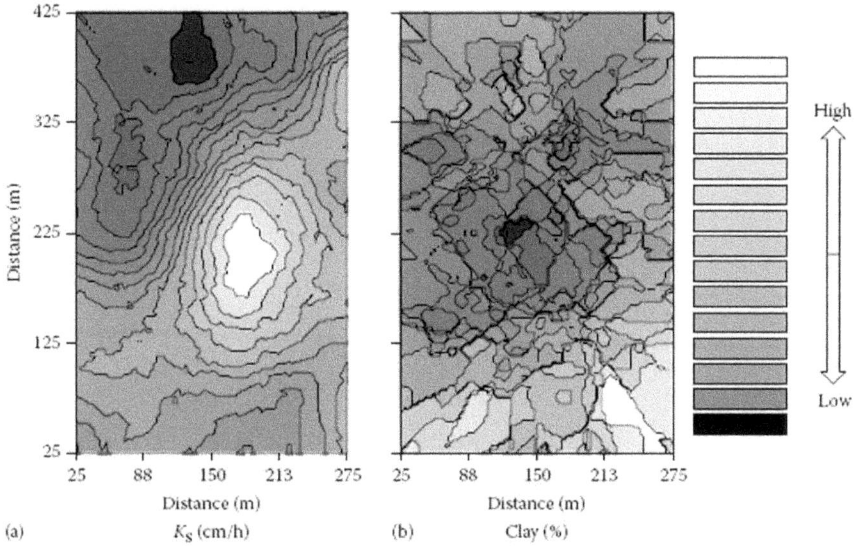

FIGURE 5.3 The kriged maps of (a) saturated hydraulic conductivity and (b) clay content for sandy desert soil located in the Chihuahuan desert west of the city of Las Cruces, New Mexico. The area was irrigated with lagoon-treated industrial wastewater. The contour maps show that samples close to each other are more similar with respect to the property than the samples away from each other.

observations separated by a lag distance, say h, as shown in Figure 5.4. The semivariance or semivariogram is expressed as

$$s^2 = \gamma(h) = \frac{1}{2m}\sum_{i=1}^{m}\left[z(x_i) - z(x_i + h)\right]^2 \tag{5.4}$$

where

$\gamma(h)$ is the semivariance for interval class h.
m is the number of pairs separated by lag distance h (separation distance between sample positions).
$Z(x_i)$ is a measured variable at spatial location i.
$Z(x_{i+h})$ is a measured variable at spatial location $i + h$.

At $h = 0$, that is, when separation distance or lag distance is zero, the semivariance [$\gamma(h = 0)$ or $\gamma(0)$] for pairs [$z(x_i)$ and $z(x_i)$] given by the equation becomes zero (Figure 5.4). As the lag distance increases from zero to one, the semivariance [$\gamma(1)$] is obtained for the pairs [$z(x)$ and $z(x_{i+1})$] separated by a distance of one. As the lag distance increases from one to two, the semivariance [$\gamma(2)$] is obtained for the pairs [$z(x_i)$ and $z(x_{i+2})$] separated by a distance of two. Similarly, $\gamma/(h)$ is obtained for additional lag distances, and a semivariogram for one direction is obtained, as shown

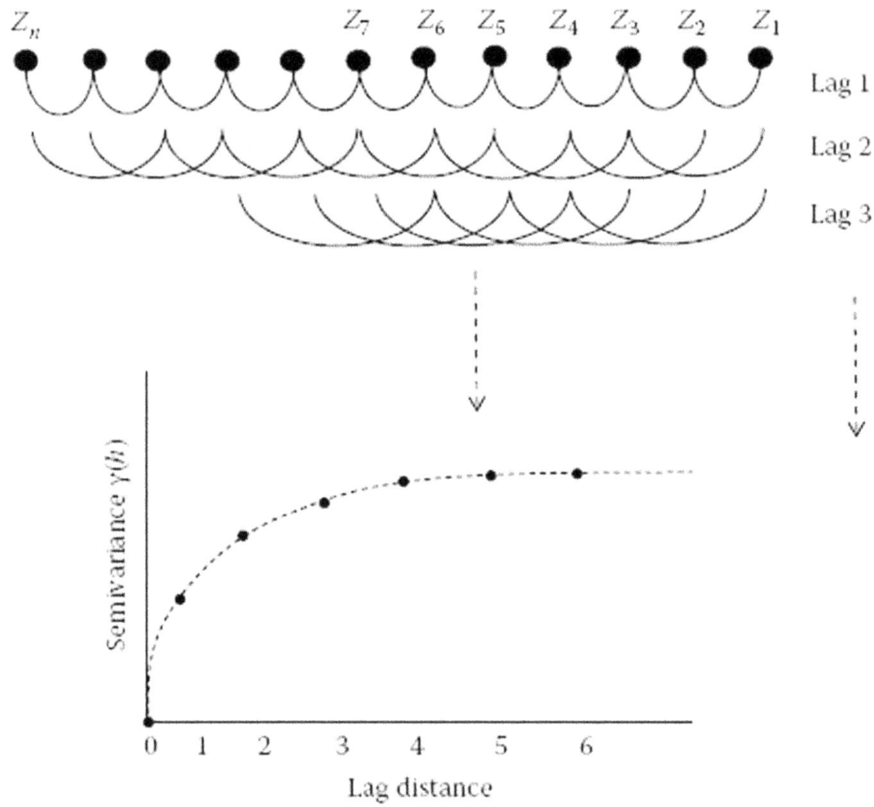

FIGURE 5.4 Soil property (Z_i) measured at a given sampling distance. Schematic of the pairs of observations for given lag distances and the calculated semivariogram (for one direction) for those lag distances.

in Figure 5.4. The variogram shown in this figure is also called an experimental variogram.

The autocorrelation coefficient for a variable separated by a lag distance of h can be calculated as follows:

$$r(h) = \frac{\text{cov}\left[z(x_i), z(x_i + h)\right]}{\sqrt{\text{var}\left[z(x_i)\right]}\sqrt{\text{var}\, z\left[(x_i + h)\right]}} \quad (5.5)$$

where

cov is the covariance function.

var is the variance function.

The covariance function is replaced in this equation, as in the next one for a dataset with m pairs of observations and the sample mean of $.\bar{z}$.

$$r(h) = \frac{1/m\sum_{i=1}^{m}\left[z(x_i)-\bar{z}\right]\left[z(x_i+h)-\bar{z}\right]}{\sqrt{\mathrm{var}\left[z(x_i)\right]}\sqrt{\mathrm{var}\,z\left[(x_i+h)\right]}} \tag{5.6}$$

The autocorrelation is one when the lag distance is equal to zero, and as the lag distance increases, the autocorrelation decreases and eventually goes down to zero. It can be negative also. Moran's *I* statistic (Moran, 1950) is also a measure of spatial autocorrelation, similar in interpretation and range to the Pearson's correlation statistic. The range for both statistics varies from –1 (perfect dispersion) to +1 (perfect correlation). The independent samples showing a zero Moran's *I* value indicate a random spatial pattern. Moran's *I* is defined as

$$I = \frac{m}{\sum_{i=1}^{m}\sum_{j=1}^{m} w_{ij}} \frac{\sum_{i=1}^{m}\sum_{j=1}^{m} w_{ij}\left(z(x_i)-\bar{z}\right)\left(z(x_j)-\bar{z}\right)}{\sum_{i=1}^{m}\left(z(x_i)-\bar{z}\right)^2} \tag{5.7}$$

where

- m is the number of spatial units.
- $z(x_i)$ is the variable of interest.
- $\bar{z}$ is the mean.

w_{ij} is the spatial weight describing the adjacency or distance between the *i*th and *j*th point.

Gajem et al. (1981) reported that information collected from autocorrelograms could potentially be used in statistical designs for sampling (obtaining truly independent samples) because autocorrelograms showed spatial dependence for distances greater than the original sampling distance.

5.10 SEMIVARIOGRAM FUNCTIONS

The shape of the semivariogram does not always have to be as shown in Figure 5.4, but depending on the type of data collected, number of samples collected, direction of sampling, heterogeneity of soils, and sampling errors, it can have different shapes (Nielsen and Wendroth, 2003). The variogram in Figure 5.4 represented mathematically is known as a theoretical variogram. A theoretical variogram provides a relation between semivariance and lag distance and can be obtained by fitting linear, spherical, exponential, power, Gaussian, or any other model form. Usually, the best model is selected based on the least residual sum of squares between experimental and theoretical semivariograms and also by the square of the correlation coefficient (r^2), also known as the coefficient of determination. The functional forms of experimental semivariograms, including linear, spherical, exponential, Gaussian, and power models, are shown here.

Linear model

$$\gamma(h) = 0; h = 0 \tag{5.8}$$

$$\gamma(h) = C_0 + C_1 \frac{h}{a}; h > 0 \tag{5.9}$$

Spherical model

$$\gamma(h) = 0; \quad h = 0 \tag{5.10}$$

$$\gamma(h) = C_0 + C_1 \left[1.5\left(\frac{h}{a}\right) - 0.5\left(\frac{h}{a}\right)^3\right]; h \le a \tag{5.11}$$

$$\gamma(h) = C_0 + C_1; \quad h > a \tag{5.12}$$

Exponential model

$$\gamma(h) = 0; \quad h = 0 \tag{5.13}$$

$$\gamma(h) = C_0 + C_1 \left[1 - \exp\left(\frac{-3h}{a}\right)\right]; h > 0 \tag{5.14}$$

Gaussian model

$$\gamma(h) = 0; \quad h = 0 \tag{5.15}$$

$$\gamma(h) = C_0 + C_1 \left[1 - \exp\left(\frac{-3h^2}{a^2}\right)\right]; h > 0 \tag{5.16}$$

Power model

$$\gamma(h) = 0; \quad h = 0 \tag{5.17}$$

$$\gamma(h) = (C_0 + C_1) h^w; \quad 0 < w < 2 \tag{5.18}$$

Pure nugget model

$$\gamma(h) = 0; \quad h = 0 \tag{5.19}$$

$$\gamma(h) = C_0; \quad h > 0 \tag{5.20}$$

The semivariance at $h = 0$ is called the nugget variance or nugget effect (C_0), $C_0 + C_1$ is the sill (or total sill), and the distance at which the semivariogram reaches the sill is called the range (a). The semivariogram value at which the semivariance attains its maximum value and then levels off is known as the sill. Sill is also expressed as overall or total sill, meaning the sum of the nugget and sill, while sill or partial sill means the difference between the total sill and nugget. Nugget variance is the height of the jump of the semivariogram at a lag distance of zero or at the discontinuity at the origin. The nugget variance represents the undetectable experimental or random variability at the sampling scale. It indicates that variation is occurring at a much smaller scale than the measured lag distance. It is usually obtained by extrapolating the semivariogram as $h \to 0$ (h tends to become 0). The errors associated with the measurement of a value at a location or the errors of assigning a location to the measurement result in a greater nugget effect. The nugget effect is usually higher for sparse data than well-distributed data over space. The range is defined as the lag distance at which the semivariogram levels off. At and beyond the range, autocorrelation, discussed in Section 5.9, is essentially zero.

Spatial dependence is reported to occur at scales ranging from a few meters to several kilometers (Trangmar et al., 1987; Ovalles and Collins, 1988; Gaston et al., 2001). Geostatistical analysis is usually carried out to understand the spatial structure and variability of the hydrological properties of soil. Geospatial analysis can also provide more insight into spatial variability of a property whether it is structured, unstructured, or directional. A detailed overview of these methods and their application on field datasets can be found in Warrick and Nielsen (1980), Webster (1985), and Nielsen and Wendroth (2003).

PROBLEMS

5.1 What are the possible sources of variability for measuring the water infiltration rate of soil?

5.2 Describe briefly how the errors due to some of the sources of variability discussed in Problem 5.1 can be reduced.

5.3 Get a dataset from any published article and calculate the mean, median, variance, skewness, and kurtosis.

5.4 Explain the differences between various semivariogram models.

5.5 Explain the variabilities of a chosen property at pore, pedon, field, and watershed scales. Explain possible sources of variability.

5.6 Collect five articles (not the ones listed here) describing saturated hydraulic conductivity measured using different methods in the field or laboratory and write at least a one-page (single-spaced) summary detailing the possible reasons for the differences in the reported values of hydraulic conductivity.

5.7 Compare variability of saturated hydraulic conductivity with soil bulk density and soil porosity using published literature. Explain why they are different.

REFERENCES

Adhikari, P., M. K. Shukla, and J. Mexal. 2012. Spatial variability of infiltration rate and sodium content of desert soils: Implications for management of irrigation using treated wastewater. *Trans. ASABE.* 55(5): 1711–1721.

Burgess, T. M. and R. Webster. 1980. Optimum interpolation and isarithmic mapping of soil properties I. The semivariogram and punctual kriging. *J. Soil Sci.* 31: 315–331.

Burrough, P. A. 1983. Multiscale sources of spatial variation in soil. I. The application of fractal concepts to nested levels of soil variation. *J. Soil Sci.* 34: 577–597.

Dasgupta, S., B. Mohanty, and J. M. Kohne. 2006. Soil hydraulic conductivities and their spatial and temporal variations in a Vertisol. *Soil Sci. Soc. Am. J.* 70: 1872–1881.

Gajem, Y. M., A. W. Warrick, and D. E. Myers. 1981. Spatial dependence of physical properties of a Typic Torrifluvent soil. *Soil Sci. Soc. Am. J.* 45: 709–715.

Gaston, L. A., M. A. Locke, R. M. Zablotowicz, and K. N. Reddy. 2001. Spatial variability of soil properties and weed populations in the Mississippi delta. *Soil Sci. Soc. Am. J.* 65: 449–459.

Isaaks, E. H. and R. M. Srivastava. 1978. *Applied Geostatistics.* Oxford University Press, New York.

Lin, H., D. Wheeler, J. Bell, and L. Wilding. 2005. Assessment of soil spatial variability at multiple scales. *Ecol. Modell.* 182: 271–290.

Moran, P. A. 1950. Notes on continuous stochastic phenomena. *Biometrika* 37: 17Y23.

Mulla, D. J. and A. B. McBratney. 2000. Soil spatial variability. In: M. E. Summer (ed.), *Handbook of Soil Science.* CRC Press, Boca Raton, FL, pp. A321–A349.

Nielsen, D. R. and O. Wendroth. 2003. *Spatial and Temporal Statistics: Sampling Field Soils and Their Vegetation.* Catena Verlag, Reiskirchen, Germany.

Nielsen, D. R., J. W. Biggar, and K. T. Erh. 1973. Spatial variability of field measured soil water properties. *Hilgardia* 42: 215–259.

Ovalles, F. A. and M. E. Collins. 1988. Evaluation of soil variability in northwest Florida using geostatistics. *Soil Sci. Soc. Am. J.* 52: 1702–1708.

Pennock, D. J. and E. de Jong. 1990. Regional and catenary variations in properties of Borolls of southern Saskatchewan, Canada. *Soil Sci. Soc. Am. J.* 54: 1697–1701.

Reynolds, W. D., B. T. Bowman, R. R. Brunke, C. F. Drury, and C. S. Tan. 2000. Comparison of tension infiltrometer, pressure infiltrometer, and soil core estimates of saturated hydraulic conductivity. *Soil Sci. Soc. Am. J.* 64: 478–484.

Shukla, M. K. 2011. Introduction to hydrology: Processes and variability of hydrological properties. In: Shukla, M. K. (ed.), *Soil Hydrology, Land Use and Agriculture: Measurement and Modeling.* CAB International, Wallingford, UK, pp. 1–23.

Shukla, M. K., R. Lal, and M. Ebinger. 2003. Tillage effects on physical and hydrological properties of a Typic Argiaquoll in central Ohio. *Soil Sci.* 168(11): 802–811.

Shukla, M. K., B. Slater, R. Lal, and P. Cepuder. 2004. Spatial variability of soil properties and potential management classification of a Chernozemic field in Lower Austria. *Soil Sci.* 169(12): 852–860.

Soil Survey Division Staff. 1993. *Soil Survey Manual.* U.S. Department of Agriculture Handbook No. 18. U.S. Government Printing Office, Washington, DC.

Trangmar, B. B., R. S. Yost, and G. Uehara. 1985. Application of geostatistics to spatial studies of soil properties. *Adv. Agron.* 38:.45–94.

Trangmar, B. B., Yost, R. S., and G. Uehara. 1987. Spatial variation of soil properties and rice yield on recently cleared land. *Soil Sci. Soc. Am. J.* 51: 668–674.

van Es, H. M. 2002. Sources of soil variability. In: J. Dane and C. Topp (eds.), *Methods of Soil Analysis, Part 4: Physical Properties.* Soil Science Society of America, Madison, WI.

Warrick, A. W. and D. R. Nielsen. 1980. Spatial variability of soil physical properties in the field. In: D. Hill (ed.), *Applications of Soil Physics.* Academic Press, Inc., New York, pp. 319–344.

Webster, R. 1985. Quantitative spatial analysis of soils in the field. *Adv. Soil Sci.* 3: 1–70.

Wilding, L. P. 1985. Spatial variability: Its documentation, accommodation, and implication to soil surveys. In. D. R. Nielson and J. Bouma (eds.), *Soil Spatial Variability.* Pudoc, Wageningen, The Netherlands, pp. 166–194.

Wilding, L. P. and L. R. Drees. 1983. Spatial variability and pedology. In: L. P. Wilding, N. E. Smeck and G. F. Hall (eds.), *Pedogenesis and Soil Taxonomy. I. Concepts and Interactions.* Elsevier, Amsterdam, pp. 83–116.

6 Fundamentals of Hydrology

6.1 INTRODUCTION

Hydrology can be defined as the branch of science pertaining to the study of water. It consists of two terms that are derived from the Greek words *hydro* and *logos* meaning "water" and "study," respectively. In general, hydrology encompasses the study of the occurrence, distribution, movement, and changes in quantity, quality, and state of atmospheric, surface, soil-borne, plant-borne, and subsurface water on earth. Water, one of the most dynamic entities on earth, can stay in all three forms—liquid, solid, and vapor—at the same time and at room temperature. The path that water takes in all three forms through the environment is given by the hydrologic cycle. Various components of the hydrologic cycle are strongly influenced by human activities, such as changing land use from natural forest to agriculture, agriculture to fallow, deforestation and afforestation, forest to urban, agriculture to urban, construction of reservoirs and dams, intensive tillage to no-tillage, and tillage or no-tillage to conservation tillage (Shukla, 2011).

Soil, the interface between the lithosphere and the atmosphere, is the most basic resource that interacts with the biosphere and the hydrosphere to support life on earth. The interaction of soil with various components of hydrology is important for the hydrologic cycle and is also identified as soil hydrology. Soil hydrology has a strong influence on the storage, transport, and release of water through soil, and water uptake and release by roots and plants during photosynthesis. Soil hydrology is closely related to agricultural water management, and it takes into account all the components of water related to irrigation and drainage, percolation and recharge to groundwater, capillary rise, root and plant water uptake, evaporation from soil and plants, and transpiration. The fate of the total amount of water applied to the soil as irrigation or natural rainfall is usually determined by the water-balance technique, which takes into account all the different components of the water cycle.

6.2 HYDROLOGIC CYCLE

The hydrologic cycle, also known as the water cycle or H_2O cycle, describes the continuous movement of water as liquid, vapor, and solid (ice) in the atmosphere, on the plant and soil surfaces, and through plant and soil bodies. Components of the hydrologic cycle are grouped into evaporation, transpiration (or evapotranspiration), condensation of water vapor, precipitation, snowfall, canopy interception, snowmelt, surface runoff, infiltration, percolation, deep percolation, subsurface flow, and return flow (Figure 6.1).

Solar energy (sun) is the driving force for the hydrologic cycle. In general, the total overall mass of water on earth is fairly constant over time. However, water in

DOI: 10.1201/9780429264849-6

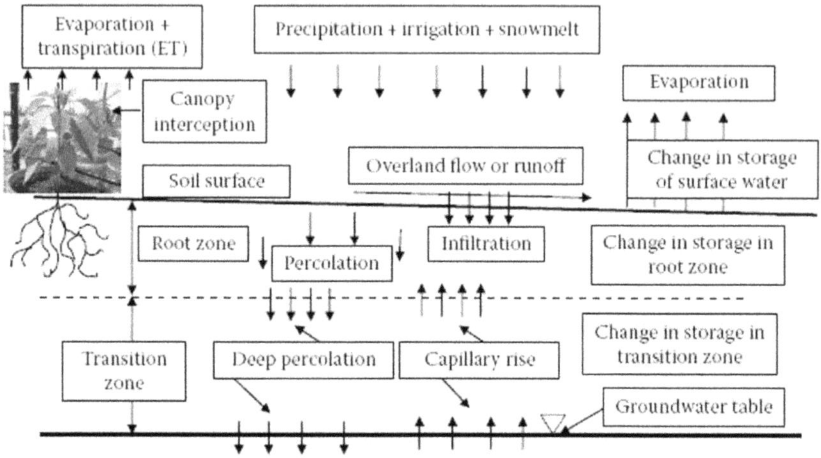

FIGURE 6.1 The hydrologic cycle and components of water balance. (Modified from Shukla, M. K., ed., *Soil Hydrology, Land Use and Agriculture: Measurement and Modeling*, Wallingford, UK, p. 434, 2011.)

TABLE 6.1
Average Residence Time of Water in Different Stores

Storage	Residence Time
Antarctica	20,000 years
Deep aquifers (may be confined)	10,000 years
Oceans	3,200 years
Shallow aquifers (unconfined)	100–200 years
Lakes	50–100 years
Glaciers	20–100 years
Snow cover (mostly seasonal)	2–6 months
Rivers	2–6 months
Soil moisture	1–2 months
Atmosphere	9 days

Source: PhysicalGeography.net, Chapter 8: Introduction to the hydrosphere, retrieved on February 8, 2010.

different storage components, such as soil, plants, lakes, mountains, rivers, seas, and groundwater, stays from a few days (water vapor) to several thousand years (deep groundwater, ice on poles) (Table 6.1). According to Table 6.1, water spends on average 9 days in the atmosphere, condenses and falls on earth as precipitation, and is stored for an average of 1–2 months as soil moisture and 2–6 months in rivers. Ice in Antarctica is reported to date back 800,000 years with an average residence time of 20,000 years (Jouzel et al., 2007). The residence time is estimated either by using the conservation of mass theory or the isotopic technique.

6.3 COMPONENTS OF THE HYDROLOGIC CYCLE

6.3.1 Evaporation and Transpiration

The sun heats the water in rivers, lakes, oceans, ice, snow, and surfaces of soil and plants, causing water to evaporate and move into the atmosphere as water vapor. The water vapor condenses into clouds due to the cooler temperatures as it moves up in the atmosphere. Air currents then move these clouds around. The energy responsible for water evaporation is known as solar radiation. Water needs about 10 kcal/mol of heat to convert into vapor. Thus, 1 mol or 18 g or 18 mL (assuming the density of water as 1 g/cm^3) of water requires 10 kcal of heat to convert it into vapor.

Transpiration from plants is a process similar to evaporation and can take place from several parts of plants, such as leaves, stems, flowers, and roots. Leaf transpiration happens through stomata and is an important process for diffusion of CO_2 from the atmosphere, changes in osmotic potential, and mass flow of water and minerals from roots. Leaves transpiring water rapidly are cooler than the surrounding air. Vegetation or canopy cover shades the soil and thus reduces evaporation from soil. Evaporation and transpiration are commonly presented together as evapotranspiration.

6.3.2 Advection

Advection is the process by which water, in different forms, moves through the atmosphere. Advection is an important process, leading to the formation of orographic clouds. Orographic clouds are formed through the uplifting of air mass followed by adiabatic (no input or release of heat) cooling to increase the relative humidity to 100%.

6.3.3 Condensation

Condensation is a process of change of state from vapor to liquid phase (opposite of vaporization) and commonly occurs when vapor cools to a temperature causing condensation. Condensation changes water vapor into droplets of water that can stay in the atmosphere as clouds or as fog.

6.3.4 Precipitation

After condensation, when water droplets become large (attain a critical size), they can no longer stay in the atmosphere and fall back on the earth due to gravity as precipitation. Precipitation, a major component of fresh water on the earth, occurs once the atmosphere is saturated with water vapor. It can occur in several different forms, such as a thunderstorm, drizzle, sleet, hail, or snow. When water vapor does not condense enough to precipitate, it is known as fog or mist.

6.3.5 Canopy Interception

A part of the precipitation during fall is intercepted by vegetation before reaching the ground. The interception of the precipitation can also occur at tree branches and forest floor. The amount of snow interception varies with the type of vegetation, for example, hardwood trees intercept less snow than conifers.

6.3.6 Infiltration

The process of entry of water into the soil is known as infiltration. The snowmelt also infiltrates into the soil. Infiltration is influenced by gravity and capillary action, discussed later in the book. The amount of infiltration depends upon soil type, depth of soil, intensity and amount of precipitation, and antecedent moisture content of the soil. The infiltrated water fills up soil pores and then moves downward in the soil profile, saturating underlying layers. The depth of water penetration depends on the volume of soil pores, the amount of precipitation or snowmelt, and the antecedent moisture content of soil. Infiltration is an important part of the water cycle and is immensely important for plant survival, growth, and development.

6.3.7 Percolation

The infiltrated water moves through the soil profile to deeper depths by first filling up the pore spaces in the soil and then letting the excess water percolate downward due to gravity or capillary action. Once water moves below the root zone, the process of water movement is known as deep percolation.

6.3.8 Surface Runoff

Part of the water falling on earth either due to precipitation or snowmelt infiltrates into the soil, and the remaining water flows on the land surface as overland flow until it joins a channel, river, or some other water body. This portion of water is known as runoff. The simplest definition of runoff is "infiltration excess" or the difference between precipitation and infiltration. It can also be called "saturation excess" when no more water can flow into the soil. Water can also flow below the ground and is sometimes known as subsurface runoff or groundwater runoff.

6.3.9 Soil Water

The infiltrated water can stay in the soil as soil water before it is taken up by plant roots to meet their transpiration needs.

Example 6.1

How much water can be present in a typical soil? Will that be enough to grow a crop in your area?

In a typical soil, almost half of the total soil volume consists of pores.

If we assume that half of the total volume of pores is filled with water, then in 100 cm^3 of soil volume, 25 cm^3 of water could be present.

Thus, in a 100 cm depth of soil, 25 cm of water is present (or 1 m soil depth contains 0.25 m of water).

Thus, in 1 ha of soil, there could be 10,000 $m^2 \times 0.25$ m = 2500 m^3 = 2.5×10^6 L of water.

This seems to be a lot of water; however, water lost from soil due to evapotranspiration is also high. For example, a cotton crop grown in Las Cruces, New Mexico, can require about 63 cm of water during the growing season, which is equivalent to 6.3×10^6 L of water. Therefore, transpiration losses from cotton could be as much as 2.5 times the water stored in the soil.

6.4 WATER BALANCE

It is a well-known fact that deficit irrigation is practiced in many parts of the world, including the southwestern United States. Deficit irrigation is a result of unavailability of the required amounts of water, which could be exacerbated by poor irrigation scheduling, poor water delivery systems, and poor accounting of the available water. It is not always possible to account for all the processes of the hydrologic cycle accurately.

Using the law of conservation of mass, the total water in a system can be accounted for by including various components of the hydrologic cycle (Figure 6.1). Water balances can be conducted at multiple scales, for a pedon, a farm, a water district, a watershed, or a river basin. Accounting of the water incorporates all the inputs, outputs or losses, and change in storages, and is known as water balance. Thus, a simple water-balance equation can be written as

$$\text{Water in} = \text{Water out} + \text{Change in water storage} \tag{6.1}$$

Water balance is important for conserving water and using it judiciously. A water-balance equation consists of various components related to soil, plant, and atmosphere. The previous equation can be expanded by including some of the processes associated with the hydrologic cycle; a simple water-balance equation for dry land areas can be written as

$$\text{Precipitation} = \text{Evaporation} + \text{Runoff} + \text{Change in water store of soil} \tag{6.2}$$

It is important to know that the change of water storage in the soil could be either positive, indicating the water is added to the soil to fill up some empty pore spaces, or negative, indicating that water was removed from the soil by emptying some of the pore spaces. Water balance can be subdivided into several categories depending on the purpose, such as surface water balance, groundwater balance, root zone water balance, as well as the user, namely, hydrologists, soil scientists, agronomists, and groundwater hydrologists.

6.4.1 Surface Water Balance

Surface water consists of all the water that runs off the soil surface; moves through drains and rivers; and is stored in lakes, ponds, depressions, and oceans. The water input could be from precipitation, snowmelt, irrigation, and return flow; and water-out components related to soil, atmosphere, and plants could be runoff and soil storage, evaporation, and transpiration, respectively. The surface water balance can be expressed as follows (Figure 6.1):

$$\begin{aligned}&\text{Precipitation} + \text{Irrigation} + \text{Snowmelt}\\&= \text{Evaporation} + \text{Transpiration} + \text{Surface runoff} + \text{Infiltration}\\&+\text{Change in storage of surface water above ground surface}\end{aligned} \tag{6.3}$$

6.4.2 Root Zone Water Balance

The root zone is the maximum depth at which the roots of the plant of interest can be found. The root zone is sometimes defined as the zone where the majority of the roots are found. The root zone water balance includes all the water that can move in and out of the root zone. The movement of water below the root zone is known as deep percolation. The upper boundary for conducting the root zone water balance for an agricultural farm could be the crop canopy. If the farm is fallow, then the upper boundary could be the soil surface (Burt, 1999). The lower boundary for the root zone water balance from a farm would be the bottom of the root zone. Root zone water balance is expressed as follows (Figure 6.1):

$$\begin{aligned}&\text{Infiltration}\left(\text{into root zone}\right) + \text{Capillary rise into the root zone}\\&= \text{Evaporation} + \text{Transpiration} + \text{Deep Percolation}\\&+\text{Change of water storage of root zone}\end{aligned} \tag{6.4}$$

6.4.3 Transition Water Balance

The vadose zone extends from the soil surface to the water table. The transition zone can be defined as the part of the vadose zone between the root zone and the water table. The upper boundary for conducting the transitional zone water balance could be the bottom of the root zone into which the capillary rise is taking place, while the lower boundary could be the water table. The water-balance equation for the transition zone involves all the water that goes below the root zone and can be written as (Figure 6.1)

$$\text{Deep percolation}\left(\text{below root zone}\right)+\text{Seepage}\left(\text{into transition zone}\right)$$
$$+\text{Capillary rise}\left(\text{from water table}\right)=\text{Capillary rise}\left(\text{into root zone}\right)$$
$$+\text{Subsurface drainage}$$
$$+\text{ Deep drainage }\left(\text{into ground water}\right)$$
$$+\text{Change of water storage of the transition zone} \quad (6.5)$$

6.4.4 Aquifer Water Balance

Similarly, water-balance equations can be written for confined and unconfined aquifers by taking into account the water that moves in and out of aquifers. The input of water into the aquifer could be the applied water (irrigation, precipitation, or seepage) that percolates through the vadose zone into the aquifer. The output or the part of the water that leaves aquifers could be via pumping or seepage.

Example 6.2

Calculate the deep percolation (DP) using the water-balance equation from known amounts (in mm) of irrigation (IR), precipitation (R), runoff (RO = 0), change in storage (ΔS), and evapotranspiration (ET) for a pecan orchard in Las Cruces, New Mexico (Deb et al., 2013).

ΔS can be estimated as

$$\left(\sum_{i=1}^{n}(\theta_2-\theta_1)\right)\Delta Z_i$$

where ΔS is the average change in soil water content in the upper 100 cm of the soil profile and is determined by subtracting the water content measured at the onset of irrigation (θ_2) from the peak soil water content (θ_1) after the irrigation measured at each depth, ΔZ_i is the thickness of each layer, and n is the number of layers (Deb et al., 2012).

DP can be estimated as

$$\text{DP}=\text{IR}+\text{R}-\Delta\text{S}-\text{RO}-\text{ET}$$

Date	IR	R	ΔS	RO	ET	DP
July 1, 2009	174	7.1	125.95	—	9.65	45.49
July 16, 2009	179	0.65	140.83	—	8.89	29.93
July 27, 2009	178	2.58	124.35	—	8.89	47.34
August 11, 2009	176	0	145.47	—	8.46	22.08
October 26, 2009	165	0	110.11	—	5.87	49.02
December 1, 2009	85	0	54.54	—	2.54	27.92

6.5 RAINFALL-RUNOFF RELATIONSHIPS

Rainfall on the surface of the earth is partitioned into infiltration and runoff. The runoff can be generated if the rainfall intensity exceeds the infiltration rate of the soil. The runoff can also be generated even if the rainfall intensity is low, but the soil is already saturated and cannot take any more water. Water initially flows on the surface as overland flow; if the rainfall intensity is not high, the water flows as a sheet. Some of the runoff water is lost as depression storage and intercepted by the vegetation, including litter, during overland flow. This water subsequently infiltrates into the soil. As the flowing water approaches a channel or a ditch, it becomes shallow concentrated flow, and subsequently water moves as channel flow (Figure 6.2) until it reaches the outlet of the watershed. The rest of the water appears as runoff in a hydrograph. The hydrograph is a plot between the rate of flow, discharge, or depth of water and a period of time at a specific point in a river or a watershed channel. A unit hydrograph is the direct runoff generated by one unit of effective rainfall that occurred over the entire watershed uniformly for a unit period of time.

Subsurface flow or runoff occurs when water flows laterally or horizontally, usually above a soil layer of lower hydraulic conductivity. This type of flow is also known as a quick return flow because it can appear as a seep or spring and can contribute to the shape of the hydrograph after a storm. If the flow is relatively steady and occurs from natural storage in soil or an aquifer replenished by infiltrated water, it is known as baseflow. The runoff in watersheds of arid land is mostly surface runoff with occasional baseflow. The runoff in watersheds of humid regions consists of surface, subsurface, and base flows.

The runoff depth can be obtained by using the US Department of Agriculture (USDA) Natural Resources Conservation Service (NRCS) curve number method (Markus, 1949):

$$Q = \frac{\left(P - I_{\mathrm{a}}\right)^2}{\left(P - I_{\mathrm{a}}\right) + S} \cdots P > I_{\mathrm{a}} \tag{6.6}$$

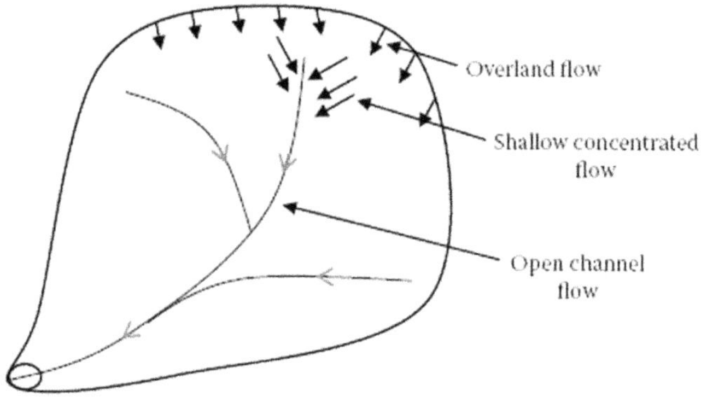

FIGURE 6.2 Process of formation of surface runoff.

$$Q = 0 \cdots P \leq I_a \tag{6.7}$$

where

Q is the depth of runoff.
P is the depth of rainfall.
I_a is the initial abstraction.
S is the maximum potential retention (USDA NRCS, 2004).

The direct surface runoff is dependent on the volume of rainfall, topography of the area, land use, and management. It decreases if the land is flat; land use is forest, agriculture, or grassland where the infiltration rate of soil is high and interception of the runoff could also be high. The subsurface runoff can be higher in areas, especially those located in humid areas. The direct surface runoff is higher from lands with higher slopes, fallow lands, and areas where agriculture or forestlands have been converted to urban land use. There are several rainfall-runoff models described using model structural categories of empirical, conceptual, and physical; and spatial structural categories of lumped, semidistributed, and distributed (Sitterson et al., 2017).

There are several publicly available programs for modeling various components of the hydrologic cycle. Some of these models are capable of evaluating the rainfall-runoff relations, infiltration, deep percolation, channel flow, soil water storage, and soil erosion under different land use and management, ecosystem, and geographic locations. Some of these models are APEX (Agricultural Policy Environmental eXtender), EPIC (Environmental Productivity Impact Calculator), SWAT (Soil and Water Assessment Tool), WEPP (Water Erosion Prediction Project), and RZWQM (Root Zone Water Quality Model). For a detailed overview of these models, readers are referred to the documentation provided on the appropriate model websites and the chapters in Shukla (2011).

PROBLEMS

6.1 Describe different components of hydrological cycle.
6.2 From a literature search find out the most common reason to conduct a water balance.
6.3 Collect refereed journal articles on water balance and identify how deep percolation and change in storage are measured in studies on agricultural farms.
6.4 Write a brief note on the SWAT model (see Chapter 18).
6.5 List various factors that influence infiltration rate into the soil.

REFERENCES

Burt, C. M. 1999. Irrigation water balance fundamentals. In: Proceedings of the Conference on Benchmarking Irrigation System Performance Using Water Measurement and Water Balances. San Luis Obispo, CA. March 10, Denver, CO, pp. 1–13. http://www.itrc.org/papers/irrwaterbalance/irrwaterbal.pdf, accessed on April 16, 2012.

Deb, S. K., M. K. Shukla, and J. G. Mexal. 2012. Simulating deep percolation in flood-irrigated mature orchards with RZWQM2. *Trans. ASABE* 55(6): 2089–2100.

Deb, S. K., M. K. Shukla, P. Sharma, and J. G. Mexal. 2013. Patterns of soil water depletion in irrigated mature pecans of arid southern New Mexico. *Irrig. Sci.* 31: 69–85.

Jouzel, J. et al. 2007. Orbital and millennial Antarctic climate variability over the past 800,000 years. *Science* 317(5839): 793–796.

Mockus, V. 1949. Estimation of total (and peak rates of) surface runoff for individual storms. In: *Exhibit A of Appendix B, Interim Survey Report*. Grand (Neosho) River Watershed, USDA Soil Conservation Service.

Shukla, M. K. (ed.). 2011. *Soil Hydrology, Land Use and Agriculture: Measurement and Modeling*. CAB International, Wallingford, UK, p. 434.

Sitterson, J. et al. 2017. *An Overview of Rainfall-runoff Model Types*. EPA Report Number EPA/600/R-14/152, September 2017.

USDA NRCS. 2004. Estimation of direct runoff from storm rainfall, Chapter 10. In: *Part 630 Hydrology National Engineering Handbook*. http://directives.sc.egov.usda.gov/OpenNonWebContent.aspx?content=17752, accessed on April 18, 2012.

7 Properties of Water

7.1 INTRODUCTION

Water in the vadose zone is very vital for sustainable food production as well as for supporting life on earth (Shukla, 2011). Water occupies about 70% of the earth and is the most widely available compound on earth. It is estimated that about 1.36 billion km^3 of water, mostly in liquid form, is available on earth. Water exists in liquid, vapor, and solid states, but the word "water" refers to the liquid state. Water is sometimes called a universal solvent because of its polarity and because it can dissolve more substances than other liquids. That is why water is seldom pure; it is almost always a solution. Water is present in lakes, oceans, and rivers as surface water; inside the earth as groundwater and as soil water; and in all living things. Water could constitute as much as 60%–70% of a livestock animal's body, 55%–60% of a young adult's body, and about 75% of an infant's body. In this chapter, properties of water are discussed to understand why water is such an important compound and to lay the foundation for understanding soil water content and its relationship to soil water potential in the vadose zone in the subsequent chapters.

7.2 PROPERTIES OF WATER

Pure water at room temperature is available as a tasteless, odorless, and colorless liquid. Water can easily dissolve a number of substances and gases, and that is why its availability in pure form is always questionable. There are three common states of matter, and water can occur in all three states—liquid, vapor, and solid—at room temperature. An example is ice cubes in a beaker at room temperature. The liquid phase is the most common state of water; in the solid state, water appears as ice and snow; and in the vapor state, water appears as water vapor or clouds. Water can also appear in a fourth state known as a supercritical fluid. A fluid subjected to pressure and temperature above critical levels is known as a supercritical fluid. At this state, distinct liquid and vapor states no longer exist. The supercritical state of water occurs in places where water is subjected to supercritical temperature (647 K) and supercritical pressure (22.06 MPa).

Naturally occurring water is mostly composed of a neutronless hydrogen isotope (protium), and only a very tiny part of natural water consists of a hydrogen isotope with one (deuterium) or two neutrons (tritium). Isotopes are natural departures of a given chemical element or compound. Light water and heavy water are essentially related to the deuterium content being low or high, respectively.

On the periodic table (presentation of elements based on properties), elements are arranged in increasing order of their atomic numbers. The atomic number is the number of protons found in the nucleus of an atom. For a neutral element, the atomic number is also equal to the number of electrons. The sum total of the number

DOI: 10.1201/9780429264849-7

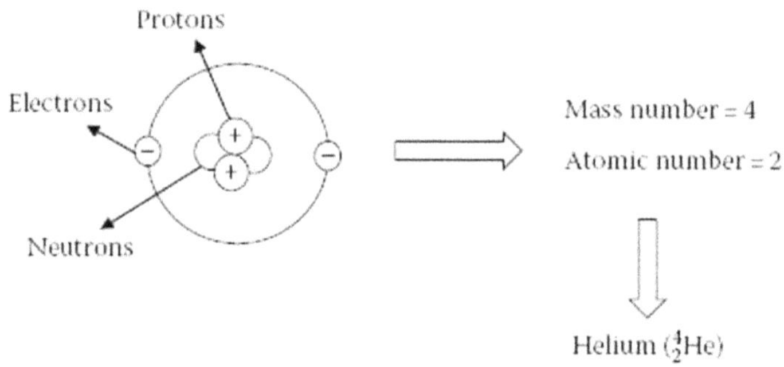

FIGURE 7.1 Atomic number and mass number for an element in the periodic table; as an example, helium is presented here.

of protons and electrons gives the mass number. For example, helium (He) has two protons, two neutrons (in the nucleus), and two electrons (Figure 7.1). Therefore, for He, the atomic number is two and the mass number is four. Hydrogen, on the other hand, has a single proton in its nucleus, no neutrons, and one electron. Therefore, the atomic number as well as mass number for hydrogen is one.

7.2.1 Polarity

Water consists of two atoms of hydrogen and one atom of oxygen, known as a binary compound with the chemical formula H_2O. Water is not electroneutral. The two hydrogen atoms are attached at an angle of about 105° to the oxygen, and the side where hydrogen atoms are attached is electropositive and the side with the oxygen atom is electronegative (Figure 7.2). Therefore, water is a dipole molecule. The distance between the two hydrogen atoms is 1.54 Angströms (Å), whereas the distance between the oxygen and hydrogen atoms is 0.97 Å (1 Å = 1.0×10^{-10} m).

A chemical bond is formed due to the attraction of atoms of opposite charges, namely, protons (+ve charge) in the nucleus and electrons (–ve charge) orbiting the nucleus. The bonds vary with respect to their strengths. Covalent bonds and ionic bonds are generally much stronger than dipole–dipole or hydrogen bonds. The bond between hydrogen and oxygen in water is a covalent bond and is the result of sharing of pairs of electrons between atoms that balances the attractive and repulsive forces (Figure 7.2). Since water is a dipole molecule, a charge opposite in nature exists on each of the atoms, positive on the hydrogen atom and negative on the oxygen atom. Water attracts other water molecules by hydrogen bonding between positively charged hydrogen and negatively charged oxygen, and the bond continues similar to a polymer type of grouping. The hydrogen bond provides structural strength and is stronger than the force of gravity. The hydrogen bond depends on the temperature of water; the higher the temperature, the weaker the hydrogen bond. Cations (Ca^{2+}, Na^+, K^+) and anions (Cl^-, NO^-, Br^-) also get bonded at negative and positive sides of the water molecule, respectively (Figure 7.3).

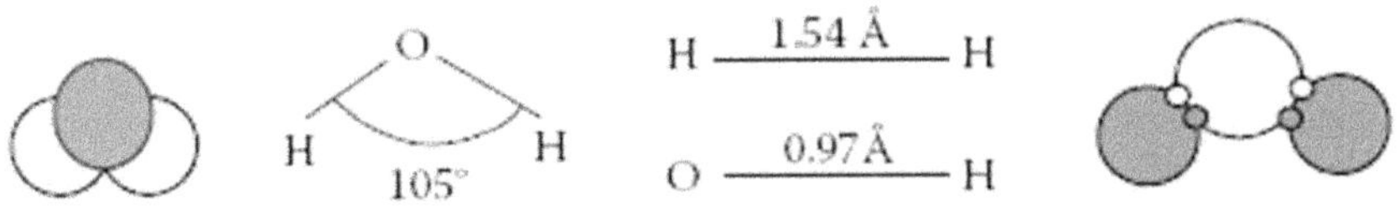

FIGURE 7.2 From left, water molecule; arrangement; distances between H and H, and O and H molecules; and covalent bonding between oxygen and hydrogen atoms by sharing of electrons.

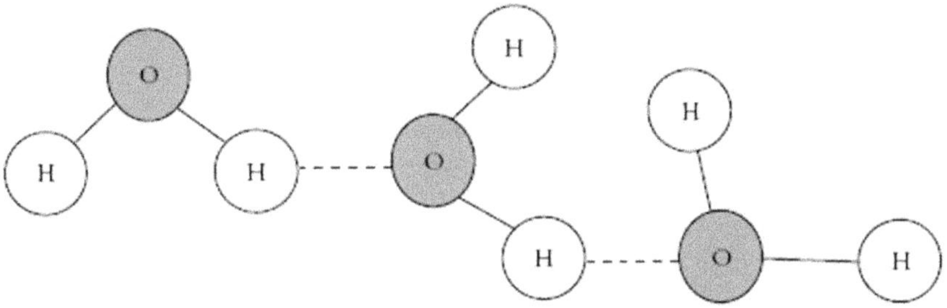

FIGURE 7.3 Hydrogen bonding between water molecules.

A practical example of the use of dipole moments in our daily life is the use of microwaves to warm food. Microwaves add energy to water molecules causing the food to get warm. If there are molecules with no dipole moments, microwaves will not interact with them.

7.2.2 Freezing Point and Boiling Point

The freezing point for pure water is 0°C (32°F or ~273 K) and the boiling point is 100°C (212°F or ~373 K). However, the freezing and boiling points of water depend on pressure. At higher elevations, the boiling point of water could be less than 100°C. Impurities in water can also change the freezing and boiling points. Ocean water contains large amounts of salts (average salinity 35 ppm), which can lower the freezing point of seawater by about 20°C and increase the boiling point by about 5°C higher than pure water. Take a look at some other H^+ compounds, such as hydrogen sulfide (H_2S), hydrogen selenide (H_2Se), hydrogen telluride (H_2Te), and water (H_2O); we find that molecular weights of these compounds follow the trend $H_2O < H_2S < H_2Se < H_2Te$. However, the freezing and boiling points of these compounds vary as $H_2S < H_2Se < H_2Te < H_2O$. Thus, as molecular weight increases, the freezing and boiling points also increase for H_2S, H_2Se, and H_2Te, but water does not follow this rule. In addition, the boiling point for water is much higher than the rest. This is due to the hydrogen bonding between water molecules, which requires more energy to change the state of water from liquid to gas. If water was not a dipole with weak bonding between its molecules, the boiling and freezing points for water would have been below those for hydrogen sulfide.

Example 7.1

Calculate the molecular weights of water (H_2O), hydrogen sulfide (H_2S), hydrogen selenide (H_2Se), and hydrogen telluride (H_2Te).

$$\textbf{Water}(\mathbf{H_2O}) = 2\times1+16=18$$

$$\textbf{Hydrogen sulfide}(\mathbf{H_2S}) = 2\times1+32=34$$

$$\textbf{Hydrogen selenide}(\mathbf{H_2Se}) = 2\times1+78=80$$

$$\textbf{Hydrogen telluride}(\mathbf{H_2Te}) = 2\times1+128=130$$

Example 7.2

Convert 25°C (T_c) into the units of Fahrenheit (T_f) and Kelvin (K).

$$T_f = \frac{9}{5}T_c + 32$$

$$K = T_c + 273.15$$

However, for most conversions 273 is added to the value in Celsius:

$$T_f = \frac{9}{5}\times 25 + 32 = 45 + 32 = 77°\mathbf{F}$$

and

$$K = 25 + 273.15 = 298.15\mathbf{K}$$

7.2.3 Density of Water

The density of water is nearly 1 g/cm^3 at 4°C. When the temperature of water is higher or lower than 4°C, the density of pure water is less than 1 g/cm^3. The density–temperature relationship as presented in Figure 7.4 is not linear, but it is unimodal because there is only one local maximum at 4°C. As the temperature of water decreases from an initially high temperature, the density of water increases up to 4°C, and further cooling of water decreases the density of water due to expansion. Water in the solid state is less dense than its liquid state, and

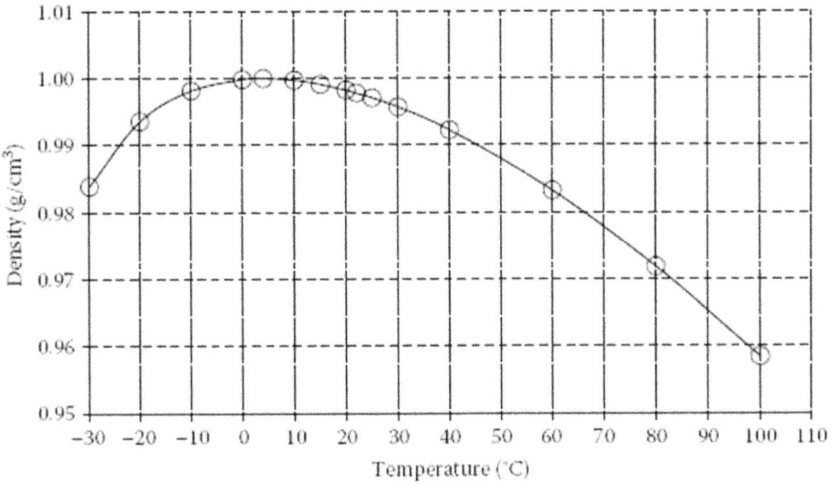

FIGURE 7.4 Variation of the density of liquid water with temperature.

that is why, unlike other substances, solid water in the form of ice floats in liquid water.

7.2.4 Viscosity

Viscosity of a fluid is the measure of its resistance to flow. Viscosity is higher for molecules that cause more internal friction during flow. Low viscosity fluids flow easily. Viscosity is defined as the ratio of the drag force and the velocity gradient perpendicular to the direction of flow. The viscosity of water can be expressed as dynamic (μ; N s/m^2 or Pa s or poise [P]) or kinematic (v; m^2/s or stokes [St]) viscosity. Kinematic viscosity is the ratio of dynamic viscosity and the density of the fluid. Viscosity decreases with increasing temperature. At 20°C, water has a dynamic viscosity of 1.002 cP (centipoize) or a kinematic viscosity of 1 cSt (centistokes). The reciprocal of viscosity is known as fluidity.

7.2.5 Specific Heat Capacity

The specific heat capacity of a substance is defined as the amount of heat required to change the temperature of a unit mass of a substance by a unit degree. The specific heat capacity of water is considered to be very high primarily due to the hydrogen bonding between water molecules. The dipolar interaction between water molecules is a large amount of energy, primarily due to the random and disordered motion of molecules. The specific heat capacity of water at a given pressure decreases as temperature increases from 0°C to about 30°C, and beyond that specific heat capacity starts to increase (Figure 7.5). The specific heat capacity of ice is 2.11 J/g/K at −10°C, water is 4.16 J/g/K at 25°C, and water vapor is 2.08 J/g/K at 100°C. The high

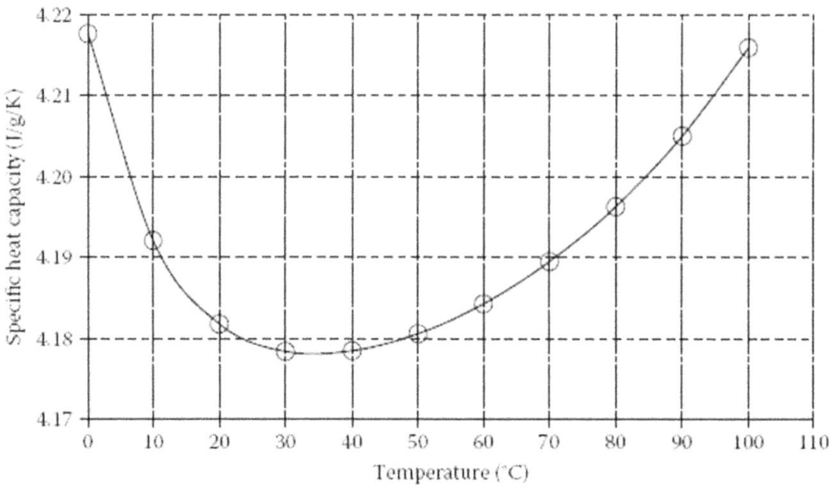

FIGURE 7.5 Specific heat capacity of water with respect to temperature at a given constant pressure.

specific heat capacity of water indicates that a lot of energy is required to change the temperature of water. Water also has a high heat of vaporization (energy required to change the state from liquid to vapor) of 2.26 kJ/g (540 cal/g) at 100°C and high specific enthalpy of fusion (energy required to change the state from solid to liquid) of 333.6 kJ/kg (79.7 cal/g) at 0°C.

7.2.6 Bulk Modulus

The bulk modulus of a substance is the measure of its resistance to the applied uniform compression and is the inverse of its compressibility. The bulk modulus of water is 2.2 GPa (2.2×10^9 Pa), and therefore the compressibility of water is very low and water is usually treated as an incompressible fluid. Under a pressure of 40 MPa, which can be observed at about 4 km deep in the ocean, the decrease in water volume is only 1.8%.

Example 7.3

What could be the pressure in bars and mega pascal (MPa) at 2 and 4 km deep in the ocean?

$$10\,\text{m of water column} = 1\,\text{bar pressure and } 10\text{ bar} = 1\text{ MPa}$$

$$2000\,\text{m}\left(\text{or } 2\text{km}\right)\text{of water column} = 2000/10 = 200\text{ bar or } 20\text{ MPa}$$

$$4000\text{ m}\left(\text{or } 4\text{ km}\right)\text{of water column} = 4000/10 = 400\text{ bar or } 40\text{ MPa}$$

7.2.7 Electrical Conductivity

Pure water is a good insulator because it does not contain charged particles. It is also a very good solvent, but is seldom free of ions. The ions could be dissolved, suspended, or floating in water. For example, water and ethanol are miscible; on the other hand, water and oils are immiscible. Even deionized water is not completely free of ions. A very small amount of impurity or ions in water will make it conduct electricity rapidly.

7.2.8 Wetting and Nonwetting

When a drop of water is placed on a clean glass, it quickly wets it, and the drop does not retain its shape. The reason behind quick wetting is the presence of oxygen on the surface of the glass and the availability of unpaired electrons that quickly form a hydrogen bond with water molecules. On the other hand, when the surface of the glass is coated with grease, water does not wet the glass because the surface of grease does not have oxygen and free electrons available for hydrogen bonding.

7.3 FORCES ON WATER MOLECULES

Several different forces act on water molecules at any given time. The attraction of water molecules for each other is known as cohesion, whereas the attraction of water molecules for solid surfaces is known as adhesion. Water molecules hold each other away from the solid surface by the force of cohesion, whereas solids hold water molecules rigidly at the solid–water surface by the force of adhesion. Together, it is possible for solids (which could be the surface of a glass tube or a soil pore) to retain water and control its movement. In addition, the force due to gravity also always acts on the water molecules in the downward direction.

Figure 7.6 shows water stored in a glass cup. The image on the left-hand side shows water molecules at the surface or at the air–water interface (point A), at the water–water interface (point B), and at the solid–water interface (point C). The water molecules at each of these interfaces are subjected to various forces that are trying to pull them up, down, or sideways from their existing position. At point B, all the cohesive forces acting on the water molecule are equal in all directions because attraction of water molecules to other water molecules is equal in all directions. At point C, the attraction of the solid surface for water molecules (adhesion) is much stronger than the attraction of water molecules for other water molecules (cohesion), and they stay attached to the solid surface. In contrast, at point A, at the air–water interface, the attraction of air for water molecules (upward) is much less that the attraction of water molecules to other water molecules (downward) as well as gravitational pull (vertically downward). Consequently, there is an imbalance of forces with a net downward force on the surface molecules, and the result is a compressed film (picture on right side) at the water–air interface. The film gets compressed until upward and downward forces become equal, and this phenomenon is called surface tension. Surface tension is a property of liquids and a result of unbalanced forces

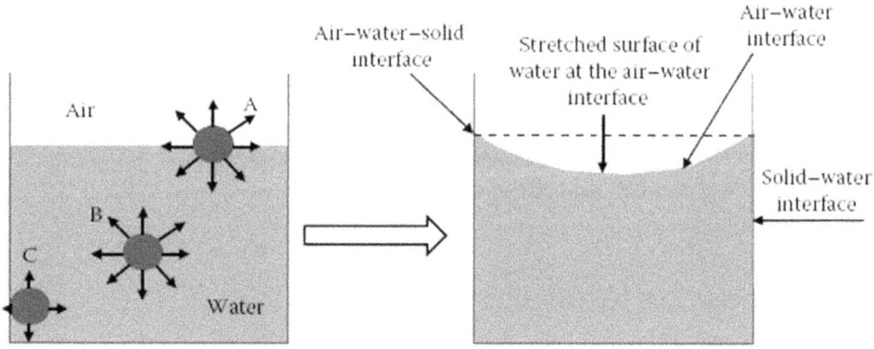

FIGURE 7.6 (Left) Forces acting on water molecules at the air–water interface (point A), water–water interface (point B), and solid–water interface (point C). (Right) Stretching of the water at the surface is caused by surface tension.

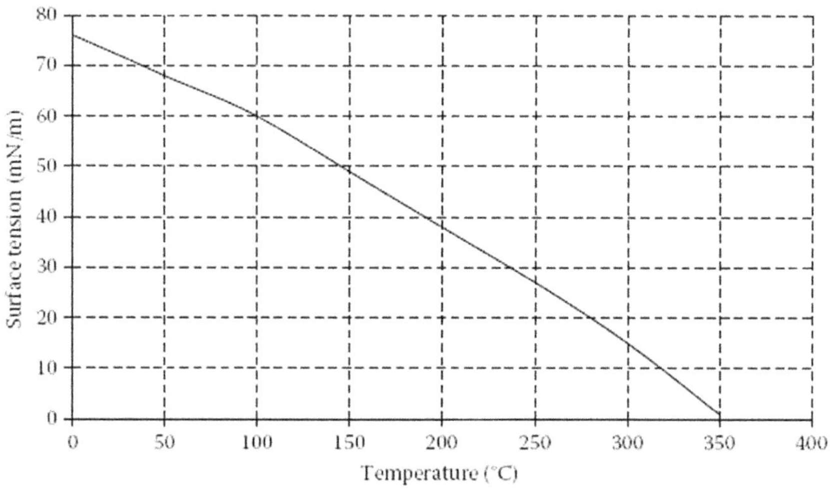

FIGURE 7.7 Schematic of the surface tension of pure water at saturation pressure as a function of temperature.

near the surface that cause the surface to be stretched. The liquid surface contracts to the minimal area. The dimensions of surface tension are given as force per unit length (MLT^{-2}/L or MT^{-2}) or energy per unit area (ML^2T^{-2}/L^2 or MT^{-2}). The SI unit for surface tension is N/m (newton per meter) and the CGS unit is dyne/cm (dyne per centimeter), with 1 dyne/cm being equal to 0.001 N/m. Surface tension decreases as temperature increases (Figure 7.7). At 25°C, water has a surface tension of 72.8 dyne/cm (or 0.0728 N/m).

Surface tension can be visually observed by dropping a paper clip in a bucket filled with water. Due to surface tension, the paper clip floats and does not submerge.

Surface tension is the phenomenon behind the shape of a liquid drop that tends to become spherical with minimum surface area-to-volume ratio due to cohesive forces. The water drop on a waxed surface is nearly spherical because of the surface tension.

7.4 CONTACT ANGLE

Surface tension is a property of the liquid as well as a property of the interface in contact with the liquid. Liquid in a glass container has several interfaces, for example, air–water, water–solid, water–water, and air–water–solid interfaces (Figure 7.6). Surface tension between a solid and water (or at a solid–water interface) is smaller than that at the water–air interface. That is why stretching of the water surface increases with increasing distance from the wall (solid surface). Figure 7.6 shows that stretching of the water meniscus is at its maximum at the center of the tube, and as the distance to the solid wall decreases, stretching also decreases. Surface tension depends on the impurities in water, not on how much the water surface is stretched. Wherever these surfaces meet, they form an angle known as the contact angle. The contact angle is the angle at which an air–liquid surface meets a solid surface. The angle is usually measured through the liquid at a point where all three interfaces meet. The contact angle is drawn as a tangent to the surface of water meeting the solid surface. If the surface tension is γ_{sa} at the solid–air interface, γ_{sw} at the solid–water interface, and γ_{wa} at the water–air interface, then according to Young (1805), for a perfectly smooth surface,

$$\gamma_{sa} = \gamma_{sw} + \gamma_{wa} \cos \alpha \tag{7.1}$$

$$\cos \alpha = \frac{\gamma_{sa} - \gamma_{wa}}{\gamma_{wa}} \tag{7.2}$$

where α is the equilibrium contact angle. Remember that even in a perfectly smooth surface, a drop of water can have many different contact angles, generally receding with increasing time (Figure 7.8).

This equation indicates that whenever $\gamma_{sa} > \gamma_{sw}$, cos α is positive, indicating that the value of the contact angle is <90°. The angle <90° is known as an acute angle. Thus, if the contact angle for a liquid is an acute angle, that liquid will wet the solid. On the other hand, whenever $\gamma_{sa} < \gamma_{sw}$, cos α is negative indicating that the value of the contact angle is >90°. This is known as an obtuse angle. Thus, if the contact angle for a liquid is an obtuse angle (between 90° and 180°), that liquid will not wet the solid.

When adhesive forces between water molecules and a surface (a solid or a soil pore) are weaker than the cohesive forces between water molecules, the surface repels water and is known as the hydrophobic surface. Hydrophobic soils restrict the entry of water into the soil, and water can sit on the surface as droplets for a long time. The wetting angle >90° causes capillary repulsion preventing water from entering soil pores. In contrast, when cohesive forces are weaker than adhesive forces, capillary

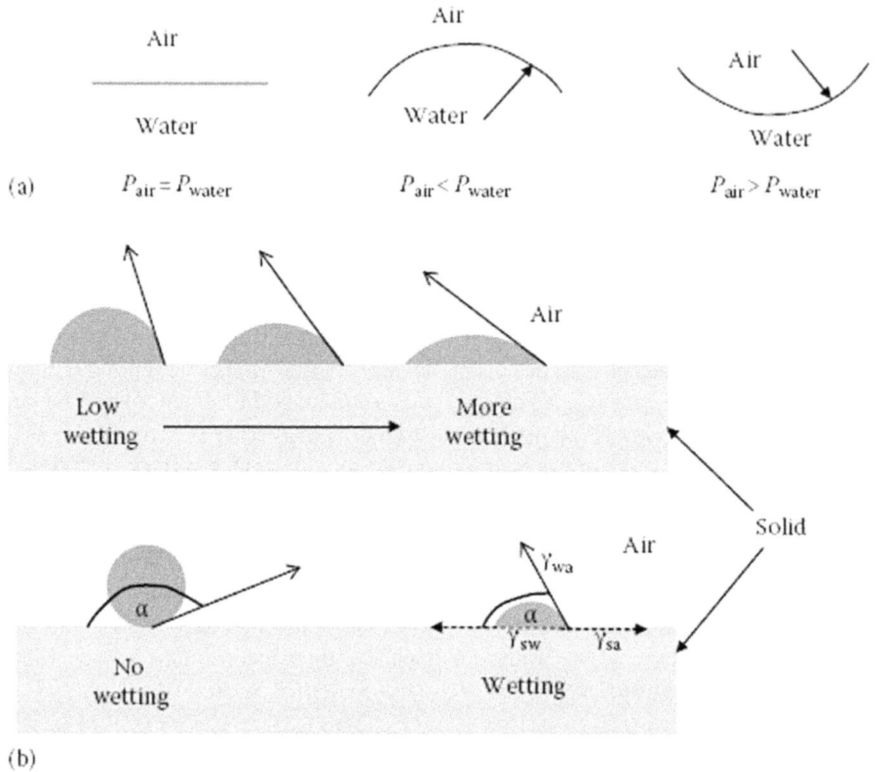

FIGURE 7.8 (a) Pressure difference across three air–water interfaces: horizontal, concave to water, and concave to air. (b) Contact angles for fluids displaying low to high wetting. *Note*: Obtuse and acute contact angles display different amounts of wetting.

attraction allows entry of water in the pore space. Such types of surfaces (or soils) are known as hydrophilic soils.

Example 7.4

If α = 0°, 45°, 90°, 180°, 270°, and 360°, calculate cos α and sin α.

α	**cos α**	**sin α**
0°	1	0
45°	0.707	0.707
90°	0	1
180°	−1	0
270°	0	−1
360°	1	0

7.5 CAPILLARITY

Capillarity, capillary rise, or wicking is the ability of a liquid to flow through small spaces, known as capillaries, without any external support and against gravity. Capillary action can be visualized by pouring a drop of water on a paper, and the water quickly moves through the pore spaces in the paper. The capillary rise is the result of adhesive and cohesive forces acting against the gravitational pull. The capillary rise can be seen by dipping a glass tube of narrow diameter in a container with water. How high the water could rise due to capillary action can be mathematically calculated by taking into account all the upward and downward forces acting on water molecules.

Let us assume a capillary of diameter $2r$ (r being the radius), as shown in Figure 7.9. The water in the tube rises due to the capillary action until the upward forces are balanced by the downward force of gravity. The force of gravity can be calculated as follows using the definition of force (Mass × Acceleration):

$$\text{Mass of water in the tube} = \text{Density of water} \times \text{Volume of water in the cylindrical tube}$$

$$= \rho_w \times \left(\pi \times r^2 \times h\right) \tag{7.3}$$

$$\text{Force} = F_g = \text{Mass of water column in the tube} \times \text{Acceleration due to gravity}$$

$$= \rho_w \times \pi \times r^2 \times h \times g \tag{7.4}$$

Total cohesive and adhesive or capillary forces acting on the water (F_c) is

$$F_c = \text{Surface tension}\left(\gamma\right) \times \text{Perimeter of the circle} \times \cos\alpha$$

$$= \gamma \times 2 \times \pi \times r \times \cos\alpha \tag{7.5}$$

At equilibrium, $F_c = F_g$

$$\gamma \times 2 \times \pi \times r \times \cos\alpha = \alpha = \pi \times r^2 \times h \times g \times \rho_w \tag{7.6}$$

Therefore, the height of the rise of water in a capillary can be given by the following formula:

$$h = \frac{2\gamma\cos\alpha}{\rho_w g r} \tag{7.7}$$

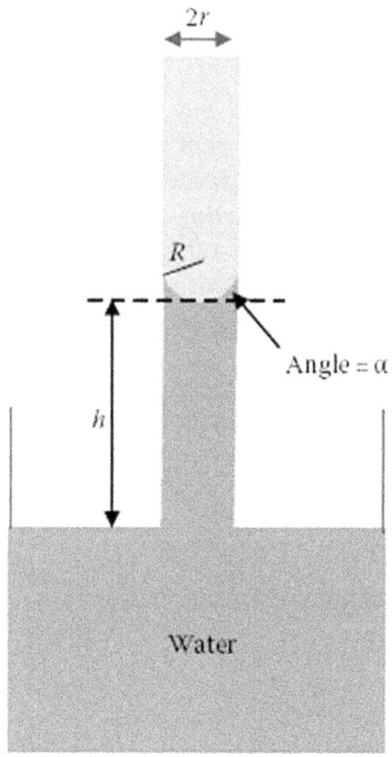

FIGURE 7.9 Capillary rise of water (h) in a glass tube with radius (r). The contact angle of the curved surface with the glass tube is α and the radius of the sphere is R.

If there are two principal radii r_1 and r_2, the height of capillary rise can be calculated as follows:

$$h = \frac{2\gamma \cos\alpha}{\rho_w g}\left(\frac{1}{r_1} + \frac{1}{r_2}\right) \tag{7.8}$$

Assuming that in Figure 7.9 there are two solutes with pure water in the tube (capillary) and a solution in the tank (instead of pure water) separated by a solute membrane and that the contact angle is zero ($\cos \alpha = \cos 0 = 1$), the capillary rise can be given by the following formula:

$$h = \frac{2\gamma}{\rho_w g r} - \frac{\varphi_o}{\rho_w g} \tag{7.9}$$

where φ_o is the osmotic potential. Since ions present in the solute will attract the water molecules on the capillary side of the membrane, the capillary rise will be smaller than when only pure water is present in the capillary as well as the tank (Jury et al., 1991).

Example 7.5

If the density of water is 0.9982 g/cm³, acceleration due to gravity is 980 cm/s², surface tension is 72.75 dyne/cm, and water completely wets the surface, what is the relationship between the height of capillary rise and the radius of the capillary?

$$\boldsymbol{h} = \frac{2\gamma}{\rho_w \boldsymbol{gr}} = \frac{2 \times 72.75}{0.9982 \times 980 \times \boldsymbol{r}} = \frac{0.1487}{\boldsymbol{r}}$$

Note that the preceding equation for units of capillary rise and radius are in centimeters (cm). There is an inverse relationship between the height of capillary rise and the radius of capillary. For a capillary with a radius of 0.2 cm, the height of rise of water in the capillary will be 0.743 cm (or 7.43 mm).

The pressure difference across a flat air–water interface is zero. However, the interface is always curved, and pressure is greater on the concave side of the air–water interface (Figure 7.9). For a spherical interface, the difference in pressure can be calculated as follows:

$$\Delta P = \frac{2\gamma}{R} = P_{\text{air}} - P_{\text{water}} \tag{7.10}$$

Thus, the drop in pressure across the air–water interface is balanced by the surface tension force. If the contact angle (α) is >0 and depends on the properties of the liquid and the surface in contact, the radius of the sphere will be a function of the contact angle. If the radius of the sphere is R, then

$$R = \frac{r}{\cos\alpha} \tag{7.11}$$

The height of rise of water in a capillary can be given by the following formula:

$$h = \frac{2\gamma\cos\alpha}{\rho_w g r} \tag{7.12}$$

In Equation 7.12, the capillary pressure is balanced by the change in the height of capillary rise and hydrostatic equilibrium is obtained.

Example 7.6

Assume that the surface tension is 0.07275 N/m, the density of water is 998.2 kg/m³, and acceleration due to gravity is 9.8 m/s². If instead of fully wetting the surface, there is a contact angle of 20°, calculate the height of rise of water for a 0.002 and 0.0002 m wide capillary.

Capillary radius = 0.002/2 = 0.001 m = 1×10^{-3} m

$$h = \frac{2\gamma\cos\alpha}{\rho_w g r} = \frac{2 \times 0.07275 \times \cos 20}{998.2 \times 9.8 \times r} = \frac{1.398 \times 10^{-5}}{r} = \frac{1.398 \times 10^{-5}}{1 \times 10^{-3}}$$

$$= 0.01398\ \mathbf{m} = 13.98 = 14\ \mathbf{mm}$$

$$\textbf{Capillary radius} = 0.0002/2 = 0.0001\textbf{m} = 1\times10^{-4}\,\textbf{m}$$

$$h = \frac{1.398\times10^{-5}}{r} = \frac{1.398\times10^{-5}}{1\times10^{-4}} = 0.1398\textbf{m} = 139.8 = 140\textbf{mm}$$

As the capillary radius decreased ten times, the height of rise of water in the capillary increased ten times (Figure 7.10). The capillary rise inside a soil mass takes place via a network of soil capillaries or soil pores (Figure 7.11). Sometimes, the capillary also contains entrapped air bubbles, and water generally follows a tortuous path, therefore, the height of capillary rise inside a soil mass is often different from the rise in an equivalent-size glass tube.

7.6 EMPIRICAL APPROACH

A semiempirical formula combined with laboratory measurements is commonly used to estimate hydraulic properties of soils under unsaturated conditions (Rajkai et al., 1996; Arya et al., 1999; Lu and Likos, 2004). Theoretically, the radii of pores can be estimated from geometrical features (Arya et al., 1999):

$$r_i = 0.816 \cdot R_i \cdot \sqrt{e \cdot n_i^{(1-\alpha_i)}} \tag{7.13}$$

where

r_i is the pore radius of soil (cm).
e is the void ratio of the soil sample (cm³/cm³) determined as $(\rho_s - \rho_b)/\rho_b$.
ρ_s is the particle density (g/cm³).
ρ_b is the bulk density of the sample (g/cm³).
R_i is the mean particle radius for the ith particle size fraction (cm).
α_i is the scaling parameter.

The scaling parameter is $\alpha_i = \log N_i/\log n_i$, where $\log N_i = a + b$, $\log(w_i/R_i^3)$, w_i is the mass fraction of particles (g/g), a and b are constants, and n_i is the number of equivalent spherical particles in the fraction and is given as $n_i 3w_i / 4\pi\rho_s R_i^3$. The empirical constants a and b, input values used to determine α_i, in Equation 7.13, can be obtained from Table 2 in Arya et al. (1999: 1065) for sand, sandy loam, loam, and clay.

The flux rate is a useful parameter to evaluate the capillary rise of water in soil. Lu and Likos (2004) modified Gardner's (1958) equation discussed later to incorporate the hydraulic gradient responsible for capillary rise to determine the flux or rate of capillary rise (Dung et al., 2011). The theoretical maximum capillary rise has a close relationship with pore radius of soil (Singh, 1999) and is given by the capillary rise equation (Equation 7.12).

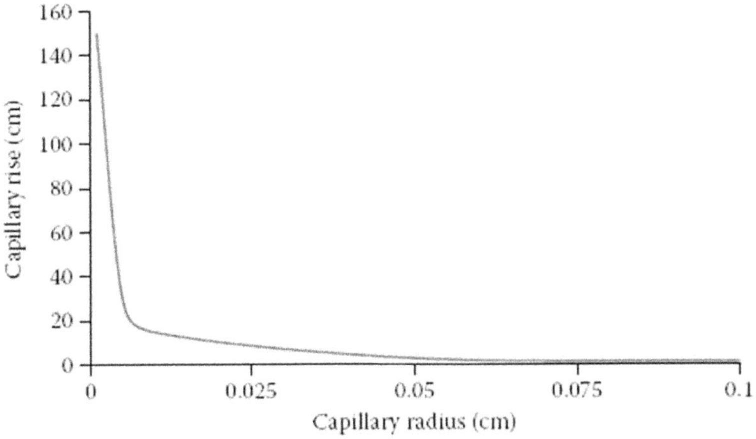

FIGURE 7.10 As the diameter of capillary decreases, the height of rise of water in the capillary increases.

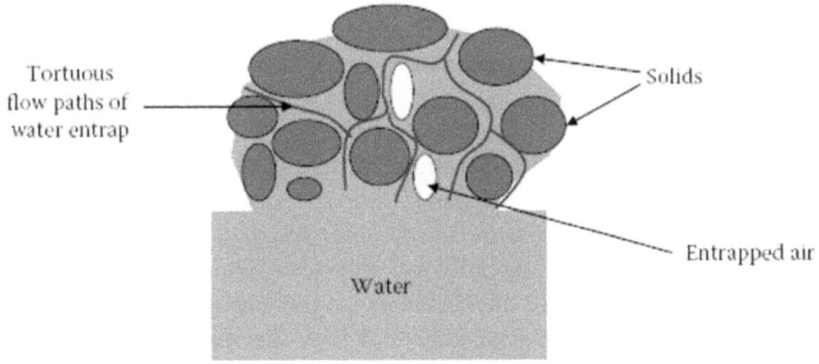

FIGURE 7.11 Water flow inside a soil mass is through a network of tortuous capillary flow paths. See entrapped air inside; therefore, capillary rise may be different from that in a glass tube of equivalent diameter.

7.7 IMPORTANCE OF CAPILLARITY

The capillary rise is caused by the hydraulic head gradient that exists across the curved air–water interface. The behavior of soil with respect to rate, height, and amount of capillary rise is a complex function of soil and pore properties (Lu and Likos, 2004). Hydrological alterations and the river system operations in arid regions worldwide have adversely impacted the riparian vegetation and the efforts to restore those (Dung et al., 2011). Both surface and groundwater, important for plant sustenance in the riparian system, have seen alterations detrimental to the vegetation. The capillary rise mechanism was the basis in the pole planting technique for certain species within the riparian regions of New Mexico, and poles were planted deep into

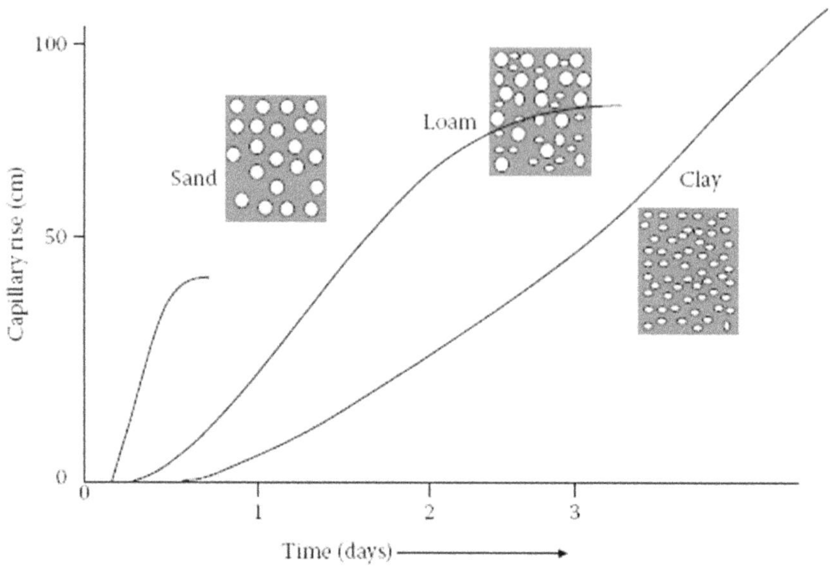

FIGURE 7.12 Schematic of capillary rise in different types of soils. Note the change in slope from sand to clay. The time in the *x*-axis is for illustration purpose and not from actual data (not to scale).

holes so that they can wick sufficient moisture into the root zone of plants (Taylor and McDaniel, 1998).

Zeolite was also shown to be a useful soil amendment to wick shallow groundwater to the root zone of plants, especially in sandy soils (Dung et al., 2011). Capillary rise was monitored in 5 and 10 cm diameter acrylic Plexiglas columns filled with clinoptilolite-type zeolite and sand collected from the banks of the Rio Grande River. A constant head of 10 cm was applied at the bottom of the column, and a capillary rise of more than 2 m was observed (Dung et al., 2011).

The capillary rise being a function of size of pore space is different for different soils. In general, among sand, silt, and clays, as particle size decreases from sand to clay, the height of capillary rise increases (Figure 7.12). The exact amount of capillary rise is dependent on the pore sizes and pore connectivity, and can range from 20 to 50 cm in coarse (sand), 50 to 80 cm in medium (loam) to more than 80 cm in fine (clay) soils. However, the time to attain the maximum height of capillary rise is much longer for clay soils than for silt and sands.

PROBLEMS

7.1 Describe some of the properties of water.

7.2 Calculate capillary rise when a glass tube with a 1 mm diameter is inserted in a container filled with alcohol and sea water. Temperatures of both liquids are 20°C and both liquids are assumed to wet the surface completely. Get all other required data using a handbook of chemistry or other sources.

7.3 Plot temperature versus capillary rise when water temperature increases from 4°C to 64°C at an increment of 20°C.

7.4 Plot time versus capillary for pure water, sea water, and alcohol on the same graph.

7.5 Explain hydrophobicity using different concepts.

7.6 In Example 7.5, calculate capillary rise if the contact angle is 45° and 75°.

REFERENCES

Arya, L. M., F. J. Leji, P. J. Shouse, and M. T. van Genuchten. 1999. Relationship between the hydraulic conductivity function and the particle-size distribution. *Soil Sci. Soc. Am. J.* 63(6): 1063–1070.

Dung, T., S. Bawazir, M. K. Shukla, and P. Bandini. 2011. Hydraulic properties of St. Cloud zeolite and zeolite soil mixtures. *Appl. Eng. Agric.* 27(6): 955–967.

Gardner, W. R. 1958. Some steady state solutions of the unsaturated moisture flow equation with application to evaporation from a water table. *Soil Sci.* 85: 228–232.

Jury, W. A., W. R. Gardner, and W. H. Gardner. 1991. *Soil Physics*. John Willey & Sons, New York, p. 328.

Lu, N. and W. J. Likos. 2004. Rate of capillary rise in soil. *J. Geotech. Geoenviron. Eng.* 130(6): 646–650.

Rajkai, K., S. Kabos, M. Th. van Genuchten, and P. E. Jannson. 1996. Estimation of water retention characteristics from the bulk density and particle size distribution of Swedish soil. *Soil Sci.* 161(12): 832–845.

Shukla, M. K. 2011. Introduction to soil hydrology. In M. K. Shukla (ed.), *Soil Hydrology, Land Use and Agriculture: Measurement and Modeling*. CAB International, Wallingford, UK, pp. 1–23.

Singh, A. 1999. *Modern Geotechnical Engineering*, 3rd edn. CBS Publishers and Distributors, New Delhi, India.

Taylor, J. P. and K. C. McDaniel. 1998. Riparian management on the Bosque del Apache National Wildlife Refuge. *New Mexico J. Sci.* 38: 219–232.

Young, T. 1805. An essay on the cohesion of fluids. *Phil. Trans. R. Soc. Lond.* 95: 65–87. doi:10.1098/rstl.1805.0005

8 Water in the Vadose Zone

8.1 INTRODUCTION

The vadose zone is defined as the zone from the soil surface (soil–air interface) to the water table (soil–water interface); in the absence of the water table, it is the space between the soil surface and the bedrock (rock–soil interface). The vadose zone is the portion of the aquifer where pore spaces are occupied with water and air. It is also known as the unsaturated zone. The vadose zone consists of various horizons and is highly spatially variable due to the soil-forming factors including parent material, climate, biota, drainage, and time. It is highly important for root growth, biomass, and grain productivity, as well as water conservation and quality of groundwater (Shukla, 2011). Water in the vadose zone is very important for sustainable food production as well as supporting life on earth. In this chapter, relationships between soil water content and soil water potential are discussed to understand water retention in soil and availability of water to plants.

Soil water is an important component of the hydrologic cycle and has a wide range of applications in soil science, horticulture, environmental science, range science, forestry, agricultural engineering, chemical engineering, and civil engineering. Since water inside the soil is bound by capillary forces, it is also referred to as soil moisture. One of the questions vastly important for supporting biota is how much water is available in the rooting zone. This chapter deals with different types of soil water content and measurement devices, and their relation to different soil water potentials, and lays the foundation for understanding soil water movement through saturated and unsaturated porous media discussed in Chapter 9.

8.2 SOIL WATER CONTENT

Soil water content is the fractional amount or fractional volume of water present at any given time in the soil. Water can be evaporated by heating to 100°C. Soil water can be evaporated by drying soil in an oven where the temperature is maintained at 105°C until the weight of the soil becomes constant. The constant weight is usually obtained within 24 h of continuous drying. All the free and held water present inside the pores due to surface tension is removed, although adsorbed or bound water present as thin films on the soil particle surfaces may not be completely removed by drying at 105°C. The bound water as well as water that is present in the soil in the vapor state is much less than the free water and is not included in the calculation of soil water content (Gardner et al., 2001). Most bound water is immobile and can be removed from the soil by drying it at temperatures around 800°C.

DOI: 10.1201/9780429264849-8

Soil water content is dynamic and changes with time. It increases due to irrigation and precipitation to the level where all pores are filled with water and soil is considered saturated. At the cessation of irrigation or precipitation, the soil water content decreases due to the progressive drying of soil by evaporation and removal of water by plants (transpiration and root water uptake) (Figure 8.1). The soil water content at any given time can be expressed as mass fraction or volume fraction. Water content when expressed as mass fraction is known as gravimetric water content (*w*) and is expressed as a ratio of mass of water (g or kg) per unit mass of dry soil (g or kg). When expressed as a volume fraction, water content is known as volumetric water content (θ) and is calculated from the ratio of volume of water (cm^3 or m^3) and total volume of soil (cm^3 or m^3). The total volume of soil is the sum of the volume of soil solids and the volume of pores filled with water or air. The volumetric water content can be calculated from known bulk density of soil, density of water, and gravimetric water content as follows:

$$\theta = \frac{\rho_b}{\rho_w} \times w \tag{8.1}$$

Another way to express soil water content is as degree of saturation. It is also known as relative saturation or saturation ratio. Degree of saturation indicates the fraction of the pore spaces filled with water and ranges from zero to one (Topp and Ferre, 2002). At complete saturation, all the pores are filled with water; therefore, the degree of saturation is equal to one. On the other hand, when all soil pores are devoid of water, the degree of saturation is equal to zero.

Accurate measurement of the soil water content is difficult, especially in a field or at a scale greater than a point scale. Some of the difficulties are related to the variability and heterogeneity of soils. Soil water is also dynamic and is spatially and temporally variable. Vegetation on the soil increases the variability of moisture content many times because plant growth and density are spatially and temporally variable and depend on several other factors, such as availability of water, nutrients, chemicals in soil, and pest pressures. In addition, the type of tools used for measurement could increase the variability of measured soil moisture. Costs associated with the soil water measurement could also be high.

8.3 SOIL WATER CONTENT MEASUREMENT METHODS

The soil water content can be measured using destructive and nondestructive methods, which are also known as direct and indirect methods (Figure 8.2). The direct methods of soil water content measurement are gravimetric, thermogravimetric, and feel methods. The indirect methods of soil water content measurement include the radiation technique, dielectric methods, electrical methods, thermal methods, chemical methods, acoustic methods, and remote sensing methods. These methods are briefly discussed here; a detailed deposition of these methods is available elsewhere (Dane and Topp, 2002; Topp and Ferre, 2002; Robinson et al., 2008).

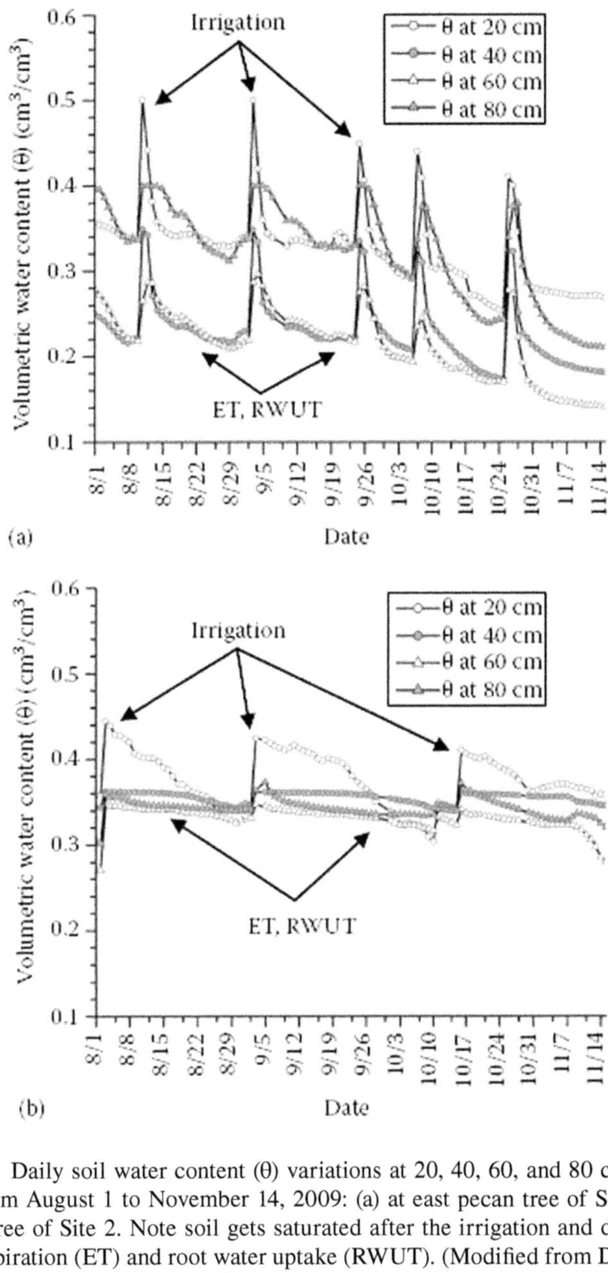

FIGURE 8.1 Daily soil water content (θ) variations at 20, 40, 60, and 80 cm deep during the period from August 1 to November 14, 2009: (a) at east pecan tree of Site 1 and (b) at north pecan tree of Site 2. Note soil gets saturated after the irrigation and desaturated due to evapotranspiration (ET) and root water uptake (RWUT). (Modified from Deb et al., *Irrig. Sci.*, 2011, doi: 10.1007/s00271-011-0293-1.)

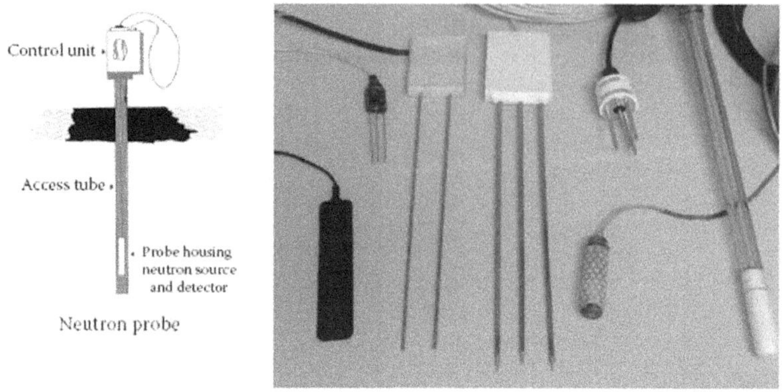

ECO probe, three TDR waveguides, Hydra probe for soil water content, soil temperature and soil electrical conductivity, and watermark and tensiometer for matric potential

FIGURE 8.2 Pictures of some indirect methods of soil water content and matric potential measurement devices and a tensiometer for direct measurement of matric potential.

8.4 DIRECT METHODS

8.4.1 Gravimetric Method Using Oven-Drying

In the gravimetric method, a representative soil sample is physically removed from an area of interest, weighed, and dried in an oven at a constant temperature of about 105°C until the weight of the dry soil becomes constant (Gardner, 1986). The gravimetric method using oven and convective heat transfer is the most common, considered the standard, and is used to calibrate other methods. The sample size for gravimetric moisture content is a function of the depth increment for which the soil water content is measured. For measuring the water content at a point in the soil profile, about 10–50 g of sample is sufficient; however, for larger depth increments, more soil must be collected to get a good average value of the gravimetric soil moisture content.

For soils containing a large amount of skeletal or gravel fractions (such as stones), more sample volume is required. For gravelly soils, the soil sample is separated into fractions smaller than and greater than 2 mm size, and the water content of these two fractions is determined separately (Gardner et al., 2001). The skeletal material or stones are generally believed to have insignificant water content, but that may not always be true. The volumetric water content of soil (θ) can be determined from the known fraction of skeletal materials (S) and the volumetric water content of the <2 mm fraction ($\theta_{<2}$) as follows (Gardner et al., 2001):

$$\theta = \theta_{<2} \times (1 - S) \tag{8.2}$$

The gravimetric method is simple, less expensive, and reliable if proper precautions are taken. A large number of samples can be simultaneously dried. The method is not dependent on soil type or soil salinity. The disadvantages of the method include the soil being physically removed from the location, and no repeat measurement is possible from the same location in space. In addition, at least 24 h are required for drying the soil, and that may be too long for certain situations. Prior knowledge of soil bulk density is required to convert gravimetric water content to volumetric water content. Soil variability can introduce errors, especially if small amounts of soil samples are collected. The oven must be maintained properly; otherwise, it can provide erroneous values of the soil water content.

8.4.2 Gravimetric Method Using Microwave Oven

In the microwave oven, heating is caused by the microwave energy using the dipole properties of water. The sample gets heated quickly and also dries fast. This method is similar to conventional convective heat transfer ovens, but the sample in the microwave oven is not dried to a constant weight. Instead, soil is dried for a certain time period sufficient for acceptable drying (Topp and Ferre, 2002). The soil moisture content is calculated as described in the previous section. The disadvantages associated with this method are similar to those associated with the convective oven method.

8.4.3 Thermogravimetric Method

In this method, the soil sample is mixed with alcohol, which burns and releases heat energy, causing evaporation of water in the soil. Burning is performed repeatedly until the weight of dry soil becomes constant. Methyl, ethyl, and propyl alcohols can be used because they can burn even when 40%–45% of water by volume is present in these alcohols. This method is rapid and can be used for determining the soil water content in the field (Bouyoucos, 1937). The disadvantages associated with this method are similar to those associated with the convective oven method, and there may be associated fire hazards.

8.5 INDIRECT METHODS

8.5.1 Radiation Techniques

Radiation techniques use a radioactive source and are based on either neutron thermalization or gamma ray attenuation.

8.5.1.1 Neutron Probe

The neutron thermalization technique is commonly known as the neutron probe method. The probe contains a long and narrow cylinder with a source and detector. The radioactive source consists of a mixture of radium–beryllium ($^{88}Ra/^{9}Be$) or americium–beryllium ($^{241}Am/^{9}Be$) high-energy, fast, epithermal neutrons. The fast-moving (kinetic energy >1 keV; one kiloelectron volt) neutrons collide with other atoms in the soil. If the collision is elastic, neutrons leave with most of the energy

TABLE 8.1
Reductions in Kinetic Energy of Fast Neutrons after Colliding with Elements in the Soil

Atom	Number of Protons	Number of Collisions Needed	Energy Loss (%)
Hydrogen	1	18	100
Oxygen	8	152	22.1
Silicon	14	252	13.8
Aluminum	13	279	13.3
Iron	26	519	6.8

Source: Modified from Gardner, C. M. K., et al., Soil water content, in: K. A. Smith and C. E. Mullins (eds.), *Soil and Environmental Analysis,* Marcel Dekker, Inc., New York, 2001.

intact. However, if the collision is inelastic, neutrons lose energy (or kinetic energy; 0.025 to 0.5 eV) and get thermalized (or slowed down) (Table 8.1; Gardner et al., 2001). Neutrons do not have charge (electrically neutral) and are not affected by electric fields. Since hydrogen (proton, H^+) has the same mass as a neutron and collisions with hydrogen are inelastic, the neutrons are radically slowed down and form a cloud in the soil. Some of the thermalized neutrons bounce into the detector that counts the thermalized neutrons. Since water is the main source of hydrogen in the soil, the count rate of thermalized neutrons is a direct function of the volume fraction of water in the soil. Although other factors can also affect the count rate, and it is always advised to calibrate the neutron probe in the field.

During calibration, the count ratio or relative count rate is determined as the ratio of the count rate in the soil and the count rate in the field and is then plotted against the volumetric water content. Calibration can be done by determining the neutron probe readings for an oven-dry soil nearly (0% soil water) and water (100%), followed by adding a known volume of water successively to the oven-dry soil and collecting neutron probe readings. Gravimetrically determined moisture content from locations near the probe can also be plotted against the count ratio to obtain the calibration curve. A schematic of the calibration curve for a neutron probe is given in Figure 8.3.

Neutron probe measurements are generally considered reliable, and one probe can record measurements at several depths by being hung inside a tube (aluminum, stainless steel, plastic, or PVC) at a desired depth. Readings are generally not influenced by soil salinity or air, and calibration curves can be developed for different soil types using the same access tube. The sphere of influence (or sample support) of a neutron probe varies with the soil water content, and it could be as low as 10 cm for wet soil and as much as 50 cm for dry soil (van Bavel et al., 1956). Thus, soil water content measurements for smaller depth increments (<10 cm) are not recommended, surface soil may seem wetter because of the water present in the deeper

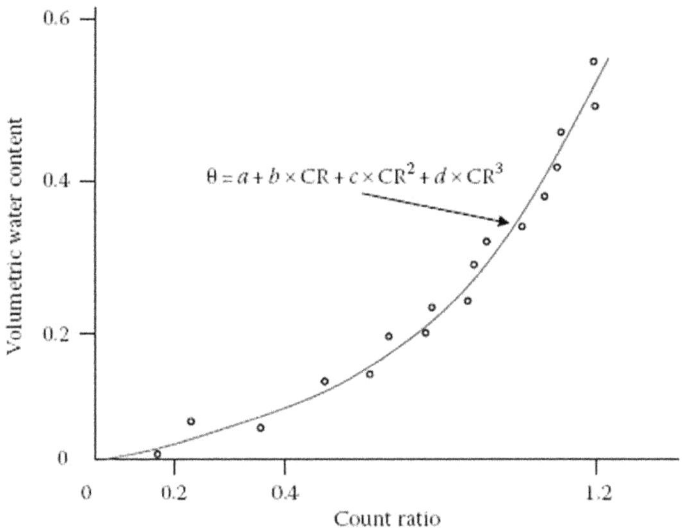

FIGURE 8.3 Schematic of a neutron probe calibration curve with count ratio (CR) and volumetric water content (θ; cm^3/cm^3). The best-fit line is a third-order polynomial.

layers, and erroneous results could be obtained for heterogeneous or layered soils. Readings close to the soil surface are difficult and unreliable and take a relatively long time between readings. The instrument cannot be left in the field unsupervised and is also expensive. The technique uses a radioactive source; therefore, it is a safety hazard. The operator must be properly trained, and warning signs must be posted at the experimental site and during transport and storage. The area where the neutron probe is housed must be accessible to only authorized personnel. The neutron probe source has a half-life of about 500 years; therefore, leak tests must be performed every 6 months, radiation badges must be kept in the storage area, and they must be worn by operators during use.

8.5.1.2 Gamma Ray Attenuation

The gamma ray attenuation method also has a radioactive source, similar to neutron thermalization, that emits a beam of gamma rays. The intensity of gamma rays moving through the soil changes with a change in the bulk density of soil. Thus, if the density of the soil is known, the change in the intensity of radiation is the result of the change in soil water content. The attenuation of the gamma ray passing through a soil volume of known density (intact soil core) is recorded by a detector. Low-energy gamma rays are more easily absorbed by water than high-energy gamma rays (e.g., cesium; ^{137}Cs) and can be counted more easily. Using ^{241}Am and ^{137}Cs together, both soil density and water content can be measured simultaneously. The gamma ray attenuation technique can determine soil water content at much smaller increments than the neutron probe method. Since the source and detector are separate, in situ measurements in the field require installation of two vertical tubes. The gamma ray technique has limited application in the field, and the limitations are similar to that of the neutron probe method.

8.5.1.3 Nuclear Magnetic Resonance

In this technique, water in the soil is subjected to both a static and an oscillating magnetic field at right angles to each other. The spin echo and free induction decays are recorded by sensors consisting of a radio detection coil, a turning capacitor, and an electromagnetic coil. The signal produced due to the adsorption of energy is linearly related to the number of protons in the sample or the soil water content.

8.5.2 Dielectric Properties

Dielectrics have a strange property of making space seem bigger or smaller than it is. When a dielectric is placed between two electric charges, it reduces the force acting between them. The dielectric constant of a medium has a strong influence on how electromagnetic waves (light waves, radio waves, etc.) move through the medium. The electromagnetic waves entering a medium with a higher dielectric constant from a medium with a lower dielectric constant require more time to travel through it because their velocity is reduced by the square root of the permittivity of the medium. The dielectric constant (K) is the relative permittivity of a dielectric material and is given as

$$K = \frac{\varepsilon}{\varepsilon_0} \tag{8.3}$$

where

ε is the permittivity of the medium.
ε_0 is the permittivity of free space (8.85×10^{-12} F/m; farads/meter).

The dielectric constant is about 80 for water, 5–7 for soil, 4 for frozen water, and 1 for air. Since soil is made up of different components, such as minerals, air, and water, the value of permittivity is made up by the relative contribution of each of the components. Since the dielectric constant for liquid water is much larger than that for soil minerals and air, the total or bulk permittivity of the soil is primarily governed by the amount of liquid water or water content of the soil. Thus, a polynomial relationship was developed by Topp et al. (1980) between gravimetrically measured soil water content using a convective oven and time domain reflectometry (TDR) measured relative permittivity or dielectric constant as follows:

$$\theta = -5.3 \times 10^{-2} + 2.92 \times 10^{-2} \times K - 5.5 \times 10^{-4} \times K^2 + 4.3 \times 10^{-6} \times K^3 \tag{8.4}$$

This equation has been widely used to measure volumetric water content for a wide range of soils. The parameters of this equation are often optimized for a site-specific calibration specifically for soils other than sandy or sandy loam soils.

8.5.2.1 Time Domain Reflectometry

In the TDR method, an electromagnetic wave (pulse) is sent to the soil between two or three metal rods, also known as waveguides. The soil bulk dielectric constant (K) can be determined by measuring the time taken by an electromagnetic wave to move between metal rods of the waveguide buried and surrounded by soil. The propagation velocity (v) is a function of K, and K is proportional to the square of the transit time (t in seconds) down and back along the transmission line (or $2L$):

$$K = \left(\frac{c}{v}\right)^2 = \left(\frac{c \times t}{2L}\right)^2 \tag{8.5}$$

where c is the velocity of electromagnetic waves in vacuum (3×10^8 m/s). This method is reliable for continuous data acquisition, but it is expensive. A TDR unit consists of a datalogger, a battery, a solar panel, waveguides, and often a multiplex to connect waveguides (more than seven) for continuous soil water content measurement. The transmission lines (waveguides) can be installed vertically, horizontally, or at an angle. Vertically installed waveguides of a given size provide the average soil water content for the length of the buried waveguide, whereas horizontally installed waveguides provide a measurement at the installed depth. Usually, horizontally installed waveguides are at a small angle (<15°). The TDR method is popular for simultaneously collecting soil water content data from several depths and locations (Groves and Rose, 2004). It is also possible to get a simultaneous measurement of soil salinity using a TDR100. For a reliable measurement, good contact between the sensor and the soil is required. Waveguides (or metal rods) must be parallel to each other after installation in soil. Some of the drawbacks of the system include high cost of the instrument, its sensitivity to high salinity and high clay content, and its relatively small sensing volume around waveguides. Site-specific calibrations are required, although manufacturers of this device provide calibration that does not always work in the field (Deb et al., 2011).

8.5.2.2 Frequency Domain Reflectometry

A capacitor is a device that can store electric charge. The soil moisture content is determined using a capacitor and the dielectric properties of water. In this method, capacitance measured between the two electrodes implanted in the soil is directly related to the soil water content, provided that the majority of the water in the soil is free water. The frequency domain capacitance sensor uses the charge time of a capacitance to measure the soil water content (Grooves and Rose, 2004). The frequency domain reflectometer (FDR) utilizes the radio frequency to obtain the dielectric constant and subsequently the soil water content. The probe is normally given a frequency excitation to permit measurements of the dielectric constant. As high-frequency radio waves, about 150 MHz, are passed through a capacitance circuit, depending upon the soil capacitance, a natural resonant frequency is established and is related to the dielectric constant by the geometry of the electric field established across the electrodes. The technique can provide good estimates of the soil water content, although site-specific calibration is required. The technique may work better in saline soils than in some other soils listed in this chapter. Similar to the TDR technique, the sensing volume is small; there is high sensitivity to clay content, bulk density, and air gaps; and good contact between the sensor and soil is vital.

8.5.2.3 Amplitude Domain Reflectometry

Impedance, measured in ohms, provides a measure of the opposition to the passing current at an applied voltage. It can be termed as resistance for the alternating current circuits because it has both magnitude and phase. When an electromagnetic wave reaches a soil, it experiences two different types of impedances: one of them is due to the electrical conductivity of the soil, and the other is due to the dielectric constant. Ignoring the effects of soil electrical conductivity on impedance, the change in the impedance of a probe can be directly related to the soil water content. The amplitude domain reflectometry (ADR) technique is almost as reliable as any other reflectometry techniques and has similar advantages and drawbacks.

8.5.2.4 Time Domain Transmission

The time domain transmission (TDT) technique measures the time that an electromagnetic wave requires to move through the waveguide. Electric connections at the beginning and at the end of the waveguide are required to get the time of travel in one direction. The measured travel time is the total time needed to travel along the connecting cables and transmission lines. For sand, the volumetric water content for TDR and TDT can be measured by the following equation (Hook and Livingston, 1996):

$$\theta = 0.1256\left(\frac{c \times t}{d} - 1.756\right) \tag{8.6}$$

where

c is the speed of light in vacuum.
t is the travel time through the medium.
d is the distance traveled by the electromagnetic signal through the medium.

For TDT, d is equal to the length of the transmission line (L), and for TDR, it is equal to $2L$ (Harlow et al., 2003). Measurements in this method are made across a larger soil volume and have similar reliability as other methods discussed in this section. Due to the design of the probe, a larger soil disturbance is caused during installation, and the generated pulse is also distorted during transmission.

8.5.3 Remote Sensing Technique

8.5.3.1 Microwave Remote Sensing

The use of microwave remote sensing to determine soil moisture began in the mid-1960s as a result of efforts by the National Aeronautics and Space Administration (NASA) to find applications for its earth-viewing satellites. The current satellite remote sensing technology used by NASA and NOAA (National Oceanic and Atmospheric Administration) for soil moisture estimation consists of microwave radiometers and microwave scatterometers. The scatterometer requires a high-power transmitter and a large antenna, and is expensive, heavy, and complex to operate. Microwave radiometers have similar characteristics. Soil moisture determination with synthetic aperture radar (SAR) is still very much in the research phase, which is part of the reason we discuss this simpler and less expensive technique (Wagner and Pathe, 2004).

Remote sensing techniques make use of electromagnetic energy emitted or reflected from the land surface. These include solar energy (e.g., with Landsat), reflected microwave energy (e.g., imaging radar), and emitted thermal radiation in microwave for soil moisture (Schmugge, 2011). Observations are usually made of the surface soil, but long-wavelength microwave sensors can sense the soil water content up to 5 cm deep. Signals with wavelengths in the range of 15–30 cm provide minimal scattering and attenuation from the atmosphere and vegetation. There is a large contrast between the emissions from water and land at microwave frequencies and at longer wavelengths (10–30 cm). Soil emissivity is a function of water content change from 0.95 for dry soil to 0.6 for wet soil at a rate of 0.01% soil water content change (Schmugge, 2011). The rate of change depends on soil texture (higher for sandy soils than for clay soils), surface roughness, and vegetation cover (Jackson and Schmugge, 1989; Schmugge and Jackson, 1994).

8.5.3.2 Ground-Penetrating Radar

A variety of ground-penetrating radar (GPR) methods have been used to estimate the soil water content (Huisman et al., 2003). Four methods can be used to determine the soil water content using GPR technology: (i) ground wave velocity, (ii) reflected wave velocity, (iii) transmitted wave velocity from boreholes, and (iv) surface reflection coefficient. In the GPR technique, the travel time for an electromagnetic wave is recorded when the wave travels between a transmitter and a receiver. Most soils have very little variability in magnetic permeability (Powers, 1997); the travel time is related to the dielectric permittivity of the soil and subsequently to the soil water content (Lunt et al., 2005). This method is similar to the TDR method but has greater potential to rapidly measure soil water content over large areas because GPR antennas are not required to be placed in contact with soil or porous media. Further details on these methods can be found elsewhere (Dane and Topp, 2002; Huisman et al., 2003).

8.5.3.3 GPS Technique

The GPS technique is under development at the moment. Remote sensing of soil moisture provides better results when performed using signals with “long” wavelengths, such as the L-band region of the electromagnetic spectrum (15–30 cm) (Jackson and Schmugge, 1991; Schmugge and Jackson, 1992). At 19 cm wavelength, the GPS signal provides an excellent opportunity to sense soil moisture remotely. GPS technology is well known for its application to locate position because of new surveying technology and the widespread availability of handheld devices as well as the use of GPS in mobile platforms (e.g., cars, planes). A recent advancement in remote sensing has demonstrated the ability to use signals from the GPS constellation of satellites for remote sensing of soil moisture (Masters et al., 2004). The use of this system as a remote sensor represents significant progress in accessible technology because the GPS signal is available everywhere continuously, which can provide for monitoring soil moisture at will in both space and time, eliminating the limitations of current in situ spot sampling. The realization and demonstration that soil-surface reflection of the GPS signal can provide quantitative information on surface soil moisture means that compact instruments can be broadly used at low cost.

The GPS method is based on the sensitivity of the soil dielectric constant to soil moisture and its subsequent effect on reflection of GPS signals (Katzberg et al., 2005). A microwave signal impinging on the soil surface will be reflected and absorbed. The soil dielectric constant determines the ratio between reflected and absorbed signals and, in addition, for the reflected portion of the signal, how far into the soil the microwave signal penetrates. The amount of water in the soil greatly influences the dielectric constant of the soil. Signals are continuously transmitted from the GPS satellites toward the surface of the earth. The GPS antenna receives both incoming and reflected solar radiation. Signals not received by a receiver are absorbed in the atmosphere, reflected back into space, or absorbed into the earth's surface.

Over land, the amplitude of the reflection is the lowest when reflecting from a bare dry soil surface (e.g., dry desert soil), increases with increasing soil moisture, and reaches a maximum over water-saturated soil or open bodies of water. Theoretically, signals reflected from saturated soils can be up to 55% of the direct signal. The signals reach 63% of the direct signal over liquid water. Other parameters affecting the reflections are surface roughness, vegetation cover, and soil properties. In the case of vegetation, the reason for conducting the experiment in the arid environment is to mitigate effects of vegetation cover because there is less in this environment. In addition, there is a wide variation in soil moisture in space and time (due to the long intervals between rainstorms) that will exercise the reflectometer over a range of effects.

The reflected and direct signals are picked up by a modified GPS receiver/reflectometer. The reflection geometry, depicted in Figure 8.4, is referred to as a bistatic configuration because the transmitter and receiver are not colocated. In our case, the GPS receiver/reflectometer is on an elevated platform and the transmitter/GPS satellite is located at a great distance. This pattern provides the prospect of varying the level of spatial details needed for different applications by changing the altitude at which the reflected signals are received.

8.5.4 Heat Dissipation Method

The principle of this method involves the measurement of temperature in a porous block before and after a heat pulse is applied to the soil. The heat flows out from the heated point outward, and the change in temperature is recorded by a thermistor at the tip of the sensor and is related to the soil water content. The change in temperature of the porous media (or soil) is inversely proportional to the soil water content,

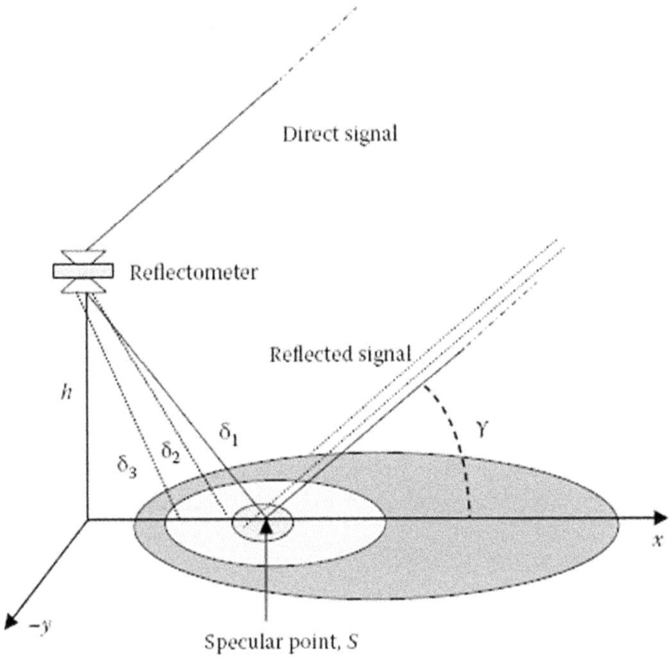

FIGURE 8.4 Illustration of GPS reflection technique geometry. The specular region (*S*; light gray) is the first Fresnel zone, while the surrounding ellipses (dark gray) represent higher-order zones of reflection, which contribute less energy to the received signal. (Courtesy Omar Torres, NASA Langley Lab, Hampton, Virginia.)

while the volumetric specific heat capacity of the soil increases with increasing soil temperature. The thermal conductivity of soil also increases with the soil water content, and thus heat flow is proportional to the soil water content.

8.5.5 Machine Learning Approaches

Alternate less expensive, reliable, rapid, repeatable, and not involving measurements in the soil are needed for soil moisture mapping across large landscapes. The soil water content can also be predicted by conducting the water balance using the already known data on irrigation, precipitation, evapotranspiration, and deep percolation. In some cases, deep percolation may be negligible. Soil moisture content can also be predicted using pedotransfer functions as well as using machine learning algorithms. Machine learning methods use existing algorithms and measured soil, water, weather, and remotely sensed data to predict trends and correlations in the datasets to predict soil moisture content. Machine learning tools can overcome issues such as outlier data, nonlinearity, heteroscedasticity, and, multicollinearity. Some examples of machine learning tools are neural network, random forest, and decision trees (Adab et al., 2020).

Soil moisture is predicted using soil temperature data using linear regression and neural network models (Altendorf et al., 1999). Some machine learning methods for predicting soil moisture used a support vector machine or neural network and meteorological data (Gill et al., 2006; Elshorbagy and Parasuraman, 2008). Other efforts focused on estimating soil moisture content from digital photographs using RGB and HSV color models (Hajjar et al., 2020; Santos et al., 2016); red, near-infrared, and shortwave infrared spectral bands (Hassan-Esfahani et al., 2015); and thermal emissivity data (Zhang et al., 2014). Machine learning methods are site-specific and the model must be calibrated for local conditions and applied locally at similar locations where no data are available.

8.6 ENERGY STATE OF SOIL WATER

There are several energy-related phenomena, for example, potential energy, kinetic energy, and electrical energy. The potential energy is due to the position of a body that allows it to do work. Potential energy (PE) of an object is provided by gravity and is given as follows:

$$\text{PE} = m \times g \times h \tag{8.7}$$

where

m is the mass of the object (kg).
g is the acceleration due to gravity (9.8 m/s^2).
h is the height above the reference level (m; could be soil surface).

The kinetic energy (KE) is due to the motion of a body and is given as

$$\text{KE} = \frac{1}{2} m \times v^2 \tag{8.8}$$

where v is the velocity of the object (m/s). In addition, there is electrical energy due to the flow of electric charge through a conductor. Electrical energy is a product of the electric potential (V) and the unit charge moved (Q). Generally, the rate of water flow through the soil is small, and kinetic energy is not considered for determining the energy state of water.

The water potential at any given location is the difference in potential energy of water at that location and a reference location containing pure water. Since water is seldom pure, especially soil water, the energy state of water is always less than the energy state of pure water. The potential energy of free pure water by definition is zero; therefore, the energy state of soil water is usually less than that of the free water and is negative. The energy state of water is important for retention and movement of water in the soil, uptake and translocation in plants, and the loss of water to the atmosphere. Water always flows from an area at high free energy to the one at low free energy. When all the soil pores are filled with water, soil is saturated, and water

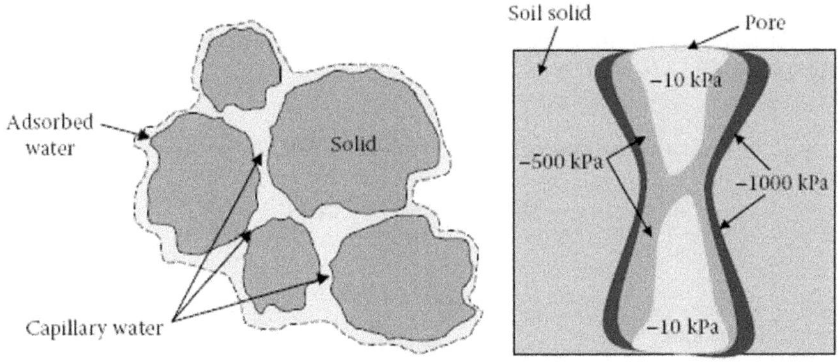

FIGURE 8.5 Schematic of adsorbed and capillary water in a given soil volume. All pores are filled with water in the diagram to the left. Schematic of the adsorbed, capillary, and free water inside at different matric potential inside a soil pore.

moves more rapidly through the soil than when the soil is dry (pores filled with both water and air; Figure 8.5) because the free energy of a saturated soil is higher than a dry (or unsaturated) soil.

Forces acting on soil water change the free energy of water. These include forces due to adhesion and cohesion, forces due to ions or solutes present in the soil, and forces due to gravity. The forces due to cohesion and adhesion (capillary and adsorptive forces) result in soil matric potential. Water flow in the vadose zone occurs from a zone of high soil water potential to a zone of low potential. The driving force for the flow is known as the gradient of flow or soil water potential gradient and is defined as the change in potential energy with distance. The driving force determines the magnitude and direction of flow, deep percolation or drainage rates through the soil, capillary rise or evaporation rate, plant water extraction rates, and solute transport rates.

8.7 DEFINITIONS AND COMPONENTS OF SOIL WATER POTENTIAL

8.7.1 Total Soil Water Potential

The difference in free energy from one contiguous site to another is of greater practical significance than the soil water content. This difference is also known as total soil water potential. A simple definition of soil water potential is the difference between the energy state of soil water and that of the pure free water. Soil water potential is always measured relative to the reference potential of zero or atmospheric pressure. Total soil water potential can be defined as the amount of work per unit quantity of pure water that must be done by external forces to transfer reversibly and isothermally an infinitesimal amount of water from a pool of pure water at a specified elevation at atmospheric pressure to the soil water at the point of interest. Water in soil varies in amount and quality, and various forces act on it; therefore, the potential energy of water differs from place to place in a contiguous soil volume.

By definition,

$$\text{Potential} = \text{force} \times \text{distance} = \text{mass} \times \text{acceleration} \times \text{distance}$$

$$= m \times g \times L = \rho_w \times V \times g \times L$$

where

m is the mass (g).
g is the acceleration due to gravity (9.8 m/s^2).
L is the distance between two points across which the soil water potential gradient is determined (cm).
ρ_w is the density of water (g/cm^3).
V is the volume (cm^3).

Soil water potential can be expressed as the potential per unit mass, per unit volume, or per unit weight as follows:

$$\text{Soil water potential per unit mass} = \frac{\rho_w \times V \times g \times L}{\rho_w \times V} = g \times L = \frac{\text{dyne cm}}{\text{g}}$$

$$= \frac{\text{ergs}}{\text{g}} = \frac{\text{N} - \text{m}}{\text{kg}}$$

$$\text{Soil water potential per unit volume} = \frac{\rho_w \times V \times g \times L}{V} = \rho_w \times g \times L$$

$$= \frac{\text{dyne}}{\text{cm}^2} = \frac{\text{N}}{\text{m}^2}$$

$$\text{Soil water potential per unit weight} = \frac{\rho_w \times V \times g \times L}{\rho_w \times V \times g} = L = \text{cm} = \text{m}$$

It is clear from the three relationships that the soil water potential does not have to be expressed as the amount of work done; instead, it can be expressed in terms of pressure and length or the height of the water units. In this book, as far as possible, the soil water potential is expressed as the height of the water column, in the units of length, or as potential per unit weight. The soil water potential can also be expressed in atmosphere (atm), pascal (P), kilopascal (kP), megapascal (MP), meter (m), centimeter (cm), pound per square inch (psi) and bar units.

Example 8.1

Convert 1 atm to m, cm, bar, Pa, kPa, and MPa. How many bars are in 1 pound per square inch (psi)? One bar is equal to how many dynes/cm²?

$$1\text{ atm} = 1\text{ bar} = 10\text{ m of water} = 1{,}000\text{ cm of water column} = 100{,}000\text{ Pa}$$

$$= 100\text{ kPa} = 0.1\text{ MPa}$$

$$1\text{ psi} = 6.8948 \times 10^{-2}\text{bar}$$

$$1\text{ bar} = 10^{6}\text{dynes / cm}^{2}$$

The total soil water potential is a relative quantity because it is measured relative to free pure water. It is mostly negative except when all pores are saturated with water. The soil water potential is always continuous across the contiguous soil domain and across different types of soils and soil layers. It is highly variable across a landscape and is a dynamic quantity because its value generally changes with time. The total soil water potential consists of pressure, matric, gravitational, osmotic, and overburden potentials among others. However, for most applications the first three potentials are considered, and sometimes the osmotic potential is also included in the calculation of the total soil water potential.

$$\text{Total soil water potential} = \text{Pressure} + \text{Matric} + \text{Gravitational} + \text{Osmotic}$$

$$+\text{Overburden} + \cdots$$

$$\Phi_T = \Phi_p + \Phi_m + \Phi_z + \Phi_o + \Phi_{ob} + \cdots \qquad (8.9)$$

8.7.2 Pressure Potential

Pressure potential is the height of water above a specific point in the soil or at the soil surface. In a flood-irrigated field with standing (ponded) water of a depth of 5 cm, the pressure potential is equal to 5 cm or the depth of the standing water on the field. The pressure potential is also defined as the water pressure exerted by the overlaying saturated column of water on a specific position within a soil. Sometimes due to an impermeable or low-permeable layer such as plow pan or hard pan, water can be standing in a field just above the impeding layer, creating a perched aquifer. The potential above the hard pan is known as the pressure potential. The pressure potential is always positive, and its lowest value could be zero.

8.7.3 Matric Potential

The water potential resulting due to capillary and adsorptive forces is known as matric potential. The matric potential is always negative, and its maximum value could be zero under saturated soil conditions. The pressure potential and matric potential do not coexist. As soil gets drier, the matric potential decreases and water in the soil is more strongly adsorbed. The matric potential is important for determining plant–water relations, plant stress, and plant survival. As soil gets drier, in order to extract water from the soil, plants require to exert more negative matric potential within the cell walls.

8.7.4 Gravitational Potential

Gravitational potential is the force of gravity that is acting on soil water exactly as it does on any other body. The gravitational attraction is always toward the center of the earth. The gravitational potential is always measured with respect to a reference, and depending on the location of the reference with respect to the point of measurement, it can be positive or negative.

8.7.5 Osmotic Potential

Osmotic potential is also known as solute potential and is a measure of the attraction of water to salts in the soil water film. It is always negative. The presence of salts in a soil volume further decreases the total soil water potential. Due to the osmotic pressure, a pure solvent (e.g., pure water) moves into a solution (containing a solute, e.g., fertilizer solution). The solute potential can be calculated as follows:

$$\Phi_{\mathrm{o}}\left(\mathrm{erg}/\mathrm{cm}^3\right) = C_{\mathrm{s}}\left(\mathrm{mol}/\mathrm{cm}^3\right) \times R\left[\mathrm{erg}/\left(\mathrm{mol\ K}\right) \times T\left(\mathrm{K}\right)\right] \tag{8.10}$$

where

C_s is the solute concentration.
R is the universal gas constant (8.32×10^7 erg/mol deg).
T is the temperature (K).

If there are two ions in the solution, the calculated osmotic potential using the previous relationship is multiplied by two. The osmotic potential can also be determined from the electrical conductivity (EC) of soil water extract (U.S. Salinity Laboratory Staff, 1954):

$$\Phi_{\mathrm{o}}\left(\mathrm{MPa}\right) = -0.036 \times \mathrm{EC}\left(\mathrm{dS}/\mathrm{m}\right) \tag{8.11}$$

According to Bresler et al. (1982), the relationship between saturated paste EC and osmotic potential works well in the range of 3–30 dS/m. The osmotic potential can also be calculated from the total dissolved solid (TDS) as follows:

$$\Phi_{\mathrm{o}}\left(\mathrm{kPa}\right) = -0.056 \times \mathrm{TDS}\left(\mathrm{mg}/L\right) \tag{8.12}$$

The osmotic potential of a solution can be determined experimentally by measuring the concentrations of dissolved ions in the solution:

$$\Phi_o(\text{psi}) = -1.12 \times \sum m_j(\text{mol/kg}) \times T(\text{K}) \tag{8.13}$$

where j denotes the number of ions present in the solution. The osmotic pressure from psi can be converted into bars by using a conversion factor of 1 psi = 6.8948 × 10^{-2} bar.

Example 8.2

If the solution concentration is 10^{-5} mol/cm^3 and the temperature is 27 K, calculate the osmotic potential of the solute and convert it to different units. If there are two ion species present, then what will be the osmotic pressure?

$$\Phi_o(\text{erg/cm}^3) = C_s(\text{mol/cm}^3) \times R[\text{erg/(mol K)}] \times T(\text{K})$$

$$\text{Universal gas constant}(R) = 8.32 \times 10^7 \text{ erg/mol/K}$$

$$\text{Temperature} = 273 + 27 = 300\text{K}$$

$$\Phi_o = 10^{-5} \times 8.32 \times 10^7 \times 300$$

$$= -2.5 \times 10^4 \text{ erg/cm}^3 = -0.25 \text{ atm} = -0.25 \text{ bar}$$

$$= -2.5 \text{ m} = -250 \text{ cm} = -25 \text{ kPa}$$

$$\text{If two ion species present;} \Phi_o = -25 \times 2 = -50 \text{ kPa}$$

Example 8.3

If the electrical conductivity of a soil solution is 4 dS/m, calculate the osmotic potential of the solution in different units.

$$\Phi_o(\text{MPa}) = -0.036 \times \text{EC}(\text{dS/m})$$

$$\Phi_o = -0.036 \times 4 = -0.144 \text{ MPa} = -1.44 \text{ bar} = -144 \text{ kPa} = -1440 \text{ cm}$$

8.7.6 Overburden Potential

The potential due to the overlaying weight of the soil is known as the overburden potential. It can be an important component of the soil water potential, especially in swelling soils. The overburden potential is always positive.

8.8 SOIL WATER POTENTIAL MEASUREMENT DEVICES

8.8.1 Piezometer

A piezometer is a device that can be used to measure the water level of groundwater or saturated soil. It can also be used to measure the rise in water level against gravity from the groundwater table or fluctuations of the groundwater table due to irrigation. The piezometer device consists of a small hollow tube that is inserted in the soil deep enough to reach the groundwater table or the saturated soil with the other end open to the atmosphere. Water slowly rises in the tube, and the level of the water in the tube corresponds to the level of groundwater for the area. Piezometers are simple devices used to measure the pressure potential or positive soil water potential. However, if the soil is unsaturated, water will not enter the piezometer tube.

Piezometers can be installed in both confined aquifers and unconfined aquifers (Figure 8.6a). A confined aquifer is bounded above and below by a low-permeable material such as clay or rocks (also known as aquitards), and water in the aquifer is mostly under positive pressure. Once a piezometer is installed in the confined aquifer, due to confinement pressure, the water rises in the tube until the pressure inside is balanced. Thus, the piezometer in this case does not measure the water level but the pressure inside the aquifer. On the other hand, water in the unconfined aquifer is not pressurized, and water in the piezometer represents the water level and fluctuates with atmospheric pressure. Usually, a piezometer tube is solid until about the depth of interest, and below it the tube is slotted and proper screens are attached to prevent sediments from entering the tube. Piezometers can be attached to pressure gauges and dataloggers to read the fluctuations of the water table continuously. One such simple device is the HOBO U20-001-01 (Onset Computer Corporation, Bourne, Massachusetts), and such a water-level logger can record the depth to the groundwater table below the soil surface (or any other reference point) continuously at any specified time interval (Figure 8.6b). Using a reference water level, the HOBOware Pro software required for the logger operation converts the pressure readings into water-level readings. The software is equipped with accounting for compensation for temperature, fluid density, and barometric pressure. The HOBO logger is launched using the HOBO software, and the logging interval and duration are entered. Using a no-stretch wire, the HOBO logger is then lowered into the pipe or well vented to the atmosphere so that pressure inside and outside the well is in equilibrium. Figure 8.6b shows the groundwater-level data recorded by HOBO U20 in a pecan orchard located in southern New Mexico (Deb et al., 2012).

8.8.2 Tensiometer

A tensiometer consists of a porous cup and a pressure measurement device. A U-shaped tube attached to the porous plate can be filled with water (Figure 8.7). The porous cup must be saturated with water at the time of installation. Initially, the water level will be at the same level or at the midlevel of the tube, as shown

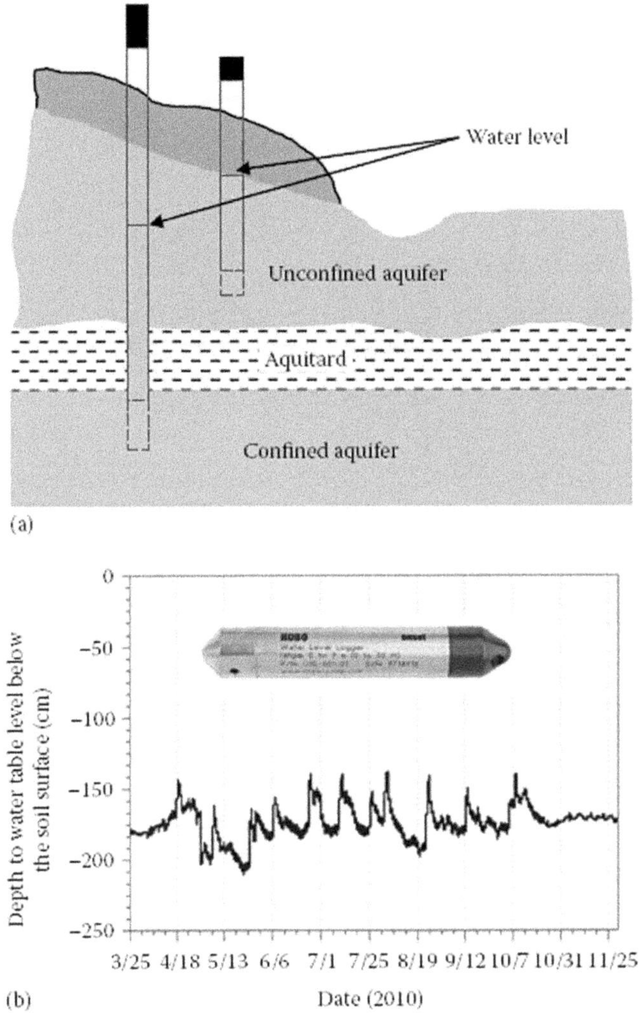

FIGURE 8.6 (a) Piezometer tube installed in a confined and an unconfined aquifer. The water level rises much higher in the piezometer installed in the confined aquifer to balance the pressure. The water level in the piezometer in the unconfined aquifer is at the water table. (b) Depth to water table measured by a HOBO U20 water-level datalogger in a pecan orchard in southern New Mexico.

in the figure for $t = 0$. However, the water will flow out of the porous cup until the soil surrounding the porous cup is saturated with water. The water level in the tube will drop and stabilize after some time ($t > 0$). The negative of the difference in the depths of water (Δh) as shown in Figure 8.7 will be the pressure potential in the soil.

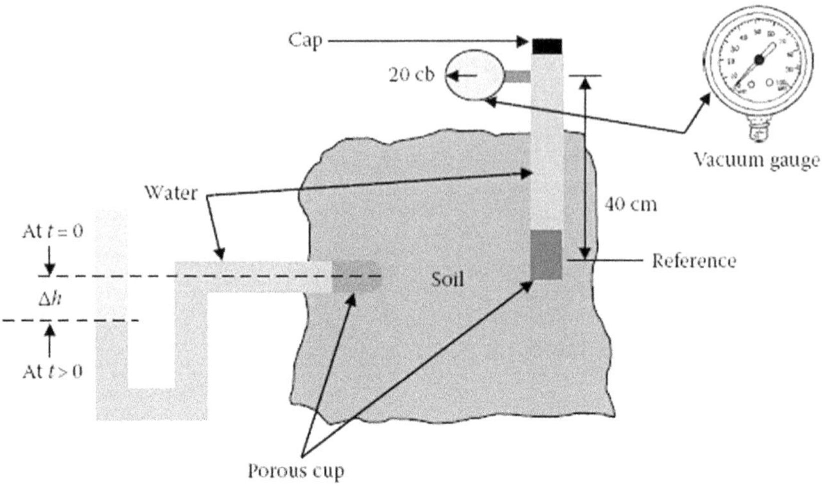

FIGURE 8.7 Tensiometers for measuring soil water potential; the tensiometer on the right has a water manometer and the one on the left uses a vacuum gauge for pressure measurement.

Example 8.4

If (Δh) is 10 cm in Figure 8.7 and the density of water is 1000 kg/m^3, calculate the soil water potential per unit weight, mass, and volume.

$$\text{Soil water potential per unit weight} = -L = -10\text{ cm} = -0.1\text{ m}$$

$$\text{Soil water potential per unit mass} = -g \times L = -9.81 \times 0.1$$

$$= -0.981\text{ N} - \text{m/kg}$$

$$\text{Soil water potential per unit volume} = -\rho_w \times g \times L = -1000 \times 9.81 \times 0.1$$

$$= -981\text{ N/m}^2$$

Instead of the water manometer, a vacuum pressure gauge can be attached to the tube for recording the soil water potential (Figure 8.7). The principle of operation stays the same, and as soon as some water flows out of the ceramic cup, the pointer in the pressure gauge moves and displays the water potential (Figure 8.7). These types of tensiometers are popular, are easy to read, and can be easily installed in a farmer's field for measuring the soil water potential to schedule irrigation. Care should be taken during field installation to make sure that (i) the porous cup on the bottom is saturated with water prior to the installation, (ii) there is no air bubble inside the tensiometer tube, (iii) the tensiometer cap is secured tightly and does not have any air leaks, and (iv) the porous cup and soil are in good contact. These tensiometers usually have a range of saturation up to 900 cm.

Example 8.5

If the vacuum pressure gauge in Figure 8.7 shows a reading of 20 cb and the distance between the center of the suction cup and the gauge is 40 cm, calculate the soil water potential. Assume the reference inside the soil as shown in the figure.

$$\text{Soil water potential} = -20 \times 10 + 40 = -200 + 40 = -160 \text{ cm} = -1.6\text{m}$$

A mercury manometer can be employed in place of a water manometer (Figure 8.8). The principle of installation and operation is similar to that of a water manometer. After equilibrium is established, the soil water potential using a mercury tensiometer in Figure 8.8 can be obtained as follows:

$$\text{Soil water potential} = -Z_{Hg} \times \frac{\rho_{Hg}}{\rho_w} + Z = -Z_{Hg} \times \frac{13.6}{1} + Z_0 + Z_{Hg} \quad (8.14)$$

$$= -Z_{Hg} \times 12.6 + Z_0 \quad (8.15)$$

where

ρ_{Hg} is the density of mercury.
ρ_w is the density of water.

Example 8.6

If the height of the rise in mercury in the manometer tube is 10 cm, and the depth from the mercury reservoir to the reference point is 40 cm, calculate the soil water potential.

$$\text{Soil water potential} = -Z_{Hg} \times 12.6 + Z_0$$

$$= -10 \times 12.6 + 40 = -126 + 40 = -86 \text{ cm}$$

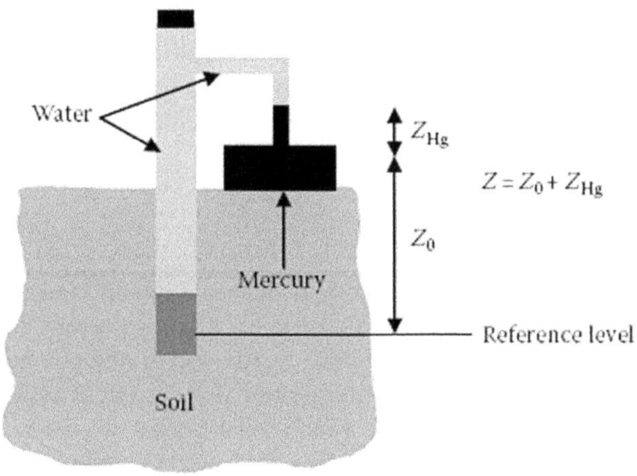

FIGURE 8.8 Mercury manometer attached to a porous cup for measuring soil water potential.

8.8.3 Dewpoint Potentiameter

The dewpoint potentiameter (WP4) is a benchtop instrument that can be used to determine the water potential for soil and plant materials. The sample is placed in a round cup about 1 cm tall and 4 cm in diameter. The principle of operation of the WP4 potentiameter centers on the measurement of the relative humidity of the soil sample. About 5–6 min are required for each measurement. A sample is placed inside the sealed chamber, and a measurement of relative humidity is made once the soil sample is in equilibrium with the vapor inside the chamber. The relative humidity is measured by the chilled-mirror method in which a tiny mirror is chilled until dew just starts to form on it. At the dewpoint, the temperature of the sample and mirror is also recorded by the WP4 potentiameter. The instrument is calibrated using saturated salt solutions. The manufacturing company reports a range of measurement between −0.1 and −300 MPa with an accuracy of ±0.05 MPa (www.decagon.com). The soil water potential can be calculated as follows:

$$\text{Soil water potential} = \frac{R \times T \times \ln\left(A_w\right)}{V_w} \tag{8.16}$$

where

R is the universal gas constant (8.314 J/mol/K).

T is the temperature (K).

V_w is the molar volume of water (1.8×10^{-5} m^3/mol).

A_w is the water activity calculated from the vapor pressure of air and the saturated vapor pressure.

The dewpoint sensor measures the dewpoint temperature of the air, and the infrared thermometer measures surface temperature of the sample. The water potential of the sample is calculated from the headspace vapor pressure and sample-surface temperature at equilibrium (Scanlon et al., 2002). According to the WP4 manual, temperature is critical in water potential determination. A 1^{0}C error in temperature difference could introduce an error of 8 MPa in water potential.

8.8.4 Resistance Blocks

Resistance blocks are indirect methods of measuring soil water potential. Two embedded electrodes inside a soil offer resistance to the current flowing between them, and the resistance is proportional to the soil water potential or the soil water content. The drier the soil, the greater the electrical resistance and the smaller the soil water potential. Most resistance blocks are made up of gypsum and granular quartz materials. Sometimes, they are covered with a porous material, such as fiberglass, ceramic, synthetic membrane, and stainless steel mesh.

These sensors must be calibrated before installation to check for the variability among the blocks. Site-specific, soil-based calibration is also recommended after field installation. The resistance can be manually read or can be attached to a datalogger for continuous monitoring of the temporal fluctuations of the soil water potential. These are simple and relatively inexpensive devices and can be used for irrigation scheduling. Care must be taken that the resistance block is completely wet and in good contact with the surrounding soil during installation. The measurement range of these devices is 0.1–2 bar.

8.8.5 Heat Dissipation Sensors

Heat dissipation sensors provide an indirect measurement of the soil water potential. The sensor buried in the soil is heated for a small time interval, and the heat dissipation of the sensor is measured over time and related to the soil water potential. The instrument typically consists of a heater and temperature sensor inside a porous granular matrix, made up of fired clay, plaster of paris, or ceramics (Scanlon et al., 2002). As the soil water potential decreases, the temperature of the soil increases rather quickly because of the high heat capacity of water (Figure 8.9). The change in temperature varies with the amount of heat applied, heating time, thermal conductivity, diffusivity of granular matrix, and matric potential of soil. The measurement range of these sensors is from −0.1 bar to −10 bar.

8.8.6 Filter-Paper Technique

The filter-paper technique can be employed by having filter paper and soil in direct contact or separated by a small vapor gap. Direct contact between soil and

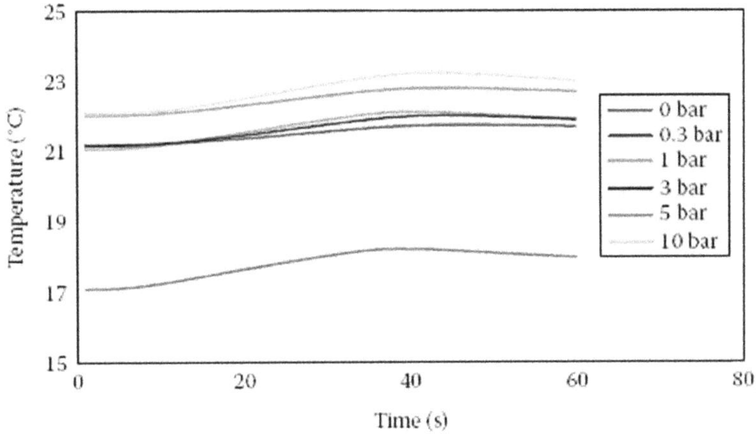

FIGURE 8.9 The heat dissipation represented by the change in soil temperature over time under different soil water potential on a soil core using a KD2 probe.

filter paper allows for liquid water and a solute to exchange freely and equilibrate after several days in sealed isothermal chambers. The change in the mass of the filter paper is related to the soil water potential by a known calibration curve. The principle is the same when there is an air gap between the soil and filter paper, but water exchange is only taking place at the vapor state. Details of the method can be found elsewhere (Scanlon et al., 2002). The method is not very accurate, and there are errors associated with incomplete equilibrium between the filter paper and soil and those associated with the measurement of the filter-paper weight.

8.9 TOTAL SOIL WATER POTENTIAL UNDER DIFFERENT CONDITIONS

There could be several different types of conditions present in a field where the soil water potential and its components can vary within the soil profile due to the upward or downward flow of water in the soil profile. These conditions can be broadly grouped into equilibrium and nonequilibrium conditions. The former is when no flow occurs, while the latter is usually accompanied by flow. Let us calculate the total soil water potential under different conditions (see Examples 8.7 and 8.8).

Example 8.7

In a pecan orchard located near the Rio Grande in an arid area of the southwestern United States, the water table is present at a depth of 100 cm from the soil surface, and water in the soil is at equilibrium with the water table. Make estimates of pressure (Φ_p), matric (Φ_m), and gravitational potential (Φ_g), and calculate the total soil water potential (Φ_T). Assume solute or osmotic potential to be zero.

Assume that the reference is at the soil surface. Assuming 20 cm deep increments for the determination of potentials, the various components of soil water potential are obtained as follows:

Depth (cm)	Φ_p (cm)	Φ_m (cm)	Φ_g (cm)	Φ_T (cm)
0	0	−100	0	−100
20	0	−80	−20	−100
40	0	−60	−40	−100
60	0	−40	−60	−100
80	0	−20	−80	−100
100	0	0	−100	−100
120	20	0	−120	−100
140	40	0	−140	−100

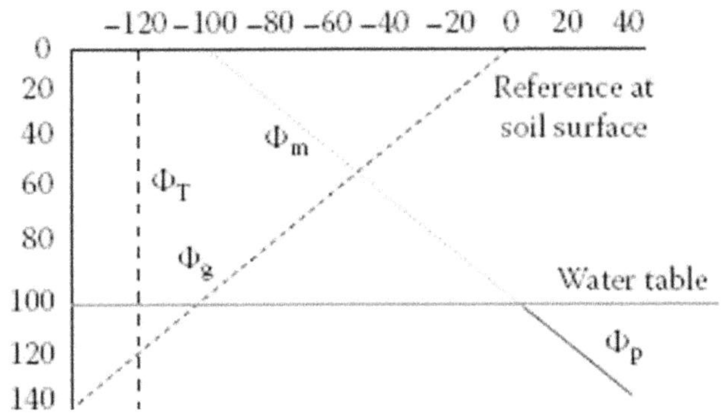

The previous calculations and the graph indicate that the gravitational potential decreases from the soil surface with depth because the reference was at the soil surface. If the reference was at the water table, the gravitational potential would have been the highest (also positive) at the soil surface and decreased toward the water table. The matric potential (negative) increases from an initial value of –100 at the soil surface to zero at the water table, and as we go deeper the matric potential stays at zero but pressure potential starts increasing (positive). When all the three components of the soil water potential are added, we see that the total soil water potential stays constant throughout the soil profile, indicating the equilibrium condition between the soil water and water table (Hanks and Ashcroft, 1980).

Example 8.8

Irrigation was applied several days ago to a cotton field, where the water table is at 50 cm deep from the surface. Assuming that the reference is at the water table and the solute potential is zero, make estimates of (Φ_p), (Φ_m), and (Φ_g), and calculate the total soil water potential (Φ_T) for up to 70 cm below the soil surface.

Depth (cm)	Φ_p (cm)	Φ_m (cm)	Φ_g (cm)	Φ_T (cm)
0	0	–500	50	–450
10	0	–300	40	–260
20	0	–140	30	–110
30	0	–80	20	–60
40	0	–30	10	–20
50	0	0	0	0
60	10	0	–10	0
70	20	0	–20	0

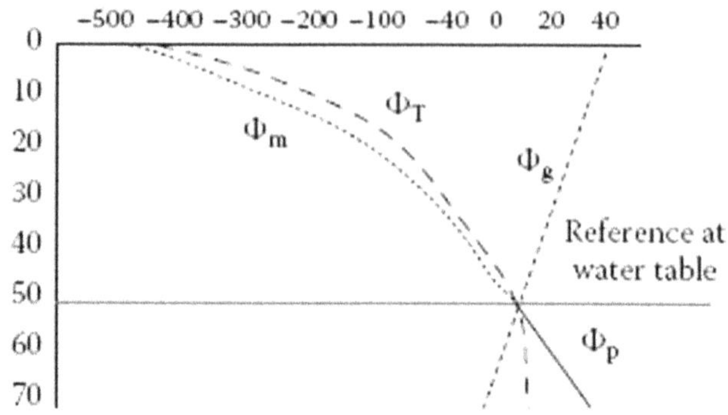

Since the soil water is not in equilibrium with the water table, the total soil water potential varies throughout the profile until it reaches the water table, and beyond that it becomes constant. The gravitational potential is zero at the water table (the reference), and it is positive upward and negative downward. The direction of flow in the soil profile can be decided by looking at the hydraulic potential at various depths; since in this table hydraulic potential increases with depth, water movement should be upward from the water table (Hanks and Ashcroft, 1980).

8.10 SOIL WATER RETENTION CURVE

The relationship between the soil water content and matric potential is known as the soil water retention curve or soil water characteristic curve (Figure 8.10). The soil water characteristic curve can be obtained by plotting the matric potential on the

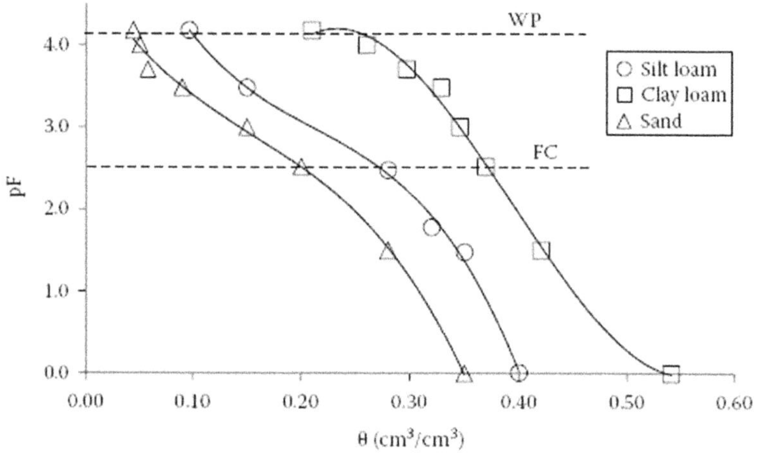

FIGURE 8.10 Soil water characteristic curves for three soils, pF = log ($-\Phi_m$), where Φ_m is matric potential in centimeters (cm) and θ is volumetric water content.

y-axis and volumetric soil water content on the x-axis or vice versa. The volumetric water content can have the units cm^3/cm^3 or m^3/m^3. On the other hand, the matric potential can have the units cm, m, bar, kPa, MPa, or it can also be expressed as pF. A soil water retention curve with the matric potential expressed as pF is also known as a pF curve:

$$\mathrm{pF} = \log\left(-\Phi_{\mathrm{m}}\right) \tag{8.17}$$

For a matric potential of –300 cm, pF is 2.5, and for a matric potential of –15,000 cm, pF is 4.2.

When all the pores are filled with water, the soil water characteristic curve or soil moisture characteristic curve is at the maximum soil water content, and the soil water potential at this point is also at the maximum or zero. The water content is known as the saturation water content. After applying a negative potential to the saturated soil, the potential at which water begins to come out of the soil is known as the bubbling pressure, air-entry pressure, or air-entry potential (water out and air in). Usually, the bubbling pressure is lower for fine-textured soils than coarse-textured soils. At or near a soil water potential of –0.33 bar (–330 cm of water; field capacity potential), the soil water content is known as the field capacity water content, and at a soil water potential of –15 bar (–15,000 cm; wilting point potential), the soil water content is known as the wilting point water content. The soil water content at field capacity and wilting point varies with the soil texture. An agriculture field must be irrigated well before the soil water content reaches the witling point water content because many plants cannot extract water from soil near the wilting point.

The soil water potential versus soil water content relationship is important for characterizing water retention and release from the soil and hydraulic properties of soil. The soil water characteristic curve is important for predicting the available water for plants, scheduling irrigation, and deciding on the amount of water that should be applied as irrigation. The soil water characteristic curve depends on soil texture. For a coarse-textured soil, a soil water characteristic curve is much more spread out with regard to the soil water content with a lower saturated ($\Phi_{\mathrm{m}} = 0$) and wilting point ($\Phi_{\mathrm{m}} = -15$ bar or 4.2 pF) water contents. However, for a fine-textured soil, the soil water characteristic curve's spread with respect to the water contents can be narrower (Figure 8.10) with higher water contents at saturation as well as at the wilting point than coarser-textured soils. Although texture is important for understanding the soil water retention and release characteristics of a soil, soil structure and pore connectivity are equally or sometimes more important for water retention and release from a soil.

The shape of a soil water retention curve shows that as the matric potential decreases (becomes more negative), more and more soil pores become empty, and therefore the soil water content decreases. The finer-textured soils of Figure 8.10 were collected from South Charleston, Ohio (Shukla and Lal, 2005), and the coarse-textured sand was collected from Las Cruces, New Mexico (Adhikari et al., 2012).

Figure 8.10 points out that as soil texture became finer (from sand to clay loam), the water content of soil at saturation as well as at the wilting point became greater. The soil water content at −15 bar of pressure is known as wilting point water content, while the water that cannot be removed from the soil is known as residual water content.

From Figure 8.10, the effective degree of saturation for a soil sample can be determined from the known values of the saturated water content (θ_s), residual water content (θ_r), and in situ water content of the sample (θ_i) as follows:

$$S_e = \frac{\theta_i - \theta_r}{\theta_s - \theta_r} \tag{8.18}$$

If a soil sample is very dry, θ_i is equal to θ_r and the degree of saturation is zero. On the other hand, when a soil sample is saturated, the numerator and denominator of Equation 8.18 are the same and the degree of saturation is equal to one. Thus, the degree of saturation of a soil varies between zero and one.

The volumetric soil water content at −1/3 bar is generally known as field capacity water content. Readers should remember that the field capacity water content is difficult to define, and water content at 1/3 bar may not represent the field capacity water content of all soils. The field capacity water content of soil in an experimental farm (or in situ) is normally the water content of soil 24 h after the cessation of a long-duration infiltration test or rainfall. The area where the field capacity water content is needed, at the end of the infiltration test or once the soil profile is saturated with water and no water is stagnant or ponded on the surface, the surface is covered with a plastic sheet to prevent evaporation of water. Water moves down into deeper soil layers until the soil water content has reached the field capacity because at that point the downward movement of water due to gravity ceases. Again, this may not be the true field capacity water content for all soils. Similarly, the wilting point water content (soil water content at −15 bar) may not be the volumetric water content beyond which plants cannot extract water from the soil. It may be true for vegetables or agronomic plants, although certain plants, depending on their stage of growth, may wilt even before the soil water content is at the wilting point. Desert vegetation, on the other hand, can extract water from soil well beyond the wilting point of the soil. The difference between the volumetric water content at field capacity and volumetric water content at wilting point is generally known as available water content or plant available water. It is also expressed in depth units as centimeters (cm) or meters (m) of water as follows:

$$\text{Plant available water}(\text{cm}) = (\theta \text{ at field capacity} - \theta \text{ at wilting point})$$
$$\times \text{ depth of soil layer}(\text{cm})$$

Example 8.9

If the saturated water content of a soil is 0.45 m^3/m^3 and the residual water content is 0.02 m^3/m^3, calculate the degree of saturation for the soil water content given in the first column of the following table.

θ_i	θ_s	θ_r	S_e
0.45	0.45	0.02	$= \dfrac{\theta_i - \theta_r}{\theta_s - \theta_r} = \dfrac{0.45 - 0.02}{0.45 - 0.02} = 1$
0.40			$= \dfrac{\theta_i - \theta_r}{\theta_s - \theta_r} = \dfrac{0.40 - 0.02}{0.45 - 0.02} = 0.88$
0.35			= 0.77
0.25			= 0.53
0.15			= 0.30
0.05			= 0.07
0.02			= 0

Example 8.10

The volumetric water contents at field capacity and wilting points are given in the following table for different soil layers from the soil surface. Calculate the available water content of each soil layer and the total soil profile as well as irrigation interval.

θ_{FC} (m^3/m^3)	θ_{WP} (m^3/m^3)	Thickness of Soil Layer (m)	Available Water Content (m)
0.40	0.18	0.3	= (0.40 – 0.18) × 0.3 = 0.066
0.35	0.14	0.4	= (0.35 – 0.14) × 0.4 = 0.084
0.25	0.08	0.25	= 0.043
0.24	0.08	0.6	= 0.096

The total available water content of the soil profile is 0.289 m or 28.9 cm. Assuming that evapotranspiration (ET) is 7 mm per day, one concern is how often does a grower need to irrigate their field without causing plant stress. Irrigation is normally done long before soil moisture content reaches the wilting point, let us assume, it is halfway between field capacity and wilting point. Assuming that the root zone depth is 70 cm, total available water content using the data in Example 8.10 will be equal to 0.075 m {[0.4 – 0.5 × (0.4 + 0.18)] × 0.3 + [0.35 – 0.5 × (0.35 + 0.14)] × 0.4} for a 0.7 m thick soil layer or 7.5 cm for 70 cm root zone depth. Since ET is 7 mm per day or 0.7 cm/day, the grower will need to irrigate approximately every 10th or 11th day [7.5 cm/(0.7 cm/d)].

8.11 SOIL WATER RETENTION MODELS

There are several commonly used empirical soil water retention models. The number of parameters is different in these models. One of the commonly used empirical equations is proposed by Brooks and Corey (1964), popularly known as the BC equation. The functional form of the BC equation is described as follows:

$$S_e = \frac{\theta_i - \theta_r}{\theta_s - \theta_r} = \left(\frac{\Phi_b}{\Phi_m}\right)^{\lambda} \qquad \Phi_m < \Phi_b,$$
$$= 1 \qquad \Phi_m < \Phi_b \tag{8.19}$$

where

Φ_b is the bubbling potential or air-entry potential.
Φ_m is the matric potential.
λ is the pore size distribution parameter.

Campbell's (1974) equation is similar to the BC equation, with the following functional form:

$$S_e = \frac{\theta_i - \theta_r}{\theta_s - \theta_r} = \left(\frac{\Phi_b}{\Phi_m}\right)^{\frac{1}{b}} \tag{8.20}$$

where b is similar to the pore size distribution parameter and is related to soil texture. Another very popular empirical equation is proposed by van Genuchten (1980), also known as the VG equation:

$$S_e = \frac{\theta_i - \theta_r}{\theta_s - \theta_r} = \left(\frac{1}{1+(\alpha\Phi_m)^n}\right)^m \qquad \Phi_m < 0,$$
$$= 1 \qquad \Phi_m \geq 0 \tag{8.21}$$

where α, n, and m are shape parameters and are the functions of the shape of the soil water retention curve. Values of m can be interpreted in the following different ways:

$$m = 1 - \frac{1}{n} \quad \text{or} \quad m = 1 - \frac{2}{n}$$

The important differences between the BC and VG equations are related to the air-entry potential and shape of the curve. The VG soil water retention curve is a smooth S-shaped curve. The BC water retention curve has a distinct air-entry potential value. Kosugi (1996) modified the van Genuchten model and suggested using a long normal distribution model with the following functional form:

$$S_e = \frac{\theta_i - \theta_r}{\theta_s - \theta_r} = \frac{1}{2}\operatorname{erfc}\left(\frac{\ln\frac{\Phi}{\alpha}}{\sqrt{2}n}\right) \qquad \Phi_m < 0$$
$$= 1 \qquad \Phi_m \geq 0 \tag{8.22}$$

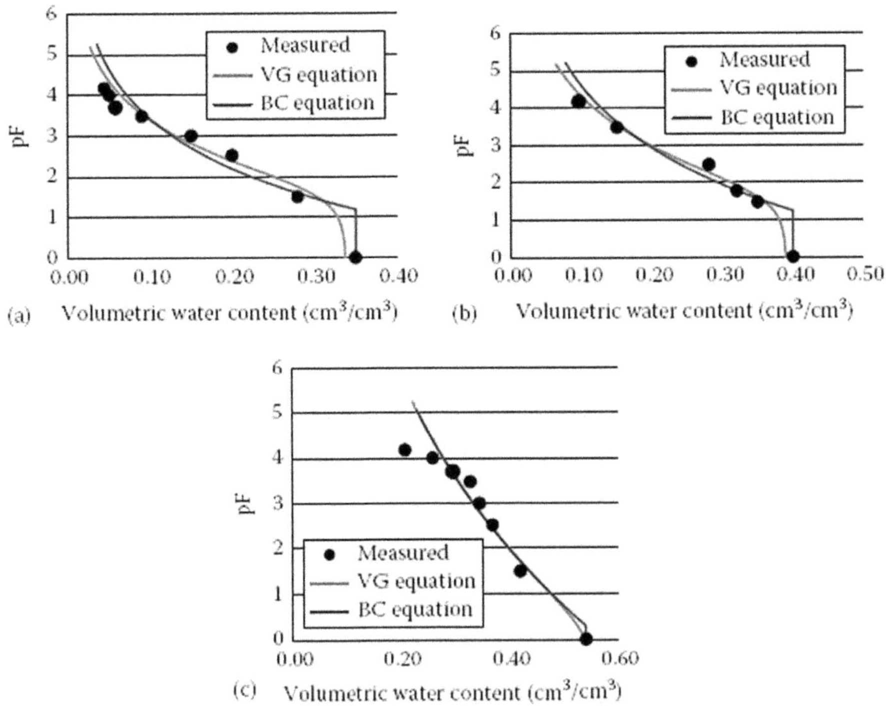

FIGURE 8.11 The soil water characteristic for (a) sand, (b) silt loam, and (c) clay loam soil. The RETC program was used to fit the BC and VG equations to measured soil water retention curve. (Panel a: From Adhikari, P., et al., *Soil Sci.*, 2012; Panels b and c: From Shukla, M. K. and Lal, R., *Soil Sci.*, 170, 445, 2005.)

Figure 8.11 presents three soil water retention curves for a sand, a silt loam, and a clay loam soil. The BC and VG equations are fitted to the measured curve using the RETC program (van Genuchten et al., 1991). Both equations fit the measured curves very well with the coefficient of correlation being greater than 0.9. The BC curve shows a distinct air-entry potential, whereas the VG curve is a smooth S-shaped curve. The fitted parameter values for three soils are presented in Table 8.2. VG model parameters for more soils can be found in the Unsaturated Soil Hydraulic Database (UNSODA; Leij et al., 1996).

8.12 HYSTERESIS PHENOMENON

The soil water content increases with soil wetting, simultaneously increasing the soil water potential. Similarly, the soil water content decreases during soil drainage or drying, with an attendant simultaneous decrease in the soil water potential. In a field soil, the wetting and drying cycles occur due to irrigation and/or rainfall. A saturated

TABLE 8.2
Van Genuchten and Brooks–Corey Model Parameters for Three Different Soil Types: Sand, Silt Loam, and Clay Loam

Parameter	Sand	Silt Loam	Clay Loam
VG Equation			
θ_s	0.338	0.390	0.562
α	0.021	0.020	0.764
n	1.310	1.225	1.078
m	0.236	0.183	0.072
BC Equation			
θ_s	0.350	0.400	0.540
α	0.064	0.056	0.499
λ	0.243	0.179	0.077

Note: The RETC program was used to fit these two models on measured soil water retention curves. The coefficient of correlation was always greater than 0.95 for both equations and all three soil moisture characteristic curves.

soil sample is dried to a fixed potential, and its water content (WC1) is determined. The same sample is then completely dried, subsequently wetted to the same fixed matric potential, and the water content of the sample (WC2) is determined. Intuition tells us that water contents (WC1 and WC2) or amounts of water in these two samples should be the same because both samples are at the same potential. It has been observed that water contents are usually different in these samples even though both are at the same fixed matric potential.

When an initial soil sample is drained at several different potentials (or suctions) while simultaneously determining the soil water content at each of the applied potentials, a plot between soil water potential and corresponding water contents at those suctions generates drainage or the drying soil moisture characteristic curve. On the other hand, if a soil sample is wetted at differential potentials while simultaneously determining the soil water content at each of the applied potentials, a plot between the soil water potential and soil water content produces a wetting soil moisture characteristic curve. It has been observed that the drainage and wetting curves do not overlay each other but there exists a gap between these two characteristic curves. This phenomenon, when the two curves do not overlay each other, is known as hysteresis (Figure 8.12).

The hysteresis phenomenon can be explained on the basis of the irregular shapes of the soil pores. The matric potential at the top of the water column inside a capillary can be estimated from the capillary rise equation (assuming contact angle is zero), repeated here for the convenience of the readers:

$$\Phi_m = -\rho_w g h = -\frac{2\gamma}{r}$$

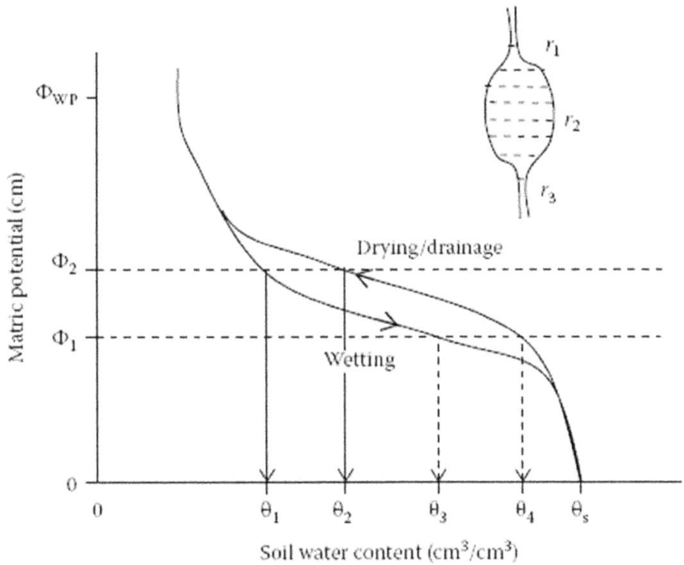

FIGURE 8.12 Schematic of hysteresis in the soil water characteristic curve. The wetting curve and the drying curve are different; thus, there are two different soil water contents (θ_1, θ_2) at the same matric potential (Φ_1). A pore with variable diameters, r_1, r_2, and r_3.

This equation indicates that for pores with uniform capillaries initially filled with water, the water content would not change and the pore will not drain with decreasing matric potential as long as $\Phi_m \geq -2\gamma/r$ (Hanks and Ashcrofts, 1980). As soon as $\Phi_m < -2\gamma/r$, all the pores will drain and the water content will change to a new value in equilibrium with the potential.

If, on the other hand, pores are of irregular shapes (see Figure 8.12), an initially filled pore when drained will retain water until $\Phi_m \geq -2\gamma/r_1$ and will lose all the water as the matric potential goes down further. During wetting, the water content will increase with increasing matric potential until $\Phi_{\mathrm{lm}} < -2\gamma/r_2$. At $\Phi_m = -2\gamma/r_2$, the entire pore will be filled with water. Soil consists of a complex network of capillaries of variable lengths and sizes, and therefore it is not possible to accurately determine the soil water content from the known soil water potential (or matric potential) because there are two different smooth soil water characteristic curves for drainage and wetting.

Hysteresis is explained by the pore geometry causing what is known as the ink bottle effect (Figure 8.13). Upon drying, a higher soil water content exists at a given matric potential than during wetting the same pore at a given potential. This is caused by the shape and size of the interconnecting pores or necks or bottlenecks, which have large cavities but narrow pints of connection to adjacent pores. Water is held in small pores that tend to fill first but do not always empty upon drying.

A soil water characteristic curve can be determined in the lab by using soil samples collected in metal cores of known weight or by repacking loose soil in small metal,

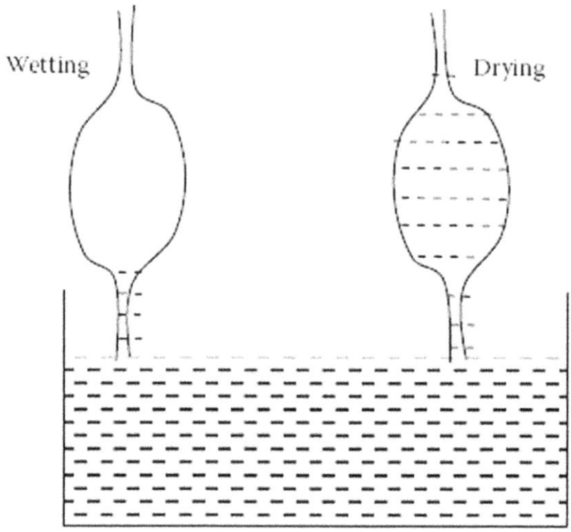

FIGURE 8.13 Ink bottle effect for an explanation of hysteresis in soil water characteristic curve.

plastic, or rubber cores. The bulk density of the soil in the core is determined from the known dry soil weight and total volume of soil (or core). These soil samples are wetted to saturation from the bottom by applying a water head and letting cores be saturated by capillary rise. The wetted cores are weighted and then placed on an already saturated pressure plate and then in the pressure plate extractor. A predetermined pressure is applied on the cores for about 24 h or until no more water is coming out of the core. Once the core is equilibrated to the applied pressure, it is removed from the extractor and weighted. The same procedure is repeated with successively higher pressures. At the end of the experiment, from the known bulk density of the soil in the core, dry soil weight in the core, and weight of core at the end of each pressure experiment, the gravimetric and volumetric water contents of soil are calculated (see Example 8.11). The water content is plotted against the pressure head to yield the soil water characteristic curve.

Example 8.11

A core soil sample (5 cm diameter and 5 cm length) was collected from an experimental area located in the Chihuahuan desert region west of the city of Las Cruces. The area was irrigated with wastewater. The core was saturated and placed inside a pressure plate apparatus, and successive pressures were applied. At equilibrium with an applied pressure, the core was weighted and placed inside the core, and again a higher pressure was applied. Calculate the amount of water, gravimetric water, and volumetric water contents at each suction. The data collected are given here and in the following table (columns 1 and 2):

Weight of empty core = 24.41 g
Weight of core + soil = 174.34

Gravimetric water content of soil = 0.03 g/g

$$\text{The dry weight of soil is determined as} = \frac{173.34 - 24.41}{1 + 003}$$

$$= \frac{149.91}{1.03} = 145.54 \text{ g}$$

$$\text{Bulk density of soil} = \frac{145.54}{98.175} = 1.48 \text{g}/\text{cm}^3$$

Weight of water = Column 2 – Dry soil wt (145.54 g) – Weight of empty core (24.41 g)
Gravimetric water content (GWC) = Column 3/145.54 g/g
Volumetric water content (VWC) = Column 4 × Soil bulk density (1.48 g/cm^3)/density of water (or 1 g/cm^3)

Suction (bar)	Core + Soil (g)	Water (g)	GWC (g/g)	VWC (cm^3/cm^3)
1	2	3	4	5
0	202.33	32.38	0.22	0.33
0.3	198.46	28.51	0.20	0.29
1	176.28	6.33	0.04	0.06
3	175.10	5.15	0.04	0.05
5	174.45	4.50	0.03	0.05
10	173.63	3.68	0.03	0.04
15	173.56	3.61	0.02	0.04

PROBLEMS

8.1 Describe various methods of measurement of the soil water content and matric potential.

8.2 An agricultural field is tile drained. If the water is flowing into a drain at 50 cm below the soil surface, assuming that the reference is at the drain level and the solute potential is zero, make estimates of Φ_p, Φ_m, and Φ_g, and calculate the total soil water potential (Φ_T) for up to 70 cm below the soil surface.

8.3 A rainfall partly wetted the soil to a certain depth; however, after several days the soil surface gets drier, whereas deeper layers still experience downward movement of water. Assuming that the reference is at the water table and the solute potential is zero, make estimates of Φ_p, Φ_m, and Φ_g, and calculate the total soil water potential (Φ_T) for up to 70 cm below the soil surface.

8.4 Plot soil moisture retention curves for when soil salinity is 1 dS and 4 dS/m.

8.5 What is the relationship between volumetric water content and gravimetric water content. Write the formula and explain each term along with their units.
8.6 What are the important distinctions between the van Genuchten model and the Brooks and Corey model?
8.7 If saturated water content is 0.50 $cm^3 cm^{-3}$ and wilting point 0.1 $cm^3 cm^{-3}$, calculate the degree of saturation for the soil water contents of 0.4, 0.3, 0.2, and 0.15 $cm^3 cm^{-3}$.

REFERENCES

Adab H., R. Morbidelli, C. Satalippi, M. Moradian, and G. A. F. Ghalhari. 2020. Machine learning to estimate surface soil moisture. *Water.* 12: 3223. https://doi.org/10.3390/w12113223.

Adhikari, P., M. K. Shukla, and J. G. Mexal. 2012. Spatial variability of soil properties in an arid ecosystem irrigated with treated municipal and industrial wastewater. *Soil Sci.* 177(7): 458–469.

Altendorf, C. T., R. Elliott, E. W. Stevens and M. L. Stone. 1999. Development and validation of a neural network model for soil water content prediction with comparison to regression techniques. *TASABE* 42: 691–700.

Bouyoucos, G. J. 1937. Evaporating the water with burning alcohol as a rapid means of determining moisture content of soils. *Soil Sci.* 44(5): 377–384.

Bresler, E., B. L. McNeal, and D. L. Carter. 1982. *Saline and Sodic Soils, Principles, Dynamics, Modeling.* Springer-Verlag, New York, p. 236.

Brooks, R. H. and A. T. Corey. 1964. *Hydraulic Properties of Porous Media, Hydrology Papers, No. 3.* Colorado State University, Fort Collins, CO.

Campbell, G. S. 1974. A simple method for determining unsaturated conductivity from moisture retention data. *Soil Sci.* 117(6): 311–314.

Dane, J. H. and G. C. Topp. 2002. *Methods of Soil Analysis, Part 4. Physical Methods.* SSSA Book Series, No. 5, Madison, WI, p. 1692.

Deb, S. K., M. K. Shukla, P. Sharma, and J. G. Mexal. 2011. Patterns of soil water depletion in irrigated mature pecans of arid southern New Mexico. *Irrig Sci.* https://doi.org/10.1007/s00271-011-0293-1.

Deb, S. K., M. K. Shukla, and J. G. Mexal. 2012. Simulating deep percolation in flood-irrigated mature orchards with RZWQM2. *Trans. ASABE*, 55(6): 2089–2100.

Elshorbagy A. and K. Parasuraman. 2008. On the relevance of using artificial neural networks for estimating soil moisture content. *J. Hydrol.* 362: 1–18.

Gardner, C. M. K., D. Robinson, K. Blyth, and J. D. Cooper. 2001. Soil water content. In: K. A. Smith and C. E. Mullins (eds.), *Soil and Environmental Analysis.* Marcel Dekker Inc., New York.

Gardner, W. H. 1986. Water content. In A. Klute (ed.), *Methods of Soil Analysis. Part 1. Physical and Mineralogical Methods*, 2nd edn. Soil Science Society of America, Inc., Madison, WI, pp. 493–544.

Gill, M. K., T. Asefa, M. W. Kemblowski, and M. McKee. 2006. Soil moisture prediction using support vector machines. *J. Am. Water Resour. As.* 42: 1033–1046.

Groves, S. J. and S. C. Rose. 2004. Calibration equations for Diviner 2000 capacitance measurements of volumetric soil water content of six soils. *Soil Use Manage.* 20: 96–97.

Hajjar, C. S., C. Hajjar, M. Esta, and Y. G. Chamoun. 2020. Machine learning methods for soil moisture prediction in vineyards using digital images. *E3S Web of Conferences* 167: 02004. https://doi.org/10.1051/e3sconf/202016702004.

Hanks, J. R. and G. L. Ashcroft. 1980. *Applied Soil Physics*. Springer Verlag, Berlin/Heidelberg/New York, p. 159.

Harlow, R. C., E. J. Burke, and T. P. A. Ferre. 2003. Measuring water content in saline sands using impulse time domain transmission techniques. *Vadose Zone J.* 2: 433–439.

Hassan-Esfahani, L., A. Torres-Rua, Jensen, and M. McKee. 2015. Assessment of surface soil moisture using high-resolution multi-spectral imagery and artificial neural networks. *Remote Sens.* 7: 2627–2646.

Hook, W. R. and N. J. Livingston. 1996. Errors in converting time domain reflectometry measurements of propagation velocity to estimates of soil water content. *Soil Sci. Soc. Am. J.* 59: 35–41.

Huisman, J. A., S. S. Hubbard, J. D. Redman, and A. P. Annan. 2003. Measuring soil water content with ground penetrating radar: A review. *Vadose Zone J.* 2: 476–491.

Jackson, T. J. and T. J. Schmugge. 1989. Passive microwave remote sensing system for soil moisture: Some supporting research. *IEEE Trans. Geosci. Remote Sens.* 27: 225–235.

Jackson, T. J. and T. J. Schmugge. 1991. Vegetation effects on the microwave emission of soils *Remote Sens. Environ.* 36: 203–212.

Katzberg, S. J., O. Torres, M. S. Grant, and D. Masters. 2005. Utilizing calibrated GPS reflected signals to estimate soil reflectivity and dielectric constant: Results from SMEX02. *Remote Sens. Environ.* 100(1): 17–28.

Kosugi, K. 1996. Lognormal distribution model for unsaturated soil hydraulic properties. *Water Resour. Res.* 32(9): 2697–2703.

Leij, F. J., W. J. Alves, M. Th. van Genuchten, and J. R. Williams. 1996. *The UNSODA Unsaturated Hydraulic Database*. EPA/600/R-96/095. U.S. Environmental Protection Agency, Cincinnati, OH.

Lunt, I. A., S. S. Hubbard, and Y. Rubin. 2005. Soil moisture content estimation using ground-penetrating radar reflection data. *J. Hydrol.* 307: 254–269.

Masters, D., P. Axelrad, and S. Katzberg. 2004. Initial results of land-reflected GPS bistatic radar measurements in SMEX02. *Remote Sens. Environ.* 92: 507–520.

Powers, M. H. 1997. Modelling frequency-dependent GPR. *The Leading Edge* 16(11): 1657–1662.

Robinson, D. A., C. S. Campbell, J. W. Hopman et al. 2008. Soil moisture measurement for ecological and hydrological watershed-scale observations: A review. *Vadose Zone J.* 7: 358–389.

Santos, J. F. C. dos, H. R. F. Silva, F. A. C. Pinto and I. R. de Assis. 2016. Use of digital images to estimate soil moisture. *Rev. Bras. Eng. Agríc. Ambient.*, 20: 1051–1056, 12.

Scanlon, B. R., B. J. Andraski, and J. Bilskie. 2002. Miscellaneous methods for measuring matric or water potential. In: J. H. Dane and G. C. Topp (eds.), *Methods of Soil Analysis—Part 4: Physical Methods*. SSSA, Madison, WI, pp. 643–670.

Schmugge, T. J. 2011 Microwave remote sensing of soil hydraulic properties. In M. K. Shukla (ed.), *Soil Hydrology, Land Use and Agriculture: Measurement and Modeling*. CAB International, Wallingford, UK, pp. 415–426.

Schmugge, T. J. and T. J. Jackson. 1992. A dielectric model of the vegetation effects on the microwave emission from soils. *IEEE Trans. Geosci. Remote Sens.* 30: 757–760.

Schmugge, T. J. and T. J. Jackson. 1994. Mapping surface soil moisture with microwave radiometers. *Meteorol. Atmos. Phys.* 54: 213–223.

Shukla, M. K. 2011. Introduction to soil hydrology. In M. K. Shukla (Ed.), *Soil Hydrology, Land Use and Agriculture: Measurement and Modeling*. CAB International, Wallingford, UK, pp. 1–23.

Shukla, M. K. and R. Lal. 2005. Erosional effects on soil properties in an on-farm study on Alfisol in west central Ohio. *Soil Sci.* 170(6): 445–456.

Topp, G. C. and P. A. Ferre. 2002. The soil solution phase. In J. H. Dane and G. C. Topp (eds.), *Methods of Soil Analysis, Part 4: Physical Methods*. SSSA Book Series, No. 5, Madison, WI, pp. 417–422.

Topp, G. C., J. L. Davis, and A. P. Annan. 1980. Electromagnetic determination of soil water content: Measurements in coaxial transmission lines. *Water Resour. Res.* 16: 574–582.

US Salinity Laboratory Staff. 1954. *Diagnosis and Improvement of Saline and Alkali Soils. Agric. Handbook No 60, USDA*. US Government Printing Office, Washington, DC.

van Bavel, C. H. M., N. Underwood, and R. W. Swanson. 1956. Soil moisture measurement by neutron moderation. *Soil Sci.* 82: 29–41.

van Genuchten, M. Th. 1980. A closed-form equation for predicting the hydraulic conductivity of unsaturated soils. *Soil Sci. Soc. Am. J.* 44: 892–898.

van Genuchten, M. Th., F. J. Leij, and S. R. Yates. 1991. *The RETC Code for Quantifying the Hydraulic Functions of Unsaturated Soils*, EPA/600/2-91/065. US Environmental Protection Agency, Ada, OK.

Wagner, W. and C. Pathe. 2004. Has SAR failed in soil moisture retrieval? Proc. Envisat & ERS Symposium (ESA SP-572), Salzburg, Austria, 6–10 September.

Zhang, D., R. Tang, W. Zhao, B. Tang, H. Wu, K. Shao, and Z.-L. Li. 2014. Surface soil water content estimation from thermal remote sensing based on the temporal variation of land surface temperature. *Remote Sens.* 6: 3170–3187.

9 Flow through the Vadose Zone

9.1 INTRODUCTION

Water movement through soil can occur when soil is fully saturated, as well as when it is unsaturated, and accordingly it is known as saturated or unsaturated flow. Movement of the applied water on the soil surface is influenced by the size and shape of pores, soil texture, soil organic matter content, soil bulk density, and soil structure. Sandy soils generally have larger pores, absorb water rapidly, and hold less water per unit soil profile depth. Thus, water penetrates the sandy soil profile faster compared to loamy, silty, or clayey soils. This is why it is recommended that sandy soils be irrigated more frequently with lesser amounts of water compared to loamy, silty, or clayey soils. Frequent irrigation with low amounts of water improves irrigation efficiency by decreasing the deep percolation of water through the sandy soil profile.

Capillary action is the primary mechanism of water movement through the soil profile. Capillary movement does not have to be upward only, and it is primarily through the smaller pores, but once larger pores are filled with water, water movement becomes faster and gravity plays a role in water movement. This chapter describes the laws governing the movement of water through saturated and unsaturated media, parameters describing flow through the porous media, sink source terms, and drainage below the root zone.

A soil matrix is considered unsaturated when some of the pores are filled with water and the remaining with air. The unsaturated vadose zone provides a continuum of water-unsaturated porous media connecting the soil–atmosphere interface and underlying groundwater. Unsaturated flow of water, although a more prevailing condition in the field than saturated flow, is more complex and often does not have direct solutions. Instead, indirect methods, approximations, and numerical methods are more commonly used in the solution of unsaturated flow problems.

9.2 LAWS GOVERNING FLOW THROUGH SATURATED POROUS MEDIA

9.2.1 Darcy's Law

When all pores in a porous medium are filled with water, it is known as a saturated porous medium, and when water flows through such a porous medium, the flow is known as saturated flow. During saturated flow, water content of the media through which flow is taking place does not change with time. Henry Philibert Gaspard

DOI: 10.1201/9780429264849-9

Darcy was the first to experimentally demonstrate the flow phenomenon and develop a law governing the flow of water through saturated porous media. The law describing flow through saturated porous media is popularly known as Darcy's law. The experiment was conducted in a vertical column filled with completely water-saturated sand. Darcy showed that water flow through the soil was proportional to the hydraulic head difference (or potential difference). The constant of proportionality that completes the relationship between flow rate and gradient of flow is known as saturated hydraulic conductivity. The volumetric flow rate of water coming out of the bottom of a saturated soil profile during one-dimensional flow is given by Darcy's law as follows (Figure 9.1):

$$Q = -K_s A \frac{H_1 - H_2}{L} = -K_s A \frac{L + h}{L} = -K_s A i \tag{9.1}$$

where

Q is the volumetric flow rate (cm^3/s).
K_s is the saturated hydraulic conductivity (cm/s).
A is the cross-sectional area through which flow is taking place.
H_1 is the head at the inlet (cm).
H_2 is the head at the outlet (can also be zero for free flow; cm).
L is the length of the soil column (cm) through which flow is taking place.
h is the head of water above the soil surface.
i is the hydraulic gradient (unitless).

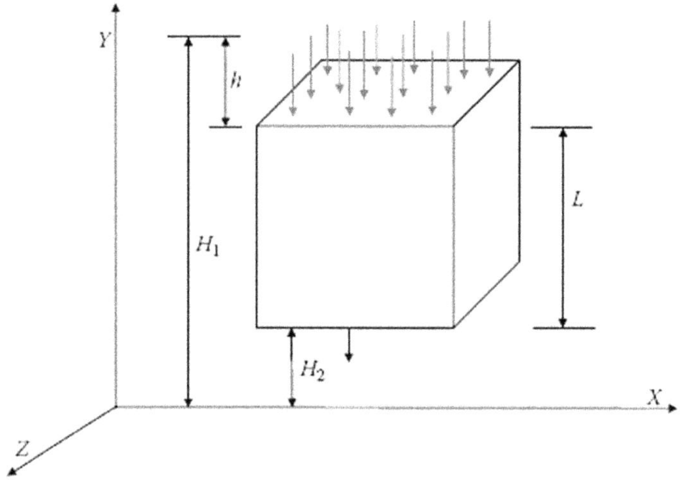

FIGURE 9.1 Flow of water through a three-dimensional profile depicting potential heads at the inlet and outlet through a soil volume of length L. A constant head of h is applied at the top of the core. Assuming that the reference is at the X-line, the total hydraulic head at the inlet (H_1) = $H_2 + L + h$, and hydraulic head at the outlet = H_2.

The negative sign is for the direction of flow and indicates that the flow of water takes place from an area with higher potential to an area with lower potential. Equation 9.1 can be rearranged as follows:

$$\frac{Q}{A} = q = -K_s i \tag{9.2}$$

where q is the rate of water flow through the soil in depth units (cm/s). q is also known as Darcy's flux and has the units of volumetric flow per unit area per unit time ($cm^3/cm^2/s$). Small q is also represented by v and is known as Darcy's velocity (cm/s). The rate of flow of water (in depth and time or velocity units) through the soil is proportional to the hydraulic gradient. If the hydraulic gradient i is equal to one, also known as the unit hydraulic gradient, then Equation 9.2 becomes

$$q = -K_s \tag{9.3}$$

Assuming that the flow in Figure 9.1 is one-dimensional, the total head at the inlet above the reference point is a net positive head, whereas the total head at the outlet (bottom of the porous medium; also called the reference point) is atmospheric or equal to zero.

The one-dimensional Darcy's equation for the flow along the y-axis can also be represented as

$$q_y = -K_{sy} \frac{\partial H}{\partial y} \tag{9.4}$$

Similarly, flow in only x and z directions can be represented as

$$q_x = -K_{sx} \frac{\partial H}{\partial x} \tag{9.5}$$

$$q_z = -K_{sz} \frac{\partial H}{\partial z} \tag{9.6}$$

Darcy's law for a three-dimensional flow can be represented by combining the flow through each direction as follows:

$$q_x + q_y + q_z = -K_{sx} \frac{\partial H}{\partial x} - K_{sy} \frac{\partial H}{\partial y} - K_{sz} \frac{\partial H}{\partial z} \tag{9.7}$$

If the porous medium or soil is isotropic, then hydraulic conductivity in the x, y, and z directions are equal, and Darcy's law can be rewritten as

$$q_x + q_y + q_z = -K_s \left(\frac{\partial H}{\partial x} + \frac{\partial H}{\partial y} + \frac{\partial H}{\partial z} \right) \tag{9.8}$$

Darcy's law is generally valid for saturated and unsaturated flow; steady and transient flow; flow in homogeneous, heterogeneous, and anisotropic systems; flow in aquifers and aquitards; and flow through granular media and rocks. Darcy's velocity assumes that flow takes place from the entire cross-section of the flow domain; however, flow only occurs through the interconnected pore channels. From the continuity equation, the total Darcian volumetric flow rate through the profile will be equal to the total flow rate through the soil pores as follows:

$$Q = Aq_{\mathrm{D}} = A_{\mathrm{p}}q_{\mathrm{p}} \tag{9.9}$$

or

$$q_{\mathrm{p}} = \frac{A}{A_{\mathrm{p}}}q_{\mathrm{D}} = \frac{A}{A_{\mathrm{p}}}q_{\mathrm{D}}\frac{L}{L} = \frac{V_{\mathrm{T}}}{V_{\mathrm{P}}}q_{\mathrm{D}} = \frac{q_{\mathrm{D}}}{\theta} \tag{9.10}$$

where

A is the total cross-sectional area of the flow domain (soil).
A_{p} is the total cross-sectional area of the pores.
q_{D} is the Darcy velocity.
q_{p} is the velocity through pores (pore water velocity is also represented as v).
L is the length of the flow domain.
θ is the volumetric water content.

It is important to remember that Darcy's law cannot always describe fluid flow through porous media. When near a gas well with high gas velocity, flow of fluid becomes turbulent. Under these circumstances, the pressure difference is much greater than those predicted by Darcy's law, and flow is known as non-Darcian flow.

Example 9.1

Assuming that Figure 9.1 represents flow in one dimension, if the length of the soil column is 10 cm, the total constant head of water at the inlet above the reference is 15 cm, the head at the outlet is 2 cm, and the saturated hydraulic conductivity of the soil is 1.5 cm/s. Calculate the rate of flow (or Darcy's velocity) at the outlet and also express it as Darcy's flux. If the volumetric water content of the soil in the column is 0.4 cm³/cm³, calculate the rate of flow through the pores:

$$q = -K_s i = -1.5 \times \frac{15-2}{10} = -1.5 \times 1.3 = -1.95 \text{ cm/s}$$

Darcy's flux = 1.95 cm³/cm²/s; the rate of flow through pores

$$= q_{\mathrm{p}} = \frac{q_{\mathrm{D}}}{\theta} = \frac{1.95}{0.4} = 4.88 \text{ cm/s}$$

9.2.2 Continuity Equation

The continuity equation takes into account all the water that is flowing through a three-dimensional porous medium. It describes the conservation of liquid mass of water during flow through a porous medium. The net rate of water flow through the soil profile changes (increases or decreases) the volumetric water content of soil as well as the net volume of water in the profile and is given on the left-hand side of the following equation. Mathematically, the net rate of flow in a three-dimensional porous medium is represented on the right-hand side of Equation 9.11. The volumetric rate of change of soil moisture content inside a soil domain is equal to the sum of the net change in volume of water per unit cross-sectional area per unit time in each direction inside the entire flow domain. These two terms are equal as given in the following equation, if there is no loss of water from the flow domain due to external gradients:

$$\frac{\partial\theta}{\partial t}\Delta x\Delta y\Delta z = -\left(\frac{\partial q_x}{\partial_x}+\frac{\partial q_y}{\partial_y}+\frac{\partial q_z}{\partial_z}\right)\Delta x\Delta y\Delta z \quad (9.11)$$

Since net volume is present on both sides of the equation, the equation reduces to the following form:

$$\frac{\partial\theta}{\partial t} = -\left(\frac{\partial q_x}{\partial_x}+\frac{\partial q_y}{\partial_y}+\frac{\partial q_z}{\partial_z}\right) \quad (9.12)$$

This equation is known as the continuity equation. For saturated flow, the rate of change of water content with respect to time is zero.

9.2.3 Laplace Equation

Transferring Darcy's equation into the right-hand side of the continuity equation for each direction of flow separately results in a three-dimensional partial differential equation known as the Laplace equation. It is named after Pierre Simon Laplace. The mathematical form of the Laplace equation is as follows:

$$\frac{\partial\theta}{\partial t} = -\left(\frac{\partial}{\partial x}\left(K_{sx}\frac{\partial H}{\partial x}\right)+\frac{\partial}{\partial y}\left(K_{sy}\frac{\partial H}{\partial y}\right)+\frac{\partial}{\partial z}\left(K_{sz}\frac{\partial H}{\partial z}\right)\right) \quad (9.13)$$

If soil is isotropic, then $K_{sx} = K_{sy} = K_{sz}$, and Equation 9.13 reduces to the following on simplification:

$$\frac{\partial\theta}{\partial t} = -K_s\left(\frac{\partial^2 H}{\partial x^2}+\frac{\partial^2 H}{\partial y^2}+\frac{\partial^2 H}{\partial z^2}\right) \quad (9.14)$$

For a saturated soil, the net rate of change of soil moisture, on the left side of Equation 9.14, is zero, and after rearranging, the Laplace equation reduces to

$$\frac{\partial^2 H}{\partial x^2}+\frac{\partial^2 H}{\partial y^2}+\frac{\partial^2 H}{\partial z^2}=0 \tag{9.15}$$

$$\nabla^2 H=0 \tag{9.16}$$

9.2.4 Hagen–Poiseuille Equation

The laminar flow through a pipe or a pore of constant radius is given by the following equation known as the Hagen–Poiseuille equation:

$$Q=\frac{h\pi r^4}{8\mu L} \tag{9.17}$$

where

Q is the volumetric flow rate (cm^3/s).
h is the pressure difference (cm).
r is the radius of the pore or pipe (cm).
μ is the dynamic viscosity [g/(cm·s)].
L is the length of the pore or pipe (cm).

9.3 SATURATED HYDRAULIC CONDUCTIVITY

Saturated hydraulic conductivity is the maximum rate of flow that can take place through a soil profile. It is one of the most important parameters for soil water movement, retention, and modeling the flow of water through both saturated and unsaturated porous media. The saturated hydraulic conductivity is highly spatially as well as temporally variable due to pedogenic, natural, or anthropogenic sources. Several factors that affect the saturated hydraulic conductivity of soil include soil texture, soil structure, size of pores, continuity of pores, presence of worm holes, biopores, cracks, chemical composition of infiltrating water, and entrapped air. There are several laboratory, field or in situ, and empirical methods (including machine learning algorithms) available in the literature to determine saturated hydraulic conductivity. An overview of various methods used to determine saturated hydraulic conductivity is given in Figure 9.2 (Deb and Shukla, 2012).

Saturated hydraulic conductivity values are higher for sandy than silty or clayey soils; however, in a watershed or a field, values of saturated hydraulic conductivity for a soil can vary by several orders of magnitude. There are several different classifications of saturated hydraulic conductivity found in the literature, and we present here the classifications suggested by Smith and Browning (1946) (Table 9.1), and Bear (1972) and Fetter (2001) (Table 9.2).

9.3.1 Laboratory Methods

The saturated hydraulic conductivity of soil can be measured directly in the laboratory. Laboratory methods of saturated hydraulic conductivity determination are grouped into constant head and falling head methods.

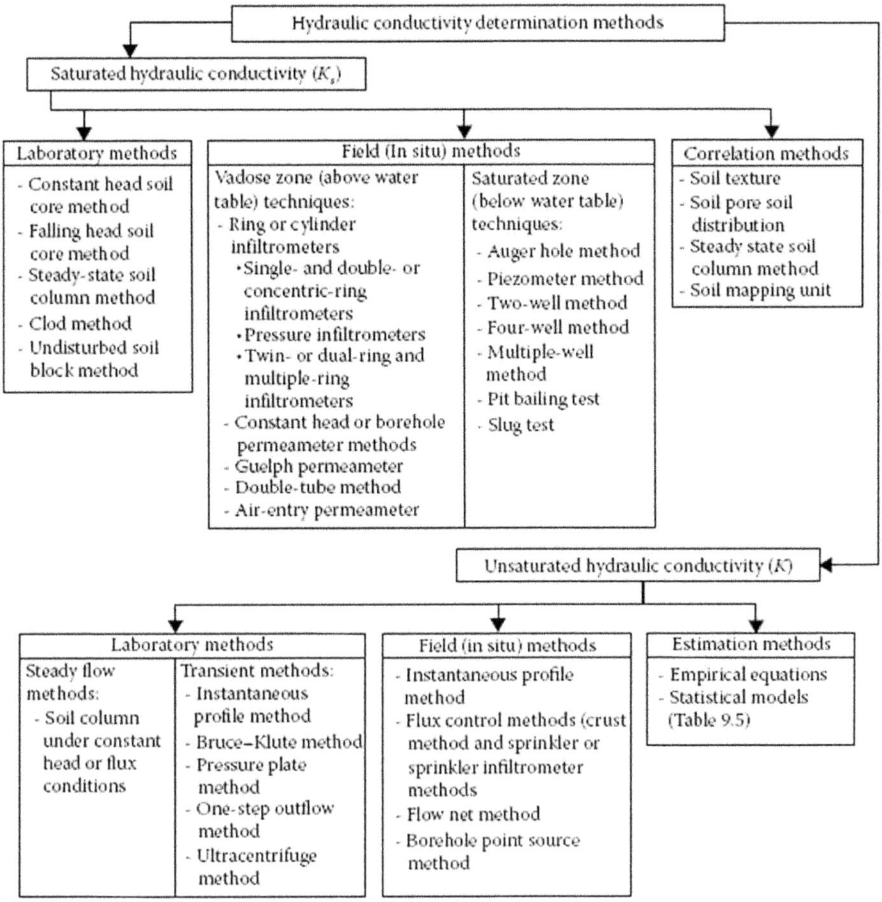

FIGURE 9.2 Overview of methods for the determination of hydraulic conductivity of soil. (From Deb, S. and Shukla, M. K., *Am. J. Environ. Sci.*, 8(5), 489, 2012.)

TABLE 9.1
Saturated Hydraulic Conductivity (K_s) of Different Soils Presented by Smith and Browning (1946)

Class	K_s (cm/day)	Comments
Extremely slow	<0.1	Very nearly impervious
Very slow	0.1–1	Poor drainage, staining, too slow for artificial drainage
Slow	1–10	Poor aeration, poor root development
Moderate rapid	10–1000	Good to excellent drainage, water holding
Very rapid	>1000	Poor water holding, excessive drainage

TABLE 9.2
Saturated Hydraulic Conductivity (K_s) of Different Soils Published in the Literature

Soil Classification	K_s (cm/s)	K_s (cm/s)
Coarse sand	10^0–10^{-3}	10^{-1}–10^{-3}
Sand	10^0–10^{-3}	10^{-1}–10^{-3}
Loamy sand	10^{-3}–10^{-7}	10^{-3}–10^{-5}
Sandy loam	10^{-3}–10^{-7}	10^{-4}–10^{-6}
Loam	10^{-2}–10^{-7}	10^{-4}–10^{-6}
Sandy clay loam	10^{-3}–10^{-7}	10^{-4}–10^{-6}
Clay	10^{-4}–10^{-11}	10^{-6}–10^{-9}

Sources: Data from Bear, J., *Dynamics of Fluids in Porous Media*, American Elsevier Publishing Company Inc., New York, 1972, p. 764; Fetter, C. W., *Applied Hydrogeology*, 4th edn., Prentice Hall, Inc., Upper Saddle River, NJ, 2001.

9.3.1.1 Constant Head Method

In the constant head method, as the name suggests, water flow is measured across saturated porous media where flow is taking place due to the constant positive potential head maintained at the inlet. The outflows of water are recorded at specific time intervals. When several readings of outflow for a given time interval are similar, with small deviations between readings, the saturated hydraulic conductivity of soil for a vertical downward flow is calculated using the rearranged Darcy's law as follows:

$$K_s = \frac{QL}{A(H_1 - H_2)} = \frac{VL}{At(L+h)} \tag{9.18}$$

where

V is the total volume of water that came out of the bottom of the saturated soil core in time t ($Q = V/t$).
h is the constant head of water maintained on the soil surface.
L is the length of the soil column.

If flow is taking place in the horizontal direction (Figure 9.3), the saturated hydraulic conductivity of the soil is calculated as follows:

$$K_s = \frac{QL}{A(H_1 - H_2)} = \frac{VL}{tAh} \tag{9.19}$$

where h is the difference of positive potential heads between the inlet and outlet. The constant head method can be used to determine the hydraulic conductivity of most soils. However, for highly permeable soil, maintaining a constant head could be difficult.

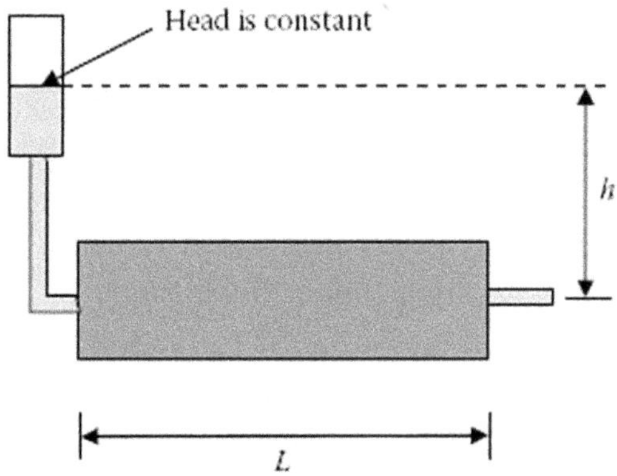

FIGURE 9.3 Flow of water through a saturated soil column in a horizontal direction at a constant head.

9.3.1.2 Falling Head Method

In the falling head method, there is usually a standpipe attached at the top of the soil core (Figure 9.4). The head at the top of the core is allowed to drop from one value to the other during a given time interval. The saturated hydraulic conductivity of soil is determined by the following equation:

$$K_s = \frac{aL}{At}\ln\left(\frac{H_1}{H_2}\right) \tag{9.20}$$

where

a is the radius of the standpipe.

H_1 and H_2 are the heads at time zero and time t.

The falling head method is useful for soils with very low hydraulic conductivity, and collection of a sufficient amount of outflow takes several hours so that evaporation might influence the accuracy of measurement. In general, the entrapped air inside the soil column influences values of saturated hydraulic conductivities. There are several options to remove entrapped air before the measurement. Some of these measures include flushing the core with carbon dioxide gas, wetting the core very slowly from the bottom, and conducting experiments under conditions of upward flux.

Example 9.2

An experiment was performed on a soil core by creating a 3 cm constant head of water to determine the saturated hydraulic conductivity of the soil. The length of the soil core is 5 cm, the diameter is 5 cm, and an outflow of 25 mL was collected in 15 min from the bottom of the core. Calculate the saturated hydraulic conductivity of the soil in the core.

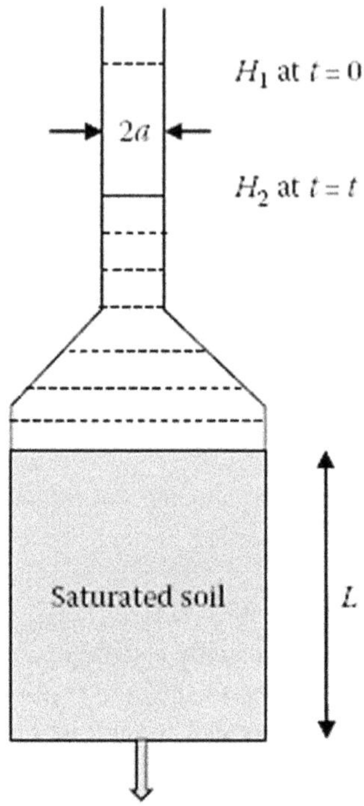

FIGURE 9.4 Flow of water through a saturated soil column in a vertical downward direction under the falling head method.

The surface area of the core = $\pi r^2 = 3.14 \times 2.5^2 = 19.63$ cm^2

$$K_s = \frac{VL}{At(L+h)} = \frac{25 \times 5}{19.63 \times 15 \times (5+3)} = 0.053 \text{ cm/min}$$

Example 9.3

Calculate the hydraulic conductivity of the soil in the core given in Example 9.2 under falling head conditions for the following dataset:

Diameter of standpipe = $2a$ = 4 cm
Head changes from 5 cm to 2 cm in 10 minutes

$$K_s = \frac{aL}{At} \ln\left(\frac{H_1}{H_2}\right) = \frac{2 \times 5}{19.63 \times 25} \ln\left(\frac{5}{2}\right) = 0.05 \text{ cm/min}$$

9.3.2 Field Methods

The common field methods for determining saturated hydraulic conductivity are the auger hole (or borehole) method, double- or single-ring infiltrometer, and Guelph permeameter.

9.3.2.1 Auger Hole Method

The auger hole method can be used to determine the saturated hydraulic conductivity of soil above as well as below the water table. The method is quick, simple, and does not require expensive equipment. Equipment required for the test include a soil auger, extension rods if deeper measurements are required, measuring tape, stopwatch, pump, and a float when the water table is shallow or a Mariotte bottle when it is deep.

For areas where the water table is at a shallow depth, after drilling a hole in the ground to a desired depth, the water is pumped out and a float attached to the bottom of a measuring tape is lowered in the hole. About ten readings on the soil tape are made at a given height at a fixed time interval.

For areas where the water table is deep, after making a borehole in the ground, the hole is filled with water to a certain depth and the head of water is maintained by a Mariotte bottle. The rate of fall of water in the Mariotte bottle is recorded at a fixed time interval. If a Mariotte bottle is not available, the rate of fall of water in the hole is measured at a fixed time interval using a float and tape device (Figure 9.5a). As many measurements should be made as possible; generally, at least ten measurements are recommended.

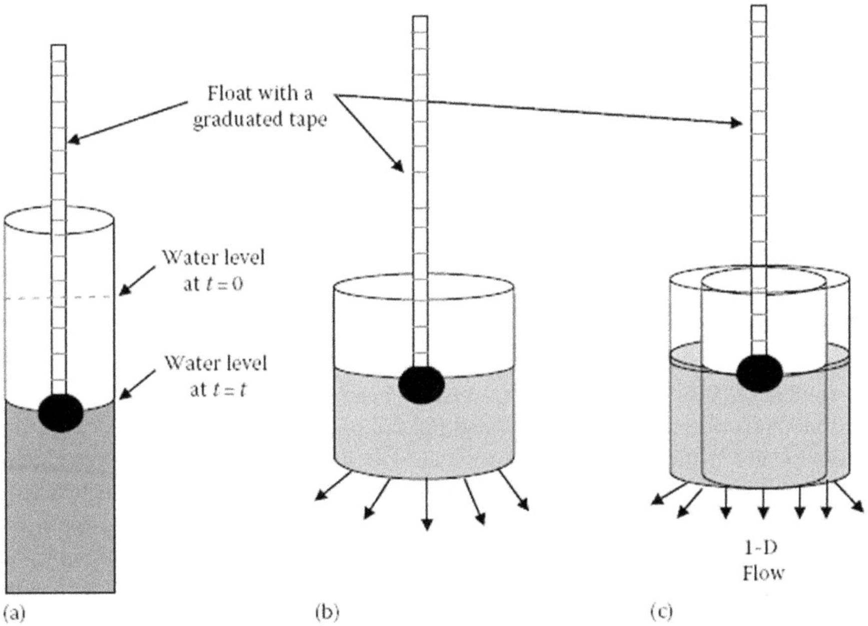

FIGURE 9.5 Schematic of (a) borehole method, (b) single-ring infiltrometer, and (c) double-ring infiltrometer. *Note*: Flow from the single ring is three-dimensional, whereas flow from the inner ring of the double infiltrometer is one-dimensional.

The saturated hydraulic conductivity can be calculated using the Hooghoudt or a functionally similar Kirkham equation (Kirkham and van Bavel, 1949) as follows:

$$K_s = C\left(\frac{\Delta y}{\Delta t}\right) \tag{9.21}$$

where C is a function of y, H, r, and the distance between the auger hole bottom and the underlying impermeable layer (Ernst, 1950).

9.3.2.2 Ring Infiltrometer Method

A ring infiltrometer is a device to measure infiltration of water into porous media. The ring infiltrometer could include just a single ring or two concentric (double) rings. A single ring is driven into the ground using a wood block (usually slightly bigger than the diameter of the outer ring) and a hammer (Figure 9.5b). An isolation layer with a small thickness consisting of mulch or any other plant residue (highly porous material) is placed on the soil surface inside the ring to prevent soil erosion when water is poured into the cylinder. A Mariotte bottle can be used to maintain a constant head of water on the soil surface, and the rate of drop of water in the Mariotte bottle is recorded with respect to time. The experiment is conducted for a long duration, usually three hours, to ensure a good saturation and steady state.

The double-ring infiltrometer technique involves two rings: an outer ring and an inner ring (Figure 9.5c). Both rings are filled with water, and care is taken that the water level in the outer ring is at the same or lower level than that in the inner ring. The primary purpose of the outer ring is to ensure that the flow from the inner ring is one-dimensional. Measurements of water depths are made only in the inner ring. A constant head in the inner ring can be maintained by a Mariotte bottle, and the rate of drop of water in the Mariotte bottle is recorded with time. The Mariotte bottle can also contain a pressure transducer, and a datalogger can record the rate of drop of the water level with respect to time. At steady state, the gradient of flow is near unity, and the rate of infiltration of water is equal to the saturated hydraulic conductivity of soil.

9.3.2.3 Guelph Permeameter

The Guelph permeameter is an easy-to-use instrument to measure saturated hydraulic conductivity in the field under constant head. It is relatively more expensive than methods previously described, and the fall of water can be recorded with time automatically using a pressure transducer. There are two reservoirs in a Guelph permeameter; however, depending upon the field conditions or the soil hydraulic conductivity, one or both reservoirs can be operated at the same time. Saturated hydraulic conductivity can be recorded using a one-constant head or two-constant head method. A vertical hole of appropriate size is dug in the field, and the Guelph permeameter is lowered in it and a constant head of water is maintained (usually 5 cm). The rate of flow of water is periodically recorded, especially after it is under steady state (or near steady state). The saturated hydraulic conductivity of soil for one head with both reservoirs can be calculated as follows:

$$K_s = \frac{C_1 Q_1}{2\pi H_1^2 + \pi a^2 C_1 + 2\pi\left(\frac{H_1}{a^*}\right)} \quad \text{where } Q_1 = 35.22\bar{R}_1 \tag{9.22}$$

where

$\bar{R}_1$ is the steady-state rate of fall of water in the reservoir (cm/s).
a^* is the macroscopic capillary length parameter (cm^{-1}).
a is the borehole diameter (cm).
H_1 is the constant head of water in the borehole (cm).
C is the shape factor.

For further details on hydraulic conductivity calculations using single and double reservoirs, a capillary length parameter, and shape factors, please refer to a Guelph permeameter user manual (Soil Moisture Equipment Corporation, 2012). In the two-head method, the experiment is repeated and steady-state outflow readings are again recorded, usually for a 10 cm head of water.

9.3.3 Empirical Methods

There are several empirical methods where one or several soil properties are combined to estimate the saturated hydraulic conductivity of soil. According to Hazen (1911), the saturated hydraulic conductivity, especially for sands, can be related to the grain diameter in millimeters (mm) for which 10% of the particles are finer ($D10$):

$$K_s = C(D10)^2 \tag{9.23}$$

where C is the Hazen coefficient and is based on the values given in Table 9.3.

Another empirical relationship for predicting saturated hydraulic conductivity is known as the Kozeny–Carman equation (Bear, 1972). It relates soil porosity and factor C to saturated hydraulic conductivity:

$$K_s = C\frac{P^3}{(1-P)^2}(D10)^2 \tag{9.24}$$

TABLE 9.3
Value of C Is Based on Different Types of Sand

Classification	Value
Very fine sand, poorly sorter	40–80
Fine sand with appreciable fines	40–80
Medium sand, well sorted	80–120
Coarse sand, poorly sorted	80–120
Coarse sand, well sorted, clean	120–150

where P is the porosity of soil. The equation was originally proposed by Kozeny and later modified by Carman.

There are several pedotransfer functions available in the literature that relate various soil properties into one- or multiple-parameter regression models for estimating saturated hydraulic conductivity of soil (Pachepsky and Rawls, 2004). There are several computer programs also available for estimating soil hydraulic parameters including saturated hydraulic conductivity (Schaap et al., 2001).

9.4 PERMEABILITY

Soil permeability is the property of soil to transmit water. It is similar to saturated hydraulic conductivity because it also describes the ability of the soil profile to conduct water through it under a hydraulic gradient. The permeability and saturated hydraulic conductivity are related as follows:

$$k = K_s \frac{\mu}{\rho_w g} \tag{9.25}$$

where

k is the soil permeability (m^2).
K_s is the saturated hydraulic conductivity (m/s).
μ is the dynamic viscosity (kg/m/s).
ρ_w is the density of the fluid or water (kg/m^3).
g is the acceleration due to gravity (m/s^2).

Transfer units of different variables in Equation 9.24 to obtain the units of permeability:

$$k = \frac{(\text{m/s})(\text{kg/m s})}{(\text{kg/m}^3)(\text{m/s}^2)} = \frac{\text{m} \times \text{kg} \times \text{m}^3 \times \text{s}^2}{\text{s} \times \text{m} \times \text{s} \times \text{kg} \times \text{m}} = \frac{\text{kg} \times \text{m}^4 \times \text{s}^2}{\text{kg} / \text{m}^2 \times \text{s}^2} = \text{m}^2$$

Therefore, the units of permeability are those of the cross-sectional area (m^2) of the flow domain. That is why permeability is also known as the property of porous media only, but not the fluid, whereas saturated hydraulic conductivity is considered the property of porous media as well as the fluid.

Example 9.4

Calculate the permeability of a soil sample with a saturated hydraulic conductivity of 0.2 m/s, and dynamic viscosity and density at 20°C of 0.001003 kg/m/s and 998.29 kg/m^3, respectively.

$$k = K_s \frac{\mu}{\rho_w g} = \frac{0.2 \times 0.001003}{998.29 \times 9.81} = 2.04E-08\text{m}^2 = 2.04E-04\text{cm}^2 = 2.04 \times 10^{-4}\text{cm}^2$$

9.5 LAWS GOVERNING FLOW THROUGH UNSATURATED POROUS MEDIA

The unsaturated soil profile is a more common scenario than saturated soil. Soil can become saturated and stay saturated for a while, especially immediately to a few hours after irrigation or precipitation, or for as long as water is ponded on the soil surface. The movement of water due to gravity, evaporation, and root water uptake causes air to enter in the soil profile as the soil profile loses water and becomes unsaturated. The unsaturated porous media usually stay at a transient state with soil water content and soil water potential changing with time. Therefore, flow through unsaturated porous media is also known as transient flow.

9.5.1 Darcy–Buckingham Equation

The equation describing the flow of water through the unsaturated soil profile is known as the Darcy–Buckingham equation. It is mathematically similar to Darcy's law for the saturated flow of water through the soil profile. The Darcy–Buckingham equation is represented as follows:

$$Q = -K(\Phi)A\frac{H_1 - H_2}{L} = K(\Phi)A\frac{\partial H}{L} = -K(\Phi)Ai \tag{9.26}$$

where
K (Φ) is the unsaturated hydraulic conductivity (cm/s).
H_1 and H_2 are the potentials across the flow domain and are negative.

Note that for vertical downward flow:

$$\partial H = L + \partial\Phi\left(\text{or} L + \partial h\right) \tag{9.27}$$

$$i = 1 + \frac{\partial\Phi}{L} \quad \text{or} \quad 1 + \frac{\partial h}{\partial x} \tag{9.28}$$

$$q = -K(\Phi)\left(\frac{\partial h}{\partial x} + 1\right) \tag{9.29}$$

where
$\partial\,\Phi$ or ∂h is the potential difference across two points inside the unsaturated soil profile.
∂x is the distance between the two points.

Unlike saturated hydraulic conductivity, unsaturated hydraulic conductivity is a function of soil water content as well as matric potential. The infiltrating water in an unsaturated soil tries to form a hydrogen bond with the already present pore water and pore walls, thus increasing the volumetric water content, matric potential (–ve potential), and unsaturated hydraulic conductivity of soil. Hydraulic conductivity is much smaller under unsaturated than saturated soil conditions. At low water content, the applied water moves through the smaller pores or through water films along the walls of larger pores, which drain first. Under unsaturated conditions, water can also move as vapor.

Example 9.5

If the water is applied at a rate slow enough to not cause ponding on the surface of an unsaturated soil column (see Figure 9.6a) and if the head at the inlet is –4 cm and the head at the outlet is –20 cm, calculate the steady-state flow through the 20 cm soil profile if the unsaturated hydraulic conductivity of the soil is 3 cm/day.

Assuming that the reference is at the bottom of the column at the outlet:

Total potential at the upper boundary = –4 + 20 = 16 cm
Total potential at the bottom of the column = –20 + 0 = –20 cm

$$q = -K(\Phi)\left(\frac{\partial H}{\partial x}\right) = -3 \times \left(\frac{16-(-20)}{20}\right) = -3 \times \frac{36}{20} = -5.4\text{cm/day}$$

The negative sign indicates that the flow takes place in the downward direction.

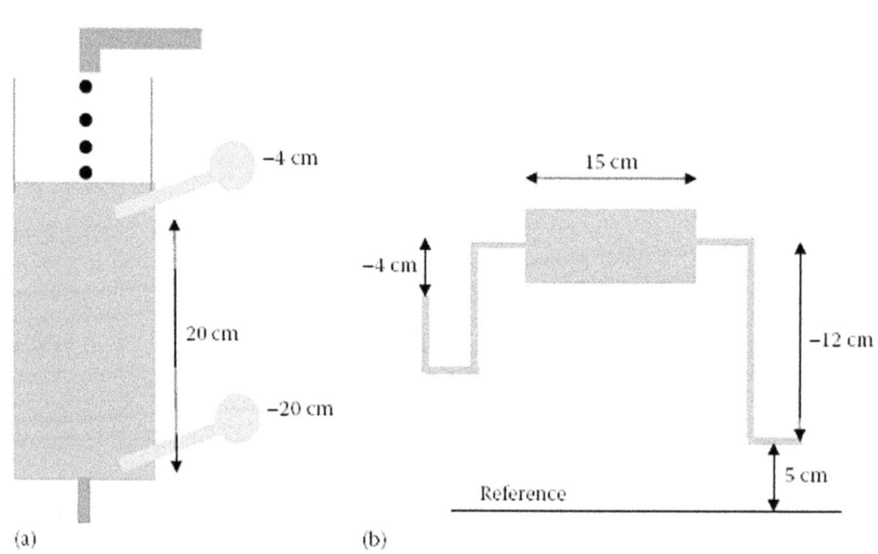

FIGURE 9.6 Steady (a) vertical and (b) horizontal water flow through an unsaturated soil column.

Example 9.6

Calculate the steady rate of flow through a 15 cm long horizontal soil column (Figure 9.6b). The unsaturated hydraulic conductivity of the soil is 3 cm/day, the head of the inlet is –4 cm and at the outlet –12 cm. The reference is 5 cm below the bottom of the outlet.

Total potential at the inlet = –4 + 17 = 13 cm
Total potential at the outlet = –12 + 17 = 5 cm

$$q = -K(\Phi)\left(\frac{\partial H}{\partial x}\right) = -3 \times \left(\frac{13-(5)}{15}\right) - 3 \times \frac{8}{15} = -1.6 \text{cm/day}$$

Note: For horizontal flow, gravity or gravitational potential is the same everywhere throughout the length of the soil column.

9.5.2 Continuity Equation

The continuity equation for the unsaturated flow of water through the soil profile is similar to that for the saturated flow, with the exception of the change in water content with time being not zero. Unsaturated soil pores are filled with air and water, and there is opportunity for water to fill the pore spaces and form hydrogen bonds:

$$\frac{\partial \theta}{\partial t} = -\left(\frac{\partial q_x}{\partial x} + \frac{\partial q_y}{\partial y} + \frac{\partial q_z}{\partial z}\right) \tag{9.30}$$

where

$$\frac{\partial \theta}{\partial t} \neq 0 \tag{9.31}$$

The conservation of mass for the unsaturated flow states that the rate of flow of water into an unsaturated soil is equal to the sum of the outflow and change in storage (Figure 9.7). The continuity equation for unsaturated flow is given as

$$\text{Inflow} = \text{Outflow} + \text{Change in storage}$$

$$q_x = (q_x + \Delta q_x) \tag{9.32}$$

or

$$\text{Net flow rate} = q_x - (q_x + \Delta q_x) \tag{9.33}$$

Writing Equation 9.32 as net volumetric flow:

$$\text{Net flow}_x = q_x \Delta y \Delta z - (q_x + \Delta q_x \Delta x)\Delta y \Delta z = -\frac{\partial q_x}{q_x} \Delta x \Delta y \Delta z \tag{9.34}$$

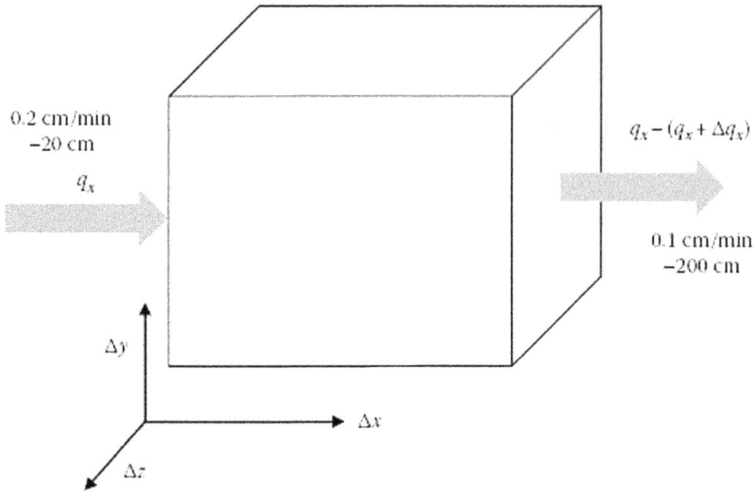

FIGURE 9.7 Continuity equation for water entering and leaving an unsaturated soil.

Equation 9.33 indicates that there is a net loss of water in the soil, indicating that the moisture content is increasing. However, there can be net gain of water that will indicate a decrease in soil water content.

Example 9.7

If water flow through a unit volume of soil (see Figure 9. 7) is only in one direction, and the inflow and outflow rate are 0.2 and 0.1 cm/min, respectively, what is the rate of change of water content?

$$\frac{\partial \theta}{\partial t}=-\frac{\partial q_x}{\partial x}=-\frac{0.1-0.2}{1}=\frac{0.1}{1}=0.01\,\text{min}^{-1}$$

Therefore, in 1 min, the change in the water content of soil is 0.1 cm^3/cm^3.

If the soil water potential at the inlet is −20 cm and at the outlet −200 cm (see Figure 9.7), what is the direction of flow?

$$q_x=-K(\Phi)\left(\frac{\partial h}{\partial x}\right)=-K(\Phi)\frac{-200-(-20)}{1-0}=-K(\Phi)\frac{-180}{1}=+180\,K(\Phi)$$

Since the sign is positive, it indicates that flow is taking place from the left to the right side of the flow domain in Figure 9.7.

9.5.3 Richards Equation for Transient Flow

The movement of water through unsaturated soils can be presented by a nonlinear partial differential equation also known as the Richards equation (Richards, 1931). The Richards equation for one-dimensional transient vertical flow of water through the unsaturated soil profile is formulated by transferring the Darcy–Buckingham equation into the continuity equation for unsaturated flow:

$$q_x = -K(\Phi)\left(\frac{\partial h}{\partial x}\right) \tag{9.35}$$

$$q_x = -K(\Phi)\left(\frac{\partial h}{\partial x} + 1\right) \tag{9.36}$$

$$\frac{\partial \theta}{\partial t} = -\frac{\partial q_x}{\partial x} \tag{9.37}$$

$$\frac{\partial \theta}{\partial t} = -\frac{\partial}{\partial x}\left(-K(\Phi)\left(\frac{\partial \Phi}{\partial x} + 1\right)\right) \tag{9.38}$$

$$\frac{\partial \theta}{\partial t} = \frac{\partial}{\partial x}\left(K(\Phi)\left(\frac{\partial \Phi}{\partial x}\right)\right) + \frac{\partial}{\partial x}\left(K(\Phi)\right) \tag{9.39}$$

Equation 9.38 describes the flow of water through unsaturated soil due to the matric potential gradient (first term on the right side of the equation) and gravity (second term on the right side of the equation). There are two forms of the Richards equation, namely, capacitance and diffusivity.

The capacitance form of the equation contains the soil water capacity function (C_w; cm^{-1}) (Lal and Shukla, 2004). C_w is the inverse of the slope of the soil water characteristic curve:

$$C_w(\Phi) = \frac{d\theta}{d\Phi} \tag{9.40}$$

$$\frac{\partial \theta}{\partial t} = \frac{\partial \theta}{\partial \Phi}\frac{\partial \Phi}{\partial t} = C_w(\Phi)\frac{\partial \Phi}{\partial t} \tag{9.41}$$

Since the left-hand side of Equation 9.40 and the Richards equation are the same, substituting the right-hand side of the Equation 9.40 into Equation 9.38 provides the capacitance form of the Richards equation as follows:

$$C_w(\Phi)\frac{\partial \Phi}{\partial t} = \frac{\partial}{\partial x}\left(K(\Phi)\left(\frac{\partial \Phi}{\partial x}\right)\right) + \frac{\partial}{\partial x}\left(K(\Phi)\right) \tag{9.42}$$

The soil water diffusivity, unsaturated hydraulic conductivity, and slope of soil water retention curve are related as follows:

$$D(\theta) = K(\theta)\frac{d\Phi}{d\theta} \tag{9.43}$$

Substituting Equation 9.42 in the first left-hand term of the Richards equation (Equation 9.38) and rearranging yields the diffusivity form of the Richards equation:

$$\frac{\partial\theta}{\partial t}=\frac{\partial}{\partial x}\left(D(\theta)\left(\frac{\partial\theta}{\partial x}\right)\right)+\frac{dK}{d\theta}\frac{\partial\theta}{\partial x} \tag{9.44}$$

The second term on the right-hand side of Equation 9.43 disappears for the horizontal flow. $D(\theta)$ is also constant for a small change in soil water content (θ). Under these conditions, Equation 9.43 reduces to

$$\frac{\partial\theta}{\partial t}=D(\theta)\frac{\partial^2\theta}{\partial x^2} \tag{9.45}$$

The Richards equation is not applicable to macropore flows, does not address hysteresis effects, and parameters K and Φ are functions of θ, making it a highly nonlinear equation and difficult to solve. For three-dimensional flows in the horizontal direction or when gravity is very small in comparison to the matric potential gradient, the Richards equation has the following form:

$$\frac{\partial\theta}{\partial t}=\frac{\partial}{\partial x}\left(K_x(\Phi)\left(\frac{\partial\Phi}{\partial x}\right)\right)+\frac{\partial}{\partial y}\left(K_y(\Phi)\left(\frac{\partial\Phi}{\partial y}\right)\right)+\frac{\partial}{\partial z}\left(K_z(\Phi)\left(\frac{\partial\Phi}{\partial z}\right)\right) \tag{9.46}$$

On the other hand, when gravity is dominant and flow is primarily taking place in the vertical direction, the one-dimensional Richards equation can be written as follows:

$$\frac{\partial\theta}{\partial t}=\frac{\partial}{\partial x}\left(K(\Phi)\left(1+\frac{\partial\Phi}{\partial x}\right)\right) \tag{9.47}$$

The sink and source term (S) can be added to the Richards equation to make it more general and applicable for situations where a subsurface drain is present in the experimental area. The S term can also take into account water uptake by plant roots. The Richards equation with the sink and source term is written as follows:

$$\frac{\partial\theta}{\partial t}=\frac{\partial}{\partial x}\left(K(\Phi)\left(\frac{\partial\Phi}{\partial x}\right)\right)+\frac{\partial}{\partial x}\left(K(\Phi)\right)-S \tag{9.48}$$

9.6 UNSATURATED HYDRAULIC CONDUCTIVITY MEASUREMENT

The unsaturated hydraulic conductivity of soil is highly variable both spatially and temporally. The unsaturated hydraulic conductivity can be measured in the laboratory and field using different methods as well as empirically using various functions available in the literature (Figure 9.2).

9.6.1 Laboratory Technique

The unsaturated hydraulic conductivity of soil can be determined on soil cores by conducting experiments similar to those for the saturated hydraulic conductivity determination. Water for the unsaturated hydraulic conductivity experiment is applied at a rate that does not cause ponding of water on the soil surface (Booltink and Bouma, 2002; Radcliff and Simunek, 2012). The total soil water potential can be measured at two places inside the soil matrix (or flow domain) using miniature tensiometers or water manometers. The volume of water applied in a given time is measured along with the potentials at two locations inside the core. The Darcy–Buckingham equation can determine the unsaturated hydraulic conductivity at different times during the experiment.

9.6.2 Field Method

In the field, unsaturated hydraulic conductivity can be determined using a tension infiltrometer, also known as a disk infiltrometer. The instrument consists of a water reservoir or tower and a disk or porous membrane that is placed on the soil surface with uniform and good contact. Water through the disk is applied to the soil surface at a rate slower than the rate at which water is ponded on the surface. This is accomplished by maintaining a negative pressure on the soil surface. A small negative pressure also ensures that the infiltration takes place through the soil matrix and not through soil cracks, biopores, or wormholes. The flow through the latter can only take place if a positive head of water (ponded water) is applied for infiltration because cracks, biopores, and wormholes are at atmospheric pressure. Reynolds and Elrick (1991) derived the unsaturated hydraulic conductivity in the middle of an interval between two applied pressure heads (Φ_1 and Φ_2), assuming α to be constant over this interval:

$$\alpha = \frac{\ln\left(Q(\Phi_2)/Q(\Phi_1)\right)}{\Phi_2 - \Phi_1} \tag{9.49}$$

Wooding's (1968) equation is used to calculate hydraulic conductivity from the steady-state infiltration rate for a homogeneous soil profile and uniform initial moisture content:

$$Q(\Phi) = \pi r^2 K(\Phi)\left(1 + \frac{4}{\pi r \alpha}\right) \tag{9.50}$$

where

- Q is the steady-state infiltration rate (L^3T^{-1}).
- r is the radius of the disk (L).

The limitations of Wooding's equation include the assumptions related to the homogeneous soil profile and uniform moisture content. Equation 9.49 has two unknown variables, $K(\Phi)$ and α, and two steady-state fluxes at different pressure heads are required. The Gardner (1958) exponential model can be used to calculate the unsaturated hydraulic conductivity for each pressure head using tension infiltrometer data:

$$K(\Phi) = K_s \exp(\alpha\Phi) \tag{9.51}$$

where

$K(\Phi)$ is the unsaturated hydraulic conductivity (LT^{-1}) for a given pressure head Φ (L).

K_s is the near saturated hydraulic conductivity (LT^{-1}).

α is the sorptive number (L^{-1}).

The unsaturated hydraulic conductivity values measured using a tension infiltrometer are presented in Table 9.4 for various potentials ranging from −30 to 0 cm. As soil becomes increasingly wet or soil water potential increases to zero, the unsaturated hydraulic conductivity also increases (Table 9.4).

9.6.3 Empirical Approach

The unsaturated hydraulic conductivity can be determined using empirical functions available in the literature. Gardner (1958) proposed the relationship between saturated and unsaturated hydraulic conductivity (repeated here):

TABLE 9.4
Unsaturated Hydraulic Conductivity Values for Chihuahuan Desert Sandy Soils Determined Using a Tension Infiltrometer at Ten Different Locations under Different Potential

			$K(\Phi)$ (cm/h)		
Test	**−30 cm**	**−20 cm**	**−10 cm**	**−5 cm**	**0 cm**
1	0.10	0.10	2.20	6.99	18.43
2	0.75	0.09	1.00	2.71	6.36
3	0.21	0.21	3.34	9.82	28.63
4	0.86	0.49	6.13	10.01	19.36
5	0.14	0.14	1.32	4.29	23.11
6	0.98	0.98	4.73	6.65	9.36
7	0.32	0.32	3.48	10.95	21.62
8	0.12	0.12	1.91	4.24	9.35
9	0.09	0.09	1.91	4.74	13.15
10	0.83	0.08	0.97	2.17	4.38

Source: Adhikari, P., et al., *Trans. ASABE,* 55(5), 1711, 2012.

$$K(\Phi) = K_s e^{\alpha\Phi} \tag{9.51}$$

α is a constant and depends on the soil type, and Φ is soil water potential (cm) and is always negative. At Φ of zero, $\exp(\alpha\ \Phi) = \exp(0) = 1$, and $K(\Phi)$ is equal to K_s. Since Φ is negative, the $K(\Phi)$ values drop as Φ decreases or becomes more negative. The drop in $K(\Phi)$ values from saturation are faster for sandy soils than clayey soils; therefore, α values are higher for clayey than sandy soils (Radcliff and Simunek, 2012).

The Campbell (1974) relationship between $K(\Phi)$ and K_s is as follows:

$$K(\theta) = K_s \left(\frac{\theta}{\theta_S}\right)^m \tag{9.52}$$

where m is an empirical constant or a fitting parameter. The Haverkamp equation has the following form (Haverkamp et al., 1977):

$$K(\Phi) = \frac{K_s}{1 + (\Phi / \alpha)^n} \tag{9.53}$$

where α (cm^{-1}) and n are fitting parameters. Some of the equations for predicting unsaturated hydraulic conductivity utilize pore size distribution. The combined van Genuchten and Mualem model for unsaturated hydraulic conductivity is also a popular model:

$$K(S_e) = K_s S_e^l \left(1 - \left(1 - S_e^{1/m}\right)^n\right)^2 \tag{9.54}$$

where m and n are empirical shape parameters. Some other empirical relationships are listed in Table 9.5 (Deb and Shukla, 2012).

There are also several computer programs available in the literature that predict the unsaturated hydraulic conductivity using the empirical equation listed in Table 9.5 and soil water content data or soil moisture characteristic curve. Mualem (1976) estimated the pore connectivity parameter (l) to be 0.5, Schaap and Leij (2000) as 1.0, and Burdine (1953) as 2.0 (Table 9.5).

Unsaturated hydraulic conductivity and soil water diffusivity are related to the macroscopic capillary length (λ_c) parameter (White and Sully, 1987; Nachabe, 1996; Radcliff and Simunek, 2012):

$$\lambda_c = \frac{1}{K(\Phi_0) - K(\Phi_1)} \int_{\Phi_1}^{\Phi_0} K(\Phi)\, d\Phi = \frac{1}{\Delta K} \int_{\theta_1}^{\theta_0} D(\theta)\, d\theta \tag{9.55}$$

where

Φ_0 is the pressure at or near saturation.
Φ_1 is the negative pressure head.

TABLE 9.5
Empirical Equations and Statistical Models for Unsaturated Hydraulic Conductivity [K (Φ) or K (θ)]

Hydraulic Conductivity Function	Reference
Empirical Equations	
$K(\Phi) = a\Phi + b$, where a and b are constants	Richards (1931)
$K(\theta) = S_e^n$, where $n = 3.5$, $S = (\theta - \theta_r)/(\theta_s - \theta_r)$	Averjanov (1950)
$K(\Phi) = a(\Phi)^{-b}$, where a and b are constants	Wind (1955)
$K(\theta) = a\theta^n$; $K(\Phi) = a/(b + \Phi^n)$; $K(\Phi) = K_s/[1 + (\Phi/\Phi_c)^n]$, where a, b, and n are constants	Gardner (1958)
$K(\Phi) = K_s$ for $\Phi \le \Phi_{aev}$; $K(\Phi) = (\Phi/\Phi_{aev})^{-\lambda}$ for $\Phi \ge \Phi_{aev}$, where λ is the pore size distribution index	Brooks and Corey (1964)
$K(\Phi) = K_s$ for $\Phi \le \Phi_{aev}$	Rijtema (1965)
$K(\Phi) = \exp[a(\Phi - \Phi_{aev})]$ for $\Phi_{aev} \le \Phi \le \Phi_1$, where a is a constant	
$K(\Phi) = K(\Phi)_1(\Phi/\Phi)^{-n}$ for $\Phi > \Phi_1$, where n is a constant	
$K(\theta) = K_s \exp[a(\theta - \theta_s]$, where a is a constant	Davidson et al. (1969)
$K(\theta) = K_s(\theta/\theta_s)^{2b+3}$; $K(\Phi) = K_s\,(\Phi_{aev}/\Phi)^{2+3/b}$, where b is the exponent of moisture release equation	Campbell (1974)
$K(\theta) = 1/\,\{\ln[e + (\Phi/A)^B]\}^C$, where A, B, and C are constants	Leong and Rahardjo (1997)
Statistical Models	
$K(S_e) = K_s S_e^l \dfrac{g(S_e)}{g(l)}, g(S_e) = \int_0^{S_e} \dfrac{1}{[\Phi(x)]^2} dx,$ where l is the pore connectivity parameter ($l = 2$)	Burdine (1953)
$K(S_e) = K_s S_e^l \left[\dfrac{f(S_e)}{f(l)}\right]^2$, in which $f(S_e) = \int_0^{S_e} \dfrac{1}{\Phi(x)} dx,$ where l is the pore connectivity parameter ($l = 0.5$)	Mualem (1976)

Notes: K_s is the saturated hydraulic conductivity, Φ is the matric suction head, Φ_{aev} is the bubbling or air-entry suction, Φ_c is the suction head at which $K = K_s/2$, Φ_1 is the soil residual suction, $K(\Phi)_1$ is $K(\Phi)$ at $\Phi = \Phi_1$, θ is the volumetric water content, θ_s is the saturated water content, and θ_r is the residual water content.

Using a tension infiltrometer, sorptivity and λ_c can be determined in the field. According to Warrick and Broadbridge (1992), capillary length can be defined as the integral of the unsaturated hydraulic conductivity with respect to the matric potential normalized by the difference between the conductivity at the two limits of integration. λ_c is an important parameter in time to reach steady-state flow during infiltration of water through the soil. The λ_c is also represented as the inverse of the sorptive number to solve the Richards equation (Woodings, 1968; Nachabe, 1996).

In the Gardner (1958) equation, λ_c is inversely related to α. λ_c is known as hydraulic conductivity weighted mean soil water potential and can also be described as a mean height of capillary rise above the water table (Raats and Gardner, 1971).

9.7 SOIL WATER DIFFUSIVITY MEASUREMENT

Soil water diffusivity can be assumed to be constant for a small range of wetness. Thus, the Richards diffusivity equation without a gravity term can be analogous to Fick's second law of diffusion as follows:

$$\frac{\partial \theta}{\partial t} = D(\theta)\frac{\partial^2 \theta}{\partial x^2} \tag{9.56}$$

Soil water diffusivity can be measured in the laboratory using outflow methods (Gardner, 1956). In the single-outflow method, a core sample is first saturated and then subjected to small increments of pressure; the volume of outflow is recorded at equilibrium for each successive pressure increment. The soil water diffusivity can be obtained by solving the diffusivity form of Richards equation neglecting gravity as follows (Passioura, 1996):

$$D(\theta) = \frac{dQ}{dW}\frac{L^2}{2} \tag{9.57}$$

where

Q is the rate of outflow.
W is the amount of water remaining in soil at any given time.
θ is the water content at $x = L$.

The experiment can be conducted in a Tempe cell, and the outflow can be recorded in a burette or by an automatic weighting system. Details of the method can be seen in Green et al. (1998). There are several numerical programs available in the literature such as RETC (van Genuchten et al., 1991) that can be used to estimate soil water diffusivity from a known soil water retention curve.

9.8 FLOW THROUGH LAYERED SYSTEM

Most of the derivations for water flow through a soil profile assume a homogeneous and isotropic flow domain. These assumptions are important simplifications of otherwise complex porous media and lead to an easier and better understanding of the fundamentals of flow through porous media. Newer soils such as Entisols could have low variability with regard to the vertical distribution of soil types. However, for well-developed soils (e.g., Mollisols, Alfisols), different types of soil occur commonly as separate layers with depth. These different layers are seldom perfectly horizontal to each other but can vary greatly with regard to their thickness below the soil surface. However, the vertical heterogeneity of soil can be simplified into

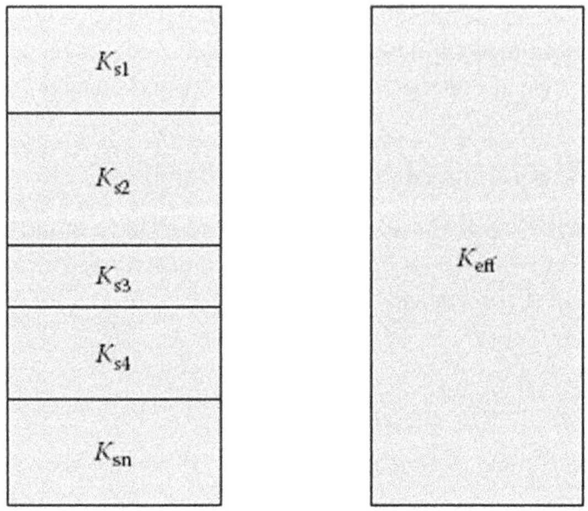

FIGURE 9.8 Effective hydraulic conductivity for a layered soil system. Note that for simplification, all layers are assumed to be parallel.

several different horizontal layers of soils (Figure 9.8). The soil hydraulic properties, for example, K_s, can be determined for each layer of known thickness (L_i) separately, and the effective K_s (K_{eff}) for the soil can be obtained as follows:

$$K_{eff} = \frac{\sum_{i=1}^{n} L_i}{\sum_{i=1}^{n} \left(L_i / K_{si}\right)} \tag{9.58}$$

Equation 9.58 provides a weighted average value of K_s, where the thickness of each corresponding layer is taken into account and provides a better estimate of the effective K_s than that obtained by the simple arithmetic average for layers of unequal thicknesses.

9.9 MODELS FOR SIMULATING FLOW THROUGH SOIL SYSTEM

There are several models available in the literature that can simulate the flow of water through saturated and/or unsaturated soil systems. Hydrus-1D, 2D, or 3D computer models (Simunek et al., 2006) can simulate water and energy transport, solute transport, and root water uptake for a variety of land use and management scenarios. The Root Zone Water Quality Model (Ahuja et al., 2000), a physically based model, can simulate the water and energy transport, chemical movement, snow dynamics, and plant growth processes. For a list of other similar programs, readers are referred to Deb and Shukla (2012) and Radcliff and Simunek (2012).

PROBLEMS

9.1 Describe Darcy's law and its limitations.
9.2 If the length of the soil column is 10 cm, constant head of water at the inlet above the reference is 15 cm, the head at the outlet is 2 cm, and saturated hydraulic conductivity of the soil is 1.5 cm/s, calculate the rate of flow at the outlet.
9.3 In Problem 9.2, if the volumetric water content of the soil in the column is 0.3 cm^3/cm^3, calculate the rate of flow through the pores.
9.4 Describe the role of biopores and micropores in water movement through soil.
9.5 Explain the difference when pores are open at the surface and not.
9.6 Calculate the saturated hydraulic conductivity of soil in vertical and horizontal directions. Length of flow is 10 cm, difference between the head at the inlet and head at the outlet is 15 cm, and 20 cm of water was collected in 20 minutes.

REFERENCES

Adhikari, P., M. K. Shukla, and J. Mexal. 2012. Spatial variability of infiltration rate and sodium content of Desert soils: Implications for management of irrigation using treated wastewater. *Trans. ASABE* 55(5): 1711–1721.
Ahuja, L. R., K. W. Rojas, J. D. Hanson, M. J. Shaffer, and L. Ma. 2000. *The Root Zone Water Quality Model: Modeling Management Effects on Water Quality and Crop Production*. Water Resources Publications, Highlands Ranch, CO.
Averjanov, S. F. 1950. *About Permeability of Subsurface Soils in Case of Incomplete Saturation. English Collection. Vol. 7, as Quoted by P. Y. Palubarinova-Kochina, 1962, The theory of ground water movement (English Translation by I. M. Roger DeWiest)*, Princeton University Press, Princeton, NJ, p. 613.
Bear, J. 1972. *Dynamics of Fluids in Porous Media*. American Elsevier, New York, p. 764.
Booltink, H. W. G. and J. Bouma. 2002. Steady flow soil column method. In: J. H. Dane and G. C. Topp (eds.) *Methods of Soil Analysis, Part 4. Physical Methods*. SSSA Book Series, no. 5, pp. 812–816.
Brooks, R. H. and A. T. Corey. 1964. *Hydraulic Properties of Porous Media*. Hydrology Paper 3, Colorado State University, Fort Collins, CO, http://digitool.library.colostate.edu///exlibris/dtl/d3_1/apache_media/L2V4bGlicmlzL2R0bC9kM18xL2FwYWNoZV9tZWRpYS8xNTIyNTc=.pdf, accessed August 1, 2013.
Burdine, N. T. 1953. Relative permeability calculation from pore–size distribution data. *J. Petroleum Technol.* 5: 71–78.
Campbell, G. S. 1974. A simple method for determining unsaturated conductivity from moisture retention data. *Soil Sci.* 117: 311–314.
Davidson, J. M., L. R. Stone, D. R. Nielsen, and M. E. Larue. 1969. Field measurement and use of soil-water properties. *Water Resour. Res.* 5: 1312–1321.
Deb, S. and M. K. Shukla. 2012. Variability of hydraulic conductivity due to multiple factors. *Am. J. Environ. Sci.* 8(5): 489–502.
Ernst, L. F. 1950. *A New Formula for the Calculation of the Permeability Factor with the Augerhole Method*. T. N. O., Groningen, the Netherlands.
Fetter, C. W. 2001. *Applied Hydrogeology*, 4th edn. Prentice Hall, Inc., Upper Saddle River, NJ.

Gardner, W. R. 1956. Calculation of capillary conductivity from pressure plate outflow data. *Soil Sci. Soc. Am. J.* 20: 317–320

Gardner, W. R. 1958. Some steady-state solutions of unsaturated moisture flow equations with application to evaporation from a water table. *Soil Sci.* 85: 228–232.

Green, T. W., Z. Paydar, H. P. Cresswell, and R. J. Drinkwater. 1998. *Laboratory Outflow Technique for Measurement of Soil Water Diffusivity and Hydraulic Conductivity.* Technical report no. 10/98. CSIRO, Highett, Victoria, Australia.

Haverkamp, R., M. Vauclin, J. Tourna, P. Wierenga, and G. Vachaud. 1977. Comparison of numerical simulation models for one-dimensional infiltration. *Soil Sci. Soc. Am. J.* 41: 285–294.

Hazen, A. 1911. Discussion of 'dams on sand foundations' by A. C. Koenig. *Trans. Am. Soc. Civ. Eng.* 73: 199–203.

Kirkham, D. and C. H. M. van Bavel. 1949. Theory of seepage into auger holes. *Soil Sci. Soc. Am. J.* 13: 75–82.

Lal, R. and M. K. Shukla. 2004. *Principles of Soil Physics*, 1st edn. Marcel Dekker Inc., New York, ISBN-10: 0824753240, p. 716.

Leong, E. C. and H. Rahardjo. 1997. Permeability functions for unsaturated soils. *J. Geotech. Geoenviron. Eng.* 123: 1118–1126.

Mualem, Y. 1976. A new model for predicting the hydraulic conductivity of unsaturated porous media. *Water Resour. Res.* 12: 513–522.

Nachabe, M. H. 1996. Macroscopic capillary length, sorptivity, and shape factor in modelling the infiltration rate. *Soil Sci. Soc. Am. J.* 60: 957–962.

Pachepsky, Y. and W. J. Rawls. 2004. Development of pedotransfer functions in soil hydrology. In: *Development in Soil Science. Vol. 30. Elsevier B. V.*, San Diego, CA, p. 512.

Passioura, J. B. 1996. Determining soil water diffusivity from one-step outflow experiments. *Aust. J. Soil Res.* 15: 1–8.

Raats, P. A. C. and W. R. Gardner. 1971. Comparison of empirical relationships between pressure head and hydraulic conductivity and some observations on radially symmetric flow. *Water Resour. Res.* 7: 921–928.

Radcliff, D. E. and J. Simunek. 2012. Water flow in soils. In P. M. Huang, Y. Li, and M. E. Sumner (eds.), *Handbook of Soil Sciences Properties and Processes*, 2nd edn. CRC Press, Boca Raton, FL.

Reynolds, W. D. and D. E. Elrick. 1991. Determination of hydraulic conductivity using a tension infiltrometer. *Soil Sci. Soc. Am. J.* 55: 633–639.

Richards, L. A. 1931. Capillary conduction of liquids through porous mediums. *Physics* 1: 318–333.

Rijtema, P. E. 1965. An analysis of actual evapotranspiration. *Agric. Res. Rep.* 659: 1–107.

Schaap, M. G. and F. J. Leij. 2000. Improved predictions of unsaturated hydraulic conductivity with Mualem-van Genuchten model. *Soil Sci. Soc. Am. J.* 64: 843–851.

Schaap, M. G., F. J. Leij, and M. Th. Van Genuchten. 2001. Rosetta: A computer program for estimating soil hydraulic parameters with hierarchical pedotransfer functions. *J. Hydrol.* 251: 163–176.

Šimůnek, J., M. Th. van Genuchten, and M. Šejna. 2006. *The HYDRUS Software Package for Simulating Two- and Three-Dimensional Movement of Water, Heat, and Multiple Solutes in Variably-Saturated Media, Technical Manual*, Version 1.0. PC Progress, Prague, Czech Republic, p. 241.

Smith, R. M. and D. R. Browning. 1946. Some suggested laboratory standards of subsoil permeability. *Soil Sci. Soc. Am. Proc.* 11: 21–26.

Soil Moisture Equipment Corporation. 2012. Guelph permeameter. Operating instruction. http://www.soilmoisture.com/pdf/82800k1.pdf, accessed on August 9, 2012.

van Genuchten, M. Th., F. J. Leij, and S. R. Yates. 1991. *The RETC Code for Quantifying the Hydraulic Functions of Unsaturated Soils, EPA/600/2-91/065*. US Environmental Protection Agency, Ada, OK.

Warrick, A. W. and P. Broadbridge. 1992. Sorptivity and macroscopic capillary length relationships. *Water Resour. Res.* 28(2): 427–431. doi: 10.1029/91WR02599

White, I. and M. J. Sully. 1987. Macroscopic and microscopic capillary length and time scales from field infiltration. *Water Resour. Res.* 23(8): 1514–1522. doi: 10.1029/WR023i008p01514

Wind, G. P. 1955. Field experiment concerning capillary rise of moisture in heavy clay soil. *Neth. J. Agric. Sci.* 3: 60–69.

Wooding, R. A. 1968. Steady infiltration from a shallow circular pond. *Water Resour.* 4: 1259–1273.

10 Water Infiltration into the Vadose Zone

10.1 INTRODUCTION

Knowledge of water infiltration characteristics of soils is important for identifying water conservation strategies and runoff and erosion control (Shukla et al., 2003b). Water applied on the soil surface as precipitation or irrigation can pond on the surface, infiltrate into the soil, or flow as surface runoff. A combination of these processes can also occur. Infiltration is the entry of water into the vadose zone through the soil–air interface. Infiltration determines how much of the total amount of applied water in the form of precipitation or irrigation on the soil surface can enter the soil storage, and how much of it flows away as surface runoff. Infiltration of water is rapid when the vadose zone soil is dry, but it rapidly decreases with time as soil becomes increasingly wet (matric potential increases) until it reaches a steady state or a nearly constant rate. High precipitation sometimes does not cause surface runoff instantly because the precipitation rate is smaller than the infiltration rate of soil. Over time, as soil becomes wetter, the infiltration rate into the soil decreases, and the rate of precipitation can become greater than the infiltration rate of soil, causing surface runoff. In addition to the antecedent soil water content and the precipitation rate, several factors can affect infiltration of water into soil. Some of these include soil structure; soil texture; soil depth; size, shape, and connectivity of pores; worm activities; soil organic matter; existing land use and management practices; and prevailing cropping system.

Infiltration into the soil is governed by two factors: capillarity and gravity. Smaller pores can move the water with or against gravity and influence the infiltration rate of the soil. The infiltration rate is sometimes also referred to as infiltration capacity. The units of infiltration rate are expressed as depth of water infiltrated (*L*; cm) in a given amount of time (*T*; s). The infiltration rate (*L*/*T*; cm/s) or infiltration capacity (*L*/*T*) is the flux of water that soil can absorb through its surface at the soil–air interface. In this chapter, the processes of infiltration, measurement techniques, management of infiltration, and some infiltration models are discussed.

10.2 PROCESS OF INFILTRATION

The process of infiltration can be classified as flux controlled or profile controlled. As long as the precipitation or irrigation rate is below the infiltration rate of soil, all the water infiltrates into the soil. The supply rate (or application rate) of water determines the infiltration rate, and the process of infiltration is known as flux controlled. On the other hand, when the rate of application of water is higher than the infiltration rate of soil, the water transmission or hydraulic properties of the soil profile control

DOI: 10.1201/9780429264849-10

the infiltration into the soil. The process of infiltration in this case is known as profile controlled.

Time-dependent infiltration is governed by the Richards (1931) equation and is dependent on the surface and lower boundary conditions of soil, antecedent soil water content, and rate of application. The initial water content varies with depth and generally increases with depth unless precipitation occurred or irrigation was recently applied. Infiltration continues until the entire soil profile becomes saturated, and at this time, surface ponding may start. These conditions can be represented as follows (ASCE, 1996):

$$-K(\Phi)\frac{\partial \Phi}{\partial x}+1=R\theta(0,t)\leq\theta_s \tag{10.1}$$

$$\Phi=\Phi_1\,\theta(0,t)=\theta_s \tag{10.2}$$

where

R is the rainfall intensity that could vary with time (L/T).

Φ_1 is positive potential indicating slight ponding of water on the soil surface (L).

θ_s is saturated water content (L^3/L^3).

t is time.

One or both conditions are generally applicable for most situations except for sloping soils, where infiltration and runoff can simultaneously occur. The lower boundary condition is dependent on the depth of the soil profile. In the presence of a water table, a constant can be assumed at the water table; however, when no shallow water table is present or the soil profile is very deep, a common assumption is to use the unit hydraulic gradient flow. These two boundary conditions are expressed as follows (ASCE, 1996):

$$\Phi(z,t)=0\;t>0 \tag{10.3}$$

$$q(z,t)=K_e(\theta,z)t>0 \tag{10.4}$$

where

q is the flux (cm/s).

K_e is the effective soil conductivity (cm/s).

z is the depth of soil (cm) at which the boundary condition is applied.

As water falls on the soil surface, it enters and starts filling the pores near the soil surface, creating a zone that is water saturated (Figure 10.1). However, the wetting front is not the boundary of the saturated zone. The liquid wetting front is ahead, and this zone is known as the transmission zone. The movement of water in this zone

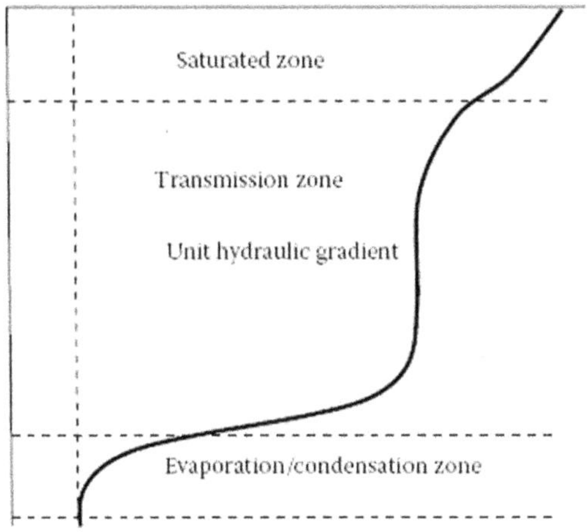

FIGURE 10.1 Infiltration zones as liquid water front moves ahead in the soil.

usually occurs under unit hydraulic gradient. The water content changes continuously in this zone. The liquid water at the end of the boundary of the transmission zone can get evaporated and move forward as vapor flow, and vapor can condense subsequently. Thus, there are evaporation and condensation zones ahead of the liquid wetting front. Temperature usually rises inside the soil profile due to the heat of wetting and the latent head of condensation. The wetting front is usually smooth in heavy textured or clayey soils, but flow could be fingered in sandy soils. The wetting front moves faster in uniform coarse-textured soils than uniform fine-textured soil. However, when soil is layered with coarse-textured soil overlying the fine-textured soil, the movement of water into and through the underlying fine-textured soil slows.

10.3 MEASUREMENT OF INFILTRATION

The infiltration rate of soil can be measured using a double-ring infiltrometer, single-ring infiltrometer, tension infiltrometer, or rainfall simulator. Double- and single-ring infiltrometers can be used to measure the one-dimensional or three-dimensional infiltration rate of soil, respectively. Use of a double-ring infiltrometer is a cheap and reliable method for determining the infiltration rate of soil under positive head created by ponding of water on the soil surface. Both cylinders are inserted into the soil to about 4–5 cm deep, and water is filled in both cylinders up to a given height. The height of water is maintained in the inner cylinder, while the outer cylinder is not allowed to drain completely to ensure one-dimensional flow from the inner cylinder.

Example 10.1

Infiltration experiments were conducted in a conventionally tilled agricultural field in Coshocton, Ohio, using a double-ring infiltrometer with an outer diameter of

27 cm and an inner diameter of 15 cm. The drop of water level in the inner ring (column III; measured using the scale attached inside the cylinder), water added to bring it back to the previous level (column IV), and the cumulative time at which the measurement is made (column I) are given in the first four columns. Calculate the cumulative infiltration and infiltration rate of soil and plot them with respect to cumulative time.

Instantaneous time is calculated by subtracting data in row$_i$ from row$_{i-1}$ of column I. The infiltration depth was calculated by subtracting data in row$_i$ from row$_{i-1}$ of column III. However, when water is added the data in row$_{i+1}$ of column III was subtracted with the amount of water added appearing in column IV. The cumulative infiltration in column VI is obtained by adding row $i + 1$ of column V to row i of column VI. The infiltration rate is calculated by dividing column V and column II.

Cumulative Time (min)	Instantaneous Time (min)	Infiltrometer Reading (cm)	Water Filled up to (cm)	Infiltration Depth (cm)	Cumulative Infiltration (cm)	Infiltration Rate (cm/min)
I	II	III	IV	V	VI	VII
				Columns		Columns V/II
0	0	1		0	0	0
2	2	1.2		0.2	0.2	0.100
5	3	1.8		0.6	0.8	0.200
10	5	2.4		0.6	1.4	0.120
15	5	3		0.6	2	0.120
20	5	3.5		0.5	2.5	0.100
30	10	4.5	2	1	3.5	0.100
45	15	3.4		1.4	4.9	0.093
60	15	4.7	2	1.3	6.2	0.087
75	15	3.5		1.5	7.7	0.100
90	15	4.5	1.5	1.3	9	0.087
120	30	4	1.6	2.5	11.5	0.088
150	30	3.6		2	14.5	0.100
180	30	6.2		2.6	17.1	0.087

The cumulative infiltration and infiltration rate with time are presented in Figure 10.2.

The tension infiltrometer can be used to obtain the infiltration rate of soil under positive, zero, and negative heads. Tension infiltrometer tests can be conducted using disks of various diameters at both positive and negative potentials. Tests are usually started at the lowest potential. Before the start of a test, the test site is made perfectly flat by removing debris and aboveground plant material or gravel. Sometimes a thin

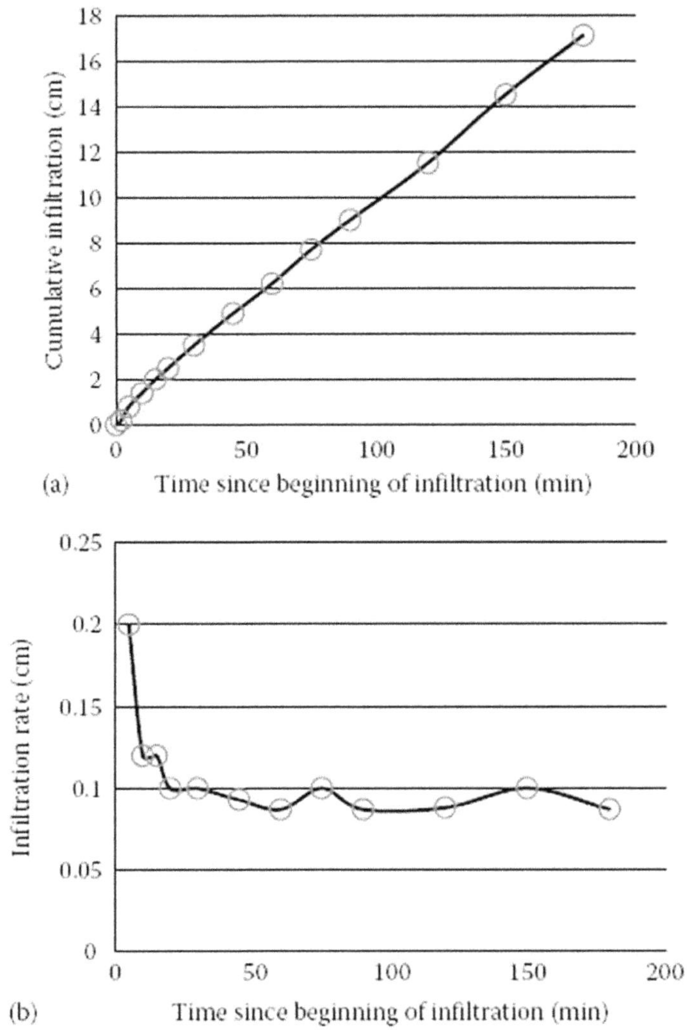

FIGURE 10.2 (a) Cumulative infiltration and (b) infiltration rate of soil for a conventionally tilled soil.

(1 cm) layer of sand is spread to ensure good contact between soil and the bottom of the disk. Details are discussed in Chapter 9.

Rainfall simulation experiments can be performed to determine the infiltration rate of soil. An experimental area is selected and water is applied via overhead sprinklers. The entire area is usually covered with plastic to reduce evaporation. The surface runoff generated is directed toward and collected at the outlet. The amount of water infiltrated into the soil is calculated by subtracting the amount of water applied via sprinklers from the amount of runoff collected at the outlet. Some of the advantages of rainfall simulation experiments are that they measure infiltration into a much larger area or volume, and therefore, values are more representative. Experiments can be conducted under a variety of precipitation rates to represent local conditions. These experiments are particularly valuable for areas with slopes where direct measurement of infiltration excess is made, while additional estimates on soil erosion are also obtained in the runoff collected. These experiments can also be conducted in laboratory settings, where different types of soils, different compaction rates, aggregation, and soil covers can be evaluated side by side for different slopes, slope lengths, and fixed or variable rainfall intensities. However, more time, energy, money, and skill are required to perform these tests reliably.

10.4 FACTORS AFFECTING INFILTRATION RATE

There are several factors, both natural and anthropogenic, that can affect the rate of infiltration of water into the soil. The duration of precipitation is an important factor affecting infiltration. The infiltration rate, unless soil is already saturated, is usually higher initially, but with time it decreases and eventually the rate of change becomes nearly constant. This behavior is due to several factors, for example, decrease in the potential gradient as soil gets wetter. Large pores also fill up first, followed by the entry of water into smaller pores. In addition, depending upon the rainfall intensity, soil erosion, surface crusting, or deposition of fine particles into the pores can happen, resulting in a decrease in the infiltration rate. Soil can also swell with increasing soil water content and reduce the size of pores.

Soil texture has an important influence on the infiltration rate of water into the soil. As the soil texture becomes finer, silt and clay content increases and the infiltration rate into the soil decreases. Soil surface conditions are important for infiltration. Good soil structure promotes infiltration, while poor soil structure diminishes it. The presence of an impeding soil layer with a different texture than the layer above it can also affect the infiltration rate greatly. If the impeding layer is predominantly clay, the lower hydraulic conductivity associated with clay slows the flow of water. On the other hand, a sandy layer can also decrease the flow due to the lower unsaturated hydraulic conductivity of soil. Flow rate into the sandy soil can only begin after there is a small and sufficient positive soil water head buildup for water to move into the sand, which could be at atmospheric (or zero) potential.

The soil infiltration rate depends on the continuity, connectivity, size, and distribution of pores. Soil physical properties, including bulk density, porosity, and hydraulic conductivity, greatly influence the infiltration rate into the soil. As soil bulk density increases, porosity and the infiltration rate decrease. If soil hydraulic conductivity is high, the final steady-state infiltration rate into the soil is also high. Soil organic matter content in the soil affects soil structure, porosity, and bulk density, and has been reported to increase the infiltration rate of soil.

The land use and management system greatly influences the infiltration of water into soil. Several accounts are available on the influence of conventional tillage on the infiltration rate. Generally, immediately after the tillage, soil porosity is high, soil water content is low, and the infiltration rate is high. However, as soil settles, the infiltration rate generally decreases. Intensive tillage can cause loss of soil structure, loss of organic matter, and breakdown of soil aggregates, resulting in reductions in water infiltration into the soil (Shipitalo and Edwards, 1996). On the other hand, no-tillage can increase the organic matter content of soil, improve soil aggregation, and increase infiltration into the soil (Shukla et al., 2003a, 2003b; Shukla and Lal, 2005). Figure 10.3 shows the infiltration–time curve for two fields—a no-till and a conventionally tilled field—in Coshocton, Ohio.

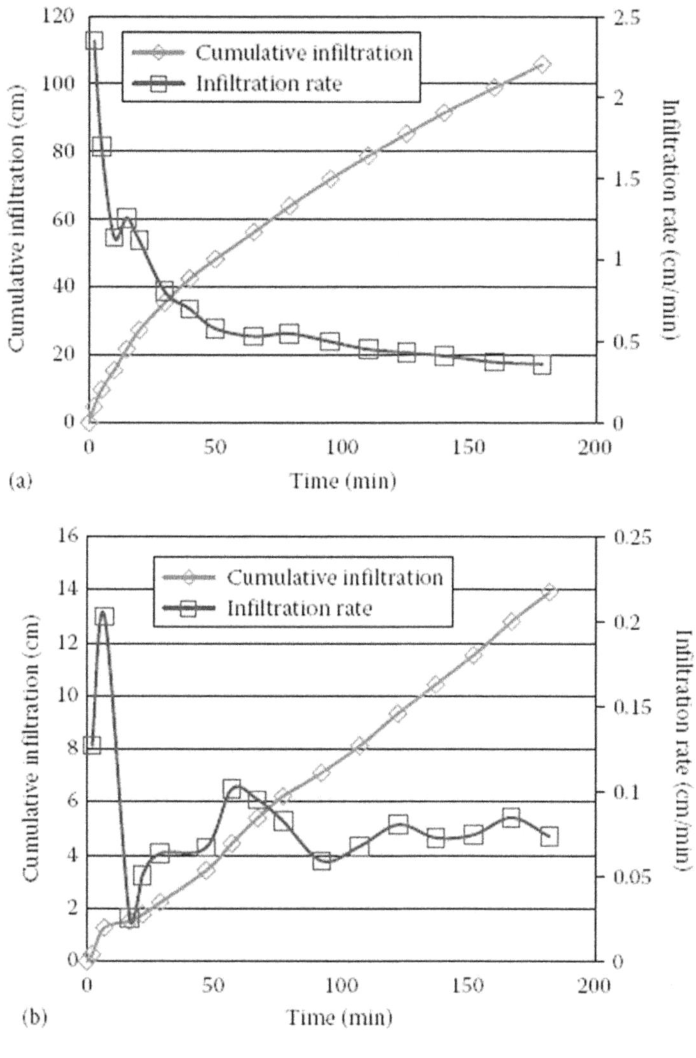

FIGURE 10.3 Cumulative infiltration and infiltration rate of (a) no-till soil and (b) a conventionally tilled soil. *Note*: The scale on the y-axis for cumulative infiltration is about seven times lower in panel b than in panel a.

The cumulative infiltration from the no-till field (106 cm) is almost 7.5 times higher than that from the conventionally tilled field (14 cm). Similarly, the steady-state infiltration rate is four times higher in the no-till field (20 cm/min) than the conventionally tilled field (5 cm/min). High cumulative infiltration is also reported for soils under forest cover with values ranging between 167.07 ± 50.6 under a canopy of piñons, 91.0 ± 34.1 under a canopy of junipers, and 91.0 ± 15.4 under the canopy of oak trees in a semiarid region in the western part of Sugarite Canyon, northeast of Raton, Colfax County, New Mexico (Shukla et al., 2006).

Drastic disturbance to soils, for example, due to surface mining activities can decrease infiltration into the soil (Shukla et al., 2004). The studies conducted to compare the infiltration rates of undisturbed and reclaimed drastically disturbed soil in Ohio have shown that the total infiltration could be as much as two times greater in undisturbed than reclaimed soil (Table 10.1). The infiltration rate at 5 min, steady-state infiltration rate, and hydraulic conductivity of soil are also usually higher for undisturbed soil likely due to the intact well-developed soil pore structure.

10.5 INFILTRATION MODELS

The infiltration of water into soil can be described by solving the Richards (1931) partial differential equation for water flow through the unsaturated soil. Several accounts are available with analytical and numerical solutions of the Richards equation for several boundary conditions (Philip, 1957; Kunze and Nielsen, 1982). Although laboratory- and field-based experimental methods are available to measure saturated hydraulic conductivity and soil water diffusivity as a function of soil water content, solutions to the Richards (1931) equation are not applied directly to the infiltration rate measurements (Parlange et al., 1982; Kutilek and Nielsen, 1994). The Burgers equation (1948) employs a nonlinear quadratic conductivity (k) – θ relationship while

TABLE 10.1
Saturated Hydraulic Conductivity (K_s), Infiltration Rate 5 min after the Start of Infiltration (i_5), Steady-State Infiltration Rate (i_c), and Cumulative Infiltration Rate (I) from Unmined Soil (UMS) and Reclaimed Minesoil (RMS) under Different Fertility Treatments (FL)

Treatment	K_s (cm/min)	i_5 (cm/min)	i_c (cm/min)	I (cm)
UMS	0.34 ± 0.19	0.63 ± 0. 14a	0.24 ± 0.1a	45.88 ± 15.58a
RMSFL1	0.15 ± 0.13	0.21 ± 0. 17b	0.05 ± 0.04b	11.85 ± 8.21b
RMSFL2	0.22 ± 0.2	0.22 ± 0. 11b	0.06 ± 0.02b	12.75 ± 4. 0b
RMSFL3	0.09 ± 0.14	0.38 ± 0.31ab	0.16 ± 0.14ab	28.78 ± 27.98ab
LSD (0.05)	NS	0.34	0.16	29.16

Source: Shukla, M. K., et al., *Soil Sci. Soc. Am. J.*, 68, 1352, 2004.
Note: Different letters indicate significant difference at alpha = 5%.

keeping D a constant and independent of θ (Clothier et al., 1981). In this section, several simple physically based and empirical infiltration equations available in the literature are described (Haverkamp et al., 1977; Shukla et al., 2003c).

10.5.1 Green and Ampt Equation

The Green and Ampt (1911) infiltration equation is based on Darcy's law. Several simplifying physical assumptions are made to develop a mathematical relationship for water infiltration calculations. The water content profile at time t during infiltration is considered to be a step-like profile, and infiltration into the soil is assumed to be a piston flow going progressively deeper with time. The wetted soil profile at time t is replaced with a uniformly wetted region of thickness L (Figure 10.4). The soil profile is considered homogeneous and isotropic; therefore, the hydraulic conductivity at saturation (K_s) is unique.

Infiltration into the soil for a uniform and homogeneous profile (L) as shown in Figure 10.4 can be given by

$$I = (\theta_0 - \theta_i)L = \Delta\theta L \tag{10.5}$$

where

θ_0 is the water content of the soil profile of thickness L (near-saturated water content) after the start of infiltration.

θ_i is the antecedent soil water content assumed to be uniform in Figure 10.4.

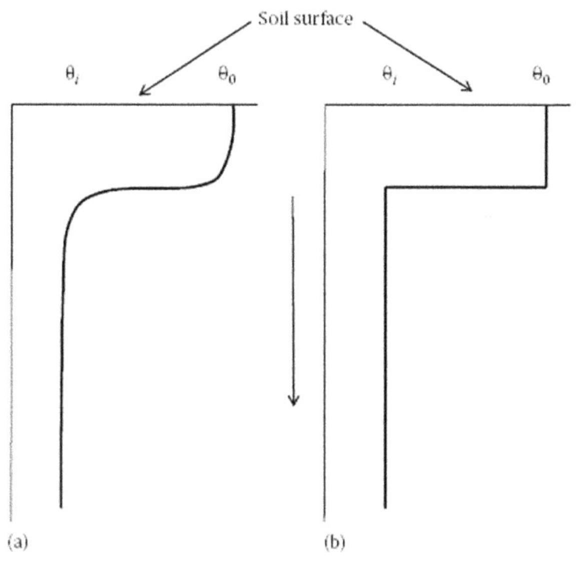

FIGURE 10.4 A soil water profile (a) during infiltration of water into the soil and (b) Green and Ampt step soil water profile.

The infiltration rate of water into the soil can be obtained by differentiating Equation 10.5 with respect to time, as follows:

$$i = \frac{d}{dt}\left[\left(\theta_0 - \theta_i\right)L\right] = \Delta\theta \frac{dL}{dt} \tag{10.6}$$

Darcy's equation for the soil profile in Figure 10.4 can be written as follows:

$$q = -K_s \frac{\Phi_{\text{wf}} - \Phi_0}{L} = K_s \frac{\Delta\Phi}{L} \tag{10.7}$$

where Φ_{wf} and Φ_0 are soil water potential at the wetting front and antecedent potential, respectively.

The right-hand sides of Equations 10.6 and 10.7 are equal:

$$\Delta\theta \frac{dL}{dt} = K_s \frac{\Delta\Phi}{L} \tag{10.8}$$

Rearranging and integrating the equation to the limits of 0 to L and 0 to t yields the following equation:

$$\frac{L^2}{2} = K_s \frac{\Delta\Phi}{\Delta\theta} t = D_0 t \tag{10.9}$$

where

$$D_0 = K_s \frac{\Delta\Phi}{\Delta\theta}$$

$$L = \left(2D_0 t\right)^{0.5} \tag{10.10}$$

or from Equation 10.5

$$I = \Delta\theta\left(2D_0 t\right)^{0.5} \tag{10.11}$$

or

$$i = \frac{dl}{dt} = \Delta\theta \sqrt{\left(\frac{D_0}{2t}\right)} \tag{10.12}$$

Equation 10.12 indicates that for horizontal infiltration in the absence of gravity, the infiltration rate is proportional to $t^{-0.5}$. According to Philip (1969), if diffusivity is a predominant factor at the wet end of the moisture range, the Green–Ampt equation corresponds to the nonlinear diffusion description of infiltration. The Green and Ampt model for vertical flow can also be written in the following form utilizing the information on soil porosity because infiltrating water can fill all the pores and the water content will be equal to the total porosity of the soil (ASCE, 1996):

$$i = K_s\left(1 + \frac{(P - \theta_i)\Phi_{\mathrm{wf}}}{I}\right) \tag{10.13}$$

The common form of the Green and Ampt model for vertical flow is as follows:

$$i = K_s + \frac{B}{I} \leftrightarrow i = i_c + \frac{B}{I} \tag{10.14}$$

where

i is the instantaneous infiltration rate of soil (cm/min).
P is the porosity ($\mathrm{cm^3/cm^3}$).
θ_i is the antecedent soil water content ($\mathrm{cm^3/cm^3}$).
I is the cumulative infiltration (cm).
i_c is the steady-state infiltration at large times usually more than 2.5 hours (cm/min).
B is a constant.

The Green and Ampt (1911) model uses an approximate description of actual flow regimes. At $t = 0$, I is zero and i is infinite; however, for large values of t, i approaches a constant value equal to i_c or K_s at natural saturation.

10.5.2 Philip Equation

Philip (1957) derived an infiltration equation using Darcy's law and making some mathematical approximations regarding the distribution of water behind the wetting front. For horizontal infiltration, Philip (1957) used the Richards (1931) equation without taking into account K (θ) and suggested that the first two terms of his series solution for vertical infiltration could be used for simulating infiltration. He proposed a concise infiltration model as follows:

$$I = St^{1/2} + At \tag{10.15}$$

$$i = \frac{1}{2}St^{-1/2} + A \tag{10.16}$$

where

S is the sorptivity ($\mathrm{cm/min^{1/2}}$).
A is a constant (cm/min).

According to Philip (1969), as t approaches infinity, i– decreases to a final asymptotic value and the coefficient A can be replaced with K_s of the upper layer, and for $t = 0$, i is infinite. The magnitude of A in Equation 10.16 is ($A_1 + K_i + \varepsilon$), where K_i is the hydraulic conductivity and ε is the truncation error. The value of S estimated from Equation 10.16 is quite reliable; however, truncation errors influence the estimated value of A. The sorptivity parameter S can be approximated as follows (Youngs, 1964):

$$S = \left(2\Phi_{\mathrm{wf}} K_e \left(P - \theta_i\right)\right)^{0.5} \tag{10.17}$$

Sorptivity can also be estimated as follows:

$$S \cong \frac{\left(\theta_s - \theta_i\right)L}{t^{0.5}} \tag{10.18}$$

where

Φ_{wf} is the soil water potential at the wetting front (cm).
P is the soil porosity (cm^3/cm^3).
θ_i is the initial soil water content (cm^3/cm^3).
θ_s is the saturated soil water content (cm^3/cm^3).
L is the thickness of the uniformly wetted region (cm).

Example 10.2

An infiltration was performed in a sandy loam soil with a saturated hydraulic conductivity of 0.05 cm/min. The antecedent soil water content was constant at 0.12 cm^3/cm^3 and saturated water content was 0.45 cm^3/cm^3. The wetting front moved to a distance of 12 cm in 15 min; calculate the values of sorptivity (S) and steady-state infiltration (A). Also, calculate horizontal and vertical infiltrations for 30, 100, 500, and 5000 min.

The sorptivity (S) can be estimated as

$$S \cong \frac{\left(\theta_s - \theta_i\right)L}{t^{0.5}} = \frac{\left(0.45 - 0.12\right) \times 12}{15^{0.5}} = 1.022\ \mathrm{cm/min}^{0.5}$$

$$A \approx K_s = 0.05\ \mathrm{cm/min}$$

The horizontal infiltration:

$$\text{At 30 min } I = St^{1/2} = 1.022 \times 30^{0.5} = 5.60\,\mathrm{cm}$$

$$\text{At 100 min } I = St^{1/2} = 1.022 \times 100^{0.5} = 10.22\,\mathrm{cm}$$

$$\text{At 500 min } I = St^{1/2} = 1.022 \times 500^{0.5} = 22.85\,\mathrm{cm}$$

$$\text{At 5000 min } I = St^{1/2} = 1.022 \times 5000^{0.5} = 72.27\,\mathrm{cm}$$

The vertical infiltration:

$$\text{At 30 min } I = St^{1/2} + At = 1.022 \times 30^{0.5} + 0.05 \times 30 = 7.10\,\text{cm}$$

$$\text{At 100 min } I = St^{1/2} + At = 1.022 \times 100^{0.5} + 0.05 \times 100 = 15.22\,\text{cm}$$

$$\text{At 500 min } I = St^{1/2} + At = 1.022 \times 500^{0.5} + 0.05 \times 500 = 47.85\,\text{cm}$$

$$\text{At 5000 min } I = St^{1/2} + At = 1.022 \times 5000^{0.5} + 0.05 \times 5000 = 322.27\,\text{cm}$$

10.5.3 Kutilek and Krejca Equation

It is well known that the truncation errors associated with the Philip equation influence the estimate of A. To overcome this problem, Kutilek and Krejca (1987) proposed to use the first three terms of Philip's series solution and proposed the following infiltration equation:

$$I = St^{1/2} + C_1 t + C_2 t^{1/2} \quad (10.19)$$

$$i = \frac{1}{2} St^{-1/2} + C_1 + C_2 t^{1/2} \quad (10.20)$$

where

C_1 is the estimate of $(A_2 + K_i)$.

C_2 the value of $(A_3 + \varepsilon_1)$, where ε_1 is the truncation error for having used three terms.

10.5.4 Swartzendruber Equation

Swartzendruber (1987) also used the adjusted Philip time series solution to derive the following infiltration model:

$$I = \frac{S}{A_0}\left[1 - \exp\left(-A_0 t^{1/2}\right)\right] + K_s t \quad (10.21)$$

$$i = \frac{S}{2} t^{-0.5} \exp\left(-A_0 t^{1/2}\right) + K_s \quad (10.22)$$

where A_0 is a constant. It provides an initial infinite infiltration at $t = 0$ and a constant equilibrium infiltration rate at large t. This equation, after substituting $4K_s/3S$ for A_0, is known as the two-parameter Stroosnijder (1976) infiltration equation. All these models provide an infinite initial infiltration rate at $t = 0$ and a finite steady-state infiltration rate at large t ($t \to \infty$).

10.5.5 Kostiakov Equation

Among the empirical models, Kostiakov (1932) was the first to propose an infiltration model, in which cumulative infiltration (I) and infiltration rate (i) are expressed as follows:

$$I = Bt^{-n} \tag{10.23}$$

$$i = B't^{-n-1}$$

where parameters B, B', and n are constants. These parameters do not have a physical meaning and can be obtained by fitting the equation to the experimental data. It can be inferred from this equation that at $t = 0$, $i \rightarrow \infty$, and as $t \rightarrow \infty$, $i \rightarrow 0$. Therefore, Equation 10.23 explains the horizontal infiltration; however, for vertical infiltration, the Kostiakov (1932) model is inadequate. To overcome the problem, Kostiakov proposed a maximum time range of application with $t_{max} = (B/K_s)^{1/n}$. Mezencev (1948) included another coefficient, i_c, which essentially shifts the axis for infiltration rate equations, and for large times, the infiltration rate approaches a finite steady-state infiltration rate:

$$I = i_c t + Bt^{-n} \tag{10.24}$$

$$i = i_c + B't^{-n-1} \tag{10.25}$$

Example 10.3

An infiltration test was conducted in a no-till field in Coshocton, Ohio, using a double-ring infiltrometer. Time of infiltration (T; min), cumulative infiltration (I; cm), and infiltration rate (i; cm/min) are as given in the following table. Fit the Green and Ampt infiltration equation to the data in the table and obtain the parameters of the infiltration equation.

T	I	i	I^{-1}	$t^{-0.5}$
I	II	III	IV	V
0	0	0		
2	3.60	1.80	0.28	0.71
5	7.60	1.33	0.13	0.45
10	12.00	0.88	0.08	0.32
15	16.30	0.86	0.06	0.26
20	19.50	0.64	0.05	0.22
30	25.50	0.60	0.04	0.18
40	31.00	0.55	0.03	0.16
50	38.00	0.70	0.03	0.14
60	43.70	0.57	0.02	0.13
75	50.40	0.45	0.02	0.12

(Continued)

90	56.40	0.40	0.02	0.11
105	62.40	0.40	0.02	0.10
120	67.80	0.36	0.01	0.09
135	74.30	0.43	0.01	0.09
150	79.20	0.33	0.01	0.08
165	85.20	0.40	0.01	0.08
180	90.10	0.33	0.01	0.07

In the Green and Ampt equation, cumulative infiltration (*I*, column II) is in the denominator. Therefore, we will calculate $1/I$ or I^{-1} (see column IV). After that we plot I^{-1} (column IV) on the *x*-axis and infiltration rate (*i*; column III) on the *y*-axis. A linear trendline ($y = a + bx$) is fitted to the curve (see Figure 10.5a), and Green and Ampt parameters are obtained as follows:

The best-fit linear equation is $y = 0.3671 + 5.6746x$.

Green and Ampt equation:

$$i = K_s + BI^{-1}$$

Thus, $K_s = 0.3671$ cm/min and $B = 5.6746$ cm²/min.

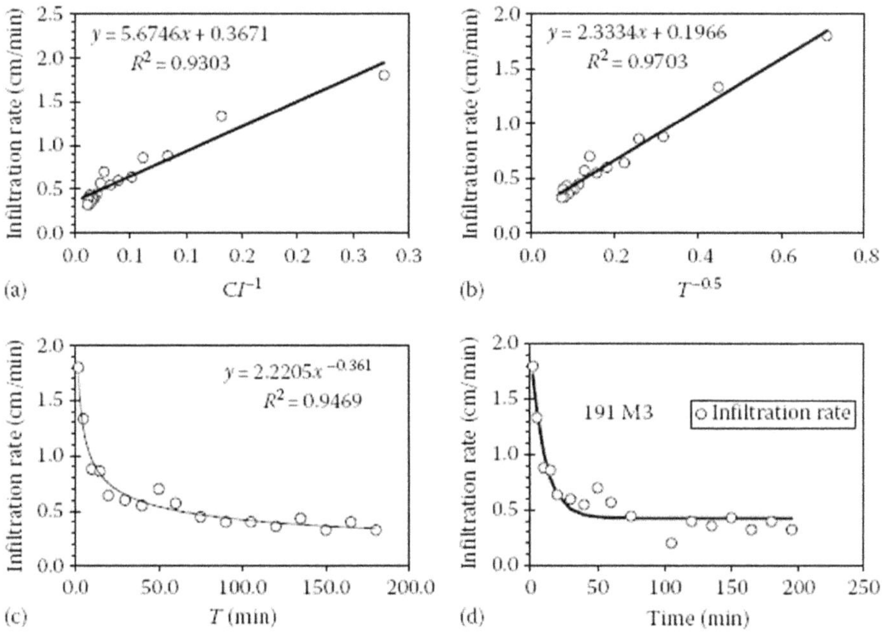

FIGURE 10.5 Measured (symbol) and fitted (line) infiltration curves using various models and the estimates of the parameters of the (a) Green and Ampt equation, (b) Philip equation, (c) Kostiakov equation, and (d) Horton infiltration equation.

Example 10.4

Using the infiltration data in Example 10.3, fit the Philip infiltration equation and obtain the parameters of the equation.

The Philip infiltration equation has square root of time (t) in the denominator (as $t^{-0.5}$). Therefore, we first calculate $t^{-0.5}$ as given in column V of the table in Example 10.3. After that, we plot $t^{-0.5}$ (column V) on the x-axis and infiltration rate (column III) on the y-axis and fit a linear equation to the curve (see Figure 10.5b). The parameters of the equation are obtained as follows:

The best-fit linear equation is $y = 2.3334x + 0.1966$ and

$$\text{Philip equationis i} = 0.5\text{St}^{-0.5} + A$$

Thus,

$$0.5\times\ \text{S} = 2.3334 \text{ or S} = 2\times 2.3334 = 4.6668 \text{ cm/min}^{0.5}$$

and

$$\text{A} = 0.1966 \text{ cm/min}$$

Example 10.5

Use the infiltration data in Example 10.3, fit the Kostiakov infiltration equation, and obtain the parameters of the equation.

The Kostiakov infiltration equation parameters can be obtained by plotting time (T, column I) on the x-axis and infiltration rate (i, column III) on the y-axis and fitting a power equation to the curve (see Figure 10.5c). The Kostiakov infiltration equation parameters are obtained as follows:

$$\text{The best}-\text{fit equation}\gamma = 2.2205x^{-0.361}$$

$$\text{Kostiakov equation}: i = B\text{'}t^{-n-1}$$

Thus,

$$B' = 2.205 \text{ and} - \text{n} - 1 = -0.361 \text{ or } n = -0.639$$

10.5.6 Horton Equation

The infiltration equation proposed by Horton (1940) has three parameters. The empirical Horton model for cumulative infiltration (I) and infiltration rate (i) is as follows:

$$I = i_c t + \frac{i_0 - i_c}{K}\left[1 - e^{-kt}\right] \tag{10.26}$$

$$i = i_c + \left(i_0 - i_c\right)e^{-kt} \tag{10.27}$$

where i_0 is the initial infiltration rate at $t = 0$. The constant k determines the rate at which i_0 approaches i_c. Unlike other models, the Horton model has a finite infiltration rate i_0 at $t = 0$. The model has three parameters, which need to be evaluated experimentally.

Example 10.6

Use the infiltration data in Example 10.3, fit the Horton infiltration equation, and obtain the parameters of the equation.

The empirical Horton model has three parameters: i_c, i_0, and k. Thus, a better option is to use the Solver program in Excel, which has the GRG nonlinear solving method for parameter estimation. The following steps are carried out:

1. From known cumulative time interval and initial estimates of the Horton model parameters—i_c, i_0, and k—the estimated infiltration rate is calculated.
2. Then, the square of the deviations between measured and estimated infiltration is calculated.
3. Then, the sum of the squared deviations is obtained.
4. The GRG nonlinear solving method for parameter estimation is employed to minimize the sum of the squared deviations from step 3.
5. The optimization process changes the initial estimates of the Horton model parameters to the final values.
6. The optimized parameters and the measured and simulated infiltration rate curves are presented in Figure 10.5d. The optimized parameters are as follows:

$$i_c = 0.428\,\text{cm/min}$$

$$i_0 = 2.009\,\text{cm/min}$$

$$k = -0.099$$

10.5.7 Holtan Equation

Holtan (1961) proposed another empirical model, which was based on a storage concept and is a two-form mathematical model as follows:

$$i = i_c + \alpha f^n \tag{10.28}$$

where a and n are constants. The exponent n dictates the steepness of the infiltration rate curve at the beginning of the infiltration, and f is the available porosity, as depleted by infiltration volumes, which can be expressed as

$$f = \mathrm{WS} - I \tag{10.29}$$

where WS is the water storage capacity (Total porosity – Antecedent water content) of the soil (to the B horizon) in depth units and

$$i = i_c \text{ for } I > WS \tag{10.30}$$

As long as $0 \leq I \leq WS$, the Holtan infiltration equation is consistent. However, for $I > WS$, $(WS - I)^n$ becomes positive or negative depending upon the value of the exponent n. Holtan (1961) did not discuss the meaning of WS in the absence of an impeding layer, and Huggings and Monk (1967) reported that effective depth is a function of land use and soil management.

Example 10.7

Use the infiltration data in Example 10.2, fit the Holtan infiltration equation, and obtain the parameters of the equation.

The Holtan parameters were obtained using Solver in the Excel program following the procedure described in Example 10.6. The optimized parameters are as follows (Figure 10.6):

$$i_c = 0.182 \text{cm/min}$$

$$\alpha = 0.002$$

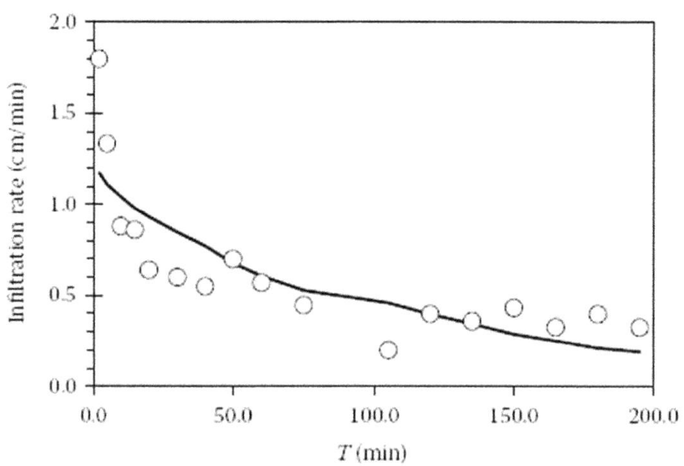

FIGURE 10.6 Measured (symbol) and fitted (line) infiltration curves using the Holtan infiltration equation.

$$k = 1.40$$

$$WS = 91.0$$

10.5.8 Smith–Parlange Equation

Analogous to the Green and Ampt equation, Smith and Parlange (1978) developed an infiltration equation for a situation where unsaturated hydraulic conductivity varies very slowly. The functional form of the equation to estimate the infiltration rate is as follows:

$$i = K_e \left(\frac{S}{K_e I} + 1 \right) \tag{10.31}$$

where

K_e is the effective hydraulic conductivity (cm/s).
S is the sorptivity (cm/s$^{0.5}$).
I is the cumulative infiltration (cm).

For conditions where hydraulic conductivity varies rapidly, the Smith–Parlange equation has the following form:

$$i = K_e \left(\frac{e^{I/S}}{e^{I/S} - 1} \right) \tag{10.32}$$

The sorptivity parameter (S) can be estimated as described previously in the Philip equation (Youngs, 1964).

Example 10.8

Use the following infiltration rate versus time curve and obtain the parameters of the Kutilek, Swartzendruber, and Mezencev infiltration equations. (*Note*: Use the procedure described in Example 10.6.)

Time (min)	Infiltration Rate (cm/min)
0	0
2	2.35
5	1.70
10	1.14
15	1.26
20	1.12
30	0.81

(Continued)

40	0.70
50	0.58
65	0.53
79	0.55
95	0.50
110	0.45
125	0.43
140	0.41
160	0.38
179	0.36

Hint: Use the Solver program in Excel (or any other parameter estimation program).

Kutilek parameters:

$$S = 13.56\,\text{cm} / \text{min}^{0.5}; C_1 = 0.11; C_2 = 0.005$$

Swartzendruber parameters:

$$S = 14.04\ \text{cm/min}^{0.5}; A_0 = -0.029; K_s = 0.058\,\text{cm/min}$$

Mezencev parameters:

$$B = 3.11; n = -0.399; i_c = 0.03\,\text{cm/min}$$

10.6 WATER REDISTRIBUTION

The applied water on the soil surface infiltrates into the soil, saturates the soil pores, and continuously moves downward. When water application is stopped, the soil water redistribution starts and water continues to move downward (Figure 10.7). The soil at the surface starts to get drier, and soil at deeper depths ahead of the wetting front gets wetter. In the absence of evaporation, the water content of the surface soil decreases to the field capacity water content, and then farther downward movement of water ceases. During soil water redistribution, changes in soil water contents at the upper soil layer as well as at deeper depths are nearly uniform for nonlayered soils over space. The total area for the wetting fronts in Figure 10.7 (near rectangle shape) at different times remains nearly equal in the absence of evaporation and root water uptake.

The gravity flow slowly decreases as the soil water content of the surface soil decreases. One should always remember that the soil water hysteresis is important during the process of redistribution, and water content can be higher for a drying soil than

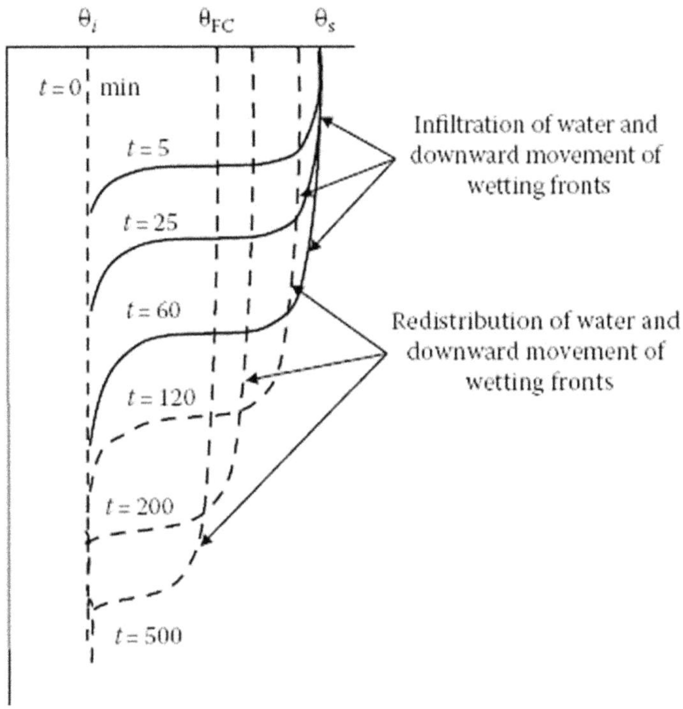

FIGURE 10.7 Movement of wetting fronts during infiltration of water (solid lines). Once the water application is stopped, soil water redistribution starts and the wetting fronts continue to move downward.

a wetting soil at any given soil water potential due to the soil water hysteresis. That is why exact analysis of redistribution gets complicated when the process of hysteresis is included in the calculation. Nevertheless, several accounts are available on the water content versus time relationship (Gardner et al., 1970; Jury et al., 1976, 1991).

PROBLEMS

10.1 Use the Philip equation for vertical and horizontal infiltration, and calculate total infiltration and the infiltration rate. Water content increases from 0.15 to 0.42 cm^3/cm^3, wetting front reaches a depth of 15 cm in 10 min, and saturated hydraulic conductivity of soil is 0.04 cm/min. Also estimate the amount of infiltration and the infiltration rate at the end of 600 and 5000 min.

10.2 Describe the processes of infiltration for a homogeneous sandy and sandy clay soils.

10.3 Describe how the wetting front movement changes for a layered soil system. Explain the differences in the movement pattern when the upper soil layer has lower conductivity than the second layer and vice versa.

10.4 What do you understand about soil water redistribution? Draw schematics with and without evaporation.

10.5 Use the data in the following table to estimate the cumulative infiltration and the parameters of the Green and Ampt, Philip, Kostiakov, Horton, and Holtan infiltration equations.

T (min)	0	2	5	10	15	20	30	40	55	70
I (cm/min)	0	0.95	1.33	0.70	0.64	0.56	0.52	0.54	0.48	0.41
T (min)	85	100	115	131	145	167	185			
I (cm/min)	0.37	0.38	0.35	0.40	0.28	0.31	0.37			

REFERENCES

ASCE. 1996. *Hydrology Handbook*, 2nd edn. ASCE Manual and Reports on Engineering Practice No. 28. American Society of Civil Engineers, New York.

Clothier, B. E., I. White, and G. J. Hamilton. 1981, Constant-rate rainfall infiltration: Field experiments, *Soil Sci. Soc. Am. J.* 45: 245–249.

Gardner, W. R., D. I. Hillel, and Y. Benyamini. 1970. Post irrigation movement of soil water I. redistribution. *Water Resour. Res.* 6(3): 851–861.

Green, W H and G. A. Ampt. 1911. Studies on soil physics: I. Flow of air and water through soils. *J. Agric. Sci.* 4: 1–24.

Haverkamp, R., M. Vauclin, J. Touma, P. J. Wierenga, and G. Vachaud. 1977. A comparison of numerical simulation models for one-dimensional infiltration. *Soil Sci. Soc. Am. J.* 41: 285–294.

Holtan, H. N. 1961. *A Concept for Infiltration Estimates in Watershed Engineering.* U.S. Department of Agriculture, Agricultural Research Service Publication, pp. 41–51.

Horton, R. E. 1940. An approach towards a physical interpretation of infiltration capacity *Soil Sci. Soc. Am. Proc.* 5: 399–417.

Huggings, L. F. and E. J. Monk. 1967. A mathematical model for simulating the hydrologic response of a watershed. *Water Resour. Res.* 4: 529–539.

Jury, W. A., W. R. Gardner, and W. H. Gardner. 1991. *Soil Physics*, 5th edn. John Wiley & Sons, Inc., New York, p. 328.

Jury, W. A., W. R. Gardner, C. B. Tammer, and P. Staffigna. 1976. Model for predicting movement of nitrate and water through a loamy sand. *Soil Sci.* 122: 36–43.

Kostiakov, A. N. 1932. On the dynamics of the coefficient of water percolation in soils and on the necessity of studying it from a dynamic point of view for purposes of amelioration. *Trans. Com. Int. Soc. Soil Sci., Moscow Part A*, 6: 17–21.

Kunze, R. J. and D. R. Nielsen. 1982. Finite-difference solutions of the infiltration equation. *Soil Sci.* 134: 81–88.

Kutilek, M. and M. Krejca. 1987. A three-parameters infiltration equation of the Philip's type solution (in Czech). *Vodohosp. Cas.* 35: 52–61.

Kutilek, M. and D. R. Nielsen. 1994. *Soil Hydrology.* Catena Verlag, Cremlingen-Destedt, Germany.

Mezencev, V. J. 1948. Theory of formation of the surface runoff (in Russian). *Meteorol. I Gidrol.* 3: 33–40.

Parlange, J.-Y., I. Lisle, R. D. Braddock, and R. E. Smith. 1982. The three parameter infiltration equation. *Soil Sci.* 133: 337–341.
Philip, J. R. 1957. Theory of infiltration: 4. Sorptivity and algebraic infiltration equations *Soil Sci* 84: 257–264.
Philip, J. R. 1969. Absorption and infiltration in two and three-dimensional systems. In R. R. Rijtema and H. Wassink (eds.), *Water in the Unsaturated Sone*, vol. 2, pp. 503–525. IASH/UNESCO Symp., Wageningen.
Richards, L. A. 1931. Capillary conductivity of liquids through porous media. *Physics*. 1: 318–333.
Shipitalo, M. J. and W. M. Edwards. 1996. Effects of initial water content on macro-pore/matrix flow and transport of surface-applied chemicals. *J. Environ Qual.* 25: 662–670.
Shukla, M. K. and R. Lal 2005. Erosional effects on soil properties in an on-farm study on Alfisol in west central Ohio. *Soil Sci.* 170: 445–456.
Shukla, M. K., R. Lal, and M. Ebinger. 2003a. Tillage effects on physical and hydrological properties of a typic Argiaquolls in central Ohio. *Soil Sci.* 168(11): 802–811.
Shukla, M. K., R. Lal, M. H. Ebinger, and C. Meyer. 2006. Physical and chemical properties of soils under some piñon-juniper-oak canopies in a semi-arid ecosystem in New Mexico. *J. Arid Environ.* 66: 673–685.
Shukla, M. K., R. Lal, L. B. Owens, and P. Unkefer. 2003b. Land use and management impacts on structure and infiltration characteristics of soils in the North Appalachian region of Ohio. *Soil Sci.* 168(3): 167–177.
Shukla, M. K., R. Lal, J. Underwood, and M. Ebinger. 2004. Physical and hydrological characteristics of minesoils in eastern Ohio. *Soil Sci. Soc. Am. J.* 68: 1352–1359.
Shukla, M. K., R. Lal, and P. Unkefer. 2003c. Experimental evaluation of infiltration models for different land use. *Soil Sci.* 168(3): 178–191.
Smith, R. E. and J.-Y. Parlange. 1978. A parameter-efficient hydrologic infiltration model. *Water Resour. Res.* 14(3): 533–538.
Stroosnijder, L. 1976. Cumulative infiltration and infiltration rate in homogeneous soils. *Agric. Res. Rep.* 847: 69–99.
Swartzendruber, D. 1987. A quasi-solution of Richards equation for the downward infiltration of water into soil. *Water Resour. Res.* 23: 809–817.
Youngs, E. G. 1964. An infiltration method of measuring the hydraulic conductivity of unsaturated porous material. *Soil Sci.* 109: 307–311.

11 Energy Flow through the Vadose Zone

11.1 INTRODUCTION

Energy flow through the vadose zone of soil is one of the important abiotic factors for any ecosystem. An ecosystem is a community consisting of nonliving and living organisms. Some other important abiotic factors in an ecosystem are water, nutrients, and air. The primary driver of energy to an ecosystem is the sun. The radiations falling on the earth's surface raise the temperature of the air and soil. Depending on the abiotic factors, such as water content of soil, soil texture, compaction, and thermal properties of soil, the changes in soil temperature take place over different depths and degrees. The type of vegetation growing in an area also influences energy flow through soils. The vegetation can completely cover the soil (known as a closed canopy), partially cover the soil (known as an open canopy), or not at all cover the soil (known as bare). Open canopy has an important influence on modifying the microclimate and soil surface conditions, characterized by localized conditions of soil water and the thermal regime within canopies and outside drip lines, thus influencing the soil water dynamics in unsaturated soils. The drip line is generally known as the area immediately outside the canopy. For a proper understanding of the soil–plant–atmospheric continuum, knowledge of soil water dynamics driven by either pressure head or temperature gradients through the vadose zone is essential. Water vapor flow is an important part of the total water flux and energy balance, particularly under bare soil and open canopy conditions in semiarid and arid regions (e.g., Milly, 1984; Scanlon and Milly, 1994; Scanlon et al., 2005). Although it is not always intuitive, vapor flow transports a substantial amount of energy as latent heat of vaporization (Cahill and Parlange, 1998; Deb et al., 2011a, 2011b) and therefore affects the movement of heat through unsaturated soils. This chapter covers water and energy transport, atmospheric energy balance, soil water dynamics, soil temperature, and measurement techniques.

11.2 ENERGY BALANCE OF SOIL

Solar radiation reaching the surface of the earth, known as extraterrestrial radiation, enters the atmosphere and reaches the soil as direct radiation or as diffused radiation. Diffused radiations reach the soil surface after reflecting from gases, aerosols, and clouds in the atmosphere. Total radiation, also known as global radiation, is the sum of direct and diffused radiations. Net radiation is the sum of all incoming minus all outgoing radiations on the earth's surface. The steady-state one-dimensional heat energy balance equation at the soil surface or crop canopy can be written as

DOI: 10.1201/9780429264849-11

$$\text{Net heat energy arriving at the surface} - \text{Net heat energy leaving the surface} = 0 \quad (11.1)$$

This equation is not applicable in the greenhouse, where all solar radiations entering the greenhouse do not leave the greenhouse. The net radiations reaching the soil surface are essentially heat energy, and once they enter the soil, soil temperature increases and soil water can evaporate. Heat transfer from the soil surface can take place in three different ways: by convective heat flux (H_c) where warm air from the soil surface and shallow depths is transferred to the atmosphere, by soil heat flux (H_s) where heat is transported into the soil, and by latent heat flux ($L \times E$) where soil water vaporizes and is transported as water vapor toward the atmosphere. Under steady-state conditions, the heat balance equation is written as follows:

$$R_N = H_c + H_s + L \times E \quad (11.2)$$

where R_N is the net radiation. During the day, the net transfer of heat is into the soil; however, during the night, net transfer of heat energy is usually away from the soil toward the atmosphere. The difference between the right-hand side and left-hand side of the equation is the net change in heat storage of soil. The convective heat flux (or H_c) is also known as sensible heat flux. Sensible heat flux is higher for a dry soil than a wet soil; however, latent heat flux ($L \times E$) is higher for a wet soil than a dry soil (Figure 11.1). The net radiations can also be estimated from the known albedo (α; also known as reflectivity coefficient), global solar radiation (R_s), and net longwave radiation (R_{nt}) as follows:

$$R_N = (1 - \alpha) R_s + R_{nt} \quad (11.3)$$

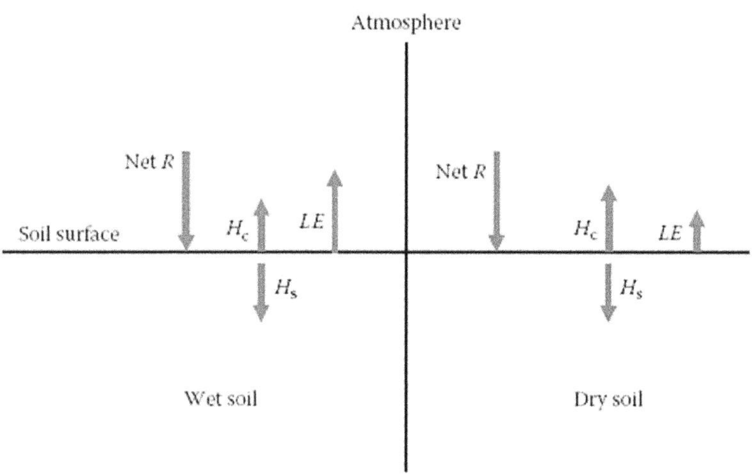

FIGURE 11.1 The net change in heat storage of wet versus dry soil. Net R is the net radiation, H_c is the convective (or sensible) heat flux, H_s is the soil heat flux, and $L \times E$ is the latent heat flux. Note the differences in H_c and LE among wet and dry soil.

The net solar radiations are also expressed as

$$R_N = R_{sky} - R_{earth} \tag{11.4}$$

R_{earth} is the longwave thermal radiation energy flux from earth and can be calculated using the Stefan–Boltzmann equation from the known temperature and emissivity coefficient. Thus, the total surface energy balance equation can be written as follows:

$$(1-\alpha)R_s + R_{nt} - (H_c + H_s + L \times E) = 0 \tag{11.5}$$

The radiations reflected by the earth's surface are slow-moving energy known as infrared radiations. Greenhouse gases, such as water vapor, carbon dioxide, nitrous oxide, methane and other hydrocarbons, and ozone, absorb it and keep the earth warm by about 30°C.

Albedo is the Latin word for "white" and denotes the reflection coefficient of an object. It is the fraction of solar energy (shortwave radiation) reflected from the earth back into space. Snow on top of ice reflects most of the solar energy, while water absorbs more and reflects less. The albedo values for some surfaces are given in Table 11.1. The global mean energy balance at the earth's surface is given in Table 11.2.

Example 11.1

Calculate the net radiation for a dry and wet soil with an albedo of 0.35 and 0.09, respectively, if the global and net radiation are 700 and –110 Ly/day, respectively.

$$R_N = (1-\alpha)R_s + R_{nt}$$

$$\text{For a dry soil, } R_N = (1-0.35)\times 700 + (-110) = 455 - 110 = 345 \text{ Ly/day}$$

$$\text{For a wet soil, } R_N = (1-0.09)\times 700 + (-110) = 637 - 110 = 527 \text{ Ly/day}$$

11.3 FACTORS AFFECTING ENERGY BALANCE

Different surfaces reflect solar energies at different strengths, intensities, and wavelengths. There are several factors influencing the absorption and reflection of incoming solar radiation, such as ground cover, clouds, pollution, greenhouse gases, climate, and latitude. Most of the incoming solar radiations entering the earth's atmosphere are dissipated before striking the soil surface.

Ground cover is an important factor in the absorption or reflection of shortwave radiations striking the earth's surface. The lighter the color of the ground cover or soil, the higher the reflection with attendant lower absorption of the incoming

TABLE 11.1
Albedo for Different Surfaces

Surface	Typical Values
Fresh snow	0.80–0.90
Clouds (thick)	0.60–0.90
Clouds (thin)	0.30–0.50
Ocean ice	0.50–0.70
Concrete (new)	0.55
White painted concrete	0.45
Grass	0.24–0.26
Alfalfa	0.25
Barley	0.21–0.22
Corn	0.16–0.17
Deciduous tree	0.15–0.18
Conifer forest (summer)	0.09–0.15
Bare soil	0.17
Dry sand	0.35–0.45
Wet sand	0.20–0.30
Dry silt loam	0.23–0.15
Dry clay loam	0.18
Wet clay loam	0.11
Fresh asphalt	0.04

Sources: Gates, D. M. and Hanks, R. J., *Plant Factors Effecting Evapotranspiration. Irrigation of Agricultural Lands*, ASA, Madison, WI, pp. 506–521, 1967; Aase, J. K., et al., *Agron. J*, 60, 234, 1968; Hanks, J. R. and Ashcroft, G. L., *Applied Soil Physics*, Springer-Verlag, New York, p. 159, 1980.

TABLE 11.2
Global Mean Energy Balance at the Earth's Surface

Energy Component	Amount (W/m^2)
Net shortwave radiation received	142–174
Net longwave radiation lost	40–72
Sensible heating lost to the air	16–27
Latent heating lost to the air	78–90

Source: Adapted from Kiehl, J. T. and Trenberth, K. E., *Bull. Am. Meteorol. Soc.*, 78, 197, 1997.

radiations (Table 11.1). The albedo values for different surfaces show that snow on ice can reflect almost 90% of all incoming radiations, whereas asphalt can only reflect 4% (absorb 96%). Thus, the surface of asphalt becomes hot quickly as compared to other lighter surfaces. The soil water content has an important influence on absorption of solar radiations; in general, wetter soils absorb more than drier soils.

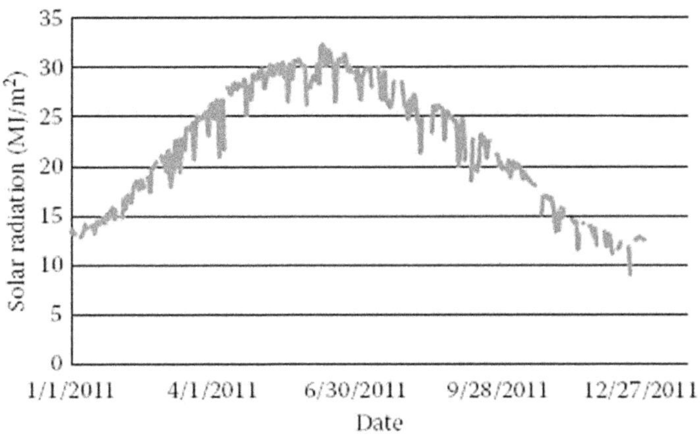

FIGURE 11.2 Solar radiation data for Las Cruces, New Mexico, for the year 2011.

Tillage influences absorption of shortwave radiation; a freshly tilled soil absorbs less because it is generally drier than the same soil type that is not tilled. Type of vegetation also influences the absorption and reflection of incoming radiation, with grass reflecting more radiation than forest (Table 11.1).

Cloud cover is another important factor in the energy balance of the atmosphere because it can absorb as well as reflect radiation. Clouds can reflect incoming shortwave radiations from the sun back to space and absorb longwave radiations reflected by the earth's surface. Usually, thicker clouds near the earth reflect incoming shortwave radiations from the sun and trap longwave radiations. These clouds retain heat from the trapped longwave radiations, making cloudy nights warmer than clear nights. Greenhouse gases also trap longwave radiations and keep the atmosphere warm. Pollutants in the atmosphere can absorb or release heat and can affect energy balance. Climate and latitude also influence net longwave thermal radiation. The equator is the line around the center of the earth. Solar radiations strike the equator directly, but radiations reach the higher latitudes at diffuse angles. The diffuse radiations spread out the same amount of solar radiations across much larger surface areas of the earth away from the equator. Solar radiation values increase from January to June and July, then subsequently go down. Figure 11.2 presents the solar radiation data for Las Cruces, New Mexico. The differences in the amounts of solar radiations across the earth's surface result in differing climate, with colder climates in higher altitude areas that receive fewer radiations than warmer climates in lower altitude areas.

11.4 HEAT FLOW PROCESSES

Heat transfer through a soil profile essentially represents movement of heat from the soil surface to the deeper layers. Heat transfer normally occurs from warmer soil layers to the cooler soil layers. During summer, heat transfer is from upper soil layers to deeper soil layers, and during winter, it is usually from deeper soil layers to upper soil layers. Heat transfer changes direction during the day and night, and depending on the air or surface soil temperature, the heat transfer can be downward during the day and upward during the night. Heat transfer continues until thermal equilibrium is reached. Heat transfer through the soil is dependent on the heat capacity and heat conductivity of different soil layers. Since both capacity and conductivity of soil are dependent on the soil water content, heat transfer is strongly influenced by soil water content. In addition, it is influenced by the soil compaction, presence of snow, type of vegetation, and topography. Heat transfer through the soil generally increases with increasing soil water content because water is a good conductor of heat. Heat transfer through dry soil is slow and less because conductivity of soil air is low. In dry soil, heat transfer predominantly takes place through conduction between soil particles in contact with each other. Heat transfer usually increases with increasing soil bulk density or compaction. Knowledge of heat-transfer processes is important for regulating soil temperature.

There are three principal heat-transport processes: radiation, conduction, and convection. The following sections describe these processes in more detail.

11.4.1 Radiation

Radiation is the act of spreading energy outward through a central source. All matter above absolute zero temperature possesses thermal energy that can be radiated by matter as electromagnetic waves. Thermal radiations can be transferred through a vacuum because they do not require a carrier. Thermal radiations propagate primarily due to the random thermal motion of molecules. That is why thermal radiations can be concentrated to a source by using reflecting mirrors to generate solar power for various purposes, for example, heating water and cooking food in a solar cooker. Thermal energy travels as discrete photons, also known as quanta. According to the Stefan–Boltzmann law, the total energy (J_t) emitted by a body is given by the following equation:

$$J_t = \sigma A T^4 \tag{11.6}$$

where

A is the cross-sectional area of the body (m^2).
T is the absolute temperature.
σ is the Stefan–Boltzmann constant, also known as Stefan's constant (5.675×10^{-8} W/m^2/K^4; 5.675×10^{-8} J/m^2/s/K^4; 5.675×10^{-5} erg/cm^2/s/K^4).

According to the Stefan–Boltzmann law, the total energy radiated by a blackbody per unit surface area in unit time is proportional to the fourth power of the temperature (K). The Stefan–Boltzmann constant is essentially a constant of proportionality in the Stefan–Boltzmann law. If the energy transfer takes place through a medium with an emissivity coefficient of ε, the Stefan–Boltzmann law can be written as

$$J_t = \varepsilon \sigma A T^4 \tag{11.7}$$

The emissivity coefficient is defined as the ratio of radiant energy emitted by soil to the maximum amount of radiant energy emitted ($\varepsilon_s/\varepsilon_b$). A blackbody transmits maximum energy and is a perfect emitter. For a perfect emitter, the emissivity coefficient is equal to one. Soils generally emit much less radiant energy. The emissivity coefficient varies as a function of the wavelength of radiation and serves as an indicator of the efficiency of natural resources. The actual intensity distribution as a function of the wavelength (λ) and temperature (T) is described by Planck's law as follows:

$$E_\lambda = \frac{C_1}{\lambda^4}\left(e^{C_2/\lambda T} - 1\right) \tag{11.8}$$

where

E_λ is the energy emitted in a given wavelength range.

C_1 and C_2 are constants.

The temperature of the soil or a body including the sun (T in K) is inversely proportional to the maximum radiation intensity (λ_m). This relationship is also known as Wien's law and is represented as follows:

$$\lambda_m = \frac{2900}{T} \tag{11.9}$$

Example 11.2

Calculate the maximum radiation intensity for a soil temperature of 300 K and for the sun at the temperature of 6000 K.

$$\text{For soil } \lambda_m = \frac{2900}{T} = \frac{2900}{300} = 9.6\,\mu\text{m}$$

$$\text{For sun } \lambda_m = \frac{2900}{T} = \frac{2900}{6000} = 0.5\,\mu\text{m}$$

11.4.2 Conduction

Conduction is the process of heat transfer in which transfer of heat from one molecule to another or within a substance or from one substance to another occurs due to the direct contact of atoms or molecules. Conduction of heat through soil also takes place through soil particles that are in contact with each other. The movement of heat is always from warmer areas to cooler areas. Heat flux (q_h) through soil due to conduction is represented by the following equation:

$$q_h = -K_t A\left(\frac{\partial T}{\partial x}\right) \tag{11.10}$$

where

K_t is the thermal conductivity (J/m/s/K).

A is the area of cross section (m^2).

T is the temperature in K.

$\partial T/\partial x$ is the temperature gradient in degrees per unit length or the slope of the temperature–distance curve.

The negative sign in Equation 11.10 indicates that the heat flow takes place from warmer areas to colder areas.

The thermal conductivity is also known as the proportionality constant in Equation 11.10, similar to the saturated hydraulic conductivity in Darcy's law. Thermal conductivity is defined as the amount of heat transferred through a unit surface area per unit time under a unit temperature gradient. K_t is dependent on different soil fractions, such as solid, air, and water. It is also dependent on the interfacial contact between solids and between solids and liquids, as well as on the size and arrangement of soil solids. An increase in porosity or air content of soil decreases thermal conductivity, while an increase in soil compaction (or bulk density) improves thermal contact between solids and thus increase K_t. Similarly, increasing soil water content increases thermal conductivity of soil. Thermal conductivity can vary from 1 to 100 J/m/s/K for solids, from 0.01 to 1.0 J/m/s/K for liquids, and from 0.001 to 0.1 J/m/s/K for gases. Dividing heat flux with the surface area provides heat flux density and is expressed as follows:

$$\frac{q_h}{A} = -K_t A\left(\frac{\partial T}{\partial x}\right) = \frac{\mathrm{J}}{\mathrm{msK}}\frac{\mathrm{K}}{\mathrm{m}} = \mathrm{J/m^2/s} \tag{11.11}$$

Heat flux density can also be expressed in W/m^2. It is defined as the rate of heat energy transfer per unit surface area. The volumetric heat capacity is the amount of heat required to change the temperature of a unit volume of soil by 1°C. The ratio of soil's thermal conductivity and volumetric heat capacity is also known as thermal diffusivity:

$$D_t = \frac{K_t}{C_v} \tag{11.12}$$

Example 11.3

The soil temperature at 5 cm depth from the surface is 25°C and at 15 cm depth from the surface is 21°C. If thermal conductivity of the soil is 0.005 cal/cm/s/°C, calculate the heat flow between these two points with an area of 100 cm^2 in 6 h and the direction of heat flow. Also, calculate the heat flux density and thermal diffusivity if volumetric heat capacity of soil is 0.32 cal/cm^3/°C.

$$q_h = -K_t A\left(\frac{\partial T}{\partial x}\right) = -0.005 \times 100 \times \frac{(25-21)}{-5-(-15)} = -0.2 \text{ cal/s}$$

The total heat flow in 6 h = −0.2 × 6 × 60 × 60 = −4320 cal.

The sign is negative, and it indicates that heat flow is downward.

$$\text{Heat flux density} = \frac{q_h}{A} = -\frac{0.2}{100} = -0.002\ \text{cal/cm}^2/\text{s}$$

$$\text{Thermal diffusivity } D_t = \frac{K_t}{C_v} = \frac{0.005}{0.32} = 0.016\ \text{cm}^2/\text{s}$$

11.4.3 Convection

Convection is the dominant process during heat transfer due to the movement of fluids. Transmission of heat energy from one place to another in the soil takes place by the physical displacement of fluids, such as air or water within the pores. In liquids and gases, convection is the most dominant form of heat transfer. Often heat transfer is the combination of the processes of conduction (heat diffusion) and convection (bulk fluid movement). The transfer of heat from moving fluid to solids and vice versa involves diffusion as well as fluid movement; thus, the convection process is a combination of both. Heat transfer exclusively due to the mass flow of fluid is more precisely expressed as advection. By the process of advection, heat transfer within a fluid takes place by the movement of heated molecules from one place to another. Newton's first law of cooling can be used to calculate energy fluxes (J/s) in and out of the system.

$$q_v = C_v A v \left(T_s - T_0\right) \tag{11.13}$$

where

q_v is the heat flux of convection (J/s or W).
C_v is the volumetric heat capacity ($J/m^3/K$).
v is the velocity of a fluid (m/s).
T_s is the temperature of the soil in contact with a fluid (K).
T_0 is the temperature of the fluid far away from the surface (K).

The convection phenomenon is probably more important in the atmosphere, where there is a consistent circulation of warm and cold air and heat exchange. In soils, the heat convection phenomenon is less important in general; however, during infiltration and redistribution of water in the soil profile, soil water which is cooler than the incoming water, convectional heat energy transport becomes important.

The soil temperature can change either by the adsorption of heat energy or by the release. Heat capacity of the soil, earlier defined as the heat energy required to change the temperature of soil by 1°, can be expressed as volumetric or gravimetric heat capacity. The relationships between gravimetric and volumetric heat capacities are as follows:

$$\text{For a dry soil}: C_v = \rho_b \times C_g \tag{11.14}$$

$$\text{For a wet soil}: C_v = \rho_b' \times C_g = \rho_b\left(1+w\right)\times C_g = \left(\rho_b + \theta\right)\times C_g \tag{11.15}$$

where

C_v is the volumetric heat capacity of soil (J/m³/K).
C_g is the gravimetric heat capacity of soil (J/kg/K).
ρ_b is the dry soil bulk density (kg/m³).
ρ_b is the wet soil bulk density (kg/m³).
θ is the soil water content (m³/m³).

The volumetric heat capacity is dependent on the composition of the soil solid phase, including organic matter and mineral matter, and air and water contents. The volumetric heat capacity can be expressed as follows (de Vries, 1975; Hillel, 1980):

$$C_v = \sum_{i=1}^{n}\left(f_{si}C_{si}\right) + f_W C_W + f_a C_a \tag{11.16}$$

$$C_v = f_m C_m + f_o C_o + f_W C_W + f_a C_a \tag{11.17}$$

where

f is the volume fraction of each component.
i is the number of components.
s is solid.
w is water.
a is air.
m is mineral phase.
o is organic matter phase.

Volumetric heat capacities and densities of different soil materials at 10°C are given in Table 11.3 (Hillel, 1980).

TABLE 11.3
Volumetric Heat Capacities and Densities of Different Soil Material at 10°C

Material	Volumetric Heat Capacity (cal/cm³/K)	Density (g/cm³)
Air	0.003	0.0013
Liquid water	1	0.98
Ice	0.45	0.92
Quartz	0.48	2.66
Organic matter	0.3	1.3

Source: Hillel, D., *Fundamentals of Soil Physics,* Academic Press, New York, p. 413, 1980.

Example 11.4

If the volumetric heat capacity of the soil is 0.4 cal/cm³/K, the temperature of the soil in contact with the fluid is 8°C and fluid far away is 5°C, the surface area is 20 cm², and the fluid velocity is 1 cm/s, calculate the energy fluxes in and out of the system:

$$q_v = C_v A v(T_s - T_0) = 0.4 \times 20 \times 1 \times (281 - 278) = 24.0 \text{cal/s}$$

Example 11.5

If volumetric heat capacity of air, liquid water, solid, and organic matter is 0.003, 1, 0.44, and 0.3 cal/cm³/K, respectively, and their respective volumetric proportions are 25%, 25%, 45%, and 5%, calculate the volumetric heat capacity of the soil:

$$C_v = f_m C_m + f_o C_o + f_W C_W + f_a C_a$$

$$C_v = 0.45 \times 0.44 + 0.05 \times 0.3 + 0.25 \times 1 + 0.25 \times 0.003 = 0.464 \text{cal}^{-3}\text{K}^{-1}$$

11.5 HEAT FLUX THROUGH SOIL

Heat flow in the soil can take place due to both conduction through solid particles and advection or convection due to flowing water, air, or latent heat. The net heat flux through soil can be represented by the following equation:

$$q_h = -K_t\left(\frac{\partial T}{\partial x}\right) + L \times E \tag{11.18}$$

The second term in the equation refers to the convective transport of latent heat in water vapor. The change in phase due to vaporization or condensation results in release or consumption of heat in the soil.

11.6 HEAT CONSERVATION EQUATION

The heat conservation equation accounts for all the heat energy entering and leaving a soil volume, as well as change in the heat energy storage inside the soil volume. The heat conservation equation can be written as follows:

$$\begin{aligned}
&\text{Amount of heat energy entering a soil volume in time } t \\
&= \text{amount of heat energy leaving the soil system} \\
&+(\text{or} -)\text{amount of heat energy change in the soil volume during time } t \\
&+\text{amount of heat energy lost due to chemical reaction}
\end{aligned} \tag{11.19}$$

In this equation, if all the compartments of heat transfer are known precisely, the energy balance will close, or total heat energy entering the soil will be fully accounted for, and the heat energy balance will be equal to zero. For one-dimensional heat flow in the x-direction, the previous equation can be represented by the following mathematical relationship:

$$\frac{\partial H}{\partial t} + \frac{\partial H_S}{\partial x} + H_r = 0 \qquad (11.20)$$

where

H is the heat content per unit volume.
H_s is the soil heat flux.
H_r is the amount of heat lost by reactions
t is the time.
x is the distance.

Heat can be lost or added to the soil due to natural factors or by reactions in the soil. The heat content per unit volume can also be written as a function of the volumetric heat capacity of soil and soil temperature as follows:

$$H = C_v \times (T - T_{ref}) \qquad (11.21)$$

where T_{ref} is the arbitrary reference temperature for which $H = 0$.

11.7 MEASUREMENT OF THERMAL PROPERTIES

Thermal conductivity of soil can be measured by the steady-state method or by the transient method. In the steady-state method, a soil sample is placed inside a cylinder and two glass plates are placed on both ends, as shown in Figure 11.3. Several simplified assumptions are made, including that the heat flux through soil and glass plates is equal, and the lateral movement of heat energy is negligible as compared to the vertical movement. During the experiment, good contact between the soil and glass plates is also ensured. The heat flux across the plate and the soil core can be obtained by the following two equations, respectively:

$$q_h = -K_{tg} \frac{(T - T_0)}{d} \qquad (11.22a)$$

$$q_h = -K_{tg} \frac{(T_2 - T_1)}{L} \qquad (11.22b)$$

where

T_0 is the temperature of the plate at the upper end.
T_1 is the temperature at the glass–soil upper interface.

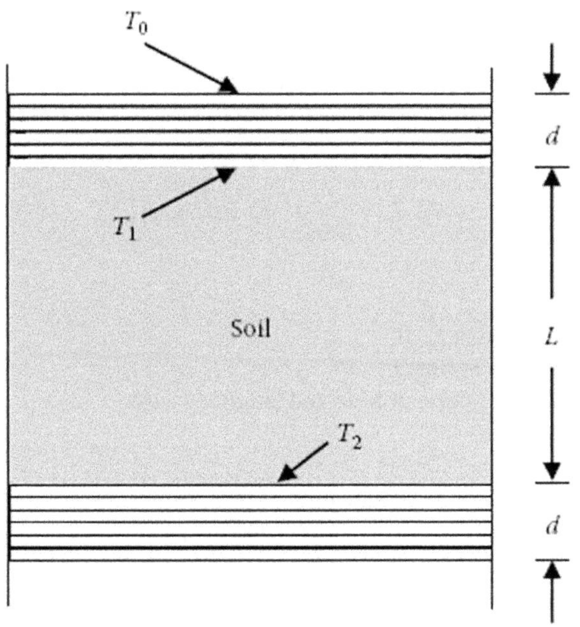

FIGURE 11.3 Thermal conductivity measurement under steady-state one-dimensional flow of heat energy.

T_2 is the temperature at the glass–soil lower interface.
d is the thickness of plates.
L is the thickness of soil core.
K_{tg} is the thermal conductivity of glass plates.

Since heat fluxes through soil and glass are equal, these two equations are also equal and can be solved for the effective thermal conductivity of the porous media (K_{te}) as follows:

$$K_{te} = \frac{K_{tg}L}{d}\frac{\left(T_1 - T_0\right)}{\left(T_2 - T_1\right)} \tag{11.23}$$

Drawbacks of the steady-state method include creation of a nonuniform profile within the column primarily due to the redistribution of water under the influence of the steady-state temperature gradient (Jury and Miller, 1974).

The transient method can be used to avoid the limitation of the steady-state method. According to de Vries and Peck (1968), in the transient method a thin metal wire is used as the heat source and a temperature-measuring device is attached to the instrument to record the change in soil temperature. The heat flux is calculated from the amount of current applied to the wire (I) and the measured resistance per unit length of the wire. The thermal conductivity can be calculated by the following equation:

$$T - T_0 = \frac{q_h}{4\pi K_t}\left[d + \ln\left(t + t_0\right)\right] \tag{11.24}$$

where
- T_0 is the temperature at $t = 0$.
- T is the temperature at time t.
- q_h is the heat flowing per unit time and length of wire.
- d and t_0 are constants.

The constant d depends on the location of the thermocouple, and t_0 on the dimensions of the probe and properties of the soil and the probe (Jury et al., 1991). If $t >>> t_0$, the equation reduces to the following:

$$K_t = \frac{q_h}{4\pi m} \tag{11.25}$$

where
- m is the slope and can be calculated by plotting T versus $\ln(t)$.
- q_h is calculated as the ratio of the current applied (I) and the resistance per unit length of the wire.

Soil thermal properties can be measured in the laboratory and field using a probe, for example, the KD2 probe developed by Decagon Devices, Inc. The KD2 probe can be used to measure soil thermal conductivity, resistivity, diffusivity, and volumetric specific heat capacity. There are single- and double-needle probes. The single-needle probe measures soil thermal conductivity and resistivity, whereas the double-needle probe also provides the estimates of volumetric heat capacity and thermal diffusivity.

Several soil core samples were collected near Las Cruces, New Mexico. The soil textural analysis identified these sampling locations as loamy sand, sandy loam, and sandy clay loam soils. These samples were brought to the laboratory and were saturated with tap water. Using a KD2 dual-needle probe, thermal properties were determined. The cores were equilibrated at different tensions inside a pressure plate apparatus, and thermal properties were measured at the end of each pressure increment when no more water was coming out of the core at the applied pressure (Tables 11.4 and 11.5). It is clear from both tables that with decreasing soil water content or decreasing matric potential, thermal conductivity and volumetric heat capacity are also decreasing.

TABLE 11.4
Thermal Conductivity, Thermal Diffusivity, and Volumetric Heat Capacity for Three Soils Collected around Las Cruces, New Mexico

Soil Type	0 bar	−0.3 bar	−1 bar	−3 bar	−5 bar	−10 bar	−15 bar
Mean Thermal Conductivity (W/m/K)							
Loamy sand	1.49	1.43	1.14	0.96	0.85	0.85	0.82
Sandy loam	1.11	0.91	0.75	0.69	0.59	0.57	0.52
Sandy clay loam	0.8	0.74	0.59	0.57	0.59	0.53	0.52
Mean Thermal Diffusivity (m^2/s)							
Loamy sand	0.58	0.62	0.63	0.61	0.59	0.56	0.56
Sandy loam	0.44	0.49	0.43	0.41	0.4	0.43	0.4
Sandy clay loam	0.26	0.27	0.27	0.28	0.27	0.25	0.28
Mean Volumetric Specific Heat Capacity ($J/m^3/K$)							
Loamy sand	2.58	2.31	1.82	1.57	1.44	1.2	1.48
Sandy loam	2.15	1.86	1.78	1.6	1.53	1.38	1.31
Sandy clay loam	3.18	2.7	2.4	2.01	2.14	2.14	2.01

Note: The soil cores after saturation were subjected to successive increments of pressure in a pressure plate apparatus, and soil thermal properties were determined using a KD2 probe (Decagon Devices, Inc.). (Unpublished data, Adhikari and Shukla, 2013.)

TABLE 11.5
Thermal Conductivity, Thermal Diffusivity, and Volumetric Heat Capacity for Three Soils Collected around Las Cruces, New Mexico

Soil Type	0 bar	−0.3 bar	−0.5 bar	−1 bar	−3 bar	−5 bar	−10 bar	−15 bar
Mean Thermal Conductivity (W/K/m)								
Silty clay loam	1.28	1.25	1.23	1.22	1.22	1.18	1.15	1.09
Loam	1.29	1.26	1.24	1.22	1.23	1.19	1.16	1.08
Silt loam	1.28	1.26	1.24	1.21	1.20	1.19	1.16	1.06
Mean Thermal Diffusivity (m^2/s)								
Silty clay loam	0.46	0.49	0.56	0.57	0.59	0.44	0.43	0.54
Loam	0.59	0.44	0.63	0.56	0.59	0.52	0.51	0.52
Silt loam	0.51	0.50	0.56	0.57	0.60	0.52	0.48	0.57
Mean Volumetric Specific Heat Capacity ($J/m^3/K$)								
Silty clay loam	2.65	2.40	2.08	2.02	1.88	2.47	2.48	1.98
Loam	2.07	2.80	1.92	2.12	2.01	2.30	2.28	2.23
Silt loam	2.46	2.45	2.09	2.07	1.94	2.26	2.35	1.96

Note: Soil thermal properties were determined using a KD2 probe (Decagon Devices, Inc.) in undisturbed soil cores subjected to successive pressure increments after saturation. (Unpublished data, Adhikari and Shukla, 2013.)

Example 11.6

A soil column contains 50 cm of dry sand over 30 cm of dry loam, the top of the column is at 28°C and bottom at 5°C. The K_t of sand is 0.45 and K_t of loam soil is 0.28 mcal/cm/s/°C. Calculate the steady-state heat flux through the two layers and the temperature at the sand–loam interface assuming a unit cross-sectional area.

The equivalent thermal conductivity for the two layers is

$$\frac{50+30}{K_{te}} = \frac{50}{0.45} + \frac{30}{0.28} = 111.11 + 107.14 = 218.25$$

$$K_{te} = \frac{80}{218.25} = 0.37 \text{mcal / cm / s / °C}$$

The steady-state heat flux across the soil column is

$$q_h = -K_{te}\frac{T_1 - T_2}{L_1 + L_2} = -0.37\frac{28-5}{50+30} = \frac{0.37\times 23}{80} = -0.11 \text{mcal/s}$$

The temperature across the interface is

$$q_h = -K_{te}\frac{T_1 - T_2}{L_1 - L_2} = -0.37\frac{T-5}{30}$$

$$-0.11 = -0.37\frac{T-5}{30}$$

$$T = 13.92°\text{C}$$

11.8 SOIL TEMPERATURE

Temperature is a measure of the thermal state of a body with respect to its ability to transfer heat. Soil temperature plays an important role in the germination of seeds and survival and growth of roots. Soil temperature below a certain degree can prevent or delay seed germination. For example, chili seeds can germinate and grow normally at soil temperatures of 15°C–25°C, with an optimum of around 20°C. Gerson and Honma (1978) reported that chili peppers can tolerate moderately cold soil temperatures of 13°C–15°C. Onion grows the best between air temperatures of 21°C and 27°C and can be planted in a wide range of soil temperatures from 20°C to 35°C, with the optimum at 24°C. Pecan grows the best during the summer mean air temperature of 27°C but can tolerate high temperatures ranging from 41°C to 46°C with slight yield reductions (McEachern and Stein, 1997). Soil temperatures generally follow the change (increase or decrease) in air temperature, but the relationship is often nonlinear and depends on the soil type and soil cover (Sharma et al., 2010). Soil temperature varies in response to exchange processes at soil–air interfaces and varies with soil depth, depending on the soil thermal properties such as thermal conductivity, thermal diffusivity, and specific heat capacity of the constituents of the soil. Soil temperature also influences the water movement, water availability, and water evaporation from the soil. It can also affect the properties of soil water such as, density, viscosity, and surface tension.

Bouyoucus (1915) was among the first to observe water movement caused by a soil temperature gradient. He conducted an experiment where temperature gradients were imposed across a column with two different soils at two different water contents. Differences in water content were observed and were dependent upon the soil type and the initial soil temperature. The soil temperature also depends on soil water content, and a dry soil warms up quicker than a wet soil. This is due to the much higher specific heat capacity of water than that of soil particles. Therefore, more heat energy is required to change the temperature of a wet soil than a dry soil. The reported values for specific heat capacity are 0.24 cal/g/°C (1.004 J/g/K) for air and 1 cal/g/°C (4.181 J/g/K) for water. The specific heat capacity for sand is 0.835 J/g/K.

11.9 EFFECTS OF SOIL TEMPERATURE

Soil temperature affects plant growth and development, water movement in soils, aeration, plant water uptake, availability and uptake of plant nutrients, microbial and enzyme activity, and decomposition of plant residues.

Seed germination is influenced by soil temperature but also by the pathogens at low soil temperature. Hart and Wells (1965) showed that emergence of forage grasses in sterilized soil was much higher (91%) than in field soil (58%) at 16°C, indicating that the low seed emergence in field soil was likely caused by pathogens. Growth and development of plants is influenced by soil temperature. Wierenga and co-workers conducted experiments at different constant temperatures and showed a gradual increase in shoot yield of beans with soil temperature up to 30°C.

Plant water uptake depends partially on the metabolic activity, which depends on soil temperature. In addition, plant water uptake can fluctuate the changes in the properties of water at low temperature, for example, increases in viscosity of water can increase resistance to water flow into roots. Soil temperature can change the solubility of mineral nutrients, for example, phosphorus (P) solubility can decrease at high soil temperature, and the rate of immobilization and chemical fixation of P in the soil can increase, making P unavailable to plants (Power et al., 1964).

The rate of infiltration of water into the soil is influenced by the viscosity of water, which at 25°C is about half of that at 0°C. Infiltration rate and hydraulic conductivity of soil increase with increasing water temperature. Water vaporizes at the warm end of a soil profile and condenses at the cold end. Thus, the water content at the cold end increases and creates a pressure gradient across cold and warm ends of the profile. Surface tension holding water in the soil pore decreases with increasing soil temperature. The decrease in surface tension drives water from the warm end to the cold end of the soil profile. Through its effect on soil moisture content, the increase in soil temperature results in increases in soil aeration.

11.10 SOIL TEMPERATURE VARIATIONS

Air temperature continuously changes during the day and over the year, and soil temperature also nearly follows that pattern. There are two soil temperature cycles:

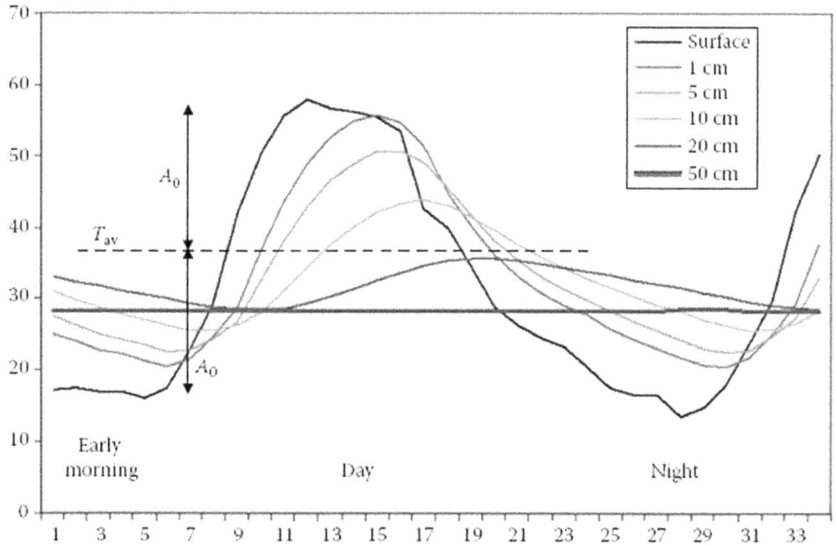

FIGURE 11.4 Soil temperature (°C) versus time (h) for 34 consecutive hours recorded at various depths in Jornada Experimental Range, near Las Cruces, New Mexico, during the month of June. See the decrease in the amplitude with increasing soil depth. A_0 is the amplitude above the mean temperature (T_{av}), known as maxima, and below the mean temperature, known as minima.

diurnal and annual. The diurnal cycle corresponds to the temperature wave in 24 h, whereas the annual temperature wave corresponds to average daily or monthly temperature for 1 year.

The change in soil temperature in the diurnal cycle is due to the heating of soil during the day and cooling of the soil during the night. Soil temperature before sunrise may be the lowest at the soil surface, and as the depth of the soil increases, temperature goes up. Even as the surface soil cools down, the temperature of deeper soil layers continues to increase. However, the maximum and minimum temperatures decrease with increasing depth. The amplitude is calculated as the deviation between the maximum temperature and the mean temperature for any given day or as the difference between the minimum and the mean temperature. Figure 11.4 presents the data of soil temperature at different depths using thermocouples from Jornada Experimental Range near Las Cruces, New Mexico. The amplitude of the diurnal temperature wave decreases with increasing soil depth as the difference between maximum and minimum temperature decreases (Figure 11.4). Below 50 cm deep, amplitudes were nearly zero, and no changes in soil temperatures during the day and the night were observed (data not shown).

The annual variations in soil temperature are due to the variations in shortwave radiations that occur throughout the year. The annual variations at the equator are the smallest because air-temperature variations are the least. However, as one moves away from the equator, the annual variations in soil temperature

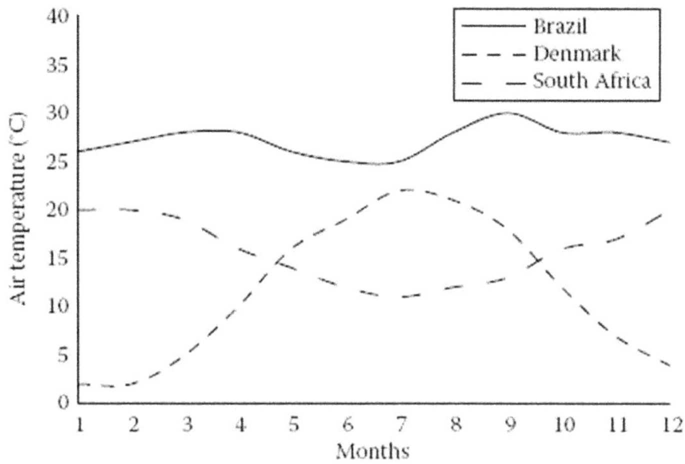

FIGURE 11.5 Annual temperature cycle presented as average monthly air temperature for a country near the equator (Brazil), in the northern hemisphere (Denmark), and in the southern hemisphere (South Africa). See the change in amplitudes and maxima and minima between the equator and away from it in the northern and southern hemispheres.

become large. June and July are the summer months in the northern hemisphere and represent the peak of global radiations and soil temperatures (Figure 11.5). During summer months, the soil temperature at the surface is greater than the soil temperature at deeper depths. The winter months have effects similar to nocturnal daily temperatures. The annual variations in the southern hemisphere also follow global radiations, but December and January are the summer months, with maximum soil temperatures (Figure 11.5).

11.11 MATHEMATICAL REPRESENTATION

The diurnal and seasonal variations of soil temperatures in Figures 11.4 and 11.5 can be mathematically represented as a sinusoidal function of time around an average temperature. Assuming that soil temperature oscillates as a pure harmonic (sinusoidal) function of time around an average temperature (T_{ave}) and the starting soil temperature is 0°C, the soil temperature at the soil surface at any time can be written as

$$T(0,t) = T_{ave} + A_0 \sin\omega t \tag{11.26}$$

where

A_0 is the amplitude of surface temperature fluctuation.

ω is the angular or radial frequency measured in radians per second (1 rad = 57.3°), which is 2π times the actual frequency.

t is the time (Jury et al., 1991).

The temperature equation for a soil at a depth of z can be written as follows (Carlaw and Jaeger, 1959):

$$T(z,t) = T_{\text{ave}} + A_z e^{-(z/d)} \sin\left(\omega t - \frac{z}{d}\right) \quad (11.27)$$

where
A_z is the amplitude at soil depth z.
d is the damping depth also known as characteristic depth.

The damping depth is defined as the depth at which temperature amplitude decreases to 0.37× amplitude at the soil surface [$1/e \times A_0$ or $(1/2.718) \times A_0$]. The damping depth can be calculated as

$$d = \sqrt{\frac{2K_t}{C_v \omega}} = \sqrt{\frac{2D_t}{\omega}} \quad (11.28)$$

As heat energy moves through the soil profile, it is absorbed or released, depending upon the decrease or increase in soil temperature of the profile. Between diurnal temperature fluctuation and annual fluctuation, the damping depth is $\sqrt{365} = 19$ times larger for a given soil (Hillel, 1980).

Example 11.7

If the daily maximum soil temperature is 42°C, the daily minimum soil temperature is 12°C, and the damping depth is 12 cm, calculate the soil temperatures at noon and midnight at 0, 10, and 15 cm depths. The soil temperature curve is symmetrical and the mean temperature is equal at each depth:

$$\text{The mean tempreture} = T_{\text{ave}} = \frac{42+12}{2} = \frac{54}{2} = 27°\text{C}$$

The amplitude is the difference between maximum or minimum temperature and mean:

$$A_0 = 42 - 27 = 15°\text{C}\left(\text{alternately } A_0 = 27 - 12 = 15°\text{C}\right)$$

$$\text{Radial frequency} = \omega = 2\pi / 24 = \pi / 12$$

$$T(z,t) = T_{\text{ave}} + A_z e^{-(z/d)} \sin\left(\omega t - \frac{z}{d}\right)$$

Soil temperature at soil surface $z = 0$ and at noon $t = 6$. (*Note*: $\sin(\pi/2) = 1$.)

$$T(0,6) = 27 + 15e^{-(0/12)} \sin\left(\frac{\pi}{12} \times 6 - \frac{0}{12}\right)$$

$$T(0,6) = 27 + 15 \times 1 \times \sin\left(\frac{\pi}{2}\right) = 27 + 15 = 42°C$$

Soil temperature at soil surface $z = 0$ and at midnight $t = 18$. (*Note*: sin (3π/2) = −1.)

$$T(0,18) = 27 + 15e^{-(0/12)}\sin\left(\frac{\pi}{12} \times 18 - \frac{0}{12}\right)$$

$$T(0,18) = 27 + 15 \times 1 \times \sin\left(\frac{3\pi}{2}\right) = 27 + 15 \times 1(-1) = 27 - 15 = 12°C$$

Soil temperature at depth $z = 10$ cm and at noon $t = 6$. Note: π = 3.14.

$$T(10,6) = 27 + 15 \times 0.435 \times \sin\left(\frac{\pi}{12} \times 6 - 0.83\right)$$

$$T(10,6) = 27 + 15 \times 0.435 \times \sin(1.57 - 0.83)$$

$$T(10,6) = 27 + 15 \times 0.435 \times \sin(0.74) = 27 + 15 \times 0.435 \times 0.674 = 31.4°C$$

Soil temperature at depth $z = 10$ cm and at midnight $t = 18$.

$$T(10,18) = 27 + 15e^{-(10/12)}\sin\left(\frac{\pi}{12} \times 18 - \frac{10}{12}\right)$$

$$T(10,18) = 27 + 15 \times 0.43 \times \sin\left(\frac{3\pi}{2} - 0.83\right) = 27 + 15 \times 0.43 \times -0.68 = 22.58°C$$

Similarly, soil temperature can be estimated at 15 cm depth for noon and midnight:

$$T(15,6) = 28.35°C$$

$$T(15,18) = 25.64°C$$

As depth increases, the time lag between maximum and minimum soil temperature values, also known as phase shift, also increases (Hillel, 1980). Looking at the soil temperature at midnight, one can conclude that as depth increases, the soil temperature at midnight also increases. At the depth of 20 cm, the midnight temperature goes up to 27.27°C, which is more than the average soil temperature. The temperature pattern with depth is due to the increasing phase shift and the decrease in the amplitude with depth.

11.12 MANAGEMENT OF SOIL TEMPERATURE

Soil temperature can be managed by altering the surface soil conditions and by treating the surface soil with covers or mulches. Soil temperature management is important for controlling weeds as well as reducing evaporation of water from the soil. The surface of the soil can be covered with mulch such as straw, saw dust, wood chips, gravel, or charcoal. The surface can also be covered with white or black plastics. The color of the mulch is also important in temperature management; generally darker colored mulch absorbs more solar energy, making the soil warmer than lighter colored mulch, which reflects more, keeping the soil cooler. Cover crops and canopy cover also shade the soil and decrease soil temperature.

Soil water content can also moderate soil temperature because water has high specific heat capacity, and when the soil water content is higher, temperature changes in the soil require more heat energy input for raising soil temperature. High soil water content also increases the thermal conductivity of water, and heat can be transported downward without heating the soil surface.

Tillage can also alter the soil temperature by changing the water content and porosity of soil. Irrigation and drainage have opposite effects on the soil water content. Drainage increases soil aeration; since specific heat capacity of air is low, soil gets warm quickly. Thermal conductivity of the dry soil is also low, and thus heat does not move to deeper layers and instead is used to warm the surface soil.

The direction of the sloping surfaces is also important for temperature modulation. Slopes facing the sun receive more solar radiations and are warmer than slopes not facing the sun. In the northern hemisphere, south-facing slopes are generally warmer with less soil water than north-facing slopes. However, the opposite is true in the southern hemisphere.

PROBLEMS

11.1 Write a short note on energy balance.

11.2 List the factors that affect energy balance.

11.3 Write a brief note on the principal heat-transport processes.

11.4 Calculate the net radiation for a dry and wet soil with an albedo of 0.45 and 0.07, respectively, if the global and net radiations are 750 and –120 Ly/day, respectively.

11.5 If the daily maximum soil temperature is 45°C, the daily minimum soil temperature is 14°C, and the damping depth is 15 cm, calculate the soil temperatures at noon and midnight at 0, 15, and 25 cm depths. Make necessary and required assumptions.

11.6 A soil column contains 40 cm of dry sand over 25 cm of dry loam, and the top of the column is at 30°C and bottom at 8°C. The K_t of sand is 0.42 and the K_t of loam soil is 0.25 mcal/cm/s/°C. Calculate the steady-state heat flux through the two layers and the temperature at the sand–loam interface.

REFERENCES

Aase, J. K., W. D. Kemper, and R. E. Danielson. 1968. Response of corn to white and black ground covers. *Agron. J.* 60: 234–236.

Bouyoucus, G. T. 1915. Effect of temperature on the movement of water vapor and capillary moisture in soils. *J Agric. Res.* 5: 141–172.

Cahill, A. T. and M. B. Parlange. 1998. On water vapor transport in field soils. *Water Resour. Res.* 34: 731–739.

Carlaw, H. S. and J. C. Jaeger. 1959. *Conduction of Heat in Solids.* Oxford University Press, London.

de Vries, D. A. 1975. Heat transfer in soils. In: D. A. de Vries and N. H. Afgan (eds.), *Heat and Mass Transfer in Biosphere.* Scripta Book Company, Washington, DC, pp. 5–28.

de Vries, D. A. and A. J. Peck. 1968. On the cylindrical probe method of measuring thermal conductivity with special reference to soils. *Aust. J. Phys.* 11: 255–271.

Deb, S. K., M. K. Shukla, and J. G. Mexal. 2011b. Numerical modeling of water fluxes in the root zone of a mature pecan orchard. *Soil Sci. Soc. Am. J.* 75: 1667–1680.

Deb, S. K., M. K. Shukla, P. Sharma, and J. Mexal. 2011a. Coupled liquid water, water vapor, and heat transport simulations in an unsaturated zone of a sandy loam field. *Soil Sci.* 176(8): 387–398.

Gates, D. M. and R. J. Hanks. 1967. *Plant Factors Effecting Evapotranspiration. Irrigation of Agricultural Lands.* ASA, Madison, WI, pp. 506–521.

Gerson, R. and S. Honma. 1978. Emergence response of the pepper at low soil temperature. *Euphytica* 27(1): 151–156.

Hanks, J. R. and G. L. Ashcroft. 1980. *Applied Soil Physics.* Springer-Verlag, New York, p. 159.

Hart, R. H. and H. D. Wells. 1965. Effect of temperature and soils on emergence of summer annual forage grasses. *Agron. J.* 57: 636–637.

Hillel, D. 1980. *Fundamentals of Soil Physics.* Academic Press, New York, p. 413.

Jury, W. A., W. R. Gardner, and W. H. Gardner. 1991. *Soil Physics.* John Wiley & Sons, Inc., New York, p. 328.

Jury, W. A. and E. E. Miller. 1974. Measurement of the transport coefficients for coupled flow of heat and moisture in a medium sand. *Soil Sci. Soc. Am. Proc.* 38: 551–557.

Kiehl, J. T. and K. E. Trenberth. 1997. Earth's annual global mean energy budget. *Bull. Am. Meteorol. Soc.* 78: 197–208.

McEachern, G. R. and L. A. Stein (eds.). 1997. *Texas Pecan Handbook.* Texas Agricultural Extension Service, TAEX Hort Handbook 105. The Texas A&M University System, College Station, TX.

Milly, P.C.D.1984. A simulation analysis of thermal effects on evaporation from soil. *Water Resour. Res.* 20: 1087–1098.

Power, J. F., D. L. Grunes, G. A. Reichman, and W. O. Willis. 1964. Soil temperature and phosphorus effects upon nutrient absorption by barley. *Agron. J.* 56: 355–359.

Scanlon, B. R. and P. C. D. Milly. 1994. Water and heat fluxes in desert soils: 2. Numerical simulations. *Water Resour. Res.* 30(3): 721. Doi:10.1029/93WR03252

Scanlon, B. R., R. C. Reedy, D. A. Stonestrom, D. E. Prudic, and K. F. Dennehy. 2005. Impact of land use and land cover change on groundwater recharge and quality in the southwestern USA. *Global Change Biol.* 11: 1577–1593.

Sharma, P., M. K. Shukla, and T. Sammis. 2010. Predicting soil temperature using air temperature and soil and meteorological parameters. *Appl. Eng. Agric.* 26(1): 47–58.

12 Evaporation from Soil

12.1 INTRODUCTION

Water can be found in three different states: solid (ice), liquid, and gas (water vapor). Evaporation is the process of change of the state of water from liquid to vapor. Water in the liquid state is known to everyone because it is a necessity on a daily basis. We drink water, use it to wash utensils and clothes, and water our lawns. However, the vapor form of water is less familiar because seeing something makes it easier to believe in it, and water vapor cannot be seen. Although water vapor is also an inseparable part of our daily life, especially if we are living in humid, semihumid, or coastal areas, we commonly refer to it as humidity. In the presence of vegetation, evaporation and transpiration are together known as evapotranspiration (ET). Evaporation and evapotranspiration are important components of the water cycle in the hydrosphere. For evaporation to occur, energy is required. Meteorological conditions strongly influence evaporation rates and amounts from an area. Surface soil conditions also influence evaporation, for example, presence of mulch on the soil surface could decrease evaporation. Other factors that can influence evaporation from soil are soil depth, soil structure, and degree of homogeneity of soil. Important aspects of energy transport, other than those discussed in Chapter 11, through soil are evaporation, transpiration, and root water uptake. Actual evaporation from soil and transpiration from plants are affected by root zone soil water dynamics, and their determination requires simultaneous knowledge of water flow and root water uptake in soils. In this chapter, we discuss soil water dynamics due to evaporation, water uptake by roots, and evapotranspiration. Some of the information from the lecture notes provided by Dr. Pete Wierenga, a soil physicist at New Mexico State University, to students are incorporated in this chapter.

12.2 EVAPORATION FROM DIFFERENT WATERBODIES

Evaporation occurs from open water surfaces, soil surfaces, and plant canopies. Water is also lost due to the root water uptake and transpiration by plants. Evaporation from open bodies, such as lakes, ponds, rivers, and drains, is dependent on the surface area of water. The higher the surface area, the higher the evaporation. From bare soil, water is quickly lost to the atmosphere by evaporation. If the evaporation process is not controlled, large amounts of water could be lost, especially from irrigated areas or areas with shallow water tables. Evaporation can change with the type of land use and management treatments applied to the soil. Evaporation can increase from a recently plowed soil because of high pore spaces and convective flow of water from the atmosphere. Evaporation also increases from the soil once crops or residues are removed from the surface. The decrease in canopy cover decreases the shading of

DOI: 10.1201/9780429264849-12

soil surface and can also increase evaporation. During planting and germination of seeds, soil is mostly bare, and that can accelerate evaporation. Soil texture also influences the rate of evaporation. More water can be evaporated from saturated clayey soil than saturated sandy soils because more water is stored in a saturated clayey soil than in a saturated sandy soil.

Evaporation from soil can be sustained, provided there is enough continuous supply of water from a deeper layer. The available water can be transported upward toward the soil surface so that enough water is always available at the soil surface for evaporation. The amount of water and the supply to the surface depend on the soil water content, soil water potential, unsaturated hydraulic conductivity of soil, soil texture, bulk density, soil homogeneity or heterogeneity, and the depth to the water table. If a shallow water table is present in an area, evaporation can be sustained at a steady state for a long time. If the rate of evaporation is equal to the rate of replenishment of soil water content from the water table, soil water content does not change, and the soil profile is considered in equilibrium with the water table. However, when the evaporation rate is higher than the rate of replenishment of water from underlying soil layers, the soil moisture content changes, and soil at the surface gets drier than at deeper depths. The water table is usually deep in this situation, and there is no hydraulic continuity of water from the water table to the soil surface. Under this condition, the soil profile is not in equilibrium with the water table.

Evaporation can also be sustained, provided there is a continuous supply of energy, for example, solar energy. The direct or radiated solar energy is required in the form of latent heat (590 cal/g) for evaporation to take place. Once evaporation starts, for it to sustain, there has to be a vapor pressure gradient between the soil and atmosphere so that water vapor from soil can be transported to the atmosphere. The transport of water vapor could take place through the process of diffusion or convection. Usually, vapor transport is external to the soil and depends primarily on meteorological factors such as air temperature, wind velocity, relative humidity, and solar radiation. Environmental conditions have an important influence on evaporation. These external meteorological conditions could be regular, for example, diurnal or annual cycles of temperature and radiation; or irregular, for example, spells of warm and cold weather, or drying and rewetting of soil.

In general, the rate of evaporation from wet soil is dependent upon external meteorological conditions, whereas the evaporation rate for drier soils depends on the unsaturated hydraulic conductivity of soil that determines the rate of upward movement of water (Philip, 1957). Water evaporation can be at the maximum rate or potential rate from the wet soil, whereas it is less than potential from the dry soil. The evaporation rate can be initially controlled by meteorological factors; however, the control can be shifted to soil factors such as unsaturated hydraulic conductivity. This is particularly true for transient-state evaporation or drying.

12.3 EVAPORATION PROCESSES

As discussed earlier, the evaporation process starts with the transport of water toward the soil surface. The transport takes place due to the hydraulic gradient between the soil surface and a deeper layer that causes upward flow of water through capillary rise. This process continues as long as there is enough water in the soil. Usually, the upward movement is longer after the end of an irrigation or good precipitation event. At or near the surface, water changes its state from liquid to vapor due to the solar energy and the latent heat of vaporization. The energy balance of the soil depends on the physical and thermal properties of soil. Liquid water can also change its state due to exothermic reactions and microbial activities that can release enough heat to vaporize water. The water vapor diffuses (by the process of convection) to the atmosphere if the vapor pressure inside the soil is greater than the vapor pressure above the evaporating surface.

12.4 STAGES OF EVAPORATION

The evaporation process is divided into three stages. Stages 1 and 2 are also known as the potential evaporation stage and soil-controlled evaporation stage, respectively (Figure 12.1). For evaporation to take place at stage 1, soil must be wet and there must be enough water available for evaporation at the maximum rate. The rate of evaporation during stage 1 is controlled by atmospheric conditions but not by the hydraulic properties of soil. The evaporation rate is known as the potential evaporation rate ($E = E_p$). This stage is sustained because as the soil water content decreases, the saturated hydraulic conductivity decreases, but the gradient across the evaporating surface and soil below it increases and compensates for the reductions in the hydraulic conductivity. This situation is similar to the flux-controlled process of water infiltration into the soil. A well-structured soil can sustain evaporation at

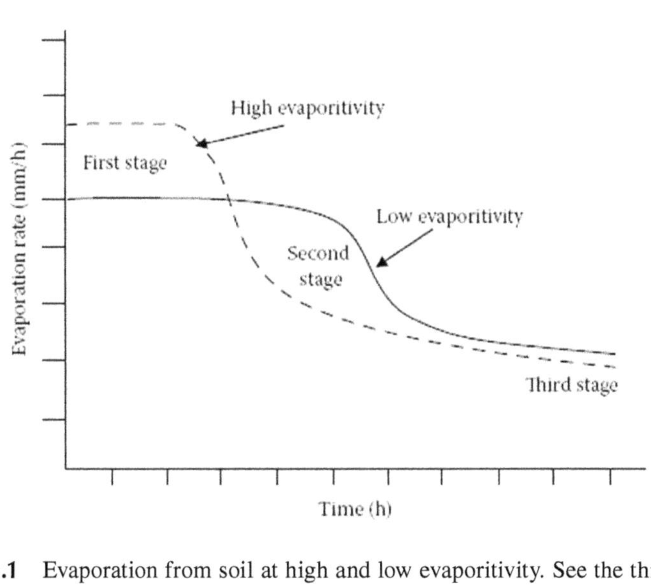

FIGURE 12.1 Evaporation from soil at high and low evaporitivity. See the three stages of drying: the first stage where evaporation is at the maximum rate, the second stage or falling rate of drying, and the third stage of drying.

stage 1 for a much longer duration than a structureless soil. Application of mulch on the soil surface increases the water content of soil, but it shortens the duration of evaporation under stage 1.

The next stage of evaporation is also known as stage 2 or the intermediate stage. At this stage, evaporation is no longer at the potential rate but starts to decrease with time. The process is no longer controlled by atmospheric condition but by the soil profile properties. The soil water content is low, and the increase in gradient could not compensate for the decrease in unsaturated hydraulic conductivity of soil. Soil water pressure at this stage is nearly equal to the partial water vapor pressure; thus, the hydraulic gradient may not increase appreciably. As soon as unsaturated hydraulic conductivity decreases to a level that an increase in gradient can no longer compensate, the second stage of drying starts. The depth of the dry soil profile continuously increases, and the ability of soil to conduct water upward decreases further.

The third stage of evaporation is usually steady, and evaporation takes place at a low rate and can continue for several days. Liquid water conductance is very low, and the stage is also known as the vapor diffusion stage because water transport is primarily by water vapor diffusion.

If there are only two stages of evaporation, then they can be mathematically represented by the following two equations:

$$E = E_P \text{ for } 0 < t < t_a \tag{12.1}$$

$$E_c = c\left(t - t_a\right)^{0.5} \text{ for } t > t_a \tag{12.2}$$

where

t_a is some arbitrary time at which stage 1 of evaporation ends.
c is a constant that denotes the decreasing evaporation rate after time t_a.

12.5 STEADY-STATE EVAPORATION

In areas where the water table is present at shallow depths, there is hydraulic connectivity between the surface and the water table. The evaporation in the presence of a shallow water table can take place under steady-state conditions. Under steady-state conditions, the water that is evaporated is quickly replenished by the water from the groundwater. Under this condition, the rate of change of flux density with depth is zero:

$$\frac{\partial q}{\partial z} = 0 \tag{12.3}$$

where

q is the flux density.
z is the distance or height above the water table.

The steady-state evaporation, although difficult to sustain for a long time, can be represented by the following equation:

$$q = K(\Phi)\left(\frac{d\Phi}{dz} - 1\right) \tag{12.4}$$

Note the gradient because water is moving upward. K (Φ) is the unsaturated hydraulic conductivity and Φ is the soil water potential. The equation can also be represented in terms of soil water diffusivity:

$$q = K(\Phi)\frac{d\Phi}{dz} - K(\Phi) \tag{12.5}$$

$$D(\theta) = K(\theta)\frac{d\Phi}{d\theta} \tag{12.6}$$

Introducing hydraulic diffusivity, $D(\theta)$, in the evaporation flux density equation yields the following equation:

$$q = D(\theta)\frac{d\theta}{dz} - K(\theta) \tag{12.7}$$

The evaporation flux density given by the equation depends on several factors. Generally, the evaporation rate and the depth to the water table are inversely related, and as the depth to the water table increases, the evaporation rate decreases. The evaporation rate is also dependent on soil texture and is generally greater from fine-textured soils than from coarse-textured soils (Gardner, 1958). The relationship between the evaporation rate and the depth to the water table can be represented by the following equation (Gardner, 1958):

$$d = \left(\frac{B}{q_{\max}}\right)^{1/n} \tag{12.8}$$

where

B is a constant.
$q_{\max}$ is the maximum rate of upward water movement.
n is a constant that depends on soil texture.

The value of n is generally smaller for fine-textured than for coarse-textured soil, and consequently evaporation is greater from the former than from the latter.

12.6 TRANSIENT-STATE EVAPORATION

Steady-state evaporation does take place for some time, but mostly evaporation is a transient process. Water table depths change with space and also with time, and therefore evaporation rates also change from steady to unsteady state. The assumption of the steady-state condition stipulates that the soil water content remains unchanged during the process, which is also not valid sometime after the start of the evaporation. Soil water content at the surface changes more than at deeper depths as evaporation progresses, hydraulic connectivity decreases, and soil is not able to fully compensate the loss of water from the upper soil layers. Another common assumption for steady-state evaporation is that the external atmospheric conditions remain unchanged, for example, atmospheric evaporativity is assumed to remain constant, which is seldom true. However, this is also a limitation for several transient-state analyses of evaporation, where an implicit assumption of constant atmospheric evaporativity is made for the sake of simplicity.

Similar to the steady state, the transient-state evaporation processes can also have three stages (Figure 12.1). In general, transition from the first to the second stage of drying is sharp, and transition from the second to the third stage is gradual and difficult to separate. The first stage of drying is directly influenced by the atmospheric evaporativity. If atmospheric evaporativity is low, the first stage of drying is usually sustained for a longer duration. As evaporation under transient conditions continues, with soil drying, soil water potential decreases with time (becomes more negative), resulting in a decreased evaporation rate with time.

If, due to evaporation, the rate of change of water content is $\partial\theta$ in time ∂t, and soil profile is uniformly wetted to a thickness L, the following equation can predict the evaporation rate from soil (Gardner, 1959; Gardner and Hillel, 1962):

$$\frac{e}{L} = \frac{\partial\theta}{\partial t} = \frac{\partial}{\partial z}\left(D(\theta)\frac{\partial\theta}{\partial z}\right) \tag{12.9}$$

where

e is the rate of water loss per unit area per unit time from a soil that is wet up to a depth of L.

z is the distance measured upward from the bottom of the wetted soil profile.

$D(\theta)$ is the soil water diffusivity.

The following exponential diffusivity relationship was used by Wagner (1952) and Gardner and Mayhugh (1958) to obtain the solutions of a one-dimensional diffusive flow equation without the influence of gravity:

$$D(\theta) = D(\theta_0)\exp\left[\beta(\theta - \theta_0)\right] \tag{12.10}$$

where

β is a constant.

$D(\theta_0)$ is the diffusivity at $\theta = \theta_0$.

The total water content (θ) at the end of the first stage of drying was approximated by Gardner and Hillel (1962):

$$\theta = \frac{L}{\beta}\ln\left(1 + \frac{e\beta L}{2D(\theta)_0}\right) \tag{12.11}$$

According to Gardner and Hillel (1962), the equation describes the region in which potential and actual evaporation are equal. After the first stage of drying, the evaporation rate depends on the soil water content. Gardner and Hillel provided some theoretical curves for the falling rate period of drying, using the exponential diffusivity function given previously. Gardner (1962) showed that using the following relationship, theoretical curves can be calculated for any known diffusivity function $D(\theta)$:

$$e = \frac{D(\theta)\theta_L\pi^2}{4L^2} = \frac{D(\theta)\theta_L\pi^2}{4L} = -\frac{d\theta}{dt} \tag{12.12}$$

where

θ is the average water content and is obtained as the ratio of the total water content of soil (θ) and depth of wetting (L).

θ_L is the total depth of water ($= \theta \times L$).

This equation can be integrated for the entire duration (or time t) to obtain the cumulative evaporation (Gardner and Hillel, 1962). Thus, the first and the second stages of drying can be predicted based on the isothermal flow equation and the properties of soil (Gardner and Hillel, 1962). During evaporation processes, the temperature of soil can also change, and thermal transport of water and vapor can take place. The nonisothermal situations are described separately.

For the second stage of drying, an equation similar to the Philip equation was proposed by Rose (1966). In the equation, sorptivity of the Philip infiltration equation is referred to as soil evaporativity (E_v) or desorptivity because soil is losing water or drying (Lal and Shukla, 2004). The rate of evaporation is given by the following equation:

$$e = 0.5E_v t^{-0.5} + B \tag{12.13}$$

where B is a constant and similar to the transmissivity coefficient in the Philip infiltration equation. The total evaporation can be obtained by integrating the equation for the total time duration. If diffusivity is constant (or average diffusivity is $D(\theta)$), the evaporation rate (e) and cumulative evaporation (E) can be given by following equations (Gardner, 1959):

$$e = (\theta_i - \theta_0) \times \left(\sqrt{D'(\theta)/(\pi \times t)}\right). \tag{12.14}$$

$$E = 2 \times (\theta_i - \theta_0) \times \left(\sqrt{D'(\theta) \times t/(\pi)}\right) \tag{12.15}$$

From the known average (or weighted average) soil water diffusivity and soil water contents during drying, the rate of evaporation and cumulative evaporation can be obtained from Equations 12.14 and 12.15, respectively.

Example 12.1

There are two chile fields, and the soil in the first field has an average volumetric water content of 0.4 cm^3/cm^3, while it is 0.12 cm^2/cm^3 in the other, and both soils are wet up to a depth of 80 cm. Calculate evaporation rates if soil water diffusivity is 200 cm^2/day. Discuss your results if the potential evaporation rate is 1.5 cm/day for both fields.

$$e = \frac{D(\bar{\theta})\theta\pi^2}{4L}$$

$$e = \frac{200 \times 0.4 \times 3.14^2}{4 \times 80} = 2.48 \text{ cm/day}$$

Thus, soil has enough water to supply at the rate of 2.48 cm/day, since the potential evaporation rate [equal to $(R_N + H_s)/L$] is only 1.5 cm/day; the energy available for evaporation can only evaporate water at the potential rate of 1.5 cm/day. The evaporation from the first chile field will continue at the maximum or potential rate or will be at stage 1. For the second chile field,

$$e = \frac{200 \times 0.15 \times 3.14^2}{4 \times 80} = 0.93 \text{ cm/day}$$

The calculated rate of evaporation from the second chile field is the actual evaporation because it is less than the potential evaporation.

12.7 SOIL WATER REDISTRIBUTION DURING EVAPORATION

Water content of the soil increases with the onset of irrigation as the waterfront moves deeper into the soil profile. Once irrigation ceases, the soil water contents of the deeper profiles continue to rise due to the downward movement of the waterfront. However, evaporation at the soil surface starts rendering the soil surface drier, thus creating an upward gradient for water movement. Initially, water can be supplied at the rate of evaporation from the underlying layers; however, as evaporation continues, the rate of water supply to the surface decreases with a decrease in the unsaturated hydraulic conductivity of soil, and the surface soil layer becomes drier (Figure 12.2).

Example 12.2

What could be the initial and boundary conditions after the cessation of the irrigation, when water is moving vertically through the soil profile and evaporation is taking place at the soil surface (refer to Figure 12.3).

The initial condition for water content will be a function of water content with soil depth $f(z)$.

The boundary condition at the surface will be initially equal to potential evaporation E_p until some time t_1. After $t > t_1$, the upper boundary condition will be

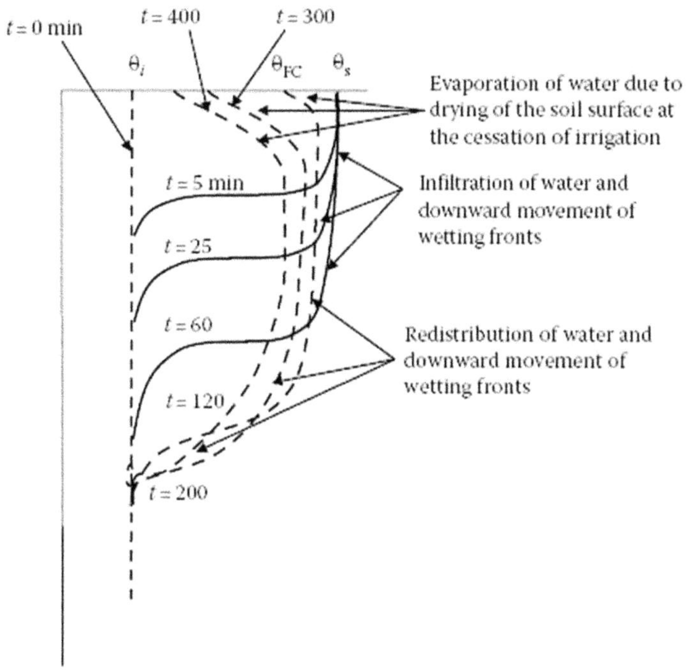

FIGURE 12.2 Redistribution of water during evaporation from the soil surface. *Note*: Units of time are in minutes and are provided to illustrate the infiltration and redistribution during evaporation at the end of an irrigation event.

$f(E)$ and will also depend upon atmospheric evaporativity. The bottom boundary condition for $t > 0$ could be free drainage for a semi-infinite soil profile or will be the saturated water content at the water table.

12.8 VAPOR FLOW THROUGH SOIL

As soon as soil desaturates, air enters the soil profile. If there is sufficient energy available inside the soil, water changes its state to vapor, and water vapor moves out of the soil profile through the air-filled pore spaces (Hanks and Ashcroft, 1980) because of the vapor density gradients. The one-dimensional vapor flux density through the soil under steady-state conditions can be given by the following equation:

$$q_v = -D_v \frac{\Delta\rho_v}{\Delta Z} \tag{12.16}$$

where

q_v is the vapor flux density (g/cm³/s).
D_v is the vapor diffusion coefficient (cm²/s).
ρ_v is the vapor density (g/cm³).
ΔZ is the separation distance or distance for vapor flow (cm).

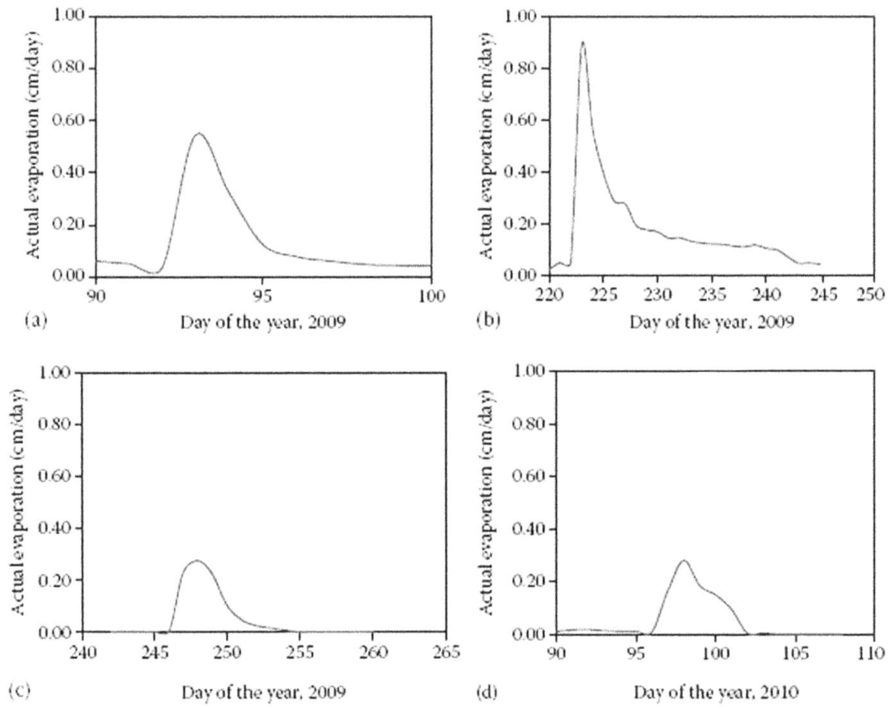

FIGURE 12.3 Average daily actual simulated evaporation from (a) a furrow-irrigated sandy loam field where onion was planted very sparsely, (b) a flood-irrigated sandy loam pecan orchard at a bare spot, (c) a flood-irrigated sandy loam pecan orchard at an under-canopy location, and (d) a flood-irrigated silty clay loam pecan orchard at an under-canopy location. The simulations were made using a numerical model with local data. (From Deb, S. K., et al., *Soil Sci. Soc. Am.* J., 75(5), 1667, 2011; Deb, S. K., et al., *Irrig. Sci*, 319–385, 2013.)

The vapor density gradients can be caused by the differences in the soil temperature at different locations or due to the differences in matric potentials (Hanks and Ashcroft, 1980). Vapor density does not vary much between a wet and a dry soil because there is only a small difference between the relative humidity of a wet soil and a dry soil (RH of 0.9889 at –15 bar). Vapor density increases with increasing soil temperature, and this effect is more pronounced than the change in the matric potential of soil.

Example 12.3

Calculate the vapor flux density between point A and point B in the soil profile with point B being 15 cm vertically below point A; the vapor pressure density is

23.05×10^{-6} g/cm³ at point A and that at point B is 14.25×10^{-6} g/cm³. Point A is 4 cm below the soil surface and diffusivity is 0.2 cm²/s.

$$q_v = D_v \frac{\Delta \rho_v}{\Delta Z}$$

$$q_v = \frac{0.2 \times \left(23.05 \times 10^{-6} - 14.25 \times 10^{-6}\right)}{-4 - (-19)} = -1.17 \times 10^{-7} \text{g}/\left(\text{cm}^2/\text{s}\right)$$

Since flux density is negative, the vapor will flow from point A to point B. Flux density is divided by the density of water to provide the depth of water lost per unit time as vapor.

12.9 CONTROL OF EVAPORATION

Evaporation from soil is influenced by meteorological factors such as wind velocity, humidity, temperature, and solar radiation; exposed surface area of the evaporating body and evaporative demand; and properties of soil such as texture, porosity, cracks, and soil water content. The evaporation of water from the soil surface can be reduced by decreasing the surface area of the water on the soil. The surface area of water applied on soil during irrigation and exposed to evaporation is greater for flood irrigation than the furrow irrigation system (Figure 12.3). Thus, changing the irrigation from a flood to a furrow system could decrease the evaporation. Another option is to irrigate when atmospheric evaporativity is low. For example, irrigating the field during the night will limit the exposure of water to solar radiations and lower the amount of evaporation. The measures discussed under control of soil temperature are also useful to lower the evaporation. Any material applied on the soil surface that can decrease the hydraulic connectivity between soil and atmosphere can help reduce evaporation. The cover crops, shading of the ground surface (Figure 12.3), and light-colored mulches do not let solar radiation directly fall onto the soil surface and can shorten the duration of the first stage of drying. Soil drying can take place at a reduced rate and can significantly decrease total evaporation from the soil surface.

12.10 TRANSPIRATION

The loss of water from plant tissues, namely, leaf, flower, stem, and root, is known as transpiration. The water from soil is taken up by roots, and it moves through the stem and branches into the leaf, and with the opening of stomata, the evaporated water moves into the atmosphere. The stomata are the openings on leaf surfaces together with guard cells, and when they open, water vapor is released to the atmosphere while the carbon dioxide is taken in for photosynthesis. The process is also known as transpiration and is important for mass flow of minerals, nutrients, and water from root to shoot; osmotic pressure in the cell membrane; and cooling the plants.

Minerals, nutrients, and water are absorbed by roots by a process known as osmosis. The movement of water from roots to leaves, especially in the upper part of the canopy, is by the capillary rise as well as due to the decrease in hydrostatic pressure or water pressure with the release of water to the atmosphere during stomata opening. Together, gravitational force is overcome, and water from soil reaches the top of the canopy to complete the transpiration process. In addition to the opening of stomata, transpiration, similar to evaporation, also depends on the evaporative demand of the atmosphere around the leaves. Humidity, wind velocity, solar radiation, air temperature, soil water content, and soil temperature can also influence the rate of transpiration from plants. In general, high relative humidity, low wind velocity, and low air temperature reduce the transpiration rate. The ratio between mass of water transpired and mass of dry matter produced is known as the transpiration ratio and is reported to range from 200 to 1000 kg of water for every kilogram of dry matter produced (Martin et al., 1976).

12.11 MEASUREMENT OF EVAPOTRANSPIRATION

Evaporation and transpiration take place simultaneously from areas under some form of vegetation (natural or grown), including agricultural areas, grasslands, forests, and areas with native vegetation. Evapotranspiration is the combined loss of water due to evaporation from soil and plant surfaces and transpiration by plants. Evaporation and transpiration processes are similar and can be assumed as only the different routes of water transport to the atmosphere. Solar energy is the driving force for evapotranspiration to occur because it provides the energy and gradient required for water movement, change of state, and release to the atmosphere. Water use by crops for meeting their transpiration needs is an important factor for deciding the amount and number of irrigations (irrigation scheduling) to be applied on a field as well as for the design of the irrigation system. Evapotranspiration can be measured in a field using several different methods described in this section.

12.11.1 SOIL WATER DEPLETION

The change in soil water content storage between two irrigation events or two precipitation events can be determined by making a water balance as follows:

$$\Delta S = (P + I) - (ET + DP + SRO) \tag{12.17}$$

where

ΔS is the change in soil water content of profile.
P is the amount of precipitation (mm).
I is the amount of irrigation (mm).
ET is the evapotranspiration (mm).
DP is the deep percolation (mm).
SRO is the surface runoff (mm).

In the soil water depletion method, ΔS is measured, and if surface runoff is zero, then the equation reduces to the following for determining evapotranspiration:

$$ET = (P + I) - (DP + \Delta S) \tag{12.18}$$

If deep percolation or the water loss below the root zone is zero (or negligible), then from the known amounts of rainfall, irrigation, and measured soil water content change, ET can be easily estimated. The changes in soil water content can be measured by various direct and indirect methods, such as the gravimetric method, neutron probe, or time domain reflectometry technique. One of the limitations of soil water depletion method is that deep percolation cannot be assumed zero all the time. Several accounts are available in the literature that show that deep percolation cannot be assumed zero under irrigated agriculture. Depending upon the initial water content of the soil profile and water application depths, deep percolation from a sandy loam, flood-irrigated alfalfa-grass field in northern New Mexico ranged from 5 to 18 cm for irrigation events (Ochoa et al., 2007). The total deep percolation throughout the growing season of two flood-irrigated pecan orchards located in southern New Mexico were 25% for sandy loam soil and 37% for silty clay loam soil of the total water applied, respectively (Deb et al., 2012). The soil water depletion method will be easily applicable where total water applied via irrigation and precipitation does not exceed soil water depletion.

12.11.2 Lysimeters

A lysimeter could be a container filled with soil from an experimental area and placed in the same location from where the soil was excavated. Lysimeter soil is not hydrologically connected with the soil around it. Lysimeters are equipped with the required instrumentation. There are several different types of lysimeters, such as drainage lysimeter, weighing lysimeter, floating lysimeter, and virtual lysimeter.

Drainage lysimeters are containers buried in the soil, and changes of soil water content in the container are measured using direct or indirect measurement devices. These lysimeters are planted with the same crop that is in the experimental field. The water inflow and outflow through the lysimeters are measured, and ET can be estimated from the known changes in soil water content, water input, and loss. These are nonweighing types of lysimeters, and they can provide estimates of ET over longer durations (for 3 days to a week). Differentiating the cumulative ET with respect to time can yield ET estimates on a daily basis.

Weighing lysimeters also contain soil from the same site and are planted with the same crop. These are attached to a weighing mechanism and with modern digital scales or transducers; the accuracy of measurement has improved substantially. Similar to the weighing lysimeters are the floating lysimeters, where weight is determined by recording the water level at which the lysimeter floats. Both lysimeters can be instrumented with various water-content measurement devices. Since the weight of these lysimeters can be recorded at much shorter time intervals, the ET estimates can be made on a much finer scale and as short as every 10 min.

Virtual lysimeters are those that are hydrologically connected to the surrounding soil. These lysimeters must be equipped with devices to measure the change in soil water content and deep drainage. The ET estimates cannot be made for as short a duration as by the weighing type of lysimeters. However, there are no boundary effects and low soil property differences within and outside the virtual lysimeter.

12.11.3 Evaporimeters

Evaporimeters are instruments that measure evaporation or the rate of evaporation. Evaporimeters can measure evaporation rates from a waterbody exposed to the atmosphere or from wet soil. Evaporation can be measured in evaporation pans, namely, the US Weather Bureau Class A pan. The pan has a diameter of 122 cm, a depth of 25 cm, and is usually kept about 14 cm above the ground. Another type of pan is known as a sunken pan. The crop evapotranspiration is obtained by multiplying the pan coefficient and the pan evaporation. Pan coefficients range from 0.2 to 0.85 for different crops (Table 12.1; Taylor and Ashcroft, 1972).

There are also several different types of atmometers available in the market. Atmometers are also simple devices to measure evaporation from a water surface. The device can contain a ceramic cup (saturated) attached with a tube dipped in a water reservoir. As evaporation continues, more water is transported toward the ceramic cup via the tube. The water level in the reservoir is measured and the change provides the value of evaporation. The ceramic cup can be covered to simulate the canopy of a given crop and for the measurement to be equal to the actual evapotranspiration. The evaporating surface can also be a filter paper. Both manual and automatic measurements of rate of fall of water in the reservoir are possible.

TABLE 12.1
Coefficients for Converting Pan Evaporation to Actual Evaporation for Some Crops at Different Stages of Growth in Logan, Utah

Crop	Conversion Coefficient
Alfalfa	
May 14–June 1	0.35
June 1–September 15	0.75
Sugar Beets	
June 3–July 15	0.2–0.07
July 16–September 15	0.79
Potatoes	
June 15–July 15	0.2–0.5
July 16–September 15	0.58

Source: After Taylor, S. A. and Ashcroft, G. L., *Physical Edaphology: The Physics of Irrigated and Nonirrigated Soils*, Freeman and Co., San Francisco, CA, 1972.

12.11.4 Micrometeorology and Climatology Data

The amount of radiation normal to the direction of the radiation reaching the earth's surface is 1.94 Ly/min, provided no atmosphere exists. For a 12 h day, it will be equal to 1400 Ly ($1.94 \times 12 \times 60$ Ly), and actually less than 700 Ly reach the earth surface on any given day. The intensity of radiation at a given location can be given as follows:

$$R = R_0 \cos\theta \tag{12.19}$$

where

R_0 is the intensity of radiation on a unit plane perpendicular to the direction of the sunlight.

θ is the angle of incidence.

$$\text{If } \theta = 90 \text{ than } \cos 90 = 0; \text{and } R = 0$$

$$\text{And if } \theta = 0, \cos 0 = 1; \text{and } R = R_0$$

Thus, intensity of radiation varies from zero at night to R_0 at noon. Latitude also affects the intensity of radiation, which is higher at the equator than at poles. The direct and diffuse solar radiation is measured by a pyranometer and net radiation is measured by a net radiometer. A net radiometer consists of two identical blackened sensing elements placed back to back in the free air. The sensing elements are enclosed in a polyethylene sphere transparent to both shortwave and longwave radiation.

Evapotranspiration can also be measured by the energy budget technique, eddy covariance technique, and empirical methods. In the energy budget approach, all the components of the energy balance equation are measured. The sensible heat flux can be measured by a scintillometer, and latent heat and sensible heat flux can be appropriately partitioned to get evaporation. The surface energy balance algorithm (SEBAL) determines the energy balance from satellite data. Using the meteorological data, SEBAL can calculate evapotranspiration in time (on a daily basis) and space.

The eddy covariance technique is the direct method of measuring evapotranspiration. In this method, convective transfer of water vapor is measured by directly measuring the properties of eddies as they pass through the sensor. Vertically upward-moving eddies carry more water vapor than vertically downward-moving eddies.

There are several empirical methods for calculating evapotranspiration, namely, Thornthwaite, Blaney–Criddle, Penman, Penman–Montieth, Hargreaves, Hargreaves and Samani, Priestley and Taylor, among others. The Penman (1948) equation for evaporation estimation involves the weighted average of energy available and the ability to remove saturated air when both are limiting. It is also known as the combined equation and is mathematically represented as follows:

$$E = \frac{m}{m+\gamma} E_r + \frac{\gamma + m}{\gamma} E_a \tag{12.20}$$

where

E is the open water evaporation estimate (mm/day).

m is the slope of the saturated vapor pressure curve.

γ is the psychrometric constant.
E_r is the evaporation ability of the system to remove moist air and is not limiting to the evaporation process.
E_a is the unlimited availability of energy for evaporation.

The psychrometric constant is given by the following equation:

$$\gamma = \frac{C_P P K_h}{0.622\ L K_w} \tag{12.21}$$

where
C_p is the specific heat at constant temperature.
P is the atmospheric air pressure.
L is the latent heat of vaporization.
K_h and K_w are the diffusivities.

The evapotranspiration can be determined by the modified Penman equation, also known as the Penman–Monteith equation (Shuttleworth, 1993). The equation, although slightly difficult to use, contains a surface resistance term, which is a measure of available water from soil surface and plant canopy (Yates and Strzepek, 1994):

$$E_v = \frac{1}{L}\left(\frac{\Delta(R_N - G) + \rho_a C_P\left((e_s - e)/r_a\right)}{\Delta + \omega\left(1 + (r_s/r_a)\right)}\right) \tag{12.22}$$

where
$R_N - G$ is the available energy.
R_N is the net radiation.
G (= H_s) is the soil heat flux.
$e_s - e$ is the vapor pressure deficit.
r_a is the aerodynamic resistance.
r_s is the surface resistance.

The Hargreaves empirical equation (Hargreaves et al., 1985) is a temperature-based equation and gives reference crop evapotranspiration:

$$ET_r = 0.0022 R_a \Delta T^{0.5}\left(T + 17.8\right) \tag{12.23}$$

$$\Delta T = T_{m\,max} - T_{m\,min} \tag{12.24}$$

where
R_a is the mean extraterrestrial radiation (mm/day).
$T_{m\,max}$ is the mean monthly maximum temperature (°C).
$T_{m\,min}$ is the mean monthly minimum temperature (°C).
T is the mean air temperature (°C).

These and other empirical equations can be used for estimating the crop water need but care must be taken to use the right values of constants and other meteorological data for any location. Guidelines for calculating the crop water requirement can be obtained from the FAO Irrigation and Drainage paper 56 (Allen et al., 1998).

PROBLEMS

12.1 Describe the evaporation processes and distinguish different stages of drying in bare soils with contrasting texture and under-canopy site with different vegetation.

12.2 List important differences between steady-state and transient-state evaporation.

12.3 Soil in a field has an average volumetric water content of 0.4 cm^2/cm^3 that decreases to 0.12 cm^2/cm^3 after 10 days. Assuming that the soil was wet up to a depth of 75 cm, calculate the evaporation rate if soil water diffusivity is 180 cm^2/day and potential evaporation is 1.6 cm/day.

12.4 Calculate the vapor flux density between point A and point B in the soil profile, with point B being 10 cm vertically below point A. The vapor pressure density is 21.14×10^{-6} g/cm^3 at point A; at point B it is 12.32×10^{-6} g/cm^3. Point A is 5 cm below the soil surface and diffusivity is 0.2 cm^2/s.

12.5 Describe different methods of evapotranspiration measurement.

12.6 Write a note on lysimeters. Do a literature search to find out some of the oldest lysimeters in the US.

12.7 What is a virtual lysimeter? How many different measurements are typically made it?

REFERENCES

Allen, R. G., L. S. Pereira, D. Raes, and M. Smith. 1998. *Crop Evapotranspiration: Guidelines for Computing Crop Water Requirements-FAO Irrigation and Drainage Paper 56.* Rome, Italy: Food and Agriculture Organization of the United Nations. http://www.fao.org/docrep/X0490E/x0490e00.HTM, retrieved August 8, 2013.

Deb, S. K., M. K. Shukla, and J. G. Mexal. 2011. Numerical modeling of water fluxes in the root zone of a mature pecan orchard. *Soil Sci. Soc. Am. J.* 75(5): 1667–1680.

Deb, S. K., M. K. Shukla, and J. G. Mexal. 2012. Simulating deep percolation in flood-irrigated mature orchards with RZWQM2. *Trans. ASABE* 55(6): 2089–2100.

Deb, S. K., M. K. Shukla, P. Sharma, and J. G. Mexal. 2013. Soil water depletion in irrigated mature pecans under contrasting soil textures for arid southern New Mexico. *Irrig. Sci.* 31: 69–85.

Gardner, W. R. 1958. Some steady state solutions of unsaturated moisture flow equations with applications to evaporation from a water table. *Soil Sci.* 85(4): 228–232.

Gardner, W. R. 1959. Solutions of the flow equation for the drying of soils and other porous media. *Soil Sci. Soc. Am. Proc.* 23: 183–187.

Gardner, W. R. 1962. Note on the separation and solution of diffusion-type equations, *Soil Sci. Soc. Am. Proc.* 26: 404.

Gardner, W. R. and D. Hillel. 1962. The relation of external evaporative conditions to the drying of soils. *J. Geophys. Res.* 67(11): 4319–4325.

Gardner, W. R. and M. S. Mayhuge. 1958. Solutions and tests of the diffusion equation for the movement of water in soil. *Soil Sci. Soc. Am. Proc.* 22: 197–201.

Hanks, R. J. and G. L. Ashcroft. 1980. *Applied Soil Physics.* Springer-Verlag, New York, p.159.

Hargreaves, L. G., G. H. Hargreaves, and J. P. Riley. 1985. Irrigation water requirements for Senegal river basin. *J. Irrig. and Drain. Engrg., ASCE* 111: 265–275.

Lal, R. and M. K. Shukla. 2004. *Principles of Soil Physics.* Marcel Dekker, Inc., New York.

Martin, J. H., W. H. Leonard, and D. L. Stamp. 1976. *Principles of Field Crop Production,* 3rd edn. Macmillan Publishing Co., Inc., New York.

Ochoa, C. G., A. G. Fernald, S. J. Guldan, and M.K. Shukla. 2007. Deep percolation and its effects on shallow groundwater level rise following flood irrigation. *Trans. ASABE* 50: 73–81; 72: 581–582.

Penman, H. L. 1948. Natural evaporation from open water, bare soil and grass. *Proc. Roy. Soc. Lond. A* 193: 120–146.

Philip, J. R. 1957. Evaporation, and soil moisture and heat fields in the soil. *J. Meteorol.* 14: 354–366.

Rose, C. W. 1966. *Agriculture Physics.* Pergamon Press, Oxford, p. 226.

Shuttleworth, W. J. 1993. Evaporation. In: D. R. Maidment (ed.), *Handbook of Hydrology.* McGraw-Hill, New York, Chapter 4, pp. 4.1–4.53.

Taylor, S. A. and G. L. Ashcroft. 1972. *Physical Edaphology: The Physics of Irrigated and Nonirrigated Soils.* Freeman and Co., San Francisco, 1972.

Wagner, C. 1952. On the solution of diffusion problems involving concentration-dependent diffusion coefficients. *J. Met. Trans. AIME* 4: 91–96.

Yates, D. and K. Strzepek, 1994. *Assessment of River Basin Runoff Under Climate Change. WP94-46.* Working paper. International Institute for Applied Systems Analysis, Laxenburg, Austria.

13 Root Water Uptake

13.1 INTRODUCTION

The efficiency with which roots are able to absorb water and nutrients is important for the growth, sustenance, and yield of crops. Roots act as a bridge between soil resources and plant tissues needing those resources for growth. They provide a pathway for water to move from soil to roots to shoots to leaves and eventually to the atmosphere. This vertical movement of water is controlled by several factors that include soil, plant, and environmental properties. Some of the important soil properties include soil texture, bulk density, hydraulic conductivity, matric potential, water-holding capacity, and plant-available water. Plant properties may include type and height of vegetation; density, distribution, and depth of root systems; physiological properties of the vegetation; and leaf stomata. Atmospheric properties include air temperature, relative humidity, wind velocity, net radiation, and atmospheric evaporativity.

A large number of microscopic and macroscopic approaches are available in the literature to model root water uptake from soil. This chapter describes various microscopic and macroscopic root water uptake models.

13.2 ROOT WATER UPTAKE MODELS

The microscopic approach is physically based, involves radial flow, and simulates water flow toward a representative individual root, which is idealized as an infinitely long, cylindrical sink of a uniform radius and water-absorbing properties (e.g., Gardner, 1960). The dynamics and detailed geometry of the rooting system at the microscopic scale are difficult to measure (Vrugt et al., 2001b). The macroscopic approach averages uptake over a large volume of roots. The oversimplified conditions of steady state or steady rate considered in most of the macroscopic models rarely exist in the field. In the macroscopic approach, water movement with plant uptake is simulated by a volumetric sink term in the Richards equation discussed in Chapter 9.

The first category of macroscopic models, hereafter referred to as "Type I," is based on steady-state or steady-rate approximations of the single-root models. The macroscopic Type I model follows the bottom-up approach, in which the water flow toward the root system is first simulated and the total water uptake by the plant is then taken as the sum of the water uptake from each soil layer. The Type I model considers root resistances and water potentials inside and at the root–soil water interface to calculate the water uptake rate (e.g., Gardner, 1964; Nimah and Hanks, 1973a; Hillel et al., 1976), which are difficult to quantify (Wu et al., 1999). Broadly,

DOI: 10.1201/9780429264849-13

resistances encountered by water moving from the soil through the plant root system include soil resistance, soil–root interface resistance, root resistance to water inflow (root radial resistance), and root axial resistance pertaining to pathways of conduction through xylem vessels.

In the second category of macroscopic models (referred to as "Type II"), root water extraction is calculated from the plant transpiration rate, rooting depth, spatial distribution of roots, and soil water potential, while the model takes into account the effect of soil moisture deficit on root water extraction by using response functions dependent on soil water potential (Molz and Remson, 1970; Feddes et al., 1976, 1978). The parameters in the second category of models are relatively easy to obtain and are generally available in many vadose zone hydrological models (e.g., van Dam et al., 1997; Ahuja et al., 2000; Fayer, 2000; Šimůnek et al., 2008). However, site-specific information on crop, soil, climate, and irrigation is required for model parameterization. Before describing different root water uptake models (Section 13.3), an overview of plant root structure is presented in Section 13.2.1.

13.2.1 Plant Root Structure

The complexity of water transport across plant roots associates with the structural variability of roots during their growth and development. Single-root functioning during water uptake is based on root structure and anatomy. Elongating roots usually possess four regions: the root cap, the meristematic region, the region of cell elongation, and the region of differentiation and maturation (Kramer, 1983; Kramer and Boyer, 1995; Figure 13.1). The tip of the root is shielded by a root cap, which is composed of loosely arranged cells and usually well defined. The root cap generally has no role in adsorption of water, as it has no direct connection with the vascular system. Following the root cap is a region of rapid cell division called the meristematic region, which typically consists of numerous small, compactly arranged, thin-walled cells almost completely filled with cytoplasm. Relatively little water enters through the meristematic region, largely because of the high resistance to movement through the cytoplasm and the lack of a conducting system (Kramer, 1983). The region of cell elongation, where the isodiametric cells elongate in the direction of the root axis, pushes the root tip forward and causes the root to extend further into the soil. Behind the region of elongation is that of differentiation and maturation. Because various types of cells and tissues are differentiated at different distances behind the root apex, it is difficult to indicate a definite zone of differentiation. The entrance of water into young roots occurs primarily in a region a few centimeters behind the root tips, approximately where root hairs are most abundant (Kramer, 1983; Kramer and Boyer, 1995).

Although sizes between monocotyledonous and dicotyledonous roots are variable, the general structure of root apices is broadly similar for many plants (Hopmans and Bristow, 2002). They contain the vascular stele and root cortex, which constitute the primary structures of roots (Figure 13.1). A cross-sectional view of concentric arrangement of the principal tissues, that is, the epidermis, cortex, endodermis, pericycle, and vascular bundles in a dicotyledonous root, is also shown in Figure 13.2.

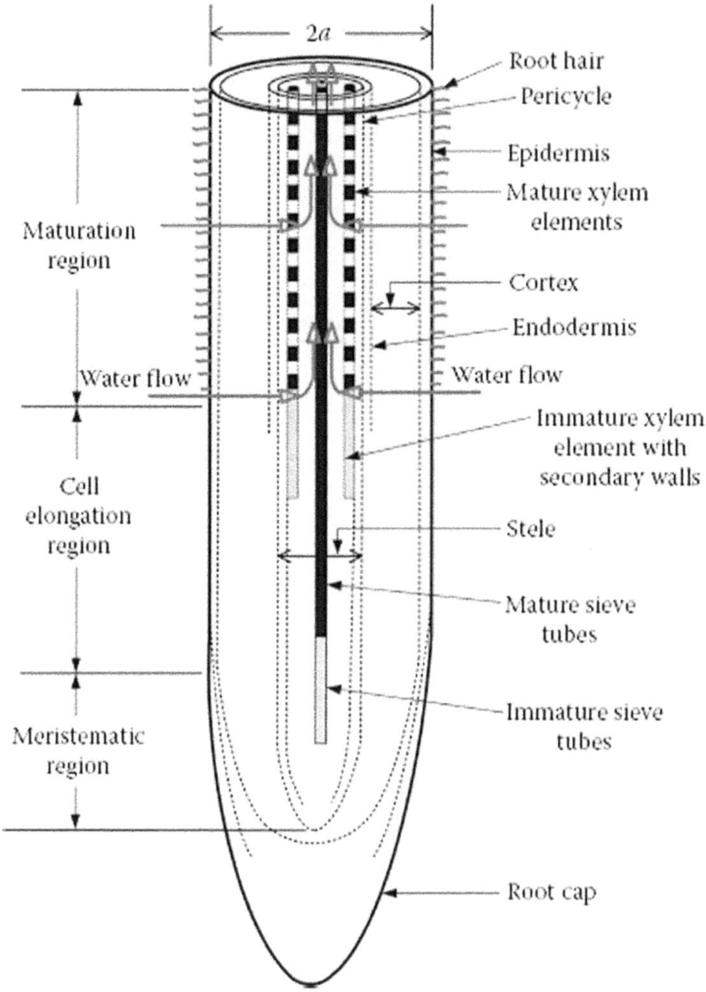

FIGURE 13.1 Illustration of four regions (i.e., the root cap, the meristematic region, the region of cell elongation, and the region of differentiation and maturation) and location of primary tissues of an elongating tobacco root tip, and radial and axial water flow across the root. (Redrawn from Kramer, P. J., *Water Relations of Plants,* Academic Press, New York, 1983.)

The inner center contains the stele, which comprises the vascular system (xylem and phloem), and the associated ground tissue (pericycle, interfascicular regions, and pith, if it occurs). Xylem is a tissue that comprises four cell types: tracheids and vessel members make up the tracheary elements that are concerned with the conduction of water and substances in water, fibers, and parenchyma cells. The conducting cells in the phloem (sieve cells) contain cytoplasm and remain metabolically active. The cortex, which makes up the bulk of the root, consists of the inner endodermis, cortex, and hypodermis and is bounded by an outer layer of epidermal cells from

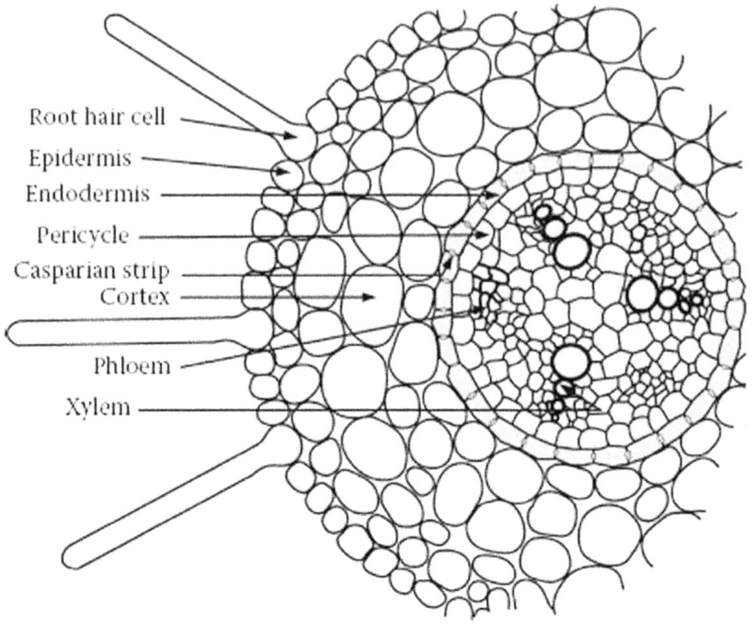

FIGURE 13.2 Cross section through a dicotyledonous root (a squash root) depicting the arrangement of principal tissues in the region where salt and water adsorption occur most rapidly. The central vascular core (stele) is surrounded by three concentric layers of cells: the endodermis, the cortex, and the epidermis. (Redrawn from Kramer, P. J., *Water Relations of Plants,* Academic Press, New York, 1983.)

where root hairs develop. Root hairs appear when the epidermis differentiates. The epidermis is usually composed of relatively thin-walled, elongated cells that form a compact layer covering the exterior of young roots. A second compact layer, called the hypodermis, may lie beneath the epidermis. Some roots include an exodermis (a specialized form of the hypodermis), which can also be a major barrier to the transport of water through suberization of cell walls and the presence of a Casparian band, which occurs in the endodermis. Roots are in contact with the surrounding soil by a film on its surface or mucigel (a sticky layer surrounding roots), which can also play a controlling role in water absorption by the plant.

Water transport from root to shoot involves both axial and radial hydraulic resistances in plants. In roots with mature vessels, it is the radial resistance rather than the axial that limits water uptake, for example, for the early metaxylem of a young maize root, axial resistance is smaller by one to two order(s) of magnitude than that for the radial flow (Steudle and Peterson, 1998). The radial pathways for water in roots are intracellular (apoplastic) and/or intercellular (symplastic pathway), and the separation of both pathways is controlled by the plasmalemma. The protoplasm of plant cells is connected through plasmodesmata, which form continuous pathways between plant cells (Hopmans and Bristow, 2002). It allows water to move along the symplastic pathway between cells. Water flow through the cortex is mostly apoplastic, which also includes symplastic flow through the endodermis, as the flow is diverted because of the presence of the Casparian band.

Water flow through the endodermis may either (1) move through the Casparian band by osmotic gradients or (2) bypass the endodermis, that is, water moves through the cell wall and plasmalemma into the symplastic pathway and returns to the apoplast once it passes the Casparian band. In both cases of water movement, for water uptake across the whole plant, hydraulic equilibrium requires that the total water potential in the apoplast and symplast are the same (Kramer and Boyer, 1995). However, component potentials may differ (Hopmans and Bristow, 2002). Generally, much smaller osmotic potentials in the symplast result in positive hydrostatic water potential. On the other hand, the high osmotic potentials in the apoplast correspond with negative matric potentials in the apoplast.

13.3 MICROSCOPIC ROOT WATER UPTAKE MODELS

13.3.1 Gardner Model

The Gardner (1960) model makes several assumptions including that (i) the roots are considered as infinitely long cylinders and a distance *2b* (b = outer radius of the cylinder) apart; (ii) the roots have a uniform radius (= a); (iii) there is uniform water absorption along the root; (iv) water moves in a radial direction only (gravity-free movement); (v) there is a uniform value for the initial soil water content ($\theta = \theta_0$), which corresponds to the matric potential in the soil for initial conditions ($\Phi = \Phi_0$). Gardner solved the following flow equation for a single root in an infinite, two-dimensional medium:

$$\frac{\partial \theta}{\partial t} = \frac{1}{r}\frac{\partial}{\partial r}\left(rD\frac{\partial \theta}{\partial t}\right) \tag{13.1}$$

where

θ is the volumetric water content of the soil.
D is the diffusivity of the soil.
t is the time.
r is the radial distance from the axis of the root.

The initial and boundary conditions are

$$\theta = \theta_0, \Phi = \Phi_0, \; t = 0$$

$$2\pi a K\left(\frac{d\Phi}{dr}\right) = 2\pi a D\left(\frac{d\theta}{dr}\right) = q, \; r = a \tag{13.2}$$

where

π is a constant (≈3.14159).
a is the radius of the root.
K is the capillary conductivity of the soil.

q is the rate of water uptake by the root as volume of water per unit length of root per unit time.

The solution to Equation 13.1 subject to Equation 13.2 for constants D and K is given by

$$\Phi - \Phi_0 = \frac{q}{4\pi K}\left(\ln\frac{4Dt}{a^2} - \gamma\right) \tag{13.3}$$

where $\gamma = 0.57722\ldots$ is Euler's constant. Because root diameters are small, it is possible to consider the root as a line source, or sink, of strength q per unit length, for which the solution of Equation 13.1 is given by

$$\Phi - \Phi_0 = \frac{q}{4\pi K}\left(\ln\frac{4Dt}{r^2} - \gamma\right) \tag{13.4}$$

When $r = a$, Equations 13.3 and 13.4 are the same (Kirkham, 2005). Equation 13.4 shows that a large part of the water being taken up by the roots comes from some distance from the roots. Equation 13.3 is compared with the steady-state solution for flow in a hollow cylinder:

$$\Phi - \Phi_0 = \frac{q}{4\pi K}\ln\left(\frac{b^2}{a^2}\right) \tag{13.5}$$

where

Φ_0 is now the matric potential at the outer radius of the cylinder $r = b$.
Φ is the matric potential at the inner radius $r = a$.

If we take $b = 2(Dt)^{1/2}$, Equation 13.5 becomes identical to Equation 13.3 except for the constant term γ. In Equation 13.3, the constant term γ is relatively small compared with the logarithmic term, so that the distribution of the matric potential in the transient case is not very different from that in the steady-state case, with all the water coming from a distance $b = 2(Dt)^{1/2}$. The maximum radius b is limited by the density of roots, which can be taken as one-half the average distance between neighboring roots. Using the matric potential instead of the terminology of soil matric potential, Equation 13.5 can be written as (Baver et al., 1972; Kirkham, 2005):

$$\Phi_b - \Phi_a = \frac{q}{4\pi K}\ln\left(\frac{b^2}{a^2}\right) \tag{13.6}$$

where

Φ_b is the matric potential midway between two roots.
Φ_a is the matric potential at the plant root–soil boundary.
q is the volume of water taken up per unit length of root per unit time.
K is the hydraulic conductivity of the unsaturated soil.

Gardner (1960) neglects any contact resistance at the soil–root interface.

13.3.2 Taylor and Klepper Model

Under steady conditions, the rate of water uptake per unit length of root (q_r) from a soil initially at uniform equilibrium water content is calculated using the following radial flow equation (Taylor and Klepper, 1975):

$$q_r = \frac{-2\pi K\left[\Phi_{\text{root surface}} - \Phi_s\right]}{\ln\left(r_{\text{cyl}} / r_{\text{root}}\right)} \tag{13.7}$$

where

π is a constant (≈3.14159).
K is the soil hydraulic conductivity.
$\Phi_{\text{root surface}}$ is the matric potential of water at the root surface.
Φ_s is the matric potential of the water at a distance $r_{\text{cyl}} = r_{\text{bulk soil}}$.
r_{root} is the root radius.
r_{cyl} is the radius of the cylinder of soil through which water is moving.
r_{cyl} is assumed to be one-half the distance between adjacent roots (Gardner, 1960).

It is difficult to measure the water potential at the soil–root surface interface for testing single-root models experimentally, particularly when root resistance to water transport is not known precisely (Taylor and Klepper, 1975). Instead of using the root surface as a boundary, Taylor and Klepper (1975) used the outer edge of the root xylem, so that sink strength could be measured with more certainty, by assigning values of $\Phi_{\text{shoot xylem}}$ to equal $\Phi_{\text{root xylem}}$. An equation that is analogous to Equation 13.7, but with different boundaries, is given to describe water movement into plant xylem from surrounding soil:

$$q_r = \frac{-2\pi K_{\text{sys}}\left[\Phi_{\text{root xylem}} - \Phi_s\right]}{\ln\left(r_{\text{cyl}} / r_{\text{stele}}\right)} \tag{13.8}$$

where

K_{sys} is the hydraulic conductivity of the combined soil–root radial pathway.
$\Phi_{\text{root xylem}}$ is a value obtained from shoot water potential measurements.
r_{stele} is the radius of the root stele.

With decreasing soil water content, decrease in hydraulic conductivity of the combined soil–root radial pathway (K_{sys}) is much greater than the decrease in the hydraulic conductivity (K) of bulk soil. For example, a comparison of the hydraulic conductivity of a soil–cotton root system (K_{sys}) at different soil depths versus the soil hydraulic conductivity (K) under various soil moisture contents is shown in Figure 13.3. The hydraulic conductivity of a soil–cotton root system (K_{sys}) is six orders of magnitude less than the hydraulic conductivity (K) of bulk soil at a water content of 0.18 cm^3/cm^3 (Figure 13.3), which indicates significant resistances to lateral water transport in root tissues. The rhizosphere (the layer of soil immediately around the

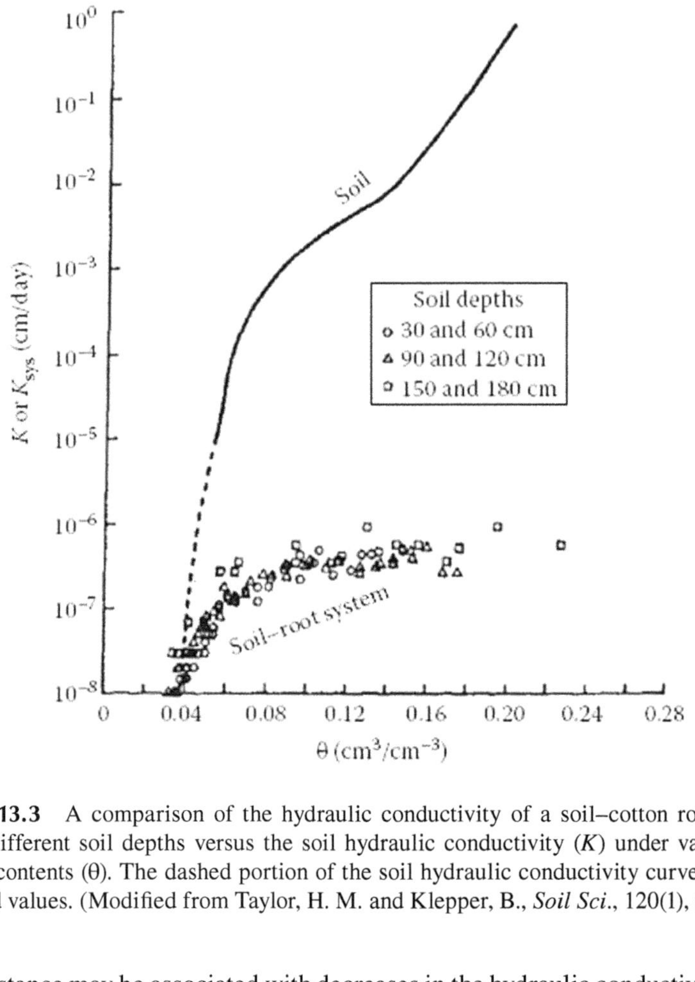

FIGURE 13.3 A comparison of the hydraulic conductivity of a soil–cotton root system (K_{sys}) at different soil depths versus the soil hydraulic conductivity (K) under various soil moisture contents (θ). The dashed portion of the soil hydraulic conductivity curve indicates calculated values. (Modified from Taylor, H. M. and Klepper, B., *Soil Sci.*, 120(1), 57, 1975.)

root) resistance may be associated with decreases in the hydraulic conductivity of the combined soil–root radial pathway (K_{sys}) at low water content, especially below 0.07 cm^3/cm^3 (Figure 13.3; Taylor and Klepper, 1975). The values of K_{sys} and K coincide at a water content of 0.04 cm^3/cm^3, which might be due to a set of compensating errors or might be real (Taylor and Klepper, 1975). As shown in Figure 13.3, soil depth appears to have no effect on the relationship between the hydraulic conductivity of a soil–cotton root system (K_{sys}) and soil water content.

13.4 MACROSCOPIC TYPE I ROOT WATER UPTAKE MODELS

13.4.1 Gardner Model

The uptake of water by a nonuniform root system is based on the assumption that the rate of water movement from soil to the plant root is proportional to the difference between the free energy of the water in the plant root and that in the surrounding soil. The equation for the volume of water taken up per unit time per unit volume of soil (S) is given by (Gardner, 1964):

$$S = \frac{\left(\Phi_{\text{root}} - \Phi\right)}{\ln\left(R_{\text{root}} + R_{\text{soil}}\right)} \tag{13.9}$$

where

Φ_{root} is the matric potential at plant roots.
Φ is the average matric potential in the soil.
R_{root} is the resistance to water movement in the plant roots.
R_{soil} is the resistance to water movement in the soil.

Gardner (1964) assumed that R_{root} may be small and can be neglected. Φ is assumed to be the sa me throughout the root system at any one time. R_{soil} is considered to be inversely proportional to the unsaturated conductivity of the soil and to the density of the root system:

$$R_{\text{soil}} = \frac{1}{BKL} \tag{13.10}$$

where

B is a constant.
K is the unsaturated conductivity of the soil.
L is the length of roots in the unit volume of soil.

The osmotic component of the total potential is not included in Equation 13.9; however, in principle, this can be included by replacing the matric potential by the total potential. The effect of gravity must be taken into account, with the uptake of water in a root system extending to a depth of more than a few centimeters. Dividing the root zone into discrete layers, Equation 13.9 can be written for the ith layer:

$$S_i = Bh\left(\Phi_{\text{root}} - \Phi_i - z_i\right)K_i L_i \tag{13.11}$$

where

S_i is the rate of water uptake per unit cross section of a layer of soil thickness h.
z_i is the distance from the soil surface to the center of the layer.

For n layers, each of thickness h, the total rate of water uptake, q, per unit cross-sectional area is given by

$$q = \sum_{i=1}^{n} S_i = Bh\sum_{i=1}^{n}\left(\Phi_{\text{root}} - \Phi_i - z_i\right)K_i L_i \tag{13.12}$$

Noting that $S_i = hd\theta_i/dt$ (where θ_i is the average water content of the ith layer), Equations 13.11 and 13.12 contain $(n + 1)$ differential equations relating the water content of the root zone to the depth and time. Therefore, the water content of the root zone can be solved as a function of time and depth from specified initial conditions if the total rate of water uptake per unit cross-sectional area (q) and the length of roots in the unit volume of soil (L_i) are known and if the average matric potential in the soil (Φ) and the unsaturated conductivity of the soil (K) are known functions of the water content. The equations must be solved simultaneously, which can be accomplished using numerical methods such as a finite difference that can be used for the time as well as depth.

Gardner (1964) derived integrated soil potential from Equation 13.12. Rearranging terms in Equation 13.12 and removing Φ_{root} from the summation sign yield

$$q = Bh\left(\Phi_{\text{root}} - \bar{\Phi}\right)\bar{K} \tag{13.13}$$

where $\bar{\Phi}$ is the integrated soil potential defined by

$$\bar{\Phi} = \frac{\left[\Sigma K_i L_i \left(\Phi_i - z_i\right)\right]}{\Sigma K_i L_i} \tag{13.14}$$

and $\bar{K}$ is the integrated conductance defined by

$$\bar{K} = \Sigma K_i L_i \tag{13.15}$$

For $\bar{\Phi}$, the Φ at each depth is weighted according to the number of roots at that depth as well as the conductivity of the soil at that depth. Different potential distributions may result in equal values for $\bar{\Phi}$; however, the effect on the plant may not be the same because the total conductance K may differ.

13.4.2 Nimah and Hanks Model

Nimah and Hanks (1973a) proposed the following root water extraction model that includes the effect due to salts in the soil:

$$S(z,t) = \frac{\left[\Phi_{\text{root}} + \left(R_{\text{root}} \times z\right) - \Phi(z,t) - \Phi_{\pi}(z,t)\right] R(z) \times K(\theta)}{\Delta x \times \Delta z} \tag{13.16}$$

where

Φ_{root} is an effective water potential in the root at the soil surface where z (soil depth) is considered zero.

R_{root} is a root resistance term equal to $1 + R_c$; R_c, a flow coefficient in the plant root system, is assumed to be 0.05.

When R_{root} is multiplied by z, the product ($R_{\text{root}} \times z$) will account for the gravity term and friction loss in Φ_{root}. Hence, the root water potential at depth z is higher than the root water potential at the surface (Φ_{root}) by a gravity term and friction loss term. The friction loss in the root is assumed to be independent of flow. In Equation 13.16, $\Phi(z,t)$ is the soil matric potential, $\Phi_\pi(z,t)$ is the salt (osmotic) potential, $R(z)$ is the proportion of total active roots in depth increment Δz, $K(\theta)$ is the hydraulic conductivity at depth z, and Δx is the distance between the plant roots at the point in the soil where $\Phi(z,t)$ and $\Phi_\pi(z,t)$ are measured. Δx is arbitrarily assumed to be one. Dividing by Δz converts the transpiration flux into change of water content per unit time (Nimah and Hanks, 1973a).

The Nimah and Hanks (1973a) model does consider hysteresis or layered soil. The soil properties, primarily the hydraulic conductivity–matric potential relationship and soil water characteristic (matric potential and water content curve), are assumed to not change with time. This situation leads to the assumption of no change in the soil structure. The proportion of total active roots in depth increment Δz, $R(z)$, generally depends on time as well as depth. The value of Φ_{root} term depends on (1) plant growth since they govern the root distribution function $R(z)$, (2) climatic conditions since they define potential transpiration, and (3) soil conditions since $\Phi(z,t)$ and $\Phi_\pi(z,t)$ are soil properties that vary greatly from wet to dry soil. In the model, a value of Φ_{root} is solved to make $S(z,t)$ over the total profile equal to potential transpiration, provided the value of Φ_{root} is higher than the potential below which wilting will occur (Φ_{wilt}). Therefore, Φ_{root} is bounded on the wet end by $\Phi_{root} = 0.0$ and the dry end by $\Phi_{root} = \Phi_{wilt}$.

The computed variation of Φ_{root} during the growing season for alfalfa is shown in Figure 13.4 (Nimah and Hanks, 1973b). Field experiments were conducted in 1970 and 1971 at a Mesa sandy loam field near Vernal, Utah. In 1970, the field was seeded with alfalfa (*Medicago sativa* L.), and by 1971, the alfalfa stand was good. The field was irrigated by a solid-set automated sprinkler system, and excess water was drained by tiles or by natural drainage. The water table was at about 165 cm throughout the year. Φ_{root} increased when irrigation or rain occurred and decreased when the water content decreased. Root water potential (Φ_{root}) for alfalfa rarely

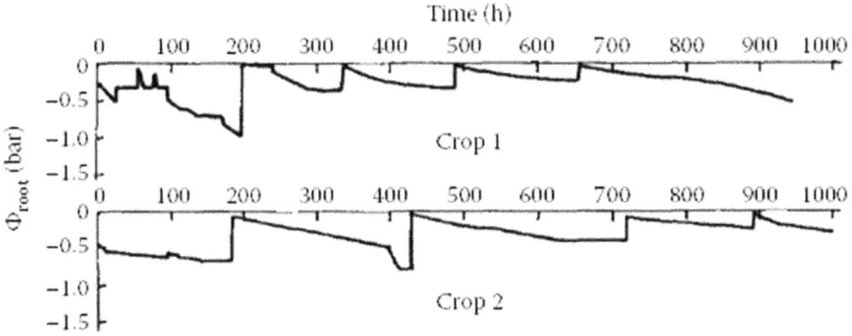

FIGURE 13.4 Variation of root water potential (Φ_{root}) of alfalfa, as computed from experiments in a mesa sandy loam field near Vernal, Utah. The increase in Φ_{root} is due to irrigation or precipitation. (Modified from Nimah, M. N. and Hanks, R. J., *Soil Sci. Soc. Am. Proc.*, 37, 528, 1973b.)

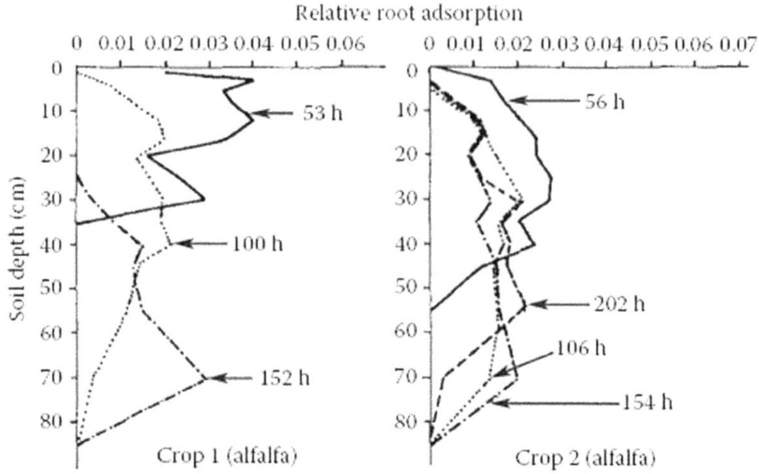

FIGURE 13.5 Relative root extraction of alfalfa as a function of soil depth and time after irrigation, as computed from experiments in a mesa sandy loam field near Vernal, Utah. (Modified from Nimah, M. N. and Hanks, R. J., *Soil Sci. Soc. Am. Proc.*, 37, 528, 1973b.)

reached –1 bar (Figure 13.4), primarily due to the increased addition of water, but could be caused in part by the relatively greater depth (≈85 cm) of the root system.

The computed relative root extraction of alfalfa as a function of soil depth and time after irrigation is shown in Figure 13.5. The relative root extraction at any depth was calculated by multiplying the root water extraction, $S(z,t)$, by the thickness of the soil layer and dividing by the total transpiration rate. During early periods (53–56 h) after irrigation, maximum root extraction occurred within the upper soil depths (0–40 cm) due to high water content (Figure 13.5). As time increased, the relative root extraction zone deepened, for example, at about 150 h, and the maximum extraction was near the bottom of the root zone. The reason was attributed to a higher water content maintained by upward flow from the water table (Nimah and Hanks, 1973b). In the Nimah and Hanks (1973a, b) model, root distribution is assumed constant with time, so that plant demand was forced to be met by an increase in water extraction from deeper layers of soil, where the water was more available.

13.4.3 Hillel Model

In the soil–plant–atmosphere continuum (SPAC) system, water flows in a "transpiration stream" down a gradient of potential energy from soil to root to stem to leaf, whence it evaporates and diffuses out to the atmosphere. Applying the analogy of Ohm's law for an electric current through a series of resistors, the transpiration stream can be represented as a catenary process through successive segments. In each of the segments, the flux is proportional to the potential difference and inversely proportional to the resistance. The Hillel et al. (1976) model considers the root system as a resistance network discharging a series of capacitors (representing water-charged soil layers). The rate of extraction of water from a unit volume of soil, S_w, is represented by

$$S_w = \frac{\Phi_{soil} - \Phi_{plant}}{R_{soil} + R_{roots}} \tag{13.17}$$

where Φ_{soil} is the total potential of soil water (the sum of the matric [Φ_m], gravitational [Φ_g], and osmotic [Φ_π] potentials). The hydraulic resistance to flow in the soil (R_{soil}) is expressed as inversely proportional to the hydraulic conductivity (K) and to the total length of active roots (L) in the unit volume of soil (Gardner, 1964): $R_{soil} = 1/BKL$, where B is an empirical constant accounting for a root-length activity factor. The term Φ_{plant} is the plant water potential at the base of the stem where all roots converge and the plant emerges from the soil with a single water potential, which is called "crown potential" (Φ_c). The hydraulic resistance of the roots (R_{roots}) is the sum of a resistance to adsorption and a resistance to conduction of water. A resistance to conduction of water is taken as a function of the depth of any particular group of roots.

The flow rate $(q_r)_i$, delivered by the roots from any particular layer i in the soil to the crown, is considered as the ratio of the difference in potential between that soil layer and the crown to the total hydraulic resistance encountered:

$$(q_r)_i = \frac{(\Phi_s)_i - \Phi_c}{(R_r)_i + (R_s)_i} \tag{13.18}$$

where

Φ_s is the soil moisture potential.
R_r is the resistance of the roots.
R_s is the hydraulic resistance of the soil.

The total extraction rate (Q) from all volume elements or layers of soil is equal to the transpiration rate, that is, a sum of the contributions of all volume elements within the root zone:

$$Q = \sum_{i=1}^{n} \frac{(\Phi_s)_i - \Phi_c}{(R_r)_i + (R_s)_i} \tag{13.19}$$

where n is the total number of volume elements or layers in the rooting zone. From Equation 13.19, Φ_c can be written as

$$\Phi_c = \left[\sum_{i=1}^{n} \frac{(\Phi_s)_i}{(R_r)_i + (R_s)_i} - Q\right] \Big/ \sum_{i=1}^{n} \frac{1}{(R_r)_i + (R_s)_i} \tag{13.20}$$

When Q, $(R_r)_i$, Φ_s, and R_s for each layer are known, the value of Φ_c can be obtained from Equation 13.20 at successive times by a process of iteration (Hillel et al., 1976). The known value of Q is first approximated for a freely transpiring plant, that is, Q

is equal to the climatically induced transpiration demand. The values of $(R_r)_i$ depend on the rooting-density distribution in the profile and the depth of the particular layer.

The Hillel et al. (1976) model for root water uptake has certain definite limitations. The model does not consider the stomatal control mechanism, by which a plant can restrict its transpiration rate during periods of high demand and may modify the pattern of its own water potential responses to the continuous transpirational demand and the dwindling supply of soil moisture. The characterization of Φ_{plant} in terms of a single-value crown potential (Φ_c) disregards the potential distribution through the root system. The root system in several of the aforementioned models is considered a fixed-resistance network allowing no growth or change in conductive properties, which is unrealistic in most real situations.

13.5 MACROSCOPIC TYPE II ROOT WATER UPTAKE MODELS

13.5.1 Molz and Remson Model

Molz and Remson (1970) devised a root water extraction term that depends on the rooting depth and plant transpiration rate. This sink term function by Molz and Remson is formulated following a hypothetical linear distribution pattern of 40%, 30%, 20%, and 10% uptake rate in each layer of equal depth over the root zone profile:

$$S(z) = -\frac{1.6T}{v}z + \frac{1.8T}{v} \quad 0 \le z \le v \tag{13.21}$$

where

$S(z)$ is the moisture extraction rate per unit volume of soil.
z is the vertical distance positive downward.
v is the vertical length of root system.
T is the transpiration rate per unit area of soil surface, which in some cases is interpreted as an average transpiration rate.

The numbers 40, 30, 20, and 10 have no special significance and should only be regarded as "reasonable" quantities (Molz and Remson, 1970). The total moisture extraction rate from a volume of soil of unit cross section, bounded by the horizontal planes $z = z_1$ and $z = z_2$ (where $z_1 < z_2$), is

$$\int_{z_1}^{z_2} S(z)\,dz \tag{13.22}$$

The extraction rate from the root zone equals the transpiration rate, and substituting Equation 13.21 into Equation 13.22 yields

$$\int_0^v S(z)\,dz = -\frac{1.6T}{v^2}\frac{z^2}{2}\Bigg|_0^v + \frac{1.8Tz}{v}\Bigg|_0^v = T \tag{13.23}$$

By integrating over the appropriate portions of the root zone, it can be verified that Equation 13.21 meets the stated percentage requirements. To account for root systems that are growing, so that $v = v(t)$, Equation 13.21 can be generalized:

$$S(z,t) = -\left(\frac{1.6T}{[v(t)]^2}\right)z + \left(\frac{1.8T}{v(t)}\right) \tag{13.24}$$

Root water extraction models, that is, Equation 13.21 or Equation 13.24, may give reasonable qualitative estimates for higher moisture contents. However, Equations 13.21 and 13.24 disregard the extraction conditions at the top and bottom of the root zone, that is, ignore the effect of soil water content on the distribution of root water uptake. For example, as the upper layers of soil dry, more of the transpiration requirement comes from deeper roots in the more moist soil, which is not accounted for in Equation 13.21 or Equation 13.24. Even if the top soil layer is desiccated by evapotranspiration, the root system always extracts 40% of the total transpiration from the top quarter of the root zone. For a steady-state condition, the moisture extraction pattern is static, and Equation 13.8 may yield reasonable results.

Root water extraction models can have an extraction term depending on moisture content, depth, and transpiration rate. In the Molz and Remson (1970) model, the transpiration demand is distributed throughout the root zone according to the product of soil water diffusivity [$D(\theta)$] and a quantity termed the "effective root density," [$R(z)$], which relates to that proportion of the roots effective in absorbing moisture. The extraction term is given by

$$S(z,\theta) = T\left[\frac{R(z)D(\theta)}{\int_0^v R(z)D(\theta)dz}\right] \tag{13.25}$$

$S(z, \theta)$ is defined so that $\int_0^v S(z,\theta)dz = T$. Molz and Remson (1970) calculated a root distribution similar in general form to that given by Gardner (1964), where effective root density distribution decreases with depth from a maximum at the soil surface However, the moisture extraction by roots in Equation 13.25 may require a better physical understanding of effective root density that is defined as the weight of roots per bulk volume of soil, $R(z)$, and the root systems with respect to the distribution of root tips and root radii.

Because soil water diffusivity may be very small in the immediate vicinity of a root as compared to the surrounding soil, a "uniform" moisture extraction may occur even in moderately dense root systems. It is well known that roots are not uniformly permeable to moisture, and for a given root radius, the root surface is most permeable a short distance behind the root tip. Therefore, $R(z)$ may not accurately reflect the actual root density (Molz and Remson, 1970).

13.5.2 Feddes Model

In general, the major shortcoming of the previously mentioned macroscopic models (i.e., Gardner, 1964; Molz and Remson, 1970; Nimah and Hanks, 1973a; and Hillel et al., 1976) is that they do not take into consideration the reduction of soil water content directly, and the reduction of soil water content is reflected in the hydraulic conductivity. To describe the sink term for actual transpiration, Feddes et al. (1976) first proposed the sink term for each time, *t*, as a function of soil water content, θ, and later (Feddes et al., 1978) as a function of soil water potential (Φ):

$$S(\Phi) = \alpha(\Phi) S_p \tag{13.26}$$

where

$\alpha(\Phi)$ is a dimensionless prescribed function of the soil water potential ($0 \le \alpha \ge 1$).
S_p is the potential water extraction by roots (cm^3/cm^3/day),

The variable S_p in Equation 13.26 is reduced by means of the factor $\alpha(\Phi)$ under nonoptimal conditions, that is, either too dry or too wet. As shown in Figure 13.6, to describe water stress, Feddes et al. (1978) proposed a linear reduction function parameterized by four critical values of the soil water potentials, $\Phi_4 < \Phi_3 < \Phi_2 < \Phi_1$:

$$\alpha(\Phi) = \begin{cases} \dfrac{\Phi - \Phi_4}{\Phi_3 - \Phi_4}, & \Phi_3 > \Phi > \Phi_4 \\ 1, & \Phi_2 \ge \Phi \ge \Phi_3 \\ \dfrac{\Phi - \Phi_1}{\Phi_2 - \Phi_1} & \Phi_1 > \Phi > \Phi_2 \\ 0, & \Phi \le \Phi_4 \text{ or } \Phi \ge \Phi_1 \end{cases} \tag{13.27}$$

Critical values of the soil water potential in Equation 13.27 describe plant stress due to dry (Φ_3 and Φ_4) and wet soil conditions (Φ_1 and Φ_2) (Figure 13.6). Water uptake above Φ_1 (oxygen deficiency point, i.e., wetter than some arbitrary anaerobiosis point) and below Φ_4 (wilting point) is assumed to be zero. Water uptake is maximal between Φ_2 and Φ_3 (reduction point). Between Φ_1 and Φ_2 and between Φ_3 and Φ_4, a linear variation is assumed. Wesseling (1991) provided critical Φ values of the sink term function $\alpha(\Phi)$ (Figure 13.4; Equation 13.27) for some agricultural crops (Table 13.1). The value of Φ_3 depends on the water demand of the atmosphere. The Feddes et al. (1978) model uses a low ($\Phi_{3,low}$) and a high or a less negative ($\Phi_{3,high}$) value of Φ_3 to differentiate between low and high atmospheric demands and T_p rates. High ($\Phi_{3,high}$) and low ($\Phi_{3,low}$) matric potential values for different crops are presented in Table 13.2 (Taylor and Ashcroft, 1972). Water should be applied at these matric potentials (Table 13.2) for maximum yields of various crops grown in deep, well-drained soil that is fertilized and otherwise managed for maximum production.

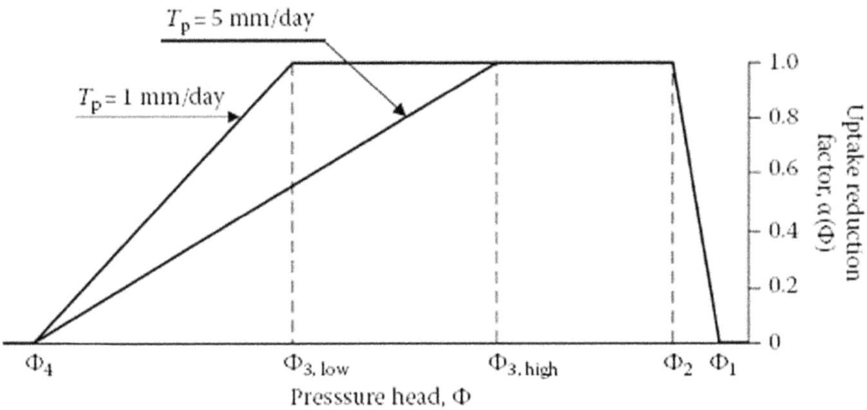

FIGURE 13.6 Schematic of the water stress uptake reduction function (Equation 13.27). (As used by Feddes, R. S., et al., *Simulation of Field Water Use and Crop Yield. Netherlands Centre for Agricultural Publishing and Documentation Simulation Monographs,* John Wiley & Sons, New York, 1978.)

TABLE 13.1
Critical Matric Potential (Φ) Values (cm) of the Sink Term Function α(Φ) (Figure 13.6; Equation 13.27) for Some Agricultural Crops

Crop	Φ_1	Φ_2	$\Phi_{3,high}$	$\Phi_{3,low}$	Φ_4
Potatoes	−10	−25	−320	−600	−16,000
Sugar beet	−10	−25	−320	−600	−16,000
Wheat	0	−1	−500	−900	−16,000
Pasture	−10	−25	−200	−800	−8,000
Corn	−15	−30	−325	−600	−8,000

Source: After Wesseling, J. G., Meerjarige simulatie van grondwaterstroming voor verschillende bodemprofielen, grondwatertrappen en gewassen met het model SWATRE, DLO-Staring Centrum, Rapport/DLO-Staring Centrum no. 152, Wageningen, the Netherlands, 1991.

Assuming a homogeneous root distribution over the rooting depth, Feddes et al. (1978) defined S_p as

$$S_p = \frac{T_p}{z_r} \tag{13.28}$$

where

T_p is the potential transpiration rate.
z_r is the root zone depth.

The potential root water extraction rate at a certain depth, $S(z)$, for nonhomogeneous root distributions may be determined by either the root mass or the root length

TABLE 13.2

High ($\Phi_{3,high}$) and Low ($\Phi_{3,low}$) Values (cm) for Critical Matric Potential Φ_3 of the Sink Term Function $\alpha(\Phi)$ (Figure 13.6; Equation 13.27) for Various Crops

Crop Type	Crop	$\Phi_{3,high}$	$\Phi_{3,low}$	Crop Type	Crop	$\Phi_{3,high}$	$\Phi_{3,low}$
Vegetative Crops	Alfalfa	−1500	−1500	**Fruit Crops**	Lemons	−400	−400
	Beans (snap and lima)	−750	−2000		Oranges	−200	−1000
	Cabbage	−600	−700		Deciduous fruit	−500	−800
	Canning peas	−300	−500		Avocadoes	−500	−500
	Celery	−200	−300		Grapes (early season)	−400	−500
	Grass	−300	−1000		Grapes (during maturity)	−1000	−1000
	Lettuce	−400	−600		Strawberries	−200	−300
	Tobacco	−300	−800		Cantaloupe	−350	−450
	Sugarcane (tensiometer)	−150	−500		Tomatoes	−800	−1500
	Sugarcane (blocks)	−1000	−2000		Bananas	−300	−1500
	Sweet corn	−500	−1000	**Grain Crops**	Corn (vegetative period)	−500	−500
	Turfgrass	−240	−360		Corn (during ripening)	−8000	−12000
Root Crops	Onion (early growth)	−450	−550		Small grains (vegetative period)	−400	−500
	Onion (bulbing time)	−550	−650		Small grains (during ripening)	−8000	−12000
	Sugar beets	−400	−600	**Seed Crops**	Alfalfa (prior to bloom)	−2000	−2000
	Potatoes	−300	−500		Alfalfa (during bloom)	−4000	−8000
	Carrots	−550	−650		Alfalfa (during ripening)	−8000	−15000
	Broccoli (early)	−450	−550		Onion (at 7 cm depth)	−4000	−6000
	Broccoli (after budding)	−600	−700		Onion (at 15 cm depth)	−1500	−1500
	Cauliflower	−600	−700		Lettuce (during productive phase)	−3000	−3000

Source: After Taylor, S. A. and Ashcroft, G. L., *Physical Edaphology: The Physics of Irrigated and Nonirrigated Soils*, Freeman and Co., San Francisco, CA, 1972.

density (RLD) (the length of roots per unit volume of soil), $L_r(z)$, as a fraction of the total RLD over the rooting depth z_r:

$$S_p(z) = \frac{L_r(z)}{\int_{z_r}^{0} L_r(z)\,dz} T_p \tag{13.29}$$

The changes in storage of water in the vegetation are assumed to be negligible and the actual rate of transpiration T at time t is given by the spatial integral of the volumetric rate of water uptake, $S(z)$, over the entire profile:

$$T(t) = \int_{z_r}^{0} S(z)\,dz \tag{13.30}$$

An example of root water uptake profile predicted by the Feddes et al. (1978) model under field conditions is presented in Figure 13.7. The Feddes et al. model, which has been implemented in the HYDRUS-1D simulation model (Šimůnek et al., 2008), was tested on a sandy loam flood-irrigated mature pecan (*Carya illinoinensis*) orchard near Las Cruces, New Mexico (Deb et al., 2011). As shown in Figure 13.5, during early periods after the irrigation, the roots extracted water at a higher rate within 0–40 cm deep, more water was being extracted from the upper depth of the soil profile (above 20 cm), and gradually less from the lower depth. Deb et al. (2011) reported that total RLD was 52%, 33%, and 13% in the 0–20, 20–40, and 40–60 cm soil depths, respectively, and the root water uptake pattern followed the RLD distribution immediately after irrigation. However, root water uptake rates gradually increased in deeper depths as the upper depths of the soil profile became more stressed (water content <0.1 cm^3/cm^3) and the soil water was still sufficient at deeper depths (water content > 0.11 cm^3/cm^3), which allowed the lower root domain to remain less stressed.

13.5.3 van Genuchten Model

In addition to a piecewise linear reduction function for $\alpha(\Phi)$ (Equation 13.27) (Feddes et al., 1978), the continuous smooth reduction function for $\alpha(\Phi)$ proposed by van Genuchten (1987) has also been commonly used. The van Genuchten model utilizes a smooth S-shaped curve (Equation 13.26) to account for water stress:

$$\alpha(\Phi) = \frac{1}{1 + (\Phi / \Phi_{50})^p} \tag{13.31}$$

where

Φ is the soil water potential at a particular spatial location.

Φ_{50} is the soil water potential at which root water uptake rate is reduced by 50% during conditions of negligible osmotic stress.

p is a dimensionless fitting parameter (Figure 13.8).

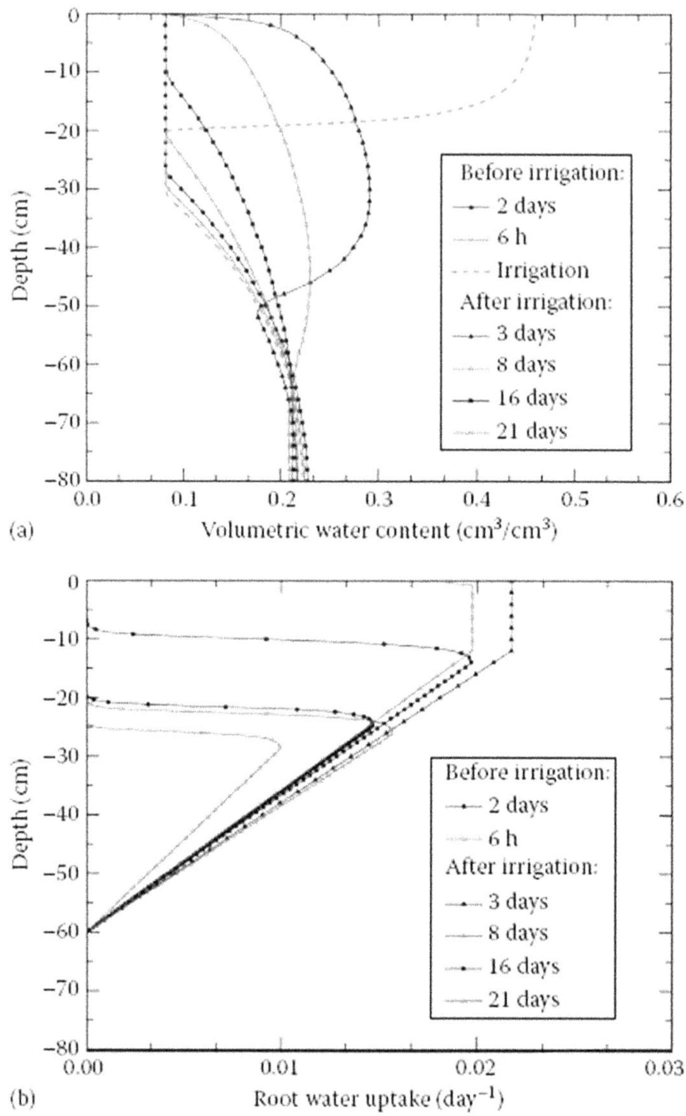

FIGURE 13.7 Simulated (a) soil water content and (b) root water uptake profiles (using the Feddes et al., 1978, model implemented in the HYDRUS-1D simulation model) at different times before and after an irrigation event in a sandy loam flood-irrigated mature pecan orchard located near Las Cruces, New Mexico. A negative sign on the y-axis indicates depth below the soil surface. (Redrawn from Deb, S. K., et al., *Soil Sci. Soc. Am. J.*, 75(5), 1667, 2011.)

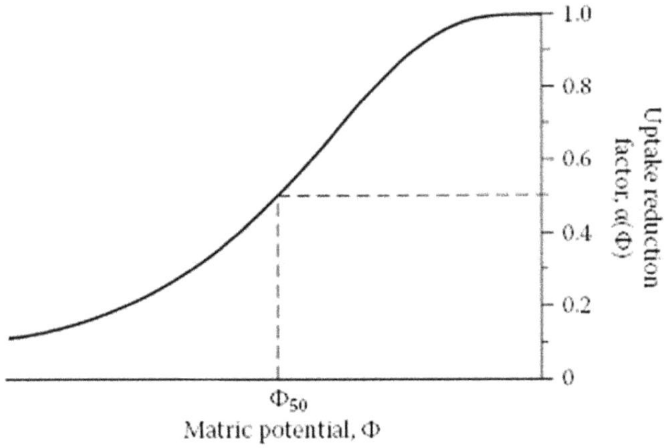

FIGURE 13.8 Schematic of the water stress uptake reduction function (Equation 13.31). (As used by van Genuchten, M. T., A numerical model for water and solute movement in and below the root zone, *Res. Rep.* 121, U.S. Salinity Laboratory, Agricultural Research Service, U.S. Department of Agriculture, Riverside, CA, 1987.)

Equation 13.31 is analogous to the S-shaped salinity stress yield reduction curve (van Genuchten and Hoffman, 1984). Water and salinity stress are assumed to have similar effects on yield and hence on transpiration. Therefore, the S-shaped function is also applicable to water-stress response data. The exponent p determines the steepness of the transition from potential to reduced uptake rates as Φ decreases.

In Equation 13.31, influence of the osmotic potential reduction can be either additive or multiplicative to the matric potential (Φ) reduction. The van Genuchten stress response function, $\alpha(\Phi)$, is appropriate for environments where saturated or near-saturated soil conditions are not expected to exist for significant periods of time, that is, when saturated or near-saturated soil conditions occur for only relatively short periods of time. The values reported in the literature are in the range of approximately −1000 to −5000 cm for Φ_{50} and 1.5–3 for p (van Genuchten, 1987; Skaggs et al., 2006a). According to Skaggs et al. (2006a), Φ_{50} may be regarded as an effective parameter, which lumps together in some unspecified way the reduction in water uptake due to reduced water potential at the root surface as well as reduced flow of water to the root surface. However, they used $\Phi_{50} = -1500$ cm and $p = 2$ to evaluate the performance of the van Genuchten (1987) S-shaped model.

13.5.4 Prasad Model

As noted earlier (Equation 13.28), the Feddes model (Section 13.5.2) assumes a constant rate of root water extraction for the entire depth of the root zone. However, when soil moisture is not limiting, the roots can mainly extract water from the upper soil layers. Prasad (1988) proposed a linear extraction model for root water uptake, with the maximum extraction at the top that reduces linearly to zero at the bottom of the root zone. For potential transpiration conditions, S_p is given by the equation of Hoogland et al. (1981):

$$S_{max} = a_j - b_j z \tag{13.32}$$

where S_{max} is the maximum root water extraction rate when soil moisture is not limiting, and a_j and b_j are the intercept and slope on the jth day, respectively. a_j and b_j depend on the rooting depth and rate of evapotranspiration in such a way that S_{max} becomes zero at the bottom, that is, the boundary condition at the bottom of the root zone is satisfied if

$$a_j - b_j z_{rj} = 0 \tag{13.33}$$

where z_{rj} is the rooting depth on the jth day. S_{max} satisfies the following equation:

$$\int_0^{z_{rj}} S_{max}\, dz = T_j \tag{13.34}$$

where T_j is the potential transpiration rate on the jth day. Equations 13.32 through 13.34 together give

$$a_j = \frac{2T_j}{z_{rj}}; b_j = \frac{2T_j}{z_{rj}^2} \tag{13.35}$$

From Equation 13.32:

$$S_{max} = \frac{2T_j}{z_{rj}}\left(1 - \frac{z}{z_{rj}}\right) \tag{13.36}$$

The variation of the model's parameters a_j and b_j over the season for different crops is shown in Figure 13.9. a_j remains constant until the maximum transpiration rate is attained. The initial constant values of a_j do not differ much among crops, which vary in the range 0.008–0.013 per day (Figure 13.9). b_j has high initial values because of the fast rate of root growth. After a day of maximum transpiration, both a_j and b_j show only a small change and eventually decrease because of the small value of T_j and the presence of the square of the root depth term in the denominator in Equation 13.35, the values of b_j at harvest are very low. The reduction in a_j at harvest from its initial values is approximately an order of magnitude, particularly when the transpiration rate does not become zero.

13.5.5 Ojha and Rai Model

The following empirical root water uptake model proposes that the extraction rate varies nonlinearly with the root depth. For potential transpiration conditions, the potential rate of soil water extraction, S_{max}, is given by (Ojha and Rai, 1996):

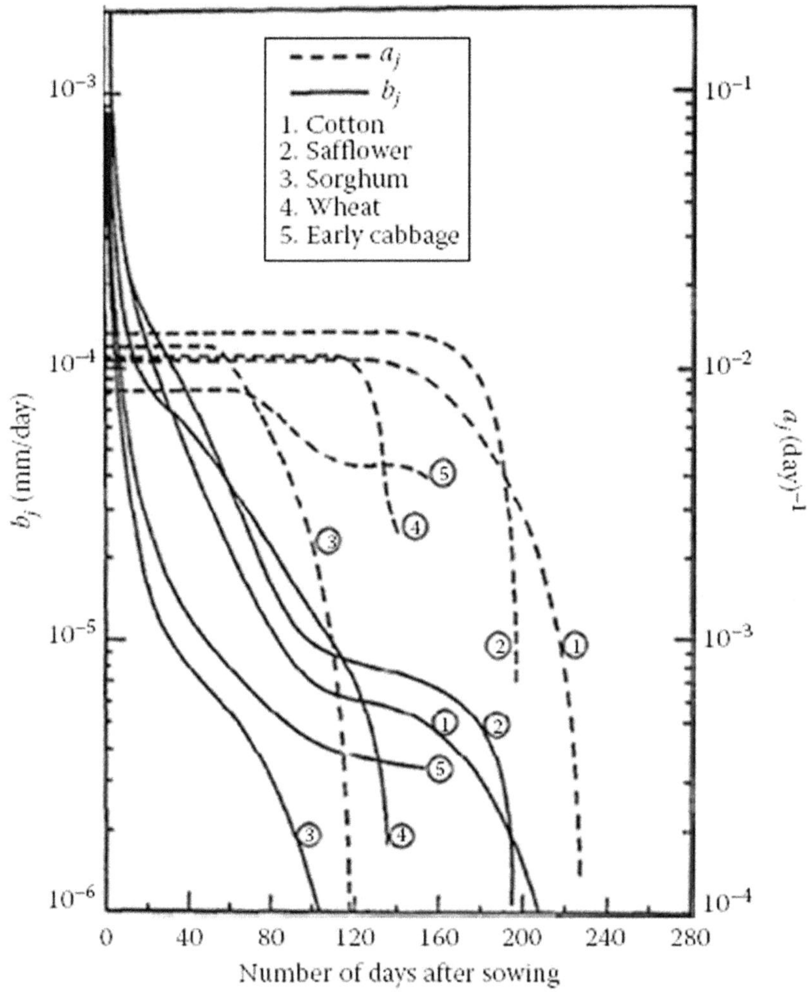

FIGURE 13.9 Variation of intercept a_j and slope b_j in the Prasad (1988) linear root water uptake model. (Reprinted from Prasad, R., A linear root water uptake model, *J. Hydrol.*, 99, 297–306, 1988. Copyright 1988, with permission from Elsevier.)

$$S_{max} = \alpha \left[1 - \left(\frac{z}{z_{rj}} \right) \right]^{\beta} \quad 0 \le z \le z_{rj} \tag{13.37}$$

where

α and β are model parameters.

z is the depth below the soil surface.

z_{rj} is the root depth on the *j*th day.

At $z = z_{rj}$, $S_{max} = 0$, and at $z = 0$ (soil surface), S_{max} attains its maximum value. Thus, extraction is maximum at the top and zero at the bottom of the root, and S_{max} has to satisfy the following equation:

$$\int_0^{z_{rj}} S_{max} dz = T_j \tag{13.38}$$

where T_j is the potential rate of transpiration on the jth day. Substituting for S_{max} from Equation 13.37 into Equation 13.38 yields:

$$T_j = \frac{\alpha z_{rj}}{(\beta + 1)} \tag{13.39}$$

Therefore, α can be obtained from Equation 13.39:

$$\alpha = \frac{T_j(\beta + 1)}{z_{rj}} \tag{13.40}$$

Using Equation 13.40 in Equation 13.37 yields

$$S_{max} = \left[\frac{T_j}{z_{rj}}(\beta + 1)\left(1 - \frac{z}{z_{rj}}\right)^{\beta}\right] \quad 0 \le z \le z_{rj} \tag{13.41}$$

From Equation 13.41, it is evident that β quantifies the nonlinearity of the root water extraction pattern. For $\beta = 0$, Equation 13.41 reduces to a constant rate extraction model (Feddes et al., 1978) with $S_{max} = T_j/z_{rj}$. For $\beta = 1$, Equation 13.41 reduces to a linear extraction model (Prasad, 1988) with $S_{max} = 2T_j / z_{rj} - 2T_j\left(z / z_{rj}^2\right)$.

Moisture-extraction patterns for different crops need to be represented by the crop-specific values of optimal β. The optimal β is obtained on the basis of the coefficient of determination (R^2), by minimizing the deviations between the field observed and the Ojha and Rai (1996) model predicted root zone soil moisture depletions. The value of β that gives maximum R^2 (equals unity for a perfect fit) for a particular crop is the optimum value of β for that crop. The optimal value of the parameter varies between 1.26 and 2.68 for different crops (Shankar et al., 2012).

13.5.6 Li et al. Model

The constant extraction rate model of Feddes et al. (1978) as well as the linear depth-dependent model of Prasad (1988) do not adequately describes the nonuniform distribution of roots in the soil profile. The nonlinear root uptake model proposed by Ojha and Rai (1996) has inherent simplicity because it requires only a root depth measurement.

Li et al. (1999) proposed that the potential transpiration should be distributed in compliance with the pattern of observed root distributions. Accordingly, based on an exponential root length distribution function, Li et al. developed a nonlinear root water uptake model:

$$S_{\max} = \frac{K_{z1-z2} T_j}{|z_1 - z_2|} \tag{13.42}$$

where

T_j is the potential rate of transpiration on the jth day.

$S_{\max}$ is the mean value of root water uptake for the compartment depth between z_1 and z_2.

The Li et al. (1999) model assumes that under optimal soil water conditions for plant growth, the fraction of water extracted between z_1 and z_2 is equivalent to its corresponding fraction of the total root length. The compartments for calculating $S_{\max}$ should correspond to those set in the flow domain for solving the water flow equation. The fraction of the total root length between depths z_1 and z_2 (K_{z1-z2}) for a given area A is

$$K_{z1-z2} = \frac{\int_{z1}^{z2} A\mathrm{RD}dz}{\int_{0}^{z_{rj}} A\mathrm{RD}dz} = \frac{\int_{z1}^{z2} \mathrm{RD}dz}{\int_{0}^{z_{rj}} \mathrm{RD}dz} \tag{13.43}$$

where

z_{rj} is the rooting depth on the jth day.

RD is the root length density (RLD).

z is the soil depth.

The exponential root distribution function is expressed as

$$\mathrm{RD} = \mathrm{RD}_0 \left[1 - \frac{1}{1+\exp(-bz)} + 0.5\exp(-bz)\right] \tag{13.44}$$

where

RD_0 is the RLD at $z = 0$.

b is an empirical extinction coefficient.

Using Equations 13.43 and 13.44,

$$K_{z1-z2} = \frac{\ln\left[\left(1+\exp(-bz_1)\right)/\left(1+\exp(-bz_2)\right)\right] + 0.5\left[\exp(-bz_1) - \exp(-bz_2)\right]}{\ln\left[2/\left(1+\exp(-bz_{rj})\right)\right] + 0.5\left[1 - \exp(-bz_{rj})\right]} \tag{13.45}$$

The extinction coefficient b and the rooting depth z_r are the only parameters that determine the distribution of water uptake by roots throughout the soil profile. The

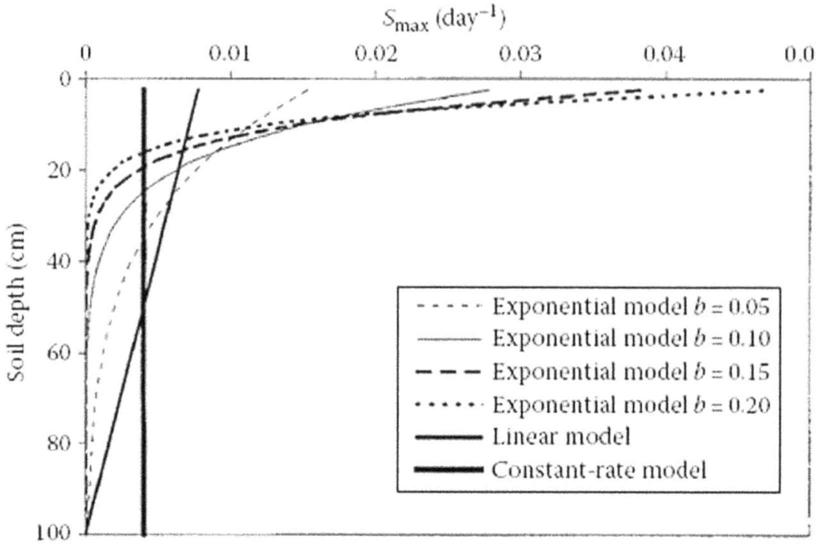

FIGURE 13.10 Estimated maximum soil water uptake with the exponential model of Li et al. (1999) (using the extinction coefficient *b* values of 0.05, 0.10, 0.15, and 0.20 cm^{-1}), linear model of Prasad (1988), and constant-rate model of Feddes et al. (1978). The potential transpiration rate was set to 0.4 cm/day. (Reprinted from Li, K. Y., et al., *Can. J. Soil Sci.*, 79, 333, 1999. With permission.)

Li et al. (1999) model, incorporated in the Soil Water Atmosphere Plant (SWAP) simulation model (van den Broek et al., 1994), is sensitive to empirical extinction coefficient *b* (Figure 13.10). The sensitivity of the Li et al. model to parameter *b* was tested on a clay loam soil cropped to soybeans and on a sandy loam soil cropped to corn, near Ottawa. For a soil with a compacted plow layer, water extraction increased at upper layers but decreased at the lower layers as coefficient *b* increased. Although different *b* values resulted in quite different patterns of S_{max} across the soil profile, the total extraction pattern over the entire soil profile remained the same. This is because the daily maximum water extraction equals the daily potential transpiration, independent of coefficient *b*. Using different *b* values (e.g., 0.05, 0.08, 0.11, and 0.17 cm^{-1}), Li et al. (1999) found that larger *b* values resulted in lower soil water content predictions for the upper layers but slightly higher water contents for the lower layers.

13.5.7 Kang et al. Model

The Kang et al. (2001) model assumes that soil water content can fully satisfy the crop root extraction, and its distribution in the root zone is uniform. The relationship at a given time between root water uptake rate at a given depth and the maximum for high root density is expressed approximately by an exponential equation:

$$S_0\left(z,t\right) = 0.9046 S_{(0,\max)} e^{-1.8z/z_r} \qquad (13.46)$$

where

$S_0(z,t)$ is the maximum water uptake rate without soil water deficit at a given soil depth z.

$S_{(0,\max)}$ is the maximum water uptake rate for high root density.

z_r is the depth of the effective root zone.

While neglecting the microchange of plant water content, the potential transpiration rate is the sum of root water extraction in the effective root zone depth, which can be expressed as

$$T_p(t) = \int_0^{z_r} S_0(z,t)\,dz \tag{13.47}$$

where $T_p(t)$ is the potential transpiration rate. After substitution of Equation 13.46 into Equation 13.47, and by integration, it yields

$$S_{(0,\max)} = \frac{T_p(t)}{0.9046} \times \frac{1.8}{\left(1-e^{-1.8}\right)z_r} \tag{13.48}$$

Substituting $S_{(0,\max)}$ in Equation 13.46, the maximum water uptake rate without soil water deficit at a given soil depth can be written as follows:

$$S_0(z,t) = T_p(t)\frac{1.8e^{-1.8z/z_r}}{\left(1-e^{-1.8}\right)z_r} \tag{13.49}$$

If the root extraction rate is limited by soil moisture content, Equation 13.49 for actual water uptake rate at depth z can be written as the Feddes et al. (1978) model:

$$S(z,t) = f(\theta)S_0(z,t) \tag{13.50}$$

where $f(\theta)$ is a dimensionless sink term ranging between 0 and 1 as a function of the soil water content, which is calculated by the following equation:

$$f(\theta) = \left[\frac{\theta(z,t)-\theta_{wp}}{\theta_{FC}-\theta_{wp}}\right]^{0.6967} \tag{13.51}$$

where

$\theta(z,t)$ is the soil water content at a given time t and a given depth z.

θ_{FC} is the field capacity of soil and is defined as the soil water content at −10 kPa soil water potential (usually θ_{FC} is defined as soil water content at −1/3 bar or −33 kPa).

θ_{wp} is the soil water content at the permanent wilting point.

13.5.8 Vrugt et al. Model

For uniform crops with a spatially uniform root water uptake pattern, root water uptake is generally considered simply to be a function of the vertical dimension only. However, a multidimensional approach would be better for row crops and tree lines where root water uptake varies in space, or for isolated trees in large monocultures where the process of water uptake is complex. The multidimensional root water uptake models allow more accurate quantification of spatial variability of soil water regime as well as water flux densities below the root zone.

Vrugt et al. (2001a, 2001b) developed one-, two-, and three-dimensional root water uptake models. As the basis of the proposed root water uptake models, they used the exponential model given by Raats (1974) and extended it with a radial component. The Raats root water uptake model is

$$\beta(z) = \frac{1}{\lambda} e^{-(z/\lambda)} \tag{13.52}$$

where

$\beta(z)$ is the spatial root water uptake distribution with depth.

λ is selected such that at depth λ (L) the cumulative root water uptake is 63% of total uptake over the whole root zone.

z is the depth in the soil profile ($z \geq 0$).

The Vrugt et al. (2001a, 2001b) model excludes the parameter λ but includes three additional parameters in the Raats (1974) root water uptake model:

$$\beta(z) = \left[1 - \frac{z}{z_m}\right] e^{-p_z / z_m \left|z^* - z\right|}; \quad z \geq 0 \tag{13.53}$$

where

$\beta(z)$ is a dimensionless shape factor describing the spatial distribution of potential root water uptake with depth (Vrugt et al., 2001b).

z_m is the maximum rooting depth.

p_z and z^* are empirical parameters.

p_z and z^* parameters were included to provide for zero root water uptake at $z \geq z_m$, to account for nonsymmetrical (or asymmetrical) root water uptake with depth, and also to allow for a maximum root water uptake rate at any depth, $z^{max}(0 \leq z^{max} \leq z_m)$. The nonsymmetry (or asymmetry) in root water uptake with soil depth is determined by the ratio between p_z for $z \leq z^*$ and the p_z value for $z > z^*$. To reduce the number of parameters, p_z is set to unity for values of $z > z^*$, whereas it is a fitted value for $z \leq z^*$. The value of $z = z^{max}$ can be calculated from the first derivative, that is, $d\beta/dz = 0$.

As the potential cumulative root water uptake must equal the potential transpiration rate (T_p), the normalized root water distribution, S_m, with depth (one-dimensional) is computed from

$$S_m(z) = \frac{\beta(z)T_p}{\int_{z=0}^{z=z_m} \beta(z)dz} \quad (13.54)$$

Since in most field studies z_m is known a priori, the uptake model of Equation 13.53 contains only two unknown parameters p_z and z^*. The actual root water uptake at any particular spatial location can be calculated from the Feddes model (1978):

$$S(\Phi,z) = \alpha(\Phi)S_m(z) \quad (13.55)$$

Vrugt et al. (2001a, 2001b) used the S-shaped reduction function [$\alpha(\Phi)$] given by van Genuchten (1987) (Equation 13.31) to account for water stress. The actual transpiration rate, T_a (one-dimensional), can be computed from

$$T_a = \int_0^{z_m} S(\Phi,z)dz \quad (13.56)$$

Information on microscopic and macroscopic root water uptake models can also be found in Molz (1981), Hopmans and Bristow (2002), Wang and Smith (2004), Feddes and Raats (2004), Green et al. (2006), and Skaggs et al. (2006b).

The roots are dynamically distributed and spatiotemporal distribution of roots influence the water uptake, specially the compensatory uptake from wetter regions in the root zone. Feddes–Jarvis model uses an empirical function, and Couvreur model uses a physical description of water flow in the soil–root system (Cai et al., 2017). Couvreur et al. (2012) developed a three dimensional macroscopic root water uptake model based on root system hydraulic structure. For crops, such as wheat, that are sown densely, the three dimensional Couvreur model can be upscaled to 1-D model that can use hydraulic conductance of soil-root system (Couvreur et al., 2014) to estimate root water uptake. Root system and root architecture are highly variable and accurate determination of the distribution using nondestructive (Rhizotrons) and destructive (large core) methods is seldom sufficient. For further information on these models, readers are referred to Cai et al. (2017).

13.5.9 Models Incorporating Soil Salinity

Homaee et al. (2002a) modified this macroscopic function using a soil salinity reduction term, $\alpha(h_0)$ instead of$\alpha(h)$:

$$S = \alpha(h_0)S_{max} \quad (13.57)$$

The piecewise linear function in the form of the Maas and Hoffman (1977) model can be written as:

$$\alpha(h_0) = 1 - \frac{\alpha}{360}(h_0^* - h_0) \quad (13.58)$$

where

h_0^*(L) is the salinity threshold value after which the uptake linearly decreases with decreasing h_0 (L).

α is the slope of yield or transpiration (decrease per one unit of salinity).
360 is the constant that converts the salinity-dependent slope into the osmotic head.

Although the Maas and Hoffman (1977) model was a significant breakthrough toward quantifying plant response to salinity, some studies also reported a nonlinear relationship under salinity stress. Therefore, van Genuchten and Hoffman (1984) were the first to propose a nonlinear function as:

$$\alpha(h_0) = \frac{1}{1 + \left[h_0 / h_{050}\right]^p} \tag{13.59}$$

where
h_{050} is the osmotic head at which the relative uptake declines by 50% (L).
p is a unitless parameter related to climate, crop, and soil.

Dirksen et al. (1993) modified Equation (13.59), incorporating the threshold value (h_0^*), and defined the following:

$$\alpha(h_0) = \frac{1}{1 + \left[(h_0^* - h_0) / (h_0^* - h_{050})\right]^p} \tag{13.60}$$

This equation can model plant response to salinity more realistically than Equation (13.59), but it is difficult to obtain h_{050} and p parameters. Therefore, Homaee et al. (2002b, 2002c, 2002d) defined the shape parameter p as a function of h_{050} and h_0^* and proposed the following:

$$p = \frac{h_{050}}{h_{050} - h_0^*} \tag{13.61}$$

To resolve the limitation of obtaining h_{050}, Homaee et al. (2002c, 2002d) used $h_{0\max}$ instead of h_{050}, beyond which relative uptake is not significantly affected by salinity increase. This indicates that at $h_0 \le h_{0\max}$, the plant is still alive with a minimal rate of biological activities. The modified nonlinear two-threshold reduction function was written as:

$$\alpha(h_0) = \frac{1}{1 + \left[(1 - \alpha_0) / \alpha_0\right]\left[(h_0^* - h_0) / (h_0^* - h_{0\max})\right]^p} \tag{13.62}$$

where $\alpha_0$is the relative uptake at $h_{0\max}$. The empirical parameter, p, similar to Equation (13.61), can be obtained from Homaee et al. (2002c, 2002d):

$$p = \frac{h_{0\max}}{h_{0\max} - h_0^*} \tag{13.63}$$

The three models of Maas and Hoffman (1977), Dirksen et al. (1993), and Homaee et al. (2002c, 2002d) demonstrated a threshold (h_0^*) after which the yield began to decline. However, Maggio et al. (2007) found that there existed another break point after which the slope of yield reduction function to salinity was reduced. To better describe plant adaptation to salinity higher than the break point and avoid the limitation of obtaining the parameters (e.g., h_{050}, h_0^*, and $h_{0\max}$), Yang et al. (2020) proposed a new model based on an inverse function as:

$$\alpha(EC_e) = \frac{1}{A(EC_e - B)} + C \tag{13.64}$$

where
- A is a shape parameter.
- B is the asymptote (parallel to the y-axis; dS/m).
- C is the asymptote (parallel to the x-axis), which is the potential lowest value for relative yield and relative ET.

When $\alpha(EC_e) = 1$, the calculated EC_e from this equation is the salinity threshold value (EC_e^*).

PROBLEMS

13.1 What is root water uptake? List what is taken up by roots during uptake.
13.2 Write a note on linear and nonlinear root water uptake models.
13.3 How does salinity affect root water uptake?
13.4 Do a literature search on root water uptake and write a summary of the different models.

REFERENCES

Ahuja, L. R., K. W. Rojas, J. D. Hanson, M. J. Shaffer, and L. Ma. 2000. *The Root Zone Water Quality Model: Modeling Management Effects on Water Quality and Crop Production*. Water Resources Publications, Highlands Ranch.

Baver, L. D., W. H. Gardner, and W. R. Gardner. 1972. *Soil Physics*, 4th edn. John Wiley & Sons, New York.

Cai, G., J. Vanderborght, V. Couvreur, C.M. Mboh, and H. Vereecken. 2017. Parameterization of root water uptake models considering dynamic root distributions and water uptake compensation. *Vadose Zone J.* 17:160125. doi:10.2136/vzj2016.12.0125.

Couvreur, V., J. Vanderborght, L. Beff, and M. Javaux. 2014. Horizontal soil water potential heterogeneity: Simplifying approaches for crop water dynamics models. *Hydrol. Earth Syst. Sci.* 18:1723–1743. doi:10.5194/hess-18-1723-2014.

Couvreur, V., J. Vanderborght, and M. Javaux. 2012. A simple three-dimensional macroscopic root water uptake model based on the hy-draulic architecture approach. Hydrol. Earth Syst. Sci. 16:2957–2971. doi:10.5194/hess-16-2957-2012.

Deb, S. K, M. K. Shukla, and J. G. Mexal. 2011. Numerical modeling of water fluxes in the root zone of a mature pecan orchard. *Soil Sci. Soc. Am. J.* 75(5): 1667–1680.

Dirksen, C., J. B. Kool, P. Koorevaar, and M. Th. van Genuchten. 1993. HYSWASOR simulation model of hysteretic water and solute transport in the root zone. In: Russo, D. and Dagan, G. (eds.), *Water Flow and Solute Transport in Soils*. Springer Verlag, New York, pp. 99–122.

Fayer, M. J. 2000. UNSAT-H Version 3.0: Unsaturated soil water and heat flow model. Theory, user manual, and examples. Pacific Northwest National Laboratory PNNL-13249.

Feddes, R. A. and P. A. C. Raats. 2004. Parameterizing the soil-water-plant-root system. In: R. A. Feddes, G. H. de Rooij and J. C. van Dam (eds.), *Unsaturated Zone Modeling: Progress, Challenges and Applications*, Vol. 6. Wageningen UR Frontis Series Kluwer Academic Publishers, Dordrecht, the Netherlands, pp. 95–144.

Feddes, R. A., P. Kowalik, K. Kolinska-Malinka, and H. Zaradny. 1976. Simulation of field water uptake by plants using a soil water dependent root extractant function. *J. Hydrol.* 31(1–2): 13–26.

Feddes, R. A., P. J. Kowalik, and H. Zaradny. 1978. Simulation of Field Water Use and Crop Yield. Netherlands Centre for Agricultural Publishing and Documentation Simulation Monographs. John Wiley & Sons, New York.

Gardner, W. R. 1960. Dynamic aspects of water availability to plants. *Soil Sci.* 89(2): 63–73.

Gardner, W. R. 1964. Relation of root distribution to water uptake and availability. *Agron. J.* 56: 41–45.

Green, S. R., M. B. Kirkham, and B. E. Clothier. 2006. Root uptake and transpiration: From measurements and models to sustainable irrigation *Agric. Water Manage* 86: 165–176.

Hillel, D., H. Talpaz, and H. van Keulen. 1976. A macroscopic-scale model of water uptake by a nonuniform root system and of water and salt movement in the soil profile. *Soil Sci.* 121: 242–255.

Homaee, M., C. Dirksen, and R. A. Feddes. 2002a. Simulation of root water uptake. I. Non-uniform transient salinity using different macroscopic reduction functions. *Agr. Water Manage.* 57, 89–109.

Homaee, M., R. A. Feddes, and C. Dirksen. 2002b. Simulation of root water uptake. II. Non-uniform transient water stress using different reduction functions. *Agr. Water Manage.* 57: 111–126.

Homaee, M., R. A. Feddes, and C. Dirksen. 2002c. Simulation of root water uptake. III. Non-uniform transient combined salinity and water stress. *Agr. Water Manage.* 57, 127–144.

Homaee, M., R. A. Feddes, and C. Dirksen. 2002d. A macroscopic water extraction model for nonuniform transient salinity and water stress. *Soil Sci. Soc. Am. J.* 66, 1764–1772.

Hoogland, J. C., R. A. Feddes, and C. Belmans. 1981. Root water uptake model depending on soil water pressure head and maximum extraction rate. *Acta Hortic.* 119: 123–136.

Hopmans, J. W. and K. L. Bristow, 2002. Current capabilities and future needs of root water uptake and nutrient uptake modeling. *Adv. Agron.* 77: 103–183.

Kang, S., F. Zhang, and J. Zhang. 2001. A simulation model of water dynamics in winter wheat field and its application in a semiarid region. *Agric. Water Manage.* 49: 115–129.

Kirkham, M. B. 2005. *Principles of Soil and Plant Water Relations.* Elsevier Academic Press, Waltham, MA.

Kramer, P. J. 1983. *Water Relations of Plants.* Academic Press, New York.

Kramer, P. J. and J. S. Boyer. 1995. *Water Relations of Plants and Soils.* Academic Press, New York.

Li, K. Y., J. B. Boisvert, and R. De Jong. 1999. An exponential root water uptake model. *Can. J. Soil Sci.* 79: 333–343.

Maas, E. V. and G. J. Hoffman. 1977. Crop salt tolerance-current assessment. *J. Irrig. Drain. Div. ASCE* IR2(103): 115–134.

Maggio, A., G. Raimondi, A. Martino, and S. De Pascale. 2007. Salt stress response in tomato beyond the salinity tolerance threshold. *Environ. Exp. Bot.* 59, 276–282.

Molz, F. J. 1981. Models of water transport in the soil-plant system. A review. *Water Resour. Res.* 17: 1245–1260.

Molz, F. J. and I. Remson. 1970. Extraction term models of soil moisture use by transpiring plants. *Water Resour. Res.* 6: 1346–1356.

Nimah, M. N. and R. J. Hanks. 1973a. Model for estimating soil water, plant, and atmospheric interactions: I. Description and sensitivity. *Soil Sci. Soc. Am. Proc.* 37: 522–527.

Nimah, M. N. and R. J. Hanks. 1973b. Model for estimating soil water, plant, and atmospheric interactions: II. Field test of model. *Soil Sci. Soc. Am. Proc.* 37: 528–532.

Ojha, C. S. P. and A. K. Rai. 1996. Nonlinear root water uptake model. *J. Irrig. Drain. Eng.*122: 198–202.

Prasad, R. 1988. A linear root water uptake model. *J. Hydrol.* 99: 297–306.

Raats, P. A. C. 1974. Steady flows of water and salt in uniform soil profiles with plant roots. *Soil Sci. Soc. Am. Proc.* 38: 717–722.

Shankar, V., K. S. Hari Prasad, C. S. P. Ojha, and R. S. Govindaraju. 2012. Model for nonlinear root water uptake parameter. *J. Irrig. Drain. Eng.* 138(10): 905–917.

Šimůnek, J., M. Th. van Genuchten, and M. Šejna. 2008. Development and applications of HYDRUS and STANMOD software packages, and related codes. *Vadose Zone J.* 7: 587–600.

Skaggs, T. H., P. J. Shouse, and J. A. Poss. 2006a. Irrigating forage crops with saline waters: 2 Modeling root uptake and drainage. *Vadose Zone J.* 5: 824–837.

Skaggs, T. H., M. Th. van Genuchten, P. J. Shouse, and J. A. Poss. 2006b. Macroscopic approaches to root water uptake as a function of water and salinity stress. *Agric. Water Manage* 86: 140–149.

Steudle, E. and C. A. Peterson. 1998. How does water get through roots? *J. Exp. Bot.* 49: 775–788.

Taylor, H. M. and B. Klepper. 1975. Water uptake by cotton root systems: An examination of assumptions in the single root model. *Soil Sci.* 120(1): 57–67.

Taylor, S. A. and G. L. Ashcroft. 1972. *Physical Edaphology: The Physics of Irrigated and Nonirrigated Soils.* Freeman and Co., San Francisco, CA.

van Dam, J. C., J. Huygen, J. G. Wesseling, R. A. Feddes, P. Kabat, P. E. V. van Valsum, P. Groenendijk, and C. A. van Diepen. 1997. *Theory of SWAP*, Version 2.0. Simulation of water flow, solute transport and plant growth in the soil–water–atmosphere–plant environment. Department Water Resources, WAU, Report 71, DLO Winand Staring Centre, Wageningen, the Netherlands, Technical Document 45.

van den Broek, B. J., J. A. Elbers, J. Huygen, P. Kabat, J. G. Wesseling, J. C. van Dam, and R. A. Feddes. 1994. *SWAP 1993—Input Instructions Manual.* Report 45, DLO- Winand Staring Centre, Wageningen Agricultural University, Wageningen, the Netherlands.

van Genuchten, M. T. 1987. A numerical model for water and solute movement in and below the root zone. *Res. Rep.* 121, U. S. Salinity Laboratory, Agricultural Research Service, U. S. Department of Agriculture, Riverside, CA.

van Genuchten, M. Th. and G. J. Hoffman. 1984. Analysis of crop salt tolerance data. In: I. Shainberg and J. Shalhevet (eds.), *Soil Salinity under Irrigation. Ecological Studies.* Springer-Verlag, New York, pp. 258–271.

Vrugt, J. A., J. W. Hopmans, and J. Šimůnek. 2001a. Calibration of a two dimensional root water uptake model for a sprinkler-irrigated almond tree. *Soil Sci. Soc. Am. J.* 65(4): 1027–1037.

Vrugt, J. A., M. T. van Wijk, J. W. Hopmans, and J. Šimůnek. 2001b. One-, two-, and three-dimensional root water uptake functions for transient modeling. *Water Resour. Res.* 37(10): 2457–2470.

Wang, E. and C. J. Smith. 2004. Modeling the growth and water uptake function of plant root systems: A review. *Aust. J. Agric. Res.* 55: 501–523.

Wesseling, J. G. 1991. Meerjarige simulatie van grondwaterstroming voor verschillende bodemprofielen, grondwatertrappen en gewassen met het model SWATRE. DLO-Staring Centrum, Rapport/DLO-Staring Centrum no. 152, Wageningen, the Netherlands.

Wu, J., R. Zhang, and S. Gui. 1999. Modeling soil water movement with water uptake by roots. *Plant Soil* 215: 7–17.

Yang H., T. Du, X. Mao, and M. K. Shukla. 2020. Modeling tomato evapotranspiration and yield responses to salinity using different macroscopic reduction functions. *Vadose Zone Journal*. https://doi.org/10.1002/vzj2.20074.

14 Airflow through the Vadose Zone

14.1 INTRODUCTION

The earth is surrounded by an 80 km thick gaseous envelope known as the atmosphere. There are several different speculations/theories on the origin of the earth's atmosphere, but it is generally accepted that the atmosphere was created after the earth cooled down. The gas composition was believed to be that of the gases normally emitted by volcanoes. The atmosphere consists of both permanent gases such as nitrogen (78.1%), oxygen (20.9%), argon, neon, helium, krypton and hydrogen, and variable gases including water vapor, carbon dioxide, methane, nitrous oxide, and ozone. Oxygen molecules combine with free oxygen atoms to form ozone, which absorbs ultraviolet radiations from the sun. Some of the variable gases are also known as greenhouse gases, which allow sunlight to enter the atmosphere but prevent outgoing ultraviolet radiations from earth, thus trapping the heat and keeping the earth warmer by about 33°C. Soil is a porous media, and gases from the atmosphere enter soil and vice versa, depending on the gas pressure and gradient. Carbon dioxide (CO_2) concentrations are usually higher in soil air than in the atmosphere, there can be an order of magnitude difference in CO_2 in soil and the atmosphere, and CO_2 concentrations are inversely related to the oxygen concentrations in the soil. Similarly, soil air content is inversely related to soil water content.

The unsaturated soil has pores filled with both water and air. As soil becomes increasingly unsaturated (dry), the attendant volume of air or gas inside the soil increases. The composition of air in the soil also changes with increasing soil drying. Air content in the soil or oxygen content in the root zone of soil is essential for root development, plant growth, and sustenance. The composition of the soil air is important because oxygen supply from leaves to the roots cannot be adequate, unless soil air has enough oxygen root growth and plant growth can be curtailed. Soil air composition can also be used to trace organic compounds, minerals, and petroleum reserves (Scanlon et al., 2002). In this chapter, terminologies associated with soil air and mechanisms of gas transport through soil are discussed.

14.2 SOIL GAS CONTENT

In unsaturated soil, liquid, gas, and solids coexist. Water is usually considered incompressible and wets the solids, while gas or air is compressible and nonwetting. In unsaturated soil, water is held inside the smaller pores, and air is generally present in the larger pores because larger pores empty first, allowing air to enter and occupy these pores. The mass of the displaced water inside a given volume of pores is much

DOI: 10.1201/9780429264849-14

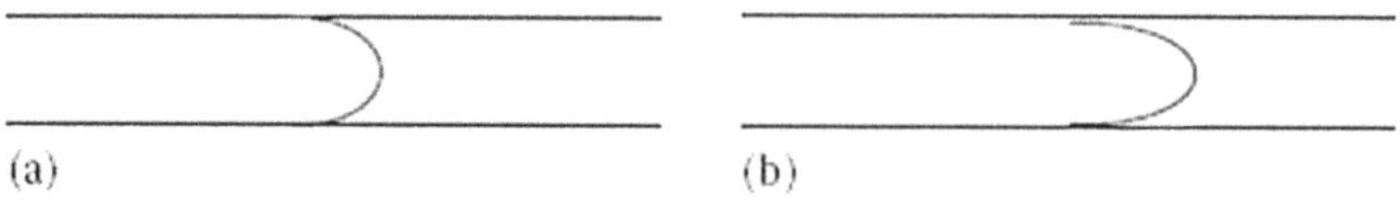

FIGURE 14.1 The flow of (a) water as compared to the flow of (b) air through a capillary. *Note*: The velocity of water is lower than air at the edges due to the higher adhesive forces between water and solid than air and solid.

larger than the mass of air entering the same pores. The soil water has strong affinity to the solids through the force of adhesion; however, air and solid have almost no affinity for each other (Figure 14.1).

The volumetric gas content (θ_g) in the soil is expressed as the ratio of total volume of gas (V_g) in the soil pores and the total volume of the soil (V_T):

$$\theta_g = \frac{V_g}{V_T} \tag{14.1}$$

When all the pores in a given volume of soil are filled with gas, soil is considered saturated with gas (S_g). S_g is expressed as the ratio of the total volume of gas in the soil pores or porosity (V_g) at any given time and the total volume of pores (V_f):

$$S_g = \frac{V_g}{V_f} \tag{14.2}$$

The degree of saturation with gas or soil air (S_g) is similar to the degree of saturation for water (S_e), and S_g also varies between zero and one. When all pores are filled with water, the degree of gas saturation is zero, and when all pores are occupied by air, the degree of saturation is equal to one. From the known soil porosity and known volumetric water content of soil, the soil air content can be calculated for any soil not contaminated with organic compounds. If total soil porosity is f, then soil air content and degree of saturation can be related as follows (Scanlon et al., 2002):

$$\theta_g = f \times S_g \tag{14.3}$$

14.3 FACTORS AFFECTING SOIL AIR COMPOSITION

There are several factors that can affect the composition of soil air, namely, soil texture, cultivation, soil organic matter, and soil temperature. The soil texture can influence the soil air composition because with decreasing size of pores in fine-textured soils, atmospheric air diffusion, and/or flow into the soil can be increasingly restricted with depth. The land use and management system also shake soil air composition. Roots and microbes are present in the cultivated soil, and they may consume more oxygen and increase the concentration of CO_2 in the soil. On the other hand, soil that is untilled for a long time may have a lack of roots, and therefore soil air composition may not change at all. The soil organic matter present in the soil can undergo chemical decomposition. Decomposition of organic matter is usually an aerobic process requiring oxygen, while CO_2 is usually released during decomposition. The speed of wind in the atmosphere generally does not influence the composition of soil air. However, increases in the soil temperature can accelerate the decomposition of organic matter through enhanced microbial activities resulting in increases in the CO_2 concentration in the soil. The water management system also influences the soil air composition; wetting (by irrigation) and drying (by drainage) can decrease or increase gas exchange between the soil and the atmosphere, respectively. Drainage of water from the soil increases the total gas or air porosity of soil and changes the soil air composition.

14.4 AIRFLOW THROUGH SOIL

The gas conservation equation can be written similar to the water or heat conservation equation, where inflow is equal to the sum of outflow and change in the concentration or composition of the soil air. Mathematically, it is presented as follows (Jury et al., 1991):

$$\frac{\partial(aC_g)}{\partial t} + \frac{\partial q_g}{\partial x} + r = 0 \tag{14.4}$$

where

a is the gas (or air) content of soil.
C_g is the soil gas concentration.
q_g is the mass flow of gas into or out of soil.
t is the time.
x is the distance.
r is the sink/source term for gas production or consumption.

14.5 AIRFLOW MECHANISMS THROUGH SOIL

The air or gas flow through the soil can take place as mass flow, convection, or diffusion. The convective flow of air from the atmosphere into the soil or vice versa occurs due to the total pressure difference between the soil pore and the atmosphere. The change in the pressure inside the soil can be due to the change in the gas composition. The gas composition can change due to the change in barometric pressure, temperature, evaporation, microbial activities, rainfall, irrigation, or drainage. The pressure difference is the main driving force during gas convection and is similar to the water flow across the flow domain. One important difference between airflow and water flow is that gravity is not a factor in vertical airflow into the soil. The convection of gas (q_x) into or out of the soil can be given by an equation similar to Darcy's law:

$$q_g = -C\frac{\partial P}{\partial x} \tag{14.5}$$

or

$$q_g = -\frac{K_g g \rho_g}{\eta_g} \frac{\partial P}{\partial x} \tag{14.6}$$

where

K_g is the air permeability coefficient.
ρ_a is the gas density.
η_g is the viscosity of soil air.
∂P is the pressure difference across a distance ∂x.

The diffusion of gas into or out of the soil system is a more important mechanism than convention. The process of diffusion is generally defined as the random thermal molecular movement from high to low concentration. By diffusion, air exchange between soil air and the atmosphere is maintained. In the presence of soil water, the air produced inside the soil can diffuse through the water films. The diffusive transport of gas through the soil can be given by Fick's law, which states that the mass rate of transfer of a diffusing gas through a unit area of flow in the soil is proportional to the concentration gradient measured normal to the surface through which diffusion is taking place. The gas diffusion through the soil due to diffusion is given as follows:

$$q_x = -D_g \frac{\partial C_g}{\partial x} \tag{14.7}$$

$$D_g = \xi D_a \tag{14.8}$$

where

D_g is the soil gas diffusion coefficient.
ξ is the gas tortuosity.
D_a is the gas diffusion coefficient in free air.
∂C_g is the concentration difference across a distance of ∂x.

D_g is also expressed as the product of ξ, volumetric air/gas content (a), and gas diffusion coefficient in free air. The gas tortuosity, ξ, accounts for increased path length for airflow; and the volumetric air content, a, accounts for the reduced cross-sectional area relative to air in the soil pores (Scanlon et al., 2002). The gas transport equation is obtained by inserting the gas diffusion equation into the gas conservation equation, an equation similar to the Laplace equation for water flow obtained by combining continuity and Darcy's equations. Since gas can be produced or taken up inside the soil, a sink or source term (r) is included in the gas transport equation as follows:

$$a \frac{\partial C_g}{\partial t} = -D_g \frac{\partial^2 C_g}{\partial x^2} + r \tag{14.9}$$

TABLE 14.1
Empirical Equations to Estimate Gas Tortuosity (ξ)

Reference	Equation
Buckingham (1904)	εf
Penman (1940a,b)	$0.66f$
Marshall (1958)	$f^{3/2}$
Millington (1959)	$a^{4/3}$
Currie (1960)	$a f^b$
Millington and Quirk (1961)	$a^{10/3}/f^2$
Lai et al. (1976)	$a^{7/3}$
Moldrup et al. (1996)	$f^2(a/f)^{2+3/b}$
Abu-El-Sha'r and Abriola (1997)	$0.435f$

Sources: Jury, W. A. et al., *Soil Physics,* John Wiley & Sons, Inc., New York, 1991; Scanlon et al., Soil gas movement in unsaturated systems, in: Warwick, A. W., ed., *Soil Physics Companion,* pp. 297–341, 2002.

Note: ε is a constant, f is porosity of soil, a is volumetric air content, and b is the Campbell pore-size distribution parameter (slope of log [matric potential] versus log [soil water content]).

where the volumetric air content, a, does not change with time and D_g does not change with space. Some of the values of ξ reported in the literature are 0.66 (Penman, 1940a, b), 0.53 and 0.89 (Flegg, 1953), and 0.58 (van Bavel, 1952). There are several models available in the literature to estimate gas tortuosity (ξ) (Table 14.1).

Example 14.1

Pecan is planted in an orchard in New Mexico. The rooting depth is 85 cm, the daily transpiration rate is 8 mm/day, and the soil respiration measured for the loamy soil is 8 g/m^2 of oxygen. Calculate the volume of air entering the soil and the oxygen supplied by convection, transport of air into the soil with change in soil water content due to the root water uptake or transpiration, and daily oxygen requirement met through convection from the unit area. The molecular weight of oxygen is 32 and the volume of 1 mol of gas at standard pressure and temperature is 22.4 L.

The depth of water lost or extracted by the roots per square meter per day = 8/10/100 = 0.008 m.

The volume of water extracted per square meter per day = 0.008 m^3.

The total volume of air removed from the soil must be equal to the volume of water withdrawn by the roots. Atmospheric air contains about 21% oxygen.

The volume of oxygen in air replacing water in soil = 0.008 × 21/100
= 0.0017 m^3
= 1.7 L (per day)

Moles of oxygen in the soil air replacing water = 1.7 [L] × 32 [g/mol]
= 54.4 [g × L/mol]
Mass of oxygen in the soil air replacing water = 54.4/22.4 [L/mol]
= 2.41 g

Daily oxygen requirement met through convection from 1 m^2 area based on soil respiration rate = 2.41/8 = 0.30% or 30%.

Example 14.2

A sandy loam soil has a moisture content of 0.28 cm^3/cm^3, total porosity of 0.45 cm^3/cm^3, and gas diffusion coefficient in free air of 0.16 cm^2/s. Calculate the soil gas diffusion coefficient using the Millington and Quirk model (Table 14.1).

Volumetric air content or air-filled porosity, $a = 0.45 - 0.28 = 0.17$ cm^3/cm^3

Millington and Quirk (1961), $D_g = D_a \times a^{10/3}/f^2 = 0.16 \times 0.17^{10/3}/0.45^2$
$= 0.00215$ cm^2/s

Example 14.3

Oxygen concentration is 21% (or 0.0003 g/cm^3) at 2 cm depth from soil surface and 10% at the depth of 42 cm (decreases to nearly half); use D_g from the Penman model and any other data needed from Example 14.2 to obtain the one-dimensional diffusive flux.

$$q_x = -D_g \frac{\partial C_g}{\partial x}$$

$$q_x = -0.0475 \times \frac{(0.0003 - 0.0003 \times 0.5)}{2 - 42}$$

$$q_x = 1.78 \times 10^{-7} \text{g cm}^2/\text{s}$$

14.6 AIR-FILLED POROSITY MEASUREMENT THROUGH SOIL

The air-filled porosity of soil can be determined in the laboratory on a soil sample collected in a metal core. The excess soil in the core is trimmed at both ends of the core, and the sample is slowly saturated with water. The saturated soil core is subsequently drained by applying a pressure of 50 cm by placing the sample in a tension table or inside a pressure plate apparatus. Once the sample is equilibrated to the applied pressure (suction) (50 cm) so that no more water can come out of the core, air permeability for the effective pore radius (0.15/50 cm = 0.003 cm) can be determined with an air pycnometer. The air pycnometer method is based on Boyle's law ($P_1V_1 = P_2V_2$), where P is pressure and V is volume. The volume of air in the soil sample is measured by obtaining the final pressure when equilibrium is reached between two chambers, one containing the soil sample and the other at higher pressure. Calibration curves are used to determine the air space at the equilibrium pressure.

Alternately, the air-filled porosity can be determined by the difference method where air-filled porosity is obtained as the difference between total porosity (or saturated volumetric water content) and water content at 50 cm (or water content of the soil sample collected from the field). The air pycnometer method is relatively quick; however, both these methods are not adequate, as considerable uncertainties exist in the measurement, and aeration dynamics inside the soil matrix remains almost untouched.

14.7 AIR PERMEABILITY

Soil aeration measured by composition of soil air and water content of soil is usually inadequate because these methods do not consider direction, rate of change, or dynamics of the soil air inside and outside the pores. The gaseous exchange between soil air and the atmosphere, and the gas transport within the soil is much more complex than can be accounted for by static methods such as pressure difference or the pycnometer method. The coefficient of proportionality in the convective flow of air through the soil is known as the air permeability coefficient. The air permeability coefficient can be determined as an alternate to measuring static soil aeration. Air permeability measurements can be made by using a constant pressure gradient method or a constant flux method.

In the constant pressure gradient method, a soil core is subjected to a pressure at one end while the other end can be left open to the atmosphere; alternately, a water manometer can be placed across the core to measure pressure difference across the soil (Figure 14.2a). In this method, the volume of airflow through the inlet of the soil

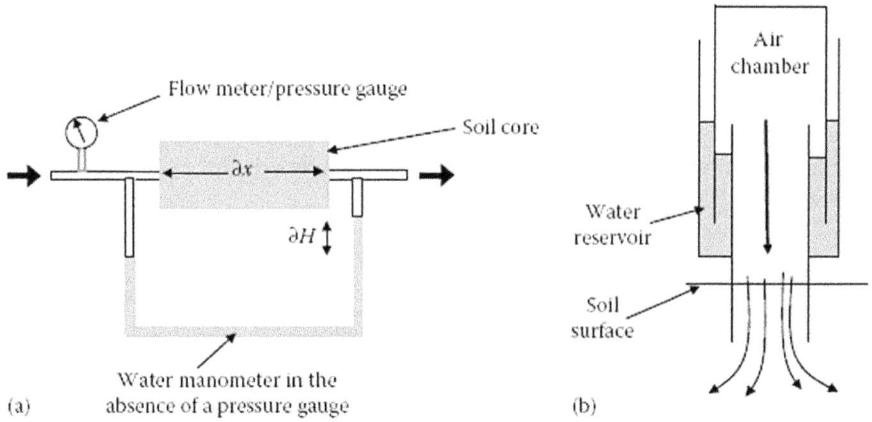

FIGURE 14.2 Measurement of air permeability with a constant pressure permeameter. The pressure difference across the soil column can be measured using a water manometer or a pressure gauge, and inflow of air can be measured by (a) a flow meter for a soil in the laboratory and (b) using a float can method in the field.

is measured with respect to time. The air permeability can be calculated using the following equation:

$$q_g = -\frac{K_g g \rho_g}{\eta_g} \frac{\partial P}{\partial x} \tag{14.10}$$

where
ρ_g is the density of air (g/cm^3).
η_g is the viscosity of air (g/cm/s).

$$K_g = \frac{q_g \eta_g \partial x}{\rho_g g \partial P} \tag{14.11}$$

$$K_g = \frac{Q_g \eta_g \partial x}{A \rho_g g \partial P} \tag{14.12}$$

where
Q_g is the volumetric flow rate (cm^3/s).
A is the cross-sectional area of flow (cm^2).

The units of air permeability are cm^2 (or L^2) and can be obtained from Equation 14.12.

The constant pressure gradient method is useful for soils where maintaining constant flux of airflow is difficult. It is also good for soils where only a small air pressure gradient is applied across the sample to ensure that soil water content does not change. Air leaks must be checked, and any flow between the soil and the core must be eliminated. The measurement should be made for short duration to avoid any soil drying due to airflow. An alternative to the pressure gauge method is the float can method, especially for measurements in the field (Figure 14.2b). In the constant flux method, a soil sample is subjected to a predetermined constant flux of air, and the resulting pressure difference is measured. Flow through the soil sample may or may not be under steady state. The constant flux method can measure air permeability under steady as well as transient state. Measurements could be made on soil with very low air permeability. This method is more complex than the pressure gradient method.

Example 14.4

A soil sample was collected from a field planted with onions every fall. The sample was equilibrated to a water content of 50 cm and at steady state; the pressure gradient method was used to determine the air permeability. The core was subjected to a pressure difference of 50 cm along a length of 5 cm. The volumetric flow rate was 50 cm^3/s, the cross-sectional area of flow was 55 cm^2, and the temperature of the lab was 20°C.

Density of air at 20°C = 1.205 kg/m^3 = 0.0012 g/cm^3

Viscosity of air at 20°C = 0.00018 g/cm/s

$$K_g = \frac{Q_g \eta_g \partial x}{A \rho_g g \partial P}$$

$$K_g = \frac{50 \times 0.00018 \times 5}{55 \times 0.0012 \times 981 \times 50} = 1.39 \times 10^{-5} \text{cm}^2$$

The acoustic method is an indirect field method for obtaining the air permeability of porous media (Moore and Attenborough, 1992). In this method, the porosity of soil is obtained by sending a sound wave from a source located close to the surface. There are two vertically separated microphones above the ground and two inside the soil. The differences in the measured sound spectra are used to calculate the air-filled porosity and air permeability.

PROBLEMS

14.1 Describe the factors that affect air or gas concentrations of soil.

14.2 Describe the various processes of air movement through soil.

14.3 A soil sample was collected from a chile field equilibrated to a water content of 50 cm. The core was subjected to a pressure difference of 100 cm along a length of 10 cm. Calculate air permeability using the steady-state pressure gradient method if the volumetric flow rate was 60 cm^3/s, the cross-sectional area of flow was 70 cm^2, and the temperature of the lab was 20°C.

14.4 A loam soil has a moisture content of 0.32 cm^3/cm^3 with a total porosity of 0.45 cm^3/cm^3, and the gas diffusion coefficient in free air is 0.16 cm^2/s. Calculate the soil gas diffusion coefficient using the models (any three) given in Table 14.1.

14.5 If the oxygen concentration is 21% (or 0.0003 g/cm^3) at 5 cm depth from soil surface and 5% at a depth of 80 cm, calculate the one-dimensional diffusive flux using at least two gas tortuosity models.

14.6 What is air permeability? Explain factors that influence air permeability of soil.

14.7 Describe different methods of measurement of air permeability. Explain their advantages and disadvantages.

REFERENCES

Abu-El-Sha'r, W. Y. and L. M. Abriola. 1997. Experimental assessment of gas transport mechanisms in natural porous media: Parameter evaluation. *Water Resour. Res.* 33: 505–516.

Buckingham, E. 1904. *Contributions to Our Knowledge of the Aeration of the Soils.* Bulletin 25, U.S. Department of Agriculture Bureau of Soils, Washington, DC.

Currie, J. A. 1960. Gaseous diffusion in porous media. II. Dry granular materials. *Br. J. Appl. Phys.* 11: 318–324.

Flegg, P. B. 1953. The effect of aggregation on diffusion of gases and vapors through soils. *J. Sci. Food Agric.* 4: 104–108.

Jury, W. A., W. R. Gardner, and W. H. Gardner. 1991. *Soil Physics.* John Wiley & Sons, Inc., New York.

Lai, S. H., J. M. Tiedje, and A. E. Erickson. 1976. In situ measurement of gas diffusion coefficient in soils. *Soil Sci. Soc. Am. J.* 40: 3–6.

Marshall, T. J. 1958. A relation between permeability and size distribution of pores. *J. Soil Sci.* 9: 1–8.

Millington, R. J. 1959. Gas diffusion in porous media. *Science* 130: 100–102.

Millington, R. J. and J. P. Quirk. 1961. Permeability of porous solids. *Faraday Soc. Trans.* 57(7): 1200–1207.

Moldrup, P., C. W. Kruse, D. E. Rolston, and T. Yamaguchi. 1996. Modeling diffusion and reactions in soil: III. Predicting gas diffusivity from the Campbell soil-water retention model. *Soil Sci.* 161: 366–375.

Moore, H. M. and K. Attenborough. 1992. Acoustic determination of air-filled porosity and relative air permeability of soils. *J. Soil Sci.* 43: 211–228.

Penman, H. L. 1940a. Gas and vapor movements in the soil. I. The diffusion of vapors through porous solids. *J. Agric. Sci.* 30: 437–462.

Penman, H. L. 1940b. Gas and vapor movements in the soil. II. The diffusion of carbon dioxide through porous solids. *J. Agric. Sci* 30: 570–581.

Scanlon, B. R., J. P. Nicot, and J. W. Massmann. 2002. Soil gas movement in unsaturated systems. In: Warwick, A. W., ed., *Soil Physics Companion*, CRC Press, Boca Raton, Florida pp. 297–341.

van Bavel, C. H. M. 1952. Gaseous diffusion and porosity in porous media. *Soil Sci.* 73: 91–104.

15 Chemical Transport through the Vadose Zone

15.1 INTRODUCTION

Water in the earth follows a cycle through the atmosphere, land surface, and open water bodies and through the vadose zone of soils. The precipitation falling on the earth's surface is seldom pure. Upon condensation of water vapor and subsequently during its descent, liquid water picks up gases and other suspended particles from the atmosphere, and the precipitation falls on the surface as a solution of varying concentrations containing dissolved gases and chemicals as well as suspended chemicals and particles. The dissolved gases could also be the gases coming out of a power plant or a factory, and the chemical composition of falling liquid water could be highly spatially and temporally variable.

Once water infiltrates into the soil, it picks up more chemicals present in the soil and in the soil water. Chemicals in the soil are naturally occurring as well as those applied by farmers or land managers. In agricultural systems, solutes may be categorized on the basis of their function, for example, nutrients, required for vegetative growth of plants and for productivity; pesticides, required to control pests in the soil; herbicides and weedicides, required to kill unwanted plants or weeds; and salts that may be present in fertilizers or other chemicals. In addition, organic compounds (natural and synthetic), manures, heavy metals, viruses, and bacteria are present in the soil system. These can be transported to deeper layers or to the groundwater table along with the percolating water through the soil profile and below the root zone. The downward movement of solutes is also known as solute leaching.

In order to reduce the solute movement through the vadose zone and leaching toward the groundwater table, knowledge of the transport mechanisms of solutes through the soil is important for designing on-site management strategies. Better understanding of the solute movement through soil, for example, can help schedule fertilizer applications to a crop for increasing plant uptake while simultaneously reducing risks of groundwater contamination or leaching of nutrients below the root zone. It can also help develop procedures for effective use of pesticides and chemicals within the root zone while minimizing their leaching into groundwater. Knowledge of transport mechanisms through soil can also help reduce risks of polluting surface water resources from runoff, salt intrusion from coastal aquifers, and seepage from chemical storages and disposal systems.

The water transport through soil is usually treated as mass or bulk flow (convection); however, solute transport through porous media is more complicated than just the mass flow (Nielsen and Biggar, 1961; Biggar and Nielsen, 1967). In this chapter, processes of solute movement, equilibrium and nonequilibrium transport mechanisms,

DOI: 10.1201/9780429264849-15

and influence of land use and management practices on solute leaching are discussed in detail.

15.2 TYPES OF SOLUTES

Several different types of solutes are present inside the soil matrix. There are several reasons for solute presence in pore water of soil, including mineralization of organic matter; saline water intrusion; atmospheric deposition; weathering of minerals; and fertilizer, herbicide, or pesticide application. Solutes can be broadly classified into two categories based on their solubility. Chemicals that mix well with water during transport through the soil column are known as miscible solutes, and the transport process is known as miscible displacement. Some examples of miscible displacement include leaching of salts through the soil profile and movement of fertilizers and pesticides through the soil. On the other hand, when a solute does not mix with water or a liquid in the soil, it is known as an immiscible solute, and the transport process is known as immiscible displacement. Common examples of immiscible displacement include movement of water containing oil and transport of contaminants like dense or light non-aqueous phase liquids (DNAPLs or LNAPLs).

Solutes can also be classified based on the reactions as reversible or irreversible according to the chemical reactions taking place during the transport through soil. Solutes remaining physically and chemically unchanged and not undergoing irreversible reactions are known as conservative solutes. These solutes are not adsorbed on the soil particle surfaces or are only weakly adsorbed. Some examples of conservative solutes are chloride, bromide, nitrate, and sulfate. On the other hand, solutes undergoing irreversible reactions with changing physical and chemical makeup are known as nonconservative solutes. Some examples of nonconservative solutes are phosphates, pesticides, herbicides, heavy metals, mercury, and arsenic.

Solutes are also classified as labile and nonlabile solutes. Labile solutes can undergo physical and chemical reactions rapidly and easily and are more prone to decomposition than nonlabile solutes. Some examples of labile solutes include nitrate, sulfate, and ammonia. Some solutes, such as pesticides, are known as reactive solutes because they react with the soil and other chemicals present in the soil and undergo chemical change by production or decay.

15.3 RELATED TERMINOLOGY

A solute or solution flowing through porous media is also termed as fluid and is defined as a liquid of a definite physical and chemical makeup, for example, water containing nitrate ions. The solution collected at the exit or outflow end of a soil column or soil matrix is known as the effluent. The fluid that is added to the soil matrix or soil column during solute transport experiments that mixes with and displaces the already existing solution in the soil is known as the displacing solute. The fluid that is removed from the soil volume in the soil matrix or soil column is known as displaced fluid.

15.4 PORE WATER VELOCITY

The velocity at which water moves through the soil pores is known as Darcy's velocity. According to Darcy's law, the rate at which water flows through the soil profile is equal to the hydraulic conductivity of soil when the hydraulic gradient causing flow is unity. Solutes move through water-filled pore spaces and through water films inside the soil pores. The velocity at which the solute travels through the porous media is known as pore water velocity. The pore water velocity (v) is defined as the ratio of the volumetric flow rate per unit cross-sectional area (Darcy flux; q) and the volumetric water content of soil (θ):

$$v = \frac{q}{\theta} \tag{15.1}$$

In general, pore water velocity accounts for straight line lengths of the flow path traversed in the soil in a given amount of time. In reality, the flow paths are not straight but are tortuous or irregular, and the property is known as tortuosity. The tortuosity relations are given in Chapter 14 describing airflow through the vadose zone; tortuosity (ξ) given by the Millington and Quirk (1961) equation after replacing volumetric air content with volumetric soil water content is presented here as follows:

$$\xi = \frac{\theta^{10/3}}{f^2} \tag{15.2}$$

15.5 SOLUTE CONSERVATION EQUATION

Similar to the water conservation equation, the solute conservation equation can be written for a given mass of solute entering and leaving a given volume of the flow domain. The solute conservation equation can be presented as follows:

Mass of solute entering a given volume of porous media in a given time (or time interval) = Mass of solute leaving the porous media during the same time interval + Decay or production of solute mass inside the soil volume during the same time interval

If the mass of solute flowing through a unit cross-sectional area of flow domain per unit time is J_x (in x-direction), the total solute concentration as the ratio of mass of solute and volume of soil inside the soil matrix is C_T, and r is the reaction rate expressed as decay or production of solute inside the soil matrix per unit volume per unit time, then the one-dimensional solute conservation equation can be expressed mathematically as follows:

$$\frac{\partial C_T}{\partial t} + \frac{\partial J_x}{\partial x} + r = 0 \tag{15.3}$$

The solute inside the soil matrix can stay at liquid state, adsorbed onto the soil particles (solid state) and at vapor state. Total solute concentration can be expressed as the sum of the solute concentrations in three phases as follows:

$$C_{\mathrm{T}} = \rho_{\mathrm{b}} C_{\mathrm{a}} + \theta C_{\mathrm{l}} + a C_{g} \tag{15.4}$$

$$\mathrm{g/cm^3} = \mathrm{g/cm^3} \times \mathrm{g/g} + \mathrm{cm^3/cm^3} \times \mathrm{g/cm^3} + \mathrm{cm^3/cm^3} \times \mathrm{g/cm^3} = \mathrm{g/cm^3}$$

where

ρ_b is the soil bulk density (g/cm^3).
C_a is the adsorbed solute concentration (g/g; gram of solute per gram of soil).
θ is the volumetric water content (cm^3/cm^3).
C_l is the dissolved solute concentration (g/cm^3).
a is the volumetric air content (cm^3/cm^3).
C_g is the solute concentration in the gaseous phase (g/cm^3).

Note: Soil bulk density, volumetric water, and air contents convert the units of solute concentrations in different phases to the units of mass per volume.

15.6 SOLUTE TRANSPORT PROCESSES

There are three processes commonly associated with solute transport: convection, diffusion, and dispersion. The solute transport through the porous media could be due to the mass flow or convection of solutes through the soil pores. The flow is also known as Darcian flow. Mathematically, the mass flow of a dissolved solute can be represented as follows:

$$J_{\mathrm{m}} = qC \tag{15.5}$$

where

J_m is the convective flux.
q is the volumetric liquid flux density.
C is the volume-averaged dissolved solute concentration.

The solute transport by convection takes place due to molecular movement. Chemicals move due to mass flow as well as due to the processes of diffusion and dispersion. When movement of solutes takes place due to the movement of ions in a solution as a result of random motion, the process is known as diffusion. During diffusion, the movement of solutes takes place from a location where concentration is higher to a location with lower concentration. This process commonly occurs within the gaseous and liquid phases in the soil due to the random thermal motion also called Brownian movement. Diffusion is a static active process, and water or solute does not need to physically move. Mathematically, the process of diffusion can be expressed as follows:

$$J_{\text{Dif}} = -\theta D_{\text{m}} \frac{\partial C}{\partial x} \tag{15.6}$$

where
- θ is the volumetric water content.
- D_{m} is the molecular diffusion coefficient.
- x is the distance.

The diffusion coefficient in soil is smaller than the diffusion coefficient in pure water (D_0) because soil contains tortuous pores or flow paths. Diffusion in soil is a relatively slow process and operates over small distances, and electrical neutrality of ions is maintained. Mathematically, the relationship between the molecular diffusion coefficient in soil and pure water is expressed as follows:

$$D_{\text{m}} = D_0 \xi(\theta) \tag{15.7}$$

where $\xi(\theta)$ is the tortuosity factor that depends on the soil water content. It is reported to vary between 0.2 and 0.7 for most soils. The process that mixes and eventually evens out the concentration gradients is known as dispersion. Dispersion is a dynamic process because liquids and solutes both move during the transport; therefore, it is also known as hydrodynamic dispersion. During dispersion, erratic flow of liquid can take place through the complex pore network. The process of dispersion is generally known as a passive process and is mathematically expressed as follows:

$$J_{\text{Dis}} = -\theta D_{\text{h}} \frac{\partial C}{\partial x} \tag{15.8}$$

where D_{h} is the mechanical dispersion coefficient. This coefficient is a function of pore water velocity and is generally expressed as follows:

$$D_{\text{h}} = \lambda v^{n} \tag{15.9}$$

where
- λ is the dispersivity (cm) and is defined as the property of porous media that combines with the pore water velocity to provide the dispersion characteristic of the porous media.
- n is a constant usually equal to one.

Dispersivity values are reported to vary between 0.5 and 2 cm for disturbed and repacked soil columns used to conduct experiments in the laboratory, between 5 and 20 cm in the field (Fried, 1975; Jury et al., 1991). The λ values were reported to be 0.389 ± 0.032 for loam soil and 0.217 ± 0.024 for sandy loam soil (Shukla et al., 2003). Sandy loam soil columns have a greater average pore size associated with sand content than loam, and convective mass transfer would likely be the dominating mechanism for solute transport. However, λ values are dependent on the size of the area over which solute and water transports are averaged and thus could be very high for regional groundwater estimates.

Molecular diffusion and hydrodynamic dispersion together are considered additive processes, and the molecular diffusion coefficient and dispersion coefficient together are known as the hydrodynamic dispersion coefficient (Bear, 1972) or apparent dispersion coefficient (Nielsen et al., 1972).

$$D = D_{\mathrm{m}} + D_{\mathrm{h}} \tag{15.10}$$

or from Equations 15.8 and 15.9

$$D = D_0\xi(\theta) + \lambda v^n \tag{15.11}$$

15.7 FICK'S LAW

The movement of ions from areas with higher concentrations to areas with lower concentrations is expressed by Fick's law. Fick's law relates diffusive flux to the concentration under steady-state conditions and states that the flux is proportional to the concentration gradient. The amount of solute that is flowing across a unit cross-sectional area in unit time can be given as follows:

$$J_{\mathrm{dif}} = D_{\mathrm{m}} \frac{\partial C}{\partial x} \tag{15.12}$$

where

J_{dif} is the solute flux density (J or g/cm^2/s).
D_{m} is the molecular diffusion coefficient (cm^2/s).
C is the solute concentration (g/cm^3).
x is the distance (cm).

Similar to the wetting front depths, the solute front depths can be approximated by $(2D_{\mathrm{m}} \times t)^{0.5}$. The mass conservation equation implies that the overall change in solute concentration is due to the overall change in the flow rate. For transient-state conditions, Fick's first law is coupled with the one-dimensional mass conservation equation as follows:

$$\frac{\partial C}{\partial t} = -\frac{\partial J_{\mathrm{dif}}}{\partial x} \tag{15.13}$$

Transferring the solute flux J (Equation 15.12) into Equation 15.13 yields

$$\frac{\partial C}{\partial t} = -\frac{\partial}{\partial x}\left(-D_m \frac{\partial C}{\partial x}\right) = -D_m \frac{\partial^2 C}{\partial x^2}. \tag{15.14}$$

Example 15.1

A soil column containing sandy loam soil has a porosity (ϕ) of 0.48 cm^3/cm^3. The dispersivity (λ) measured in the column is 1.2 cm, and the diffusion coefficient of solute in water (D_0) is 1 cm^2/day. Calculate the tortuosity factor, molecular diffusion (D_m), dispersion (D_h), and apparent dispersion (D) coefficients for the volumetric water contents (θ) and Darcy's velocity (q) given in the following table:

θ (cm^3/cm^3)	q (cm/day)	v (cm/day), (q/θ)	ξ ($\theta^{10/3}/\phi^2$)	D_m (cm^2/day), $D_0 \times \phi$	D_h (cm^2/day), $\lambda \times v$	D (cm^2/day), $D_m + D_h$
0.12	0.5	4.17	0.004	0.004	5.00	5.004
0.22	2.1	9.55	0.028	0.028	11. 45	11. 482
0.32	3.2	10.00	0.097	0.097	12.00	12.097
0.42	4.8	11.43	0.241	0. 241	13.71	13.955

15.8 SOLUTE MOVEMENT

Solute transport through the soil takes place when an incoming (or displacing) chemical enters the soil matrix containing either some of the cations and anions present in the displacing solute or none whatsoever. As soon as the displacing solution enters, the initial sharp boundary between these liquids fades due to the mixing of the solutes (Figure 15.1a). The mixing occurs at the boundary due to the mass transfer of the displacing solute as well as through the diffusion and dispersion of ions across the boundary. If mixing is taking place in a soil column, the solute moves faster near the center of the column and slower near the edge, mainly due to the high viscous drag and the adhesion of water to the walls of the column (Figure 15.1b). Sometimes, the solutes do not mix together (immiscible solutes), for example, oil and water, and therefore the sharp boundary persists during the displacement of the resident solution by the displacing fluid (Figure 15.1c).

15.9 SOLUTE BREAKTHROUGH CURVE

During the solute transport through a soil column, the displaced solution can be collected at the outlet at various time intervals, and concentrations of the ions of interest can be determined in the laboratory while simultaneously recording the volume of the displaced sample collected in a given time interval. A plot between time or volume of effluent (on x-axis) and concentration (on y-axis) is known as the breakthrough curve for that ion (Figure 15.2). For immiscible solutes, the concentration of displaced solute stays constant until the displacing solution appears in the effluent solution and the concentration jumps from an initial low value (or zero) to the maximum value. This type of displacement is also known as piston flow (Figure 15.2a). However, for miscible solutes, processes associated with mass flow, diffusion, and

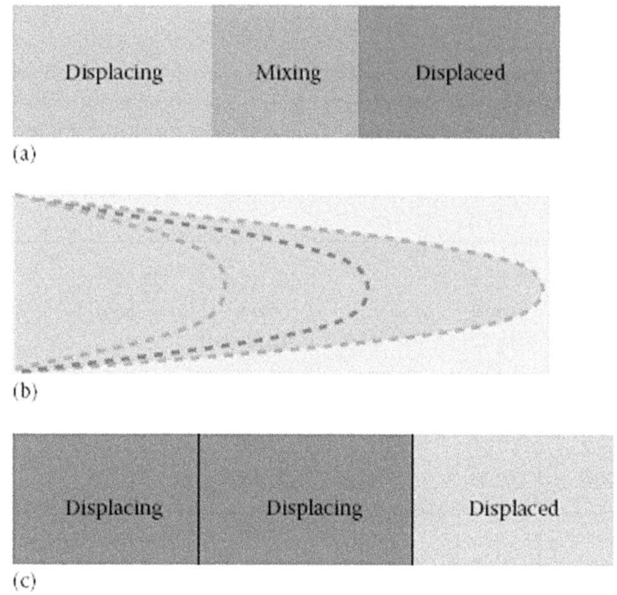

FIGURE 15.1 Mixing between displacing and displaced solutes, and miscible and immiscible displacement. (a) Displacing solute and displaced solute undergo mixing at the interface. (b) Sharp boundary fades due to mass flow, diffusion, and dispersion, and solute moves faster near the center because of high viscous drag near the wall. (c) Sharp boundary persists because displacing and displaced solutes are not mixing or immiscible.

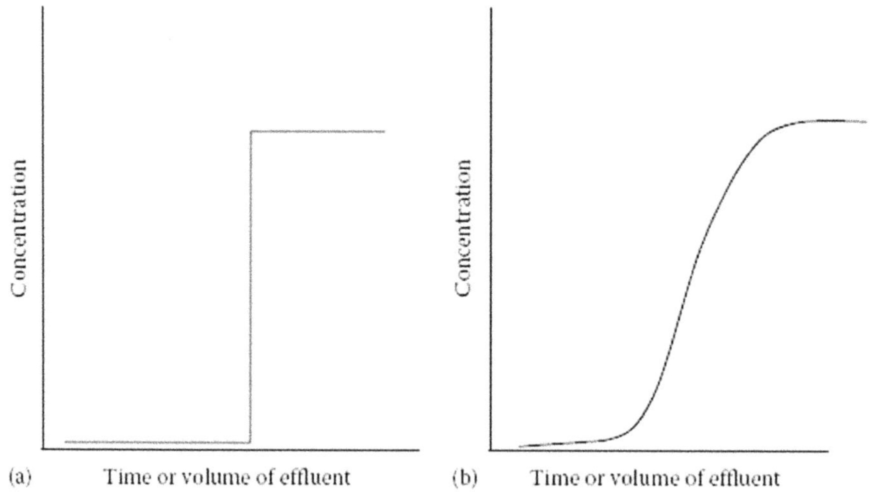

FIGURE 15.2 Schematic of (a) immiscible or piston flow where concentration jumps to its maximum value after a certain time interval or volume of effluent. The displacing solution has completely replaced the antecedent or displaced solution. (b) Due to mixing by diffusion and dispersion, displacing solute moves ahead in the soil and appears early in the effluent solution.

dispersion cause the solute to move ahead and appear early in the effluent solution, and concentration of the displacing solution increases slowly to its maximum concentration in the effluent (Figure 15.2b).

The breakthrough curves can also be plotted from the known pore volumes. Pore volumes are the ratio of the volume of water (or solution) leached through the soil column in any given time interval and the total volume of water (or solution) present inside the soil column at the beginning of the solute transport experiment. The number of pore volumes can be calculated in two different ways:

1. By dividing the volume of water (V_i) leached through the soil column in a given time interval (t_i) by the total volume of water present inside the soil volume (V_0), the number of pore volumes (T_i) can be obtained as follows:

$$T_i = \frac{V_i}{V_0} = \frac{V_i}{A\theta L} \tag{15.15}$$

where
A is the cross-sectional area of the soil.
θ is the water content of the soil.
L is the length of the soil column.

The number of pore volumes is a dimensionless number and as more and more water or solution is leached through the soil column, the number of pore volume increases. When all the water originally present in the soil column is displaced by the displacing fluid, the number of pore volumes becomes equal to one.

2. By using the known average pore water velocity (v), time interval (t_i), and length of soil column (L) as follows:

$$T_i = \frac{vt_i}{L} \tag{15.16}$$

The breakthrough curve for a solute can also be obtained by plotting the number of pore volumes on the x-axis and concentration on the y-axis (Figure 15.3). The solute can be applied as a step input so that the concentration increases with increasing time, effluent volumes, and pore volumes to the maximum concentration and then stays at that concentration (Figure 15.3a). The solute can also be applied as a pulse, that is, a predetermined volume of sample, followed by the antecedent solution to leach the applied pulse of solution completely from the porous media (Figure 15.3b).

Example 15.2

Soil is contained in a core of internal diameter of 5 cm and length of 5 cm. The volumetric water content of soil in the core is 0.40 cm^3/cm^3. The solute transport experiment was conducted through this core, and effluent solutions were collected at 10 min intervals and the volume of the effluent solution was determined

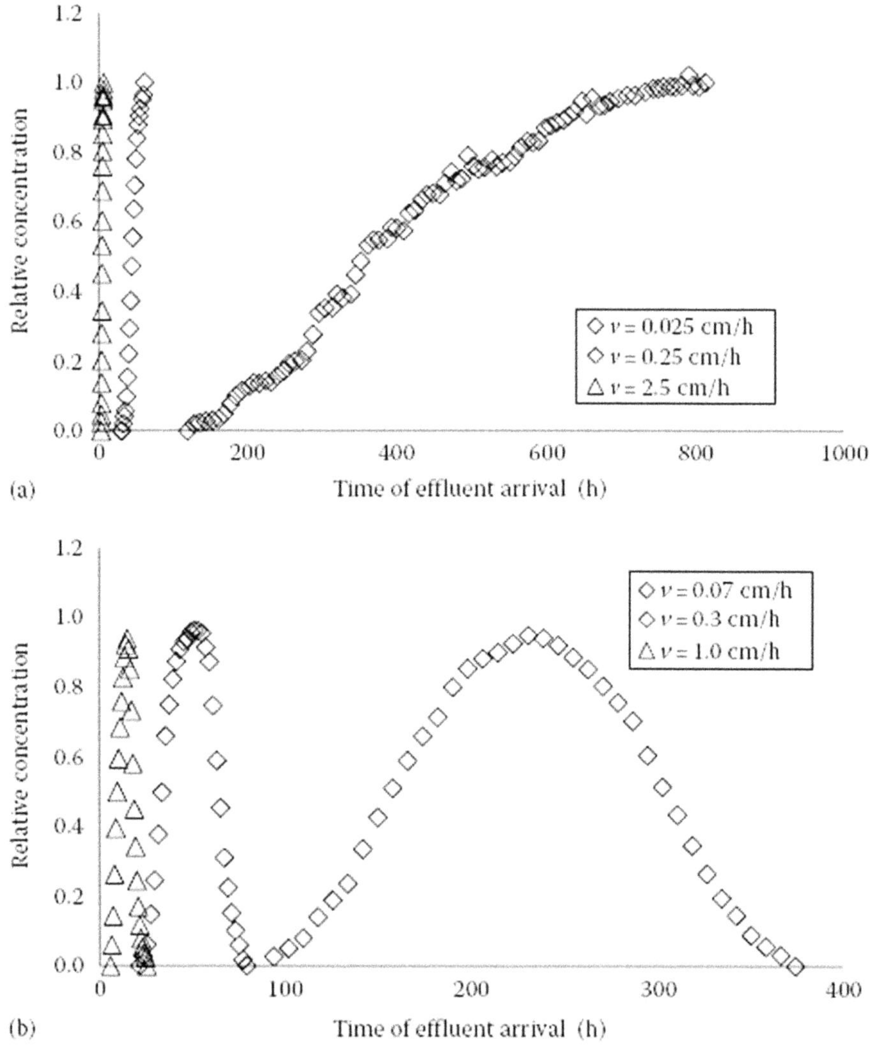

FIGURE 15.3 Breakthrough curves with time of effluent arrival on x-axis and relative concentrations on y-axis for (a) step and (b) pulse inputs of chloride for a slow, medium, and high water velocity during chloride transport through saturated repacked sandy loam soil columns. Note that as the pore water velocity increases, the breakthrough curve becomes narrower and the total time for solute to pass through the column decreases. (From Shukla, M. K., et al., *Soil Sci. Soc. Am. J.*, 67, 449, 2003.)

for each time increment. Calculate the number of pore volumes using both methods described earlier using the data from the following table.

$$V_0 = A \times L \times \theta = \pi \times r^2 \times L \times \theta = 3.14 \times (5/2)^2 \times 5 \times 0.4 = 39.27\,\text{cm}^3$$

Time Interval (min)	Cumulative Time, t_i (min)	Volume Effluent (mL)	Cumulative Effluent, V_i (mL)	Pore Volume, $v \times t_i/L$	Pore Volume, V_i/V_o
10	10	5	5	0.13	0.13
10	20	5	10	0.25	0.26
10	30	5	15	0.38	0.38
10	40	5	20	0.51	0.51
10	50	5	25	0.64	0.64
10	60	5	30	0.76	0.77
10	70	5	35	0.89	0.90
10	80	5	40	1.02	1.02
10	90	5	45	1.15	1.15
10	100	5	50	1.27	1.28

Example 15.3

Chile was planted in a sandy loam field in Las Cruces, New Mexico. Nitrogen fertilizer was applied to the field along with the irrigation water. A soil sample was collected from the field, and soil water content and solute concentrations were 0.38 cm^3/cm^3 and 4.2 mg/L. Calculate the pore water velocity for a soil water flux density of 0.1 cm^3/cm^2/day. Also, calculate the amount of nitrate–nitrogen leached per unit area of the field due to convective flow below the root zone in 4 days.

$$\text{Pore water velocity} = \text{Flux density} / \text{water content}$$

$$= 0.1/0.38 = 0.26\ \text{cm/day}$$

The flux density for convective flow of nitrate–nitrogen:

$$J_m = qC = 0.1\frac{\text{cm}}{\text{day}} \times \frac{\text{m}}{100\ \text{cm}} \times 4.2\frac{\text{mg}}{\text{L}} \times \frac{100\text{L}}{\text{m}^3} = 4.2\frac{\text{mg}}{\text{m}^2\text{day}}$$

Therefore, the amount of nitrate–nitrogen leached through the root zone in 4 days through 1 m^2 area = 4.2 (mg/m^2/day) × 1 (m^2) × 4 (day) = 16.8 mg.

15.10 INTERPRETATIONS FROM BREAKTHROUGH CURVES

The shape of the breakthrough curve can describe the transport mechanism of a solute through the porous media. When the displacing solution and displaced solution

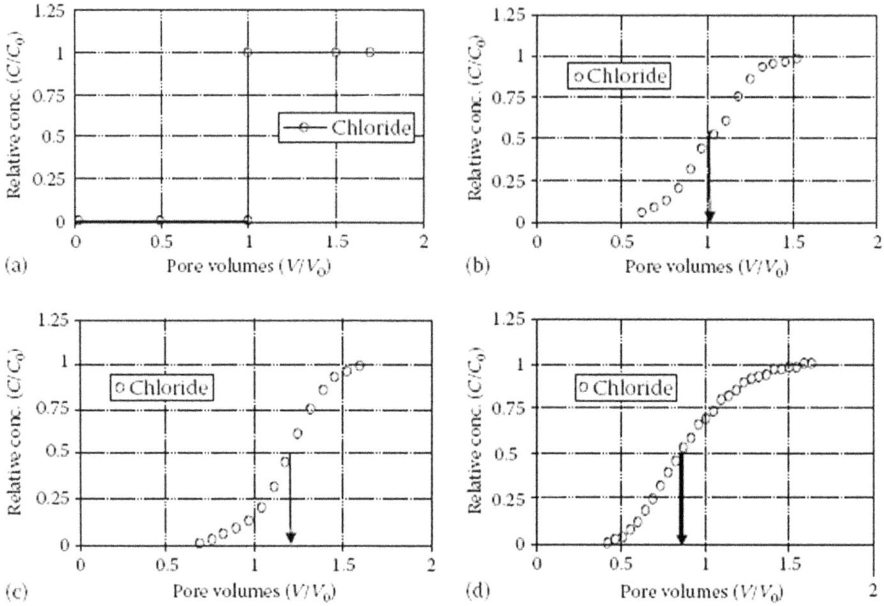

FIGURE 15.4 Some interpretations from the breakthrough curves with respect to the position at one pore volume and a relative concentration of 0.5. (a) Piston flow, (b) convective dispersion, (c) sorption or retardation, and (d) anion exclusion.

do not mix, the concentration jumps to the concentration of the displacing solution as soon as one pore volume of solution has been displaced. The transport process is known as piston flow (Figure 15.4a). When two solutions mix together during the transport, the displacing solution appears in the effluent earlier than one pore volume due to the twin processes of diffusion and dispersion. The dispersive transport results mainly due to the velocity variations compared to the average pore water velocity through the soil. The solute element near the wall of the pore moves at a slower velocity than that near the center of the pore. Thus, a velocity gradient is created inside the pore and solute elements move at different velocities relative to each other. If the breakthrough curve passes through a relative concentration of 0.5 at one pore volume, the areas below and above the relative concentration of 0.5 are equal, and the transport process is known as the convective dispersion process (Figure 15.4b).

The breakthrough curve in Figure 15.4c shows that displacing solute arrives in effluent solution before one pore volume and the breakthrough curve does not pass through a relative concentration of 0.5 at one pore volume but is translated away from the *y*-axis (or intersection of relative concentration of 0.5 and one pore volume). The solute transport process represented by this breakthrough curve is known as adsorption or sorption (Figure 15.4c). The delay in solute arrival in the effluent solution occurs due to the abrupt loss of some ions from the solute front that are being sorbed on the solid (organic and inorganic particle) surfaces. The breakthrough curve in Figure 15.4d also does not pass through the relative concentration of 0.5 at one pore volume but is translated toward the y-axis. The solute transport process represented by this breakthrough curve is known as anion exclusion (Figure 15.4d). The early solute arrival in the effluent solution could be due to the repulsion of anions from negatively charged clay surfaces.

The pore water velocity has a strong influence on the shape of the breakthrough curve. For a given displacement volume (or soil volume), as the pore water velocity increases, the breakthrough curve becomes steeper. Figure 15.3a shows three breakthrough curves for a slow, medium, and high pore water velocity ranging from 0.025 to 2.46 cm/h (two-order magnitude difference). Chloride was applied as a step input to displace water present in the soil volume. It can be observed that the time taken for the solute to reach the maximum concentration decreased with increasing pore water velocity. Similarly, when chloride was applied as a pulse input, the base of the breakthrough curves as well as the time required for complete leaching of the chloride pulse decreased with increasing pore water velocity (Figure 15.3b).

The solute displacement lengths have a strong influence on the shape of the breakthrough curves. When the same volume of solute pulse is sent through soil columns of increasing displacement lengths (or volumes), attenuation of the peak concentration takes place (Figure 15.5a; Shukla et al., 2000). The solute is not able to carry its maximum concentration all the way to the outlet because of the dilution. Peak concentration can also decrease if chemical reactions lead to the degradation of chemicals during the transport of solute through soil columns (Figure 15.5b). On the other hand, production of chemicals during the transport can lead to increases in peaks of the solute breakthrough curves.

Soil structure comprises of a complex network of pores, and it has a strong effect on the shape of a solute breakthrough curve. Soil structure is influenced by land use and management-related factors. Unlike a repacked soil core where the likelihood of macropores or preferential flow channels is low, a soil under no-tillage management where plant biomass is returned to the soil on a regular basis could have a good soil structure with a well-developed network of larger pores including macropores. The soil is more homogeneous in a repacked soil column than that in an undisturbed soil core. An undisturbed soil could be aggregated with both macro- and microaggregates present inside the soil matrix. Thus, the physical nonhomogeneity of soil associated with variable pore spaces could cause solute to appear in the effluent solution early via flow through macropores. Physical nonhomogeneity could also be due to the presence of immobile or slowly conducting pores (physical nonequilibrium), and that could result in a breakthrough curve with a long tail before the entire pulse of the solute is leached through the soil. The solute can move into the immobile zone by the process of diffusion and can slowly come out of the immobile zone creating a long tail. On the other hand, the breakthrough curve through a more homogeneous repacked soil core could have a higher relative concentration and late arrival in the effluent solution, but a smaller base than the breakthrough curve from the undisturbed core (Figure 15.6). Similar transport behavior of chloride through long, wide, undisturbed, and repacked soil columns is reported by Khan (1988).

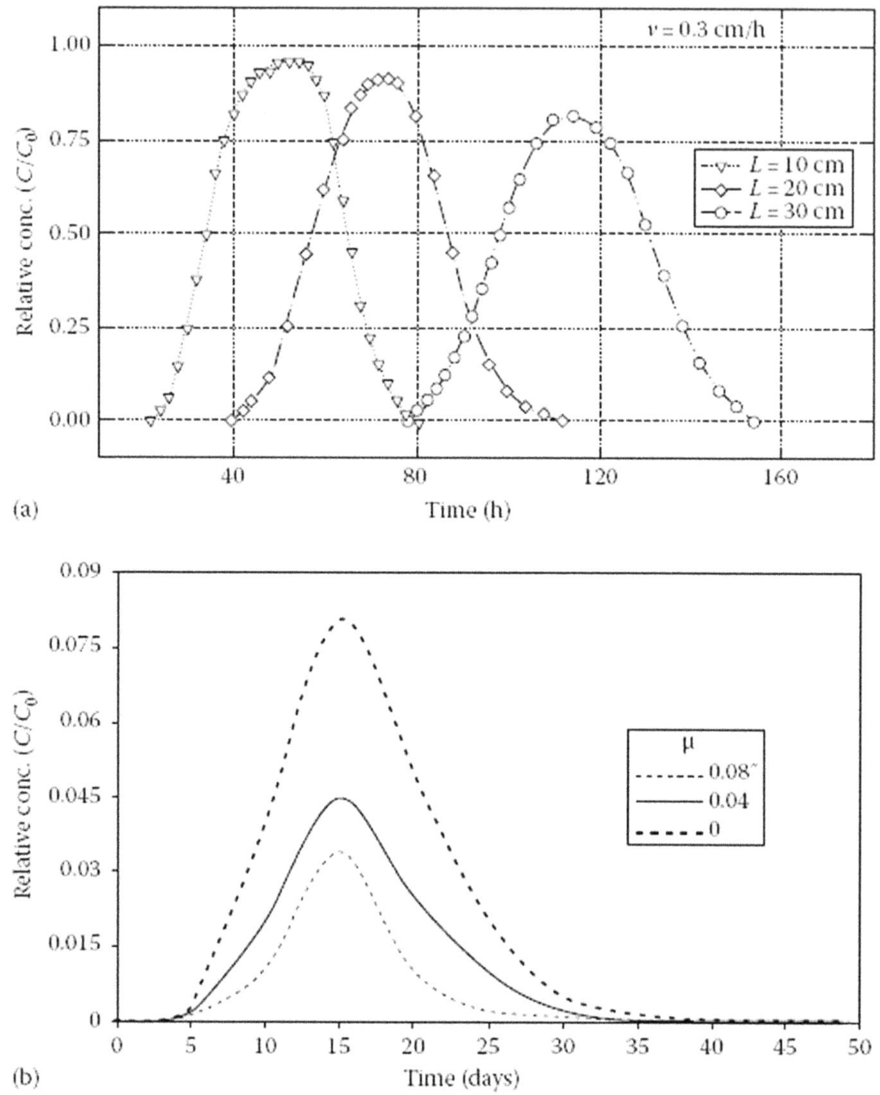

FIGURE 15.5 Effect of (a) displacement length. (Modified from Shukla, M. K., et al., *Soil Sci. Soc. Am. J.* 66(4), 1090, 2002.) (b) Schematic of chemical reactions on breakthrough curves.

The breakthrough curve can also translate away from the y-axis as water content associated with the mobile region increases with attendant decrease in water content associated with the immobile region and the breakthrough curve becomes more symmetric. Once the water content in the mobile region becomes equal to the total water content, the breakthrough curve represents a convective dispersion process.

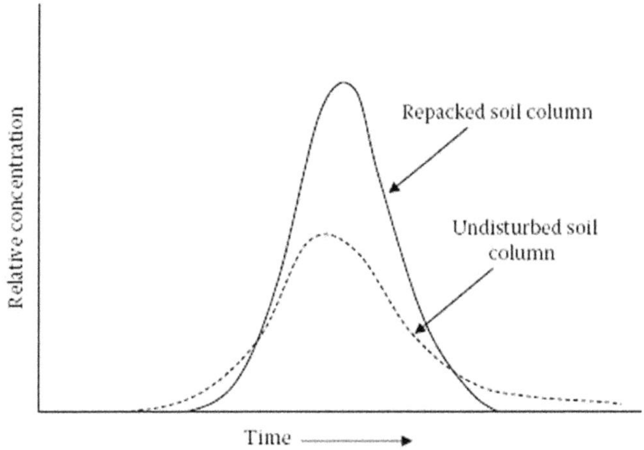

FIGURE 15.6 The schematic of breakthrough curves from an undisturbed soil core and a repacked soil core.

15.11 EQUILIBRIUM CONVECTIVE DISPERSION EQUATION

Total flux of the dissolved solute during convective–dispersive transport can be expressed as follows:

$$J = qC - \theta D_{\mathrm{m}} \frac{\partial C}{\partial x} - \theta D_{\mathrm{h}} \frac{\partial C}{\partial x} \tag{15.17}$$

or

$$J = -\theta D \frac{\partial C}{\partial x} + qC \tag{15.18}$$

or

$$J = -\theta \times \left[D \frac{\partial C}{\partial x} + vC \right] \tag{15.19}$$

where

D is the effective dispersion coefficient.
v is the pore water velocity, and the fluxes are expressed on a volumetric basis.

Transferring this equation, which combines the three solute transport processes—convection, diffusion, and dispersion (Equation 15.19)—into the mass conservation equation (Equation 15.3), the solute transport equation describing one-dimensional flow of nonadsorbing solutes is obtained as follows:

$$\frac{\partial(\theta C)}{\partial t} = -\frac{\partial J}{\partial x} = -\theta \frac{\partial}{\partial x}\left(-D \frac{\partial C}{\partial x} + vC \right) \tag{15.20}$$

$$\frac{\partial C}{\partial t} = D\frac{\partial^2 C}{\partial x^2} - v\frac{\partial C}{\partial x} \tag{15.21}$$

This equation is also known as the one-dimensional convective dispersion equation (Lapidus and Amundson, 1952). For a solute that is sorbed (S; g/g) on a soil particle surface, the equation changes as follows:

$$\frac{\rho}{\theta}\frac{\partial S}{\partial t} + \frac{\partial C}{\partial t} = D\frac{\partial^2 C}{\partial x^2} - v\frac{\partial C}{\partial x} \tag{15.22}$$

If the adsorption is taking place at equilibrium condition, $S = KC$, where K is the distribution coefficient (also expressed as K_D). Transferring S into this equation (Equation 15.22) yields

$$\frac{\rho K}{\theta}\frac{\partial C}{\partial t} + \frac{\partial C}{\partial t} = \left(\frac{\rho K}{\theta} + 1\right)\frac{\partial C}{\partial t} = D\frac{\partial^2 C}{\partial x^2} - v\frac{\partial C}{\partial x} \tag{15.23}$$

$$R\frac{\partial C}{\partial t} = D\frac{\partial^2 C}{\partial x^2} - v\frac{\partial C}{\partial x} \tag{15.24}$$

where R is the retardation factor. When during transport a solute, cation, or a positive-charged ion is adsorbed on the particle surface with a net negative charge, such as clay, the value of R is greater than one and is equal to

$$R = 1 + \frac{\rho K}{\theta} \tag{15.25}$$

The solute breakthrough time for the convective dispersion equation is given as the ratio of length of displacement and pore water velocity (L/v). Dividing the convective dispersion equation by R yields the following equation:

$$\frac{\partial C}{\partial t} = D_R\frac{\partial^2 C}{\partial x^2} - v_R\frac{\partial C}{\partial x} \tag{15.26}$$

where

D_R ($= D/R$) is the retarded dispersion coefficient.
v_R ($= v/R$) is the retarded pore water velocity.

The solute breakthrough time for this equation is given as the ratio of length of displacement and retarded pore water velocity (L/v_R or $R \times L/v$).

Example 15.4

If Darcy's water velocity (q) is 2.4 cm/day, soil water content (θ) is 0.34 cm³/cm³, and average bulk density of soil is 1.25 g/cm³, how long will it take for a conservative tracer, such as chloride, and a reactive tracer, such as a herbicide (K_D = 2.5 cm^3/g), to move to the water table 15 m below the soil surface?

Pore water velocity = q/θ = 2.4/0.34 = 7.06 cm/day
For chloride, R will be equal to one because it is weakly sorbed on soil.
Breakthrough time for chloride = L/v = 15 × 100/7.06 = 213 days
Retardation factor for herbicide = 1 + 1.25 × 2.5/0.34 = 10.19
Breakthrough time for herbicide = $L \times R/v$ = 15 × 100 × 10.19/7.06
= 2165 days

Negatively charged ions or anions are usually repelled by soil particle surfaces consisting of, for example, predominantly clays or organic matter. Thus, the solute is excluded from the water envelope surrounding the soil particle, and the process of solute transport is known as anion exclusion (Shukla and Cepuder, 2000; Lal and Shukla, 2004). For a saturated porous media with no immobile water, one-dimensional solute transport with anion exclusion for a nonreactive solute can be represented by the following equation (James and Rubin, 1986):

$$\left(1-\frac{\theta_{ex}}{\theta}\right)\frac{\partial C}{\partial t}=D\frac{\partial^2 C}{\partial x^2}-v\frac{\partial C}{\partial x} \tag{15.27}$$

where θ_{ex} is the exclusion volume. Anion exclusion is also represented by the retardation factor; exclusion occurs when the values of retardation factor are less than unity (Shukla and Cepuder, 2000).

Adsorption is the process that describes the attachment of molecules on to the particle surface. The simplest form of describing adsorption is by a linear adsorption isotherm ($S = KC$). Other common forms of describing adsorption, particularly nonlinear adsorption, include a Freundlich and a Langmuir isotherm. In addition, there are other forms of isotherms that can take into account the kinetics or time dependency of the adsorption process. The Freundlich and Langmuir isotherms are represented by the following equations:

$$S = KC^n \tag{15.28}$$

$$S=\frac{aQC}{1+aC} \tag{15.29}$$

where

a is the Langmuir energy constant or bond strength (L/g).
Q is the maximum adsorption capacity or total available number of sites per unit soil mass (mg/kg).
C is the solution concentration (g/L) (Selim and Amacher, 1997; Lal and Shukla, 2004).

During solute transport, the solute can also be produced or undergo decay, and such a process can be represented by the following one-dimensional convective dispersion equation with production and decay terms:

$$R\frac{\partial C}{\partial t} = D\frac{\partial^2 C}{\partial x^2} - v\frac{\partial C}{\partial x} - \mu C + \gamma \tag{15.30}$$

where

μ is the first-order degradation coefficient (s^{-1}).

γ is the zero-order production coefficient ($g/cm^3/s$).

Remember that the degradation and production of chemicals can take place both in solution and adsorbed phases (Parker and van Genuchten, 1984). The equations described so far describe solute transport under equilibrium conditions. The equations can be solved for known initial and boundary conditions. One of the initial conditions could be that solute concentration throughout the soil profile is constant (it can also be zero) at time zero:

$$C(x,0) = C_i \tag{15.31}$$

where $C(x, 0)$ indicates concentration in a soil column of length x at time equal to zero and is equal to C_i. At the upper end of the soil profile or a soil column ($x = 0$), two different types of boundary conditions can be applied. The constant concentration or a first-type boundary condition is as follows:

$$C(0,t) = C_0 \tag{15.32}$$

A constant flux type or third-type boundary condition is represented as follows:

$$\left(-D\frac{\partial C}{\partial x} + vC\right)_{x=0} = vC_0 \tag{15.33}$$

where C_0 is the concentration of the input solution. It is reported that unlike the constant flux boundary condition, the constant concentration type of boundary condition results in a mass balance error (van Genuchten, 1981). For the lower boundary of the porous media of finite (*L*) or semi-infinite length (∞), the following two boundary conditions can be applied:

$$\frac{\partial C}{\partial x}(L,t) = 0 \tag{15.34}$$

$$\frac{\partial C}{\partial x}(\infty,t) = 0 \tag{15.35}$$

The solutions for a semi-infinite porous media provide a solution similar to a finite porous medium of a given length (van Genuchten, 1981). The one-dimensional convective dispersion equation can be modified by introducing dimensionless variables as follows:

$$T = \frac{vt}{L}; \quad z = \frac{x}{L}; \quad P = \frac{vL}{D}; \quad C_1 = \frac{C - C_i}{C_0 - C_i}$$

Transferring the nondimensional variables changes the one-dimensional convective dispersion solute transport equation as follows:

$$R\frac{\partial C_1}{\partial T} = \frac{1}{P}\frac{\partial^2 C_1}{\partial z^2} - v\frac{\partial C_1}{\partial z} \tag{15.36}$$

where

T is the number of pore volumes.
P is the Peclet number.
C_e is the dimensionless exit concentration, which reduces to $C_e = C_1(1, T)$.

15.12 SOLUTE TRANSPORT EQUATION UNDER PHYSICAL NONEQUILIBRIUM

Solute transport through the soil profile does not always occur under equilibrium conditions but sometimes also under physical or chemical nonequilibrium conditions. The shape of the breakthrough curve for a conservative solute is symmetrical during the transport under equilibrium conditions and asymmetrical under nonequilibrium conditions. The breakthrough curves of sorbing as well as nonsorbing solutes under nonequilibrium conditions may have a long tail. The nonequilibrium conditions could be due to the physical nonequilibrium in which the soil profile consists of two (or more) regions, where one of the regions is mobile while the other region is immobile. This situation is analogous to the dual-porosity soil system in which the water in the immobile region is stagnant, likely due to the low porosity or small pore sizes. The water in the other region is flowing with greater porosities and large pore sizes, and actively participates in the transport of solution through porous media. In addition to a mobile and immobile zone concept, a two-flow domain theory, in which flow through one of the domains is higher than the other, is also proposed (Skopp et al., 1981).

In the mobile–immobile zone models, the retention mechanism for both regions is considered the same. However, during solute movement, the solute is exchanged between the mobile and immobile regions primarily by the process of diffusion. The governing transport equations for mobile and immobile regions for a nonsorbing solute are described by the following equations also known as two-region model (Coats and Smith, 1964):

$$\theta_m \frac{\partial C_m}{\partial t} + \theta_{im}\frac{\partial C_{im}}{\partial t} = \theta_m D_m \frac{\partial^2 C_m}{\partial x^2} - \theta_m v_m \frac{\partial C_m}{\partial x} \tag{15.37}$$

$$\theta_m \frac{\partial C_{im}}{\partial t} = \alpha(C_m - C_{im}) \tag{15.38}$$

The governing transport equations for mobile and immobile regions (or two-region model) for a sorbing solute are described by the following equations (van Genuchten and Wierenga, 1976):

$$\theta_m \frac{\partial C_m}{\partial t} + f\rho_b \frac{\partial C_m}{\partial t} + \theta_{im} \frac{\partial C_{im}}{\partial t} + (1-f)\rho_b \frac{\partial S_{im}}{\partial t} = \theta_m D_m \frac{\partial^2 C_m}{\partial x^2} - \theta_m v_m \frac{\partial C_m}{\partial x} \tag{15.39}$$

$$\theta_{im} \frac{\partial C_{im}}{\partial t} + (1-f)\rho_b \frac{\partial S_{im}}{\partial t} = \alpha(C_m - C_{im}) \tag{15.40}$$

where

Subscripts m and im refer to mobile and immobile fractions of water, respectively.
v_m is the pore water velocity in the mobile region.
θ_m is the mobile water.
θ_{im} is the immobile water content.
f is the mass fraction of the solid phase assigned to the mobile region.
α is the mass transfer coefficient that describes exchange between mobile and immobile regions.

Equations 15.39 and 15.40 are derived assuming that the exchange of solute between mobile and immobile regions takes place only as a diffusional process, and the adsorption around the micropores located in the immobile region is different from that taking place in mobile regions. Total adsorption (S) is given by the following formula:

$$S = fS_m + (1-f)S_{im} \tag{15.41}$$

S_m and S_{im} are the adsorbed concentration in mobile and immobile regions, respectively. Essentially for a solute that does not sorb (conservative solute), the second and the fourth terms on left-hand side of the two-region model (Equations 15.39 and 15.40) vanish:

$$v_m = \frac{q}{\theta_m} = \frac{v}{\phi_m}$$

$$\phi_m = \frac{\theta_m}{\theta}$$

$$\theta = \theta_m + \theta_{im}$$

where

q is Darcy's flux.
ϕ_m is the fraction of mobile water.
v is the pore water velocity.
v_m is the pore water velocity in the mobile region.

The following additional nondimensional variables can be introduced:

$$T = \frac{vt}{L} = \frac{v_m t \phi_m}{L}; \quad P = \frac{v_m L}{D}; \quad \beta = \frac{\theta_m + f\rho_b K}{\theta + \rho_b K} = \frac{\phi_m R_m}{R};$$

$$R_m = 1 + \frac{f\rho_b K}{\theta_m}; \quad \omega = \frac{\alpha L}{q} = \frac{\alpha L}{v_m \theta_m}; \quad C_2 = \frac{C_{im} - C_i}{C_0 - C_i}$$

The Peclet number and retardation factor for two-region models are expressed with respect to the mobile or dynamic region. Transferring the nondimensional variable to the two-region model (Equations 15.39 and 15.40) yields the following general form of equation for nonequilibrium transport:

$$\beta R \frac{\partial C_1}{\partial T} + (1-\beta) R \frac{\partial C_2}{\partial T} = \frac{1}{P}\frac{\partial^2 C_1}{\partial x^2} - \frac{\partial C_1}{\partial x} \tag{15.42}$$

$$(1-\beta) R \frac{\partial C_2}{\partial T} = \omega (C_1 - C_2) \tag{15.43}$$

Equations 15.42 and 15.43 are solved for several different initial and boundary conditions by researchers and can be found elsewhere (e.g., van Genuchten, 1981). Instead of being sorbed, a solute can be repelled by the solid phase and get excluded from the liquid zones. Assuming that anion exclusion is restricted to the immobile region, the following physical nonequilibrium equation can be used (van Genuchten, 1981):

$$\theta_m \frac{\partial C_m}{\partial t} + \theta_a \frac{\partial C_a}{\partial t} = \theta_m D \frac{\partial^2 C_m}{\partial x^2} - \theta_m v_m \frac{\partial C_m}{\partial x} \tag{15.44}$$

$$\theta_a \frac{\partial C_a}{\partial t} = \alpha (C_m - C_a) \tag{15.45}$$

where $\theta_a = \theta_{im} - \theta_{ex}$. The following additional nondimensional variables can be transferred in this equation to get the same functional form of nondimensional solute transport equation (Equations 15.42 and 15.43):

$$\phi_{ex} = \frac{\theta_{ex}}{\theta}; \quad \boldsymbol{R} = 1 - \phi_{ex}; \quad C_2 = \frac{\boldsymbol{C_a} - \boldsymbol{C_i}}{\boldsymbol{C_0} - \boldsymbol{C_i}}$$

15.13 SOLUTE TRANSPORT EQUATION UNDER CHEMICAL NONEQUILIBRIUM

The transport of solute through soil is a complicated process because of the heterogeneity associated with soils. The heterogeneity associated with nonuniform pore geometry was described in the previous section by a two-region model. In this section, we discuss the chemical heterogeneity caused by nonuniform chemical composition of soil (Selim et al., 1976). The soil matrix consists of organic matter, oxides of aluminum and iron, quartz, and other organic and mineral materials. The chemical sorption takes place differently in various constituents of the soil matrix. A two-site model assumes that the soil matrix consists of two different types of sites; the sorption in one of the sites is instantaneous, while the sorption in the other is time-dependent. The instantaneous sorption sites are also known as type-1 sites, whereas time-dependent sorption sites are known as type-2 sites:

$$\left(1+\frac{F\rho_b K}{\theta}\right)\frac{\partial C}{\partial t}+\frac{\rho_b}{\theta}\frac{\partial S_2}{\partial t}=D\frac{\partial^2 C}{\partial x^2}-v\frac{\partial C}{\partial x} \tag{15.46}$$

$$\frac{\rho_b}{\theta}\frac{\partial S_2}{\partial t}=\alpha\left[\left(1-F\right)KC-S_2\right] \tag{15.47}$$

where F is the fraction of sites occupied by type-1 sorption sites. Sorption on type-1 (S_1) and type-2 (S_2) sites is expressed as follows:

$$S_1=K_1C \quad \text{and} \quad S_2=K_2C=\left(1-F\right)KC$$

Thus, total sorption is the sum of the sorption taking place in these two sites. Introducing the following nondimensional variables leads to the same general form of nondimensional solute transport equation (Equations 15.42 and 15.43):

$$T=\frac{vt}{L};\quad P=\frac{v_m L}{D};\quad \beta=\frac{\theta+f\rho_b K}{\theta+\rho_b K}=\frac{R_m}{R};\quad R_m=1+\frac{f\rho_b K}{\theta};\quad \omega=\frac{\alpha\left(1-\beta\right)RL}{v}$$

$$C_1=\frac{C-C_i}{C_0-C_i};\quad C_2=\frac{S_2-\left(1-F\right)KC_i}{\left(1-F\right)K\left(C_0-C_i\right)}$$

The two-site chemical nonequilibrium model and two-region physical nonequilibrium model are mathematically equivalent (Nkedi-Kizza et al., 1984). These two models are reported to successfully predict nonequilibrium transport of chemicals especially when breakthrough curves are asymmetrical (e.g., Selim and Amacher, 1997; Shukla and Kammerer, 1998; González-Delgado and Shukla, 2011). There are also studies available where one model was reported to perform better than the other. For example, Rao et al. (1979) reported that the two-site model provided better predictions of 2,4-D (herbicide) and atrazine transport than the two-region model.

However, van Genuchten et al. (1974) reported that both equilibrium and kinetic models failed to predict Picloram transport at high pore water velocities.

15.14 EFFECT OF PORE WATER VELOCITY ON SOLUTE TRANSPORT PARAMETERS

The apparent dispersion coefficient increases with increasing pore water velocity (Pfannkuch, 1962; Biggar and Nielsen, 1976; Shukla et al., 2002). Relations between D/D_o and the Peclet number (vd/D_o), given as solid lines in Figure 15.7, satisfy the following equation:

$$\frac{D}{D_o} = 1 + m\left(\frac{vd}{D_o}\right)^n \tag{15.48}$$

The value of D_o can be estimated from $D_o = 0.66\ \theta\ D_m$. The relationship in Figure 15.7 was given for field soils by Biggar and Nielsen (1976), for single-grained material by Pfannkuch (1962), and for loam and sandy loam soils by Shukla et al. (2002). All these curves indicated that for low pore water velocity, molecular diffusion dominates, and as pore water velocity increases further, dispersion becomes the dominant component of apparent dispersion coefficient. Other experiments (González-Delgado and Shukla, 2011) provided similar results and showed that with decreasing average particle diameter, the values of m (see Equation 15.48) increase but values of n range from 1.02 to 2.36 (Table 15.1).

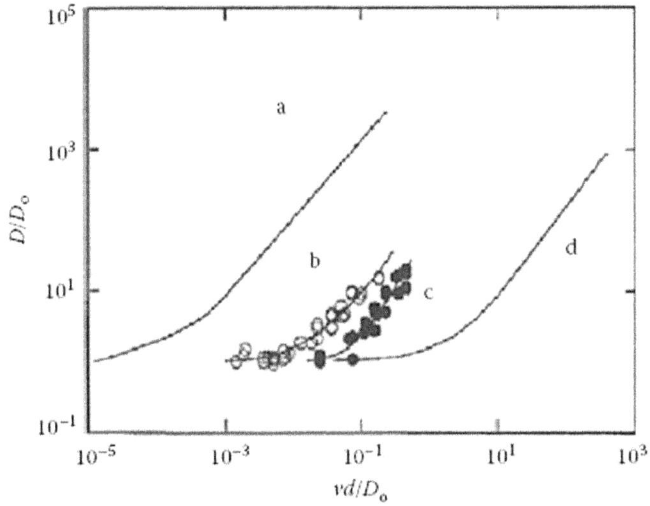

FIGURE 15.7 The relationship between DD_o^{-1} and Peclet number (vd/D_o). (a) From Biggar and Nielsen (1976). (b and c) From Shukla et al. (2002). (d) From Pfannkuch (1962). (Modified from Shukla, M. K., et al., *Soil Sci. Soc. Am. J.*, 66(4), 1091, 2002.)

TABLE 15.1
Parameters for the Equation Fitted to the Curves in Figure 15.7

Soil	m	N	d (mm)	D_o (cm²/h)
Laboratory columns	0.50	1.20	0.156	0.0220
Sand	14.72	2.36	0.062	0.0170
Sandy loam	70.50	1.71	0.051	0.0216
Loam	117.63	1.02	0.018	0.0217
Loam	141.00	1.21	0.016	0.0222
Field soil	17,780.00	1.11	0.003	0.0250

Sources: Pfannkuch, H. O., *Rev. Inst. Fr. Petrol.*, 18(2), 1962; Shukla, M. K., et al., *Soil Sci. Soc. Am. J.*, 66(4), 1087, 2002; González-Delgado, A. M. and Shukla, M. K., *Soil Sci.*, 176(7), 346, 2011; Biggar, J. W. and Nielsen D. R., *Agronomy*, 11, 254–274, 1967.

Pore water velocity is reported to have no influence to slight influence on the retardation factor, likely due to the reductions in the immobile water content of the soil (González-Delgado and Shukla, 2011). Anion exclusion volume is reported to decrease with increasing pore water velocity primarily due to the decrease in immobile water content (Shukla and Cepuder, 2000). Contradictory accounts are available on the influence of pore water velocity on mass transfer coefficient describing the exchange between mobile and immobile regions. The mass transfer coefficient is reported to show no consistent trend with increasing pore water velocity (Shukla and Cepuder, 2000; González-Delgado and Shukla, 2011). It is also reported to decrease with decreasing flow velocity, indicating a slower mixing between the two regions (van Genuchten and Wierenga, 1977). Similar observations on a direct relationship between mass transfer coefficient and pore water velocity are reported by Kookana et al. (1993) and Padilla et al. (1999).

15.15 PARAMETER ESTIMATION

There are several parameters associated with equilibrium and nonequilibrium solute transport models. In this section, we discuss how some of these parameters can be estimated.

15.15.1 From Batch Experiment

The retardation factor that provides an estimate of sorption or anion exclusion is a function of the distribution coefficient (K), soil bulk density ($_b$), and volumetric soil water content (θ). K can be calculated from the batch experiment. In a batch experiment, soil (usually air-dry and <2 mm in size) is mixed separately with the solute of several different concentrations (at least six). The mixture is shaken thoroughly and allowed to rest. It is then placed inside a centrifuge, supernatant and soil are

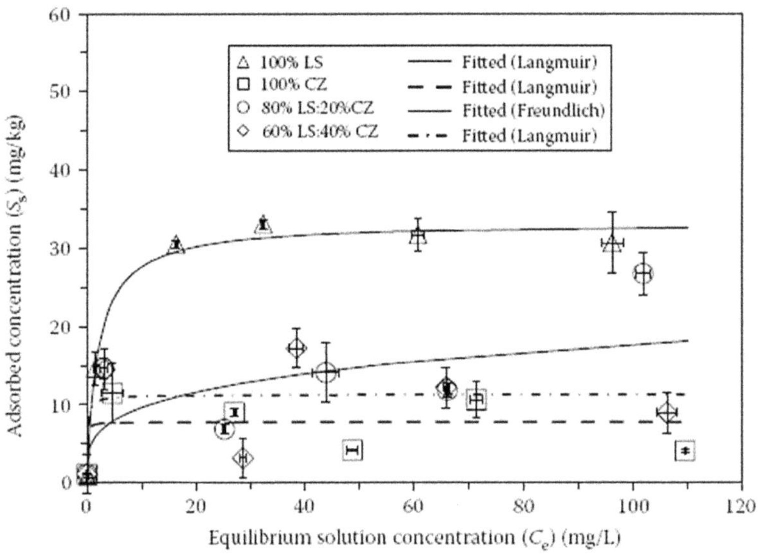

FIGURE 15.8 Mean adsorbed $NO_3^- - N$ concentration (S_s) (y-axis) as function of $NO_3^- - N$ equilibrium concentration (C_e) from a solution of UAN®32 [$CO(NH_2)_2NH_4NO3$] (y-axis), and best-fit adsorption isotherms. CZ is clinoptilolite zeolite and LS is loamy sand. (Adapted from Piñón-Villarreal, A. R., et al., *J. Irrig. Drain. Eng.*, 139(9), 760, 2013.)

separated, and the concentration of supernatant is determined. The sorbed concentration is calculated as the difference between the known solution concentration and the concentration of the supernatant. The units for sorbed concentration are appropriately converted from mass of solute per unit volume of solution (mg/L) to the units of mass of solute per unit mass of the soil (mg/g). The sorbed concentration is plotted on the y-axis and the solution concentration (mg/L) on the x-axis, and the slope of the curve provides the value of the distribution coefficient (Figure 15.8).

The batch adsorption experiments of Figure 15.8 were carried out to investigate the sorption potential of loamy sand (LS) and clinoptilolite zeolite (CZ) media to retain $NO_3^- - N$ and $NH_4^+ - N$ ions in the solid phase (Piñón-Villarreal et al., 2013). These were undertaken by mixing 36 mL of six different concentrations of $NO_3^- - N$ and $NH_4^+ - N$ fertilizer solutions with 18 g of the four soil treatments, a 2:1 ratio of liquid to soil. The liquid fertilizer UAN®32 (urea ammonium nitrate; 32–0–0; $CO(NH_2)_2NH_4NO_3$; TETRA Inc.) provided the NO_3^- and NH_4^+ ions in the prepared solutions. The soil solution (replicated three times) was shaken for 3 h with a rotary machine at 50 rpm, stored for 21 h at 0°C, centrifuged at 2,500 rpm for 15 min, and concentrations of $NH_3^- - N$ and $NH_4^+ - N$ were determined using portable spectrophotometers (models DR/2000 and DR/2010, HACH®, Loveland, CO). $NO_3^- - N$was determined with the cadmium reduction method, and $NH_4^+ - N$ was analyzed by the salicylate method (HACH®, 2000). Adsorbed solute concentration (S; mg/kg) was determined as

$$S = \frac{(C_1 - C_e)V}{M} \tag{15.49}$$

where

C_1 and C_e are the concentrations of the solution and supernatant, respectively (g/L).
V is the volume of the solution (L).
M is the mass of the media (g).

The K obtained by the batch experiment can be used to calculate the retardation factor for that solute during transport through a soil column.

15.15.2 From Breakthrough Curves

The retardation factor can be directly obtained from the breakthrough curves where relative concentration is plotted against pore volumes. The point at which a breakthrough curve crosses the relative concentration of 0.5 is obtained, and the corresponding pore volume (vertically downward) is read on the axis that provides an estimate of the retardation factor.

The apparent dispersion coefficient (D) can be estimated from the known pore water velocity of the solute (v), length of displacement (L), and the slope of the breakthrough curve at one pore volume (m) as follows:

$$D = \frac{vL}{4\pi m^2} \tag{15.50}$$

where

$$m = \frac{\partial(C/C_0)}{\partial p}(\text{at } p = 1) = \frac{(C/C_0)_1 - (C/C_0)_2}{p_1 - p_2} \tag{15.51}$$

In a similar approach, the apparent dispersion coefficient (D) can be obtained by estimating the value of the Peclet number (P). The inverse complementary error function in the solution of the solute transport equation (van Genuchten, 1981) can be obtained from a measured breakthrough curve plotted against the log of the pore volumes on the x-axis. The slope of the curve is obtained to calculate the nondimensional Peclet number (van Genuchten and Wierenga, 1986). From the known pore water velocity, length of displacement, and estimated P, the apparent dispersion coefficient can be obtained.

Example 15.5

Estimate the retardation factor for the breakthrough curves presented in Figure 15.4b through d. List the processes with respect to the retardation factor value.

Three breakthrough curves for step inputs of chloride through saturated soil columns are presented in Figure 15.4b through d (Shukla et al., 2003).

The pore volume corresponding to the breakthrough curve in Figure 15.4b when it passes through a relative concentration of 0.5 is 1. Thus, retardation factor is equal to 1, indicating no sorption.

For Figure 15.4c, the corresponding pore volume is 1.2, meaning a retardation factor of 1.2, which indicates sorption of solute on soil particle surfaces.

For Figure 15.4d, the corresponding pore volume is 0.9, meaning a retardation factor of 0.9. Since the value is less than 1, it indicates that anion exclusion is taking place during the transport of the solute through soil.

The retardation factor calculated from a batch experiment may be higher than that obtained from a breakthrough curve because in a batch experiment all the sites are fully available for sorption, which may not be the case during solute transport through a soil matrix.

Example 15.6

Assuming that the breakthrough curve in Figure 15.4b is generated for an average pore water velocity of 0.5 cm/min and length of the column (displacement) is 10 cm, estimate the slope at one pore volume and the apparent diffusion coefficient.

The slope (m) and apparent diffusion coefficient (D) can be calculated as follows:

$$m = \frac{\partial (C/C_0)}{\partial p}(\text{at}\, p = 1) = \frac{(C/C_0)_2 - (C/C_0)_1}{p_2 - p_1} = \frac{0.53 - 0.44}{1.2 - 0.95} = \frac{0.9}{0.25} = 3.6$$

$$D = \frac{vL}{4\pi m^2} = \frac{0.5 \times 10}{4 \times 3.14 \times 3.6^2} = 0.031\,\text{cm}^2/\text{min}$$

Note: The slope of the breakthrough curve in Figure 15.4b can also be calculated by first fitting a curve (polynomial or any other form) that goes through most of the points (typically a coefficient of determination >0.9) and then differentiating the curve with respect to the x-axis (or pore volume) and then transferring one in the equation for pore volume.

Suppose the best-fit equation for the breakthrough curve in Figure 15.4b is a polynomial curve as follows:

$$y = a + b \times x + c \times x^2 + d \times x^3$$

where

y is the relative concentration.
x is the number of pore volumes.

Differentiating with respect to x will yield:

$$\frac{dy}{dx} = b + 2 \times c \times x + 3 \times d \times x^2$$

Since slope is determined at pore volume of one, transferring one for x will result in the following equation for slope:

$$\frac{dy}{dx} = b + 2 \times c + 3 \times d$$

The estimates of the coefficients b, c, and d are transferred to the equation to obtain the slope.

15.15.3 From Fitting Solute Transport Equations

The solute transport equation can be solved for the appropriate boundary conditions. The solution is usually obtained in the form of complementary error functions. As an example, the equilibrium transport of a sorbing solute (Equation 15.36) for an initial soil profile concentration of zero and an infinite soil profile (concentration type BC) using dimensionless parameters can be obtained as follows (van Genuchten, 1981):

$$C_e = 0.5erfc\left(\left(\frac{P}{4RT}\right)^{0.5}(R-T)\right) + 0.5\exp(P)erfc\left(\left(\frac{P}{4RT}\right)^{0.5}(R+T)\right) \quad (15.52)$$

Initial estimates of P, R, and T can be assumed or can be obtained from the measured breakthrough curves. Using Equation 15.52, the solute concentration can be calculated at each pore volume. Using a least-square optimization program (e.g., Solver in Excel), the optimized parameter can be obtained by minimizing the sum of squared deviations between measured and predicted solute concentration curves.

The solution of Equation 15.24 (Lapidus and Amundson, 1952) for a pulse type of input is as follows:

$$\frac{C(x,t)}{C_o} = 0.5erfc\left(\frac{Rx - vt}{\sqrt{4DRt}}\right) - 0.5\exp\left(\frac{vx}{D}\right)erfc\left(\frac{Rx + vt}{\sqrt{4DRt}}\right)$$

$$+0.5erfc\left(\frac{Rx - v(t - t_0)}{\sqrt{4DRt(t - t_0)}}\right) - 0.5\exp\left(\frac{vx}{D}\right)erfc\left(\frac{Rx + v(t - t_0)}{\sqrt{4DR(t - t_0)}}\right) \quad (15.53)$$

For a continuous application of the solution of concentration C_0 at the inlet ($t < t_0$), the last two terms of Equation 15.53 are ignored.

Alternately, there are public domain computer programs by van Genuchten and coworkers available from the US Soil Salinity Research Lab in Riverside, California. Some of these programs are known as CFITIM, CXTFIT, and other versions of CXTFIT (van Genuchten, 1981; Parker and van Genuchten, 1984; Toride et al., 1995). These programs require a measured breakthrough curve as input and initial estimates of dimensional or nondimensional parameters of the equilibrium and nonequilibrium solute transport equations. Nonlinear optimization produces the estimates of solute transport equation parameters. Using CXTFIT, the pore water velocity can be optimized by using an R value equal to one because the fitted pore water velocity will be equal to the ratio of the pore water velocity and retardation factor; the ratio is also known as the retarded velocity. These programs can also fit all the dimensional or nondimensional parameters of equilibrium and nonequilibrium transport models.

15.15.4 Nonequilibrium Parameters

Some of the parameters of nonequilibrium transport equations (Equations 15.39 and 15.40), such as mobile (θ_m) and immobile (θ_{im}) water contents and mass transfer coefficient (α) can also be estimated from the known soil water content at saturation and at field capacity. The soil water content at field capacity is used as a reference for dividing total soil water content into mobile and immobile water contents. All the water in the soil between saturation and field capacity is freely mobile and therefore can be considered as mobile water content of the soil. The immobile water content is obtained as the difference between the total soil water content and the mobile water content.

Parameters of nonequilibrium transport equations (Equations 15.39 and 15.40), can also be estimated by conducting a single or multiple simultaneous solute infiltration tests. Clothier et al. (1992) utilized the solute mass balance in mobile and immobile zones to estimate water contents of these two regions. A conservative tracer was infiltrated through the soil until steady-state condition is obtained (Lal and Shukla, 2004). The soil samples were collected subsequently and the concentration of the conservative tracer in the soil was determined. The following equations provided the estimates of mobile and immobile water:

$$\theta C = \theta_m C_m + \theta_{im} C_{im} \tag{15.54}$$

If all the water is mobile, then the concentration is equal to the applied concentration (C_0) and mobile water can be obtained as follows:

$$\theta_m = \theta \frac{C}{C_0} \tag{15.55}$$

The immobile water content, assuming that the transfer of solute between mobile and immobile regions is low, can be obtained as follows:

$$\theta_{im} = \theta \left(1 - \frac{C}{C_0} \right) \tag{15.56}$$

This approach was modified by Jaynes et al. (1995) to include the estimation of the mass transfer coefficient (α). They applied a series of conservative tracers for different amounts of time (t) and used the following nonequilibrium solute transport equation:

$$ln\left(1-\frac{C}{C_0}\right)=-\frac{\alpha}{\theta_{im}}t+ln\frac{\theta_{im}}{\theta} \quad (15.57)$$

A plot between time (t; term on right-hand side) on the x-axis and the term on the left-hand side of the equation is plotted for each tracer. The intercept is equal to the second term on the right side of the equation and from the known total soil water content; immobile water content can be estimated. The difference between total soil water content and immobile water content provides the estimate of mobile soil water content. The slope term associated with t is used to calculate the mass transfer coefficient from earlier estimated immobile water content. This equation was subsequently modified by Jaynes and Horton (1998) to account for solute being moved slightly ahead than that given by t.

15.15.5 Other Methods

The nondimensional parameters of physical or chemical nonequilibrium transport equations can be estimated by using a public domain program such as CXTFIT. Once the nondimensional parameters are quantified, the dimensional parameters can then be calculated by making certain assumptions. For example, parameter β in two-region models consists of a fraction of water considered to be mobile (ϕ_m) and a fraction of sites located in the mobile region (f) as follows (Nkedi-Kizza et al., 1984):

$$\phi_m=\beta R-f(R-1) \quad (15.58)$$

There are two unknowns in this equation: ϕ_m and f. The estimate for ϕ_m can be made by generating a breakthrough curve for a conservative solute through the same soil column at the same pore water velocity (De Smedt, 1979). Alternately, breakthrough curves can be generated for two solutes with different sorption strengths to get two sets of the values of R and β. With the help of these two equations, the values of the two unknowns can be obtained (van Genuchten and Cleary, 1979). The value of f can be assumed equal to zero, indicating that sorption only occurs in the immobile region. f can also be assumed to equal ϕ_m or f can be assumed equal to one. However, assumption of f being equal to one reduces the two-region model into a one-region convective dispersion equation.

The two-region model assumes that the solute diffusion between mobile and immobile regions is by the process of diffusion and depends on the time taken for the diffusion process to complete (Lal and Shukla, 2004). Rao et al. (1980) proposed the following model based on the geometry of aggregates for estimating the mass transfer coefficient (α):

$$\alpha=\frac{D_e(1-f)\theta b}{r^{20}} \quad (15.59)$$

where

D_e is the effective diffusion coefficient.
θ is the volumetric water content.
b is a time-dependent variable.
r is the equivalent radius.

The value of mass transfer coefficient can also be estimated using the equations of van Genuchten (1985) and van Genuchten and Dalton (1986) as follows:

$$\alpha = \frac{nD_e(1-f)\theta}{b^2} \quad (15.60)$$

where

n is the geometry factor.
b is the average effective diffusion length.

Skopp et al. (1981) proposed the following equation for the estimation of α:

$$\alpha = \frac{2K_e\theta_1(\theta_1-\theta_2)^2}{gdR_p\pi} \quad (15.61)$$

where

K_e is the overall conductivity.
θ_1 and θ_2 are the flow velocities for the two regions.
d is the aggregate size.
R_p is the intra-aggregate pore size.

15.16 TRANSPORT OF REACTIVE SOLUTES

The appearance of reactive agrochemicals in surface water and groundwater is an important concern. Some of the examples of reactive solutes are pesticides, herbicides, and ammonium-based fertilizers. An evaluation of the potential of reactive solute transport to the groundwater requires an understanding of their transport mechanism through soils. Several studies over the years have reported pesticides being found in groundwater (U.S. Environmental Protection Agency, 1986, 1988; Domagalski and Dubrovsky, 1992). The agrochemicals considered fairly immobile in column studies have been reported in groundwater, for example, herbicides (Jury and Fluhler, 1992). Studies between 1970 and 1993 have reported deeper than expected movement of pesticides, primarily through preferential flow channels consisting of cracks, root channels, earthworm channels, etc. (Rao et al., 1974; Trojano et al., 1993; Flury et al., 1995).

There are several different pathways, macropore channels, burrows, cracks, etc. for the transfer of reactive solutes through the soil matrix. There are also different sites for sorption of these reactive solutes inside the soil matrix that can potentially slow the movement toward the groundwater table. For example, sorption by organic coatings on earthworm burrows can slow the leaching of organic chemicals (Stehouwer et al., 1994). Such a mechanism will result in poor mass balance or mass recovery of applied solute. The land use also influences the transport of reactive solutes, for example, fields under a no-tillage system may have a large number of macropores and preferential flow channels, and they can enhance the movement of reactive solutes to the groundwater. On the other hand, conventional tillage may break the continuity of these channels, decreasing the downward leaching. Isensee et al. (1990) conducted experiments on the movement of atrazine through conventional and no-tillage fields in Beltsville, Maryland. They reported that atrazine concentrations were 2–50 times greater under no-tillage plots than conventional tillage plots. They concluded that unusually rapid leaching to deeper layers than predicted by solute transport theories is due to the flow through macropore channels.

The solute transport equations for nonreactive solutes are often modified to predict the transport of reactive solutes. Reaction rates are important for chemical reactions, and they could be reversible and slow, reversible and fast, nonreversible and slow, or nonreversible and fast. However, there is a need to also account for the preferential flow paths to simulate the faster leaching of reactive solutes to deeper depths.

PROBLEMS

15.1 If soil water contents of a gradually drying soil are 0.44, 0.32, 0.24, and 0.15, and corresponding water velocities (Darcy) through the soil are 0.02, 0.1, 0.28, and 0.45 cm/day, respectively. Calculate pore water velocity, molecular, dispersion, and apparent dispersion coefficients.

15.2 Describe different possible mechanisms involved in the transport of a conservative and a labile solute.

15.3 List some of the differences and similarities of two-region and two-site models.

15.4 If Darcy's water velocity (q) is 1.4 cm/day, soil water content (θ) is 0.4 cm^3/ cm^3, and average bulk density of soil is 1.05 g/cm^3, how long will it take for a conservative tracer such as chloride and a reactive tracer such as a herbicide (K_D = 1.5 cm^3/g) to move to the water table 10 m below the soil surface?

15.5 List some of the differences between conservative and nonconservative solutes.

15.6 Soil water content and solute concentrations were 0.38 cm^3/cm^3 and 4.2 mg/L in an agricultural field fertilized with nitrogen fertilizer in Clovis, New Mexico. Calculate the pore water velocity for a soil water flux density of 0.1 cm^3/cm^2/day. How much nitrate–nitrogen was leached per unit area of the field due to convective flow below the root zone in 3 days?

REFERENCES

Bear, J. 1972. *Dynamics of Fluids in Porous Media*. American Elsevier, New York, p. 764.

Biggar, J. W. and D. R. Nielsen. 1967. Miscible displacement and leaching phenomenon. *Agronomy* 11: 254–274.

Biggar, J. W. and D. R. Nielsen. 1976. Spatial variability of the leaching characteristics of a field soil. *Water Resour. Res.* 12: 78–84.

Clothier, B. E., M. B. Kirkham, and J. E. Mclean. 1992. In situ measurements of the effective transport volumes for solute moving through soil. *Soil Sci. Soc. Am. J.* 56: 733–736.

Coats, K. H. and B. D. Smith. 1964. Dead end pore volume and dispersion in porous media. *SPE J.* 4: 73–84.

De Smedt, F. 1979. Theoretical and experimental study of solute movement through porous media with mobile and immobile water. Thesis, Vrije Universiteit, Brussels, Belgium.

Domagalski, J. L. and N. M. Dubrovsky. 1992. Pesticides residues in ground water of the San Joaquin Valley, California. *J. Hydrol.* 130: 299–338.

Flury, M., J. Leuenberger, B. Studer, and H. Fluhler. 1995. Transport of anions and herbicides in a loamy and a sandy field soil. *Water Resour. Res.* 31: 823–835.

Fried, J. J. 1975. *Groundwater Pollution.* Elsevier, New York.

González-Delgado, A. M. and M. K. Shukla. 2011. Coupled transport of nitrate and chloride in soil columns. *Soil Sci.* 176(7): 346–355.

Isensee, A. R., R. G. Nash, and C. S. Helling. 1990. Effect of conventional vs. no-tillage on pesticide leaching to shallow groundwater. *Soil Sci. Soc. Am. J.* 19: 434–440.

James, R. V. and J. Rubin. 1986. Transport of chloride ion in a water unsaturated soil exhibiting anion exclusion. *Soil Sci. Soc. Am. J.* 50: 1142–1149.

Jaynes, D. B. and R. Horton. 1998. Field parameterization of the mobile/immobile domain model. In: H. M. Selim and L. Ma (Eds.), *Physical Nonequilibrium in Soils: Modeling and Application.* Ann Arbor Press, Chelsea, MI, pp. 297–310.

Jaynes, D. B., S. D. Logsdon, and R. Horton. 1995. Field method for measuring mobile/immobile water content and solute transfer rate coefficient. *Soil Sci. Soc. Am. J.* 59: 352–356.

Jury, W. A. and H. Fluhler. 1992. Transport of chemicals through soil: Mechanisms, models, and field applications. *Adv. Agron.* 47: 141–201.

Jury, A. W., W. R. Gardner, and W. H. Gardner. 1991. *Soil Physics*, 5th edn. John Wiley, New York.

Khan, A U H. 1988. A laboratory test of the dispersion scale effect. PhD thesis, University of California, Riverside, CA.

Kookana, R. S., R. D. Schuller, and L. A. G. Aylmore. 1993. Simulation of simazine transport through soil columns using time-dependent sorption data measured under flow conditions. *J. Contam. Hydrol.* 14: 93–115.

Lal, R. and M. K. Shukla. 2004. *Principles of Soil Physics.* Marcel Dekker, Inc., New York.

Lapidus, L. and N. R. Amundson. 1952. Mathematics of adsorption in beds. *J. Phys. Chem.* 56: 584.

Millington, R. J. and J. P. Quirk. 1961. Permeability of porous solids. *Faraday Soc. Trans.* 57(7): 1200–1207.

Nielsen, D. R. and J. W. Biggar. 1961. Miscible displacement in soils: I. Experimental information. *Soil Sci. Soc. Am. Proc.* 15: 1–5.

Nielsen, D. R., R. D. Jackson, J. W. Cary, and D. D. Evans (eds.). 1972. *Soil Water.* American Society of Agronomy and Soil Science Society of America, Madison, WI.

Nkedi-Kizza, P., P. S. C. Rao, R. E. Jessup, and J. M. Davidson. 1984. Ion exchange and diffusive mass transfer during miscible displacement through and aggregated Oxisol. *Soil Sci. Soc. Am. J.* 46: 471–476.

Padilla, I. Y., T. C. Jim Yeh, and M. H. Conklin. 1999. The effect of water content on solute transport in unsaturated porous media. *Water Resour. Res.* 35: 3303–3313.

Parker, J. C. and M. Th. van Genuchten. 1984. Flux-averaged and volume-averaged concentrations in continuum approaches to solute transport. *Water Resour. Res.* 20(7): 866–872.

Pfannkuch, H. O. 1962. Contribution a l'etude des deplacement de fluides miscible dans un milieu poreux. *Rev. Inst. Fr. Petrol.* 18(2): 215.

Piñón-Villarreal, A. R., A. S. Bawazir, M. K. Shukla, and A. T. Hanson. 2013. Retention and transport of nitrate and ammonium in loamy sand and amended with clinoptilolite zeolite. *J. Irrig. Drain. Eng.* 139(9): 755–765.

Rao, P. S. C., R. E. Green, V. Balasubramanian, and Y. Kanehiro. 1974. Field study of solute movement in a highly aggregated oxisol with intermittent flooding: II Picloram. *J. Environ. Qual.* 3: 197–202.

Rao, P S. C., J. M. Davidson, R. E. Jessup, and H. M. Selim. 1979. Evaluation of conceptual models for describing nonequilibrium adsorption-desorption of pesticides during steady-state flow in soils. *Soil Sci. Soc. Am. J.* 43: 22–28.

Rao, P. S. C., R. E. Jussup, D. E. Rolston, J. M. Davidson, and D. P. Kilcrease. 1980. Experimental and mathematical description of nonadsorbed solute transfer by. diffusion in spherical aggregates. *Soil Sci. Soc. Am. J.* 44(4): 684–688.

Selim, H. M. and M. C. Amacher. 1997. A second-order kinetic approach for modeling solute retention and transport in soils. *Water Resour. Res.* 24: 2061–2075.

Selim, H. M., J. H. Davidson, and R. S. Mansell. 1976. Evaluation of a two-site adsorption desorption model for describing solute transport in soils In Proceedings of the Summer Computer Simulation Conference, Washington, DC, pp. 444–448.

Shukla, M. K. and P. Cepuder. 2000. Anion exclusion during transport of chloride through soil columns. *Trans. ASAE* 43(6): 1425–1430.

Shukla, M. K. and G. Kammerer. 1998. Comparison between two models describing solute transport in porous media with and without immobile water *Aust. J. Water Manag.* 50(9/10): 254–260.

Shukla, M. K., F. J. Kastanek, and D. R. Nielsen. 2000. Transport of chloride through water-saturated soil columns. The Bodenkulture. *Aust. J. Agric. Res.* 51(4): 235–246.

Shukla, M. K., F. J. Kastanek, and D. R. Nielsen. 2002. Inspectional analysis of convective dispersion equation and application on measured breakthrough curves. *Soil Sci. Soc. Am. J.* 66(4): 1087–1094.

Shukla, M. K., T. R. Ellsworth, R. J. Hudson, and D. R. Nielsen. 2003. Effect of water flux on solute velocity and dispersion. *Soil Sci. Soc. Am. J.* 67: 449–457.

Skopp, J., W. R. Gardner, and E. J. Tyler. 1981. Solute movement in structured soils, Two region model with small interactions. *Soil Sci. Am. J.* 45(5): 837–842.

Stehouwer, R. C., W. A. Dick, and S. J. Traina. 1994. Sorption and retention of herbicides in vertically oriented earthworm and artificial burrows. *J. Environ. Qual.* 23: 286–292.

Toride, N., F. J. Leij, and M. Th. van Genuchten. 1995. *The CXTFIT Code for Estimating Transport Parameters from Laboratory or Field Tracer Experiments. Research Report 137.* U.S. Salinity Lab, USDA-ARS, Riverside, CA.

Trojano, J., C. Garreston, C. Krauter, J. Brownell, and J. Hudson. 1993. Influence of amount and method of irrigation water application on leaching of atrazine. *J. Environ. Qual.* 22: 290–298.

U.S. Environmental Protection Agency. 1986. *Pesticides in Ground Water: Background Document, Report EPA 440/6–86–002.* U.S. Environmental Protection Agency, Washington, DC.

U.S. Environmental Protection Agency. 1988. *Protecting ground water: Pesticides and agricultural practices, Report EPA 440/6–88–001.* U.S. Environmental Protection Agency, Washington, DC.

van Genuchten, M. Th. 1981. Non-equilibrium transport parameters from miscible displacement experiments. *Research report 119, USDA.* U.S. Soil Salinity Lab, Riverside, CA.

van Genuchten, M. Th. 1985. A general approach for modeling solute transport in structured soils. In: Proceedings of the 17th International Congress on Hydrogeology of Rocks of Low Permeability, January 7–12, 1985, Tucson, AZ. *Mem. Int. Assoc. Hydrogeol.* 17: 512–526.

van Genuchten, M. Th. and R. W. Cleary. 1979. Movement of solutes in soils: Computer-simulated and laboratory results. In: G. H. Bolt (ed.), *Soil Chemistry, B. Physico-Chemical Models, Developments in Soil Science 5B*. Elsevier Scientific Publ. Co., Amsterdam, The Netherlands, pp. 349–386.
van Genuchten, M. Th. and F. N. Dalton. 1986. Models for simulating salt movement in aggregated field soils. *Geoderma* 38: 165–183.
van Genuchten, M. Th. and P. J. Wierenga. 1976. Mass transfer studies in sorbing porous media I: Analytical solutions. *Soil Sci. Soc. Am. Proceedings* 40(4): 473–480
van Genuchten, M. Th. and P. J. Wierenga. 1977. Mass transfer studies in sorbing porous media II: Experimental evaluation with tritium (^3H_2O). *Soil Sci. Soc. Am. Proc.* 41(2): 272–278.
van Genuchten, M. Th. and P. J. Wierenga. 1986. Solute dispersion coefficients and retardation factors. In: A. Klute (ed.), *Methods of Soil Analysis, Part 1: Physical and Mineralogical Methods*, 2nd edn. American Society of Agronomy, Madison, WI.
van Genuchten, M. Th., J. M. Davidson, and P. J. Wierenga. 1974. An evaluation of kinetic and equilibrium equations for the prediction of pesticide movement through porous media *Soil Sci. Soc. Am. Proc* 38: 29–35.
van Genuchten, M. Th. and P. J. Wierenga. 1976. Mass transfer studies in sorbing porous media I: Analytical solutions. *Soil Sci. Soc. Am. Proc* 40(4): 473–480.

16 Modeling Flow through the Vadose Zone Using the HYDRUS-1D Model

16.1 INTRODUCTION

The water flow through an unsaturated zone is a complex process. These processes can be modeled using HYDRUS-1D and HYDRUS 2D/3D software. Recently HYDRUS version 5 (5.02) was released with several new features (https://www.pc-progress.com/en/Default.aspx?h3d-new-features) including a new module for PFAS. The purpose of this chapter is to give readers hands-on experience using numerical simulation with HYDRUS-1D (Šimůnek et al., 2008). HYDRUS-1D is a public domain program, and this chapter has been written using published data (Deb et al., 2011) and available model documentation. Hydrus-1D pro includes several new add-on features including hydrus-modules Cosmic, dynamic plant uptake (DPU), C-Ride (for colloid-facilitated solute transport), and PFAS. In this chapter, the step-by-step guide on model application is explained using an example. Terminologies as they appear in the model screen have been followed. The pressure head appears in the model screen and it denotes both positive head due to ponding on the soil surface as well as the absolute of the matric potential.

HYDRUS-1D, a software package for simulating one-dimensional (1D) movement of water, heat, and multiple solutes in variably saturated media, consists of the HYDRUS computer program, and the HYDRUS interactive graphical user interface (GUI) for Microsoft Windows environments. The HYDRUS program (version 4.17) numerically solves the Richards equation for saturated–unsaturated water flow (see Chapter 9) and Fickian-based advection–dispersion-type equations for both heat and solute transport using the Galerkin-type linear finite element schemes. A sink term is also included in the flow equation to account for water uptake by plant roots as a function of both water and salinity stress The dual-porosity-type flow (i.e., one fraction of the water content being mobile while the other fraction is immobile) or the dual-permeability-type flow (i.e., two mobile regions, one representing the matrix and the other macropores) may also be considered in the flow equation.

The transport due to conduction and convection with flowing water is considered in the heat transport equation (see Chapter 11). HYDRUS-1D also considers the coupled liquid water, water vapor, and heat transport in soils, as well as mass transfer and energy balance (i.e., the balance of latent heat and sensible fluxes) at the soil surface HYDRUS-1D also provides options to estimate potential evapotranspiration using either the Penman–Monteith or Hargreaves equations (see Chapter 12).

DOI: 10.1201/9780429264849-16

The solute transport equations in HYDRUS-1D consider advective–dispersive transport in the liquid phase, as well as diffusion in the gaseous phase (see Chapter 15). The transport equations also allow considerations of nonlinear nonequilibrium reactions between the solid and liquid phases, linear equilibrium reactions between the liquid and gaseous phases, zero-order production, and two first-order degradation reactions. The physical nonequilibrium solute transport can also be accounted for by assuming a two-region, dual-porosity-type formulation that partitions the liquid phase into mobile and immobile regions (see Chapter 15). The transport equations include provisions for kinetic attachment/detachment of solute to the solid phase, which can be used to simulate transport of viruses, colloids, or bacteria.

A Marquardt–Levenberg-type parameter optimization algorithm is also implemented in HYDRUS-1D for the inverse estimation of soil hydraulic and/or solute transport and reaction parameters from measured transient or steady-state flow and/or transport data. HYDRUS-1D also includes modules for simulating CO_2 and major ion solute movement. Note that HYDRUS-1D is regularly updated with new processes and options, and new features as well as the HYDRUS-1D user/reference manual may be available publicly on the website http://www.pc-progress.com/en/Default.aspx?H1d-downloads. Readers are referred to Šimůnek et al. (2008), the user/reference manual (Šimůnek et al., 2012), and Radcliffe and Šimůnek (2010) for details on the HYDRUS-1D model.

16.2 INSTALLATION OF HYDRUS-1D

The HYDRUS-1D (version 4.17) software package requires a Microsoft Windows–based system running Microsoft Windows 95, 98, NT, 2000, XP, Vista (32/64-bit), and Windows 7, 8, or 10 (32/64-bit); 128 MB of RAM memory; VGA graphics (SVGA is recommended); and hard disk with at least 20 MB free disk space. The latest version of HYDRUS-1D can be downloaded (requires registration) from the website http://www.pc-progress.com/en/Default.aspx?downloads.

It is recommended to save the installation file "Hydrus1D_4.17.0140.exe" to a known location on your computer.

- To install, run or double-click on the installation file "Hydrus1D_4.17.0140.exe" and the setup is launched automatically.
- Follow the instructions of the InstallShield Program by clicking on the Next button to advance through the installation.
- Application installation is complete when Finish is chosen.

Note that the HYDRUS-1D user/reference manual is available in the Help menu of HYDRUS-1D interface (i.e., click Help | User Manual).

16.3 HYDRUS-1D TUTORIAL

Temporal variations of soil water content in the unsaturated soil zone due to temperature gradients may induce water fluxes in the gas and liquid phase, especially in semiarid and arid areas, and can play a key role in mass and energy transfer near soil surface. The example describes the simulation of coupled liquid water, water vapor, and heat transport in the unsaturated zone of a sandy loam furrow-irrigated onion field using HYDRUS-1D. Readers are referred to the user/reference manual (Šimůnek et al., 2012), Saito et al. (2006), and other chapters of this book for the details on theoretical description of coupled water, vapor, and heat transport in the vadose zone.

Since all furrows are irrigated simultaneously, flow through the soil profile is assumed to be adequately described as one-dimensional. The 80 cm soil profile is assumed to be vertically homogeneous. During the measurement period from day of the year (DOY) 85 (March 26) to DOY 134 (May 14) 2009, the experimental plot of 40 m length and 10 m width was irrigated once a week at a rate of 20 cm/day. The water flowmeter monitored the flow rate so that the volume applied to the experimental plot could be determined. Staff gauges were installed in the furrow to continuously measure depth of irrigation water during an irrigation event. Note that onion was planted only sparsely in the experimental plot to fill up the space, and during the measurement period onion plants were small, and thus measurements were considered to be made on bare plot.

Core and bulk soil samples were collected to determine soil physical and thermal properties down to 80 cm depth. Temporal soil water content and soil temperature variations in the furrow bed were measured using time domain reflectometry (TDR) and temperature sensors, respectively. TDR and temperature sensors were installed at soil depths of 5, 10, 20, and 50 cm to monitor volumetric soil water content and soil temperature every 15 or 30 min at two locations. Meteorological variables, including precipitation, solar radiation, maximum and minimum soil temperatures, wind speed, and relative humidity, were measured on an hourly basis at 2 m above the soil surface.

16.3.1 HYDRUS-1D Interface

Before we begin with the tutorial project, it is essential to be familiar with the HYDRUS-1D interface, which facilitates menus, submenus, sub-submenus, and toolbars; and user interaction with the simulation model and the entire process of entering input files/parameters through interactive graphics-based dialog windows/screens, spreadsheets, alert wizards/messages and guide dialogs, help, and output/result visualizations and edits.

- Start the HYDRUS-1D program (i.e., Start → All Programs → PC-Progress → HYDRUS-1D 4.xx) or double-click on the HYDRUS-1D 4.xx desktop icon or wherever it is saved on the computer.

The Microsoft Windows and interactive graphics-based user-friendly interface contains a Project Manager and both the Pre-processing and Post-processing units (Figure 16.1). You can either unzip downloaded HYDRUS projects directly into any

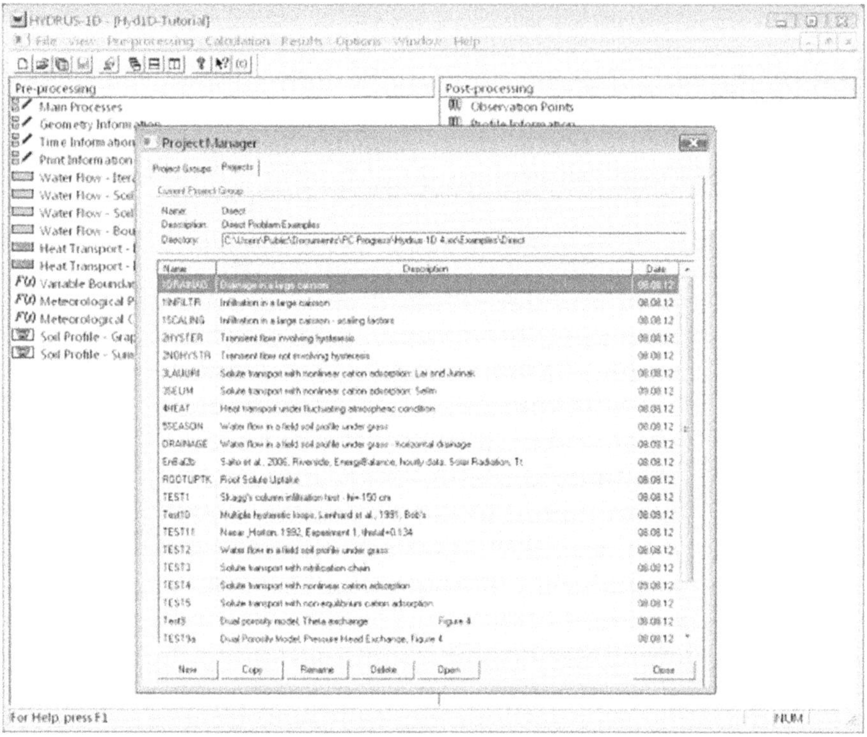

FIGURE 16.1 Main window of the HYDRUS-1D software package, including the Pre-processing pane on the left, the Post-processing pane on the right, and the Project Manager.

existing HYDRUS Project Group (e.g., Direct or Inverse) or create a new Project Group.

- Clicking File | Project Manager brings up the Project Manager window (Figure 16.1).

The Project Manager allows users to manage data of existing projects as well as to locate, open, copy, delete, and/or rename desired projects or their input or output data. Note that a "project" in the HYDRUS-1D program represents a particular problem to be solved by the program. In the Project Manager, several example projects (i.e., grouped into workspaces such as direct, inverse, etc.) are given to familiarize users with all the modules (i.e., Pre-processing and Post-processing modules) of the HYDRUS-1D interface. The Pre-processing unit includes specification of all necessary input parameters for a project to be solved/simulated by HYDRUS-1D, while the Post-processing unit consists of simulation results or output graphics. Both Pre-processing and Post-processing units are described in Sections 16.3.3 and 16.3.4.

16.3.2 Create a New HYDRUS-1D Project

- Create a directory "Hydrus1D-Tutorial" in drive C (i.e., C:\Hydrus1D-Tutorial).
  Note that you can create the "Hydrus1D-Tutorial" directory (i.e., folder) in any accessible drive or the HYDRUS-1D default directory (i.e., project group(s) in the Project Manager).
- Start the HYDRUS-1D program. The program will not open any project or it will open the last active project.
- Click File | New, which will bring up the New Project window (Figure 16.2).
- Enter the project name "Hyd1D-Tutorial," the project description "Coupled liquid water, water vapor, and heat transport simulations," and locate the directory C:\Hydrus1D-Tutorial using the Browse button.
- Click the OK button.

Note that the tutorial project directory or project folder (i.e., C:\Hydrus1D-Tutorial\Hyd1D-Tutorial) will be automatically created and all the input (*.IN) and output (*.OUT) files will be stored in the project folder "Hyd1D-Tutorial."

The "Hyd1D-Tutorial.h1d" file will also be created in the directory (C:\Hydrus1D-Tutorial\Hyd1D-Tutorial.h1d), which is recognized by the HYDRUS-1D to open the project. Note that the "Hyd1D-Tutorial.h1d" can be opened from the HYDRUS-1D interface, but not from inside the directory C:\Hydrus1D-Tutorial. For example:

- Start the HYDRUS-1D program, click File | Open, browse the "*.h1d" file, and click on the Open button.

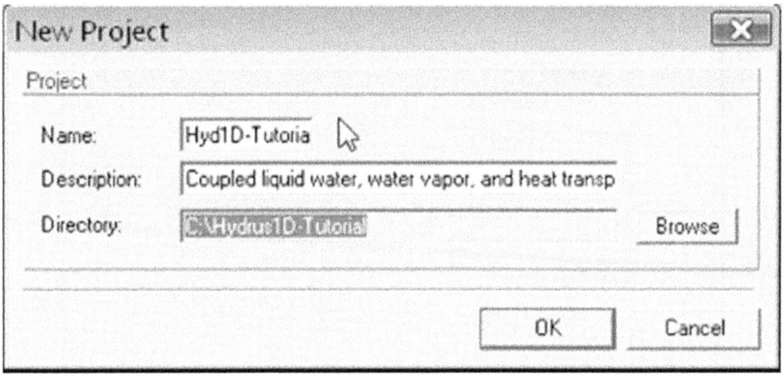

FIGURE 16.2 New Project window.

16.3.3 Working with the HYDRUS-1D Pre-Processing Unit

16.3.3.1 Main Processes (Pre-Processing | Main Processes)

- Click Pre-processing I Main processes, which will bring up the Main Processes window (Figure 16.3).
- Keep the default title "Welcome to HYDRUS-1D" under the field Heading, which will be printed into all output files.
- For the tutorial project "Hyd1D-Tutorial.h1d," specify (check box) the processes Water Flow and Vapor Flow, and Heat Transport under the option labeled Simulate.

HYDRUS-1D automatically considers transient flow when the Water Flow option is selected. For other processes such as Solute Transport, Root Water Uptake, Root Growth, and CO2 Transport options, readers are referred to the HYDRUS-1D user manual (i.e., click Help I User Manual). Once the processes are assigned in the dialog window Main Processes (Figure 16.3), users can select particular commands from a menu, or use interactive dialog windows or screens that will lead them through the entire process of entering input data and executing the simulation run. Note that dialog window(s) will not appear and submenu(s) or toolbar icon(s) will be disabled (grayed out) for the process(es) that is (are) not considered in the Main Processes (Figure 16.3).

This tutorial with HYDRUS-1D describes the direct solution of coupled liquid water, water vapor, and heat transport in the unsaturated zone of a sandy loam

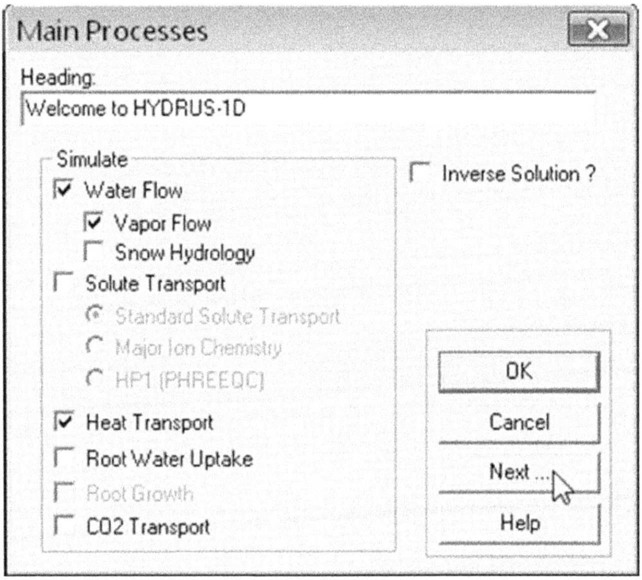

FIGURE 16.3 Main Processes dialog window.

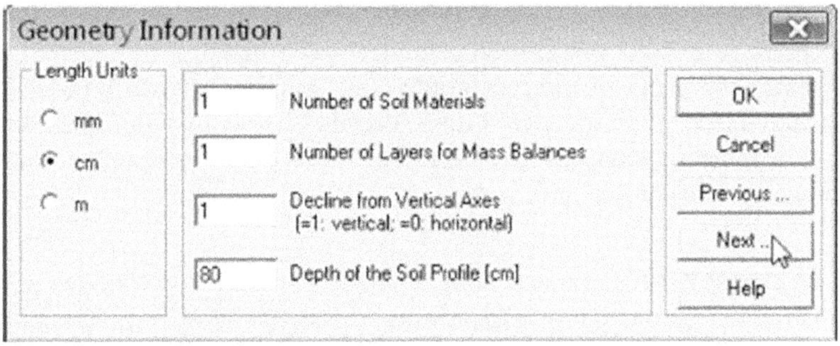

FIGURE 16.4 Geometry Information dialog window for the flow domain.

field. Note that users can check the Inverse Solution? option (Figure 16.3) to use the Levenberg–Marquardt optimization algorithm of HYDRUS-1D for inverse estimation of water and heat transport parameters using measured soil water content as well as soil temperature data at different measurement depths. Readers are referred to Šimůnek et al. (2008), the user/reference manual (Šimůnek et al., 2012), and Deb et al. (2011) for details on inverse modeling using HYDRUS-1D. Note that the HYDRUS-1D user/reference manual is available in the Help menu of the HYDRUS-1D (version 4.15.0110) interface.

- Click on the Next button (Figure 16.3), and proceed to the next dialog window Geometry Information (Figure 16.4).

16.3.3.2 Geometry Information (Pre-Processing | Geometry Information)

- For the tutorial project, select the cm radio button under the option labeled Length Units (Figure 16.4).

Note that change in the Length Units button during or after reading all input data will convert variables into the new unit. The soil profile is vertically homogenous/uniform.

- Specify by entering a value "1" (one) for both the Number of Soil Materials and Number of Layers [i.e., subregion] for Mass Balances fields.

The number of soil materials and number of subregions will be specified/identified later by material/subregion numbers in the Soil Profile-Graphical Editor (Pre-processing | Profile Information) module (see Section 16.3.3.14).

The field Decline from Vertical Axes allows users to choose any inclined direction of the soil profile (e.g., a vertical and/or horizontal soil profile). The inclination is specified as the cosine of the angle between the vertical axis and the axis of the soil profile. The value is equal to 1 for vertical soil column and zero for horizontal soil column.

- For the tutorial project, enter the value “1” (one) in the field Decline from Vertical Axes for a vertical soil column (Figure 16.4).
- Specify the depth of the soil profile (“80” cm), which will be used later in the Soil Profile-Graphical Editor module (see Section 16.3.3.14) for the profile discretization, the specification of initial conditions, and the domain property distribution.

Note that in the HYDRUS-1D interface, all the data entry dialog windows or screens have the Previous button to return to previous window(s). Also, by pressing the function key F1 or clicking on the Help button while working in any dialog window, users can access the HYDRUS-1D Application Help information about the content of the dialog window.

- Click on the Next button (Figure 16.4), which will bring up the next dialog window Time Information (Figure 16.5).

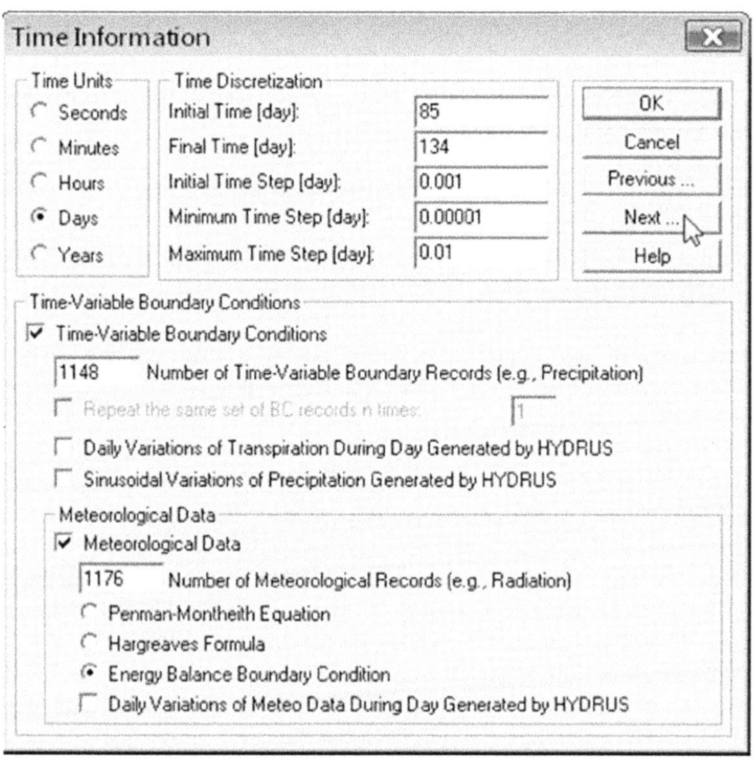

FIGURE 16.5 Time Information dialog window.

16.3.3.3 Time Information (Pre-Processing | Time Information)

- For the tutorial project, select the Days radio button under the option Time Units (Figure 16.5).

Note that a change in the Time Units button during or after reading all input data will convert variables into the new unit. The option labeled Time Discretization allows users to specify main time variables that govern the time discretization.

- Specify Initial Time (starting time of the calculation) as "85" (i.e., DOY 85 [March 26]) and Final Time (final time of the calculation) as "134" (DOY 134 [May 14]).
- Set the Initial Time Step to "0.001" day.

Note that the initial time step depends on the type of simulation and boundary conditions used. For example, when simulating a process that starts with a large initial pressure head at the boundary such as ponded infiltration, a smaller initial time step is recommended in HYDRUS-1D. A larger time step is recommended for simulating a long-term process with variable boundary conditions such as seasonal or multiyear simulation.

- Set the Minimum Time Step (i.e., minimum permitted value of the time increment) as "0.00001" day, and the Maximum Time Step (i.e., maximum permitted value of the time increment) to "0.01" day.

Note that the minimum time step must be smaller than the initial time step, interval between print times (see Section 16.3.3.4), and interval between time-variable boundary condition records (see Section 16.3.3.11). For details about Time Discretization, see the HYDRUS-1D Application Help topic on Notes on Spatial and Temporal Discretization or the HYDRUS-1D user manual.

- Check the box Time-Variable Boundary Conditions and specify the Number of Time-Variable Boundary Records as "1148" (Figure 16.5) (see Section 16.3.3.11).

Note that if the box Time-Variable Boundary Conditions is not checked, boundary conditions are assumed to be constant in time.

- Check the box Meteorological Data and specify the Number of Meteorological Records as "1176" (see Section 16.3.3.13).
- For the tutorial project, check the radio button Energy Balance Boundary Condition to simulate mass transfer and energy balances at the "bare" (i.e., no crop) soil surface.

Note that when the crop is considered, users can estimate potential evapotranspiration using the Penman–Monteith combination equation or the Hargreaves equation

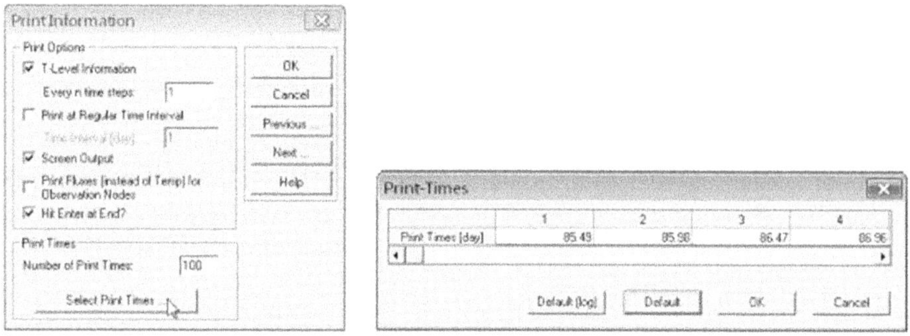

(a) (b)

FIGURE 16.6 (a) Print Information and (b) Print-Times dialog windows for the tutorial project.

as recommended by the Food and Agriculture Organization of the United Nations (FAO). Readers are referred to the user manual for these options.

- Click on the Next button (Figure 16.5), which will bring up the next dialog window Print Information (Figure 16.6a).

16.3.3.4 Print Information (Pre-Processing | Print Information)

- Check the box T-level Information under the option labeled Print Options (Figure 16.6a).

T-level Information, a group of output files (discussed later in Section 16.3.4), contains certain information concerning the mean pressure head, mean water fluxes, cumulative water fluxes, and time and iteration information, which will be printed at the end of each time step.

- Specify a value “1” (one) in the Every n time steps field after which the output is printed.

Note that you can also specify that the output is printed at regular specified time intervals.

- Check the box Screen Output to print simulation run information to the screen during the execution of the HYDRUS-1D code.

Note that the check box Print Fluxes (instead of Temp) for Observation Nodes prints output water fluxes for observation nodes instead of temperatures. For the tutorial project, do not check that box.

- Check the box Hit Enter at End?, that is, when a simulation run is completed, the DOS window remains open, waits for the command Enter to close the DOS window, then returns to the GUI.

Note that if this box is unchecked, the DOS window automatically closes after the completion of simulation run.

- Specify "100" in the Number of Print Times field under the option Print Times (Figure 16.6a).
- Click on the Select Print Times button, which will bring up the Print-Times dialog screen (Figure 16.6b).

You can specify the printing time at which detailed information about the pressure head, water content, water fluxes, and soil–water balance will be printed.

- For the tutorial project, click on the Default button (Figure 16.6b), which will redistribute printing times at equal time intervals.

Note that, as shown in Figure 16.6b, clicking on the Default (log) button redistributes printing times at logarithmically increasing time intervals.

- Click OK to return to the Print Information screen (Figure 16.6a).
- Click on the Next button (Figure 16.6a), which will bring up the next window Iteration Criteria (Figure 16.7).

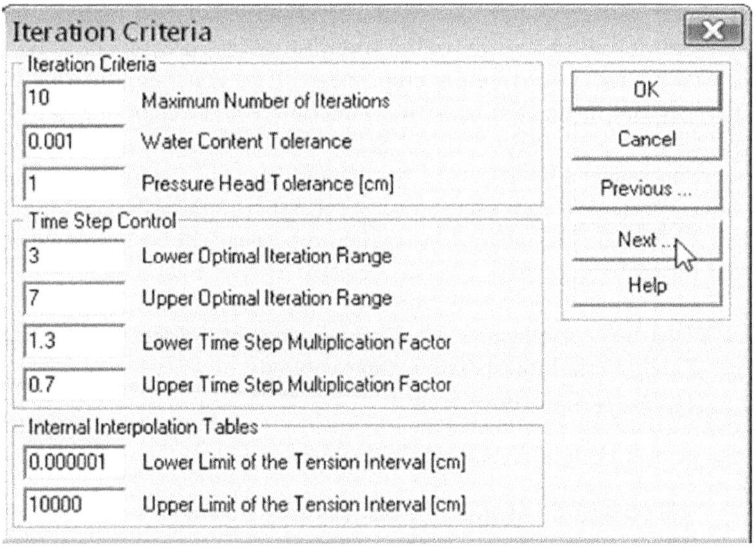

FIGURE 16.7 Iteration criteria for the solution precision, parameters for controlling the time step, and limits for internal interpolation tables.

16.3.3.5 Iteration Criteria (Pre-Processing | Water Flow | Iteration Criteria)

The Iteration Criteria window allows users to specify iteration criteria for the solution precision, parameters for controlling the time step, and limits for internal interpolation tables (Figure 16.7).

- For the tutorial project, specify Maximum Number of Iterations during any time step ("10"), Water Content Tolerance ("0.001") for nodes in the unsaturated part of the flow region, and Pressure Head Tolerance ("1") for nodes in the saturated part of the flow region.

Note that the Water Content Tolerance and Pressure Head Tolerance parameters represent the maximum desired absolute change in water content and pressure head values between two successive iterations during a particular time step, respectively.

- As shown in Figure 16.7, use the default values for all the fields under the Time Step Control option. Note that when the number of iterations necessary to reach convergence for water flow is less than the Lower Optimal Iteration Range (default value is 3), the time step is multiplied by the Upper Time Step Multiplication Factor (default value is 0.7).
- On the other hand, the time step is multiplied by the Lower Time Step Multiplication Factor (default value is 1.3) when the number of iterations necessary to reach convergence for water flow is higher than the Upper Optimal Iteration Range (default value is 7).
- Enter the default value "0.000001" in the field Lower Limit of the Tension [or matric potential when expressed as a negative number] Interval, which represents the absolute value of the lower limit of the pressure head interval for which a table of hydraulic properties will be generated internally (for each material if the number of soil materials is more than one in Section 16.3.3.2).
- Enter the value "10,000" in the field Upper Limit of the Tension Interval, which represents the absolute value of the upper limit of the pressure head interval for which a table of hydraulic properties will be generated internally for each material.

See the HYDRUS-1D Application Help or user manual for details about the Time Step Control and Internal Interpolation Tables options.

- Click on the Next button (Figure 16.7), which will bring up the next dialog window Soil Hydraulic Model (Figure 16.8).

16.3.3.6 Soil Hydraulic Model (Pre-Processing | Water Flow | Hydraulic Properties Model)

HYDRUS-1D allows users to select several single-porosity models for the soil hydraulic properties: the van Genuchten–Mualem model, the van Genuchten–Mualem

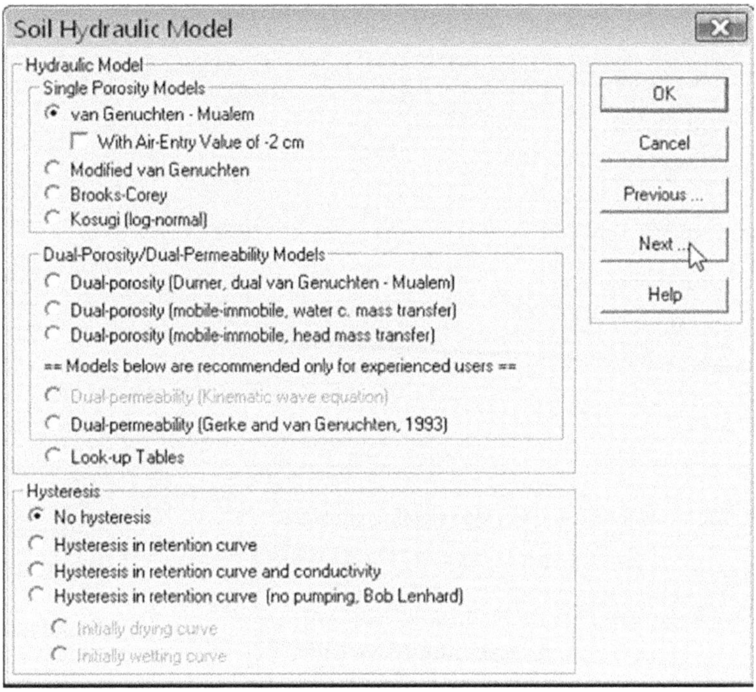

FIGURE 16.8 Soil Hydraulic Properties Model dialog window.

model with an air-entry value of –2 cm, the modified van Genuchten model, the Brooks–Corey model, and the lognormal distribution model of Kosugi (Figure 16.8). Other approaches include the dual-porosity and dual-permeability models and look-up tables. Users can select two dual-porosity nonequilibrium flow models with mass transfer between the mobile and immobile zones assumed to be proportional to either the water content or the pressure head. See the user manual for a detailed description of these models.

- Check the radio button van Genuchten–Mualem for the tutorial project. Do not check the box labeled With Air-Entry Value of –2 cm for this tutorial project (Figure 16.8).

When the van Genuchten model is used, HYDRUS-1D allows users to choose either a nonhysteretic description or a hysteretic description only in the soil water retention curve, or hysteretic descriptions in both the retention curve and the hydraulic conductivity (Figure 16.8). Note that if hysteresis in the soil hydraulic properties is considered, both the main drying and main wetting curves must be known. Users must specify whether the initial condition is associated with the wetting curve (Initially wetting curve) or drying curve (Initially drying curve). See the user manual for details.

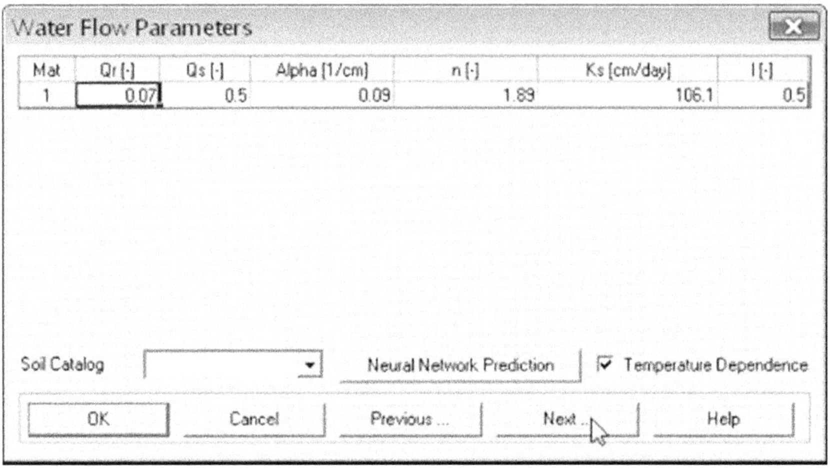

FIGURE 16.9 Water Flow Parameters dialog window.

- Check the radio button No hysteresis for the tutorial project.
- Click on the Next button (Figure 16.8), which will bring up the next window Water Flow Parameters (Figure 16.9).

16.3.3.7 Water Flow Parameters (Pre-Processing | Water Flow | Water Flow Parameters)

Since the number of soil materials has been specified as "1" (one) (i.e., vertically homogeneous/uniform; sandy loam soil) for the flow system (Section 16.3.3.2), the Water Flow Parameters window shows a data (van Genuchten–Mualem parameters) entry spreadsheet for one soil material (Figure 16.9).

- The parameters residual water content (Qr; 0.07 cm^3/cm^3), saturated water content (Qs; 0.5 cm^3/cm^3), reciprocal of the air-entry potential (Alpha [α]; 0.09 cm^{-1}), and the pore size distribution parameter (n; 1.89) were estimated by fitting the van Genuchten soil water retention model to the measured drainage curve.
- Specify the default value of "0.5" in the pore connectivity parameter l field.

Readers are referred to the RETC program (van Genuchten et al., 1991) (http://www.pc-progress.com/en/Default.aspx?retc), which can be used to fit several analytical models to measure water retention and/or unsaturated hydraulic conductivity data.

- The saturated hydraulic conductivity (Ks) of the 80 cm sandy loam soil profile is 106.1 cm/day.
- Check the box Temperature Dependence to consider the hydraulic properties to be temperature-dependent.

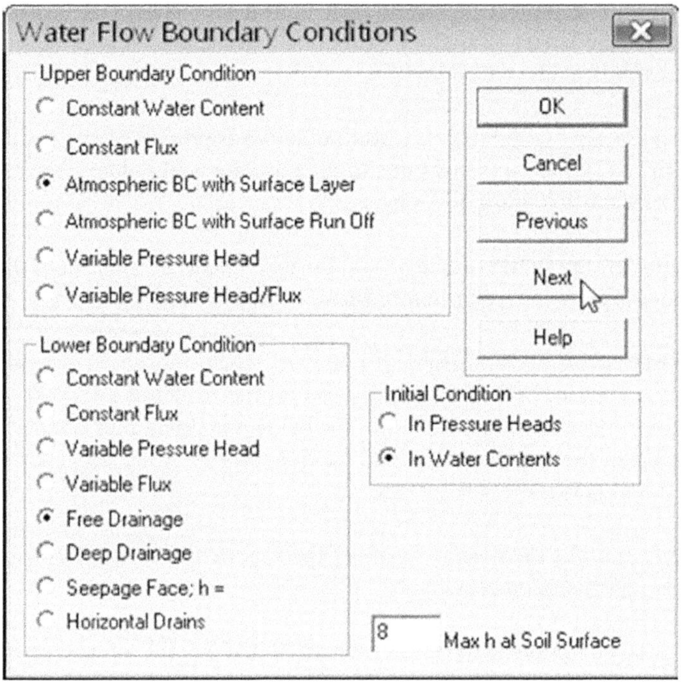

FIGURE 16.10 Water Flow Boundary Conditions dialog window.

Note that HYDRUS-1D also offers two pedotransfer function (PTF) options: (1) the hydraulic parameters of selected soils are included in the drop-down list of Soil Catalog and (2) the Rosetta Lite module to predict Ks and van Genuchten water retention parameters (Figure 16.9). To use the Rosetta Lite PTF, click on the Neural Network Prediction button and the Rosetta Lite dialog window will appear. See HYDRUS-1D Application Help or user manual, and also note that the Rosetta Lite module has its own Help features.

- Click on the Next button (Figure 16.9) and the Water Flow Boundary Conditions dialog window will appear (Figure 16.10).

16.3.3.8 Water Flow Boundary Conditions (Pre-Processing | Water Flow | Boundary Conditions)

The Water Flow Boundary Conditions dialog window allows users to specify the type of upper and lower boundary conditions (BCs) to be used (Figure 16.10). In HYDRUS-1D, the types of Upper Boundary Condition include Constant Water Content, Constant Flux, Atmospheric BC with Surface Layer, Atmospheric BC with Surface Run Off, Variable Pressure Head, and Variable Pressure Head/Flux. The types of Lower Boundary Condition are Constant Water Content, Constant Flux, Variable Pressure Head, Variable Flux, Free Drainage, Deep Drainage, Seepage Face, and Horizontal Drains. See the user manual for details about boundary conditions.

- For the tutorial project, check the radio button Atmospheric BC with Surface Layer for the Upper Boundary Condition and the button Free Drainage for the Lower Boundary Condition.

The Atmospheric BC with Surface Layer boundary condition permits water to build up on the surface, which increases due to precipitation and reduces because of infiltration and evaporation at the soil surface.

- Specify "8" cm in the Max h at Soil Surface field (i.e., maximum thickness of the surface water layer before surface runoff is initiated).

The Free Drainage (i.e., zero-gradient boundary condition) lower boundary condition is used to simulate a freely draining soil profile, which is the most appropriate for a situation where the water table lies far below the domain of interest. Note that the water table at the experimental field for this tutorial project is generally located more than 2.5 m below the soil surface.

- Specify Initial Condition in terms of the water content and check the radio button In Water Contents.

Note that initial conditions will be later specified in the Soil Profile-Graphical Editor module (see Section 16.3.3.14).

- Click on the Next button (Figure 16.10) and the Heat Transport Parameters dialog window will appear (Figure 16.11).

16.3.3.9 Heat Transport Parameters (Pre-Processing | Heat Transport | Heat Transport Parameters)

This dialog window allows users to specify the heat transport parameters for each soil material (Figure 16.11). Since the number of soil materials has been specified "1" (one) (Section 16.3.3.2), the Heat Transport Parameters window shows a data entry spreadsheet for one soil material (Figure 16.11).

- Check the radio button Chung & Horton under the Thermal Conductivity option and select Sand from the drop-down list. Default parameter values will appear in the spreadsheet for each parameter field.
- Override the values as follows: "0.5" for Solid (volume fraction of solid phase); "0" for Org.M. (volume fraction of organic matter); "6" for Disp (longitudinal thermal dispersivity); "1.47×10^{16}" for b1 (coefficient b1 in thermal conductivity function), "-1.55×10^{17}" for b2 (coefficient b2 in thermal conductivity function; W/m/°C); "3.17×10^{17}" for b3 (coefficient b3 in thermal conductivity function; W/m/°C); "1.64×10^{14}" for Cn (volumetric heat capacity of the solid phase); "1.87×10^{14}" for Co (volumetric heat capacity of organic matter); and "3.12×10^{14}" for Cw (volumetric heat capacity of the liquid phase).

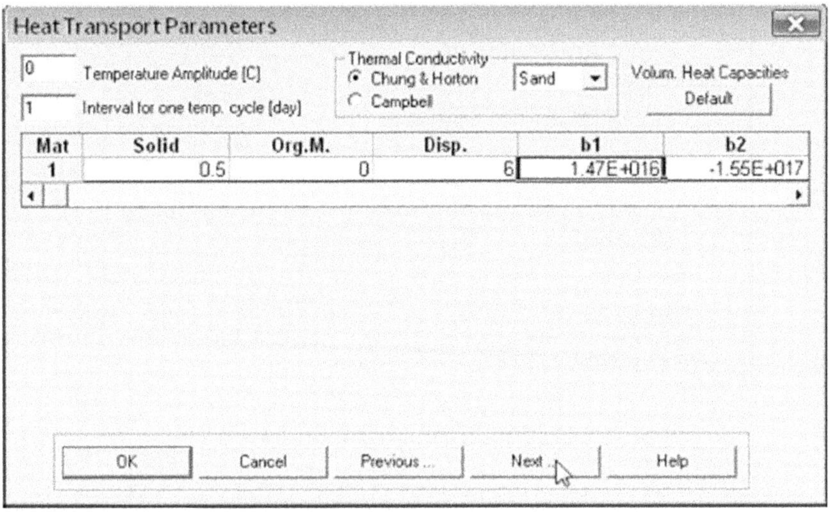

FIGURE 16.11 Heat Transport Parameters dialog window for the tutorial project.

Note that, as shown in Figure 16.11, the b1, b2, and b3 input parameters have the dimension $ML/T^3/K$ (e.g., W/m/°C), while the Cn, Co, and Cw parameters have the dimension $ML/T^2/K$ (e.g., J/m^3/°C). The nonlinear thermal conductivity equation of Chung and Horton was fitted to the measured thermal conductivity data (i.e., thermal conductivity as a function of the volumetric water content), and regression parameters (b1, b2, and b3) were determined.

- Set "0" (zero) as the Temperature Amplitude field and "1" day as the Interval for one temp. cycle field.

Readers are referred to HYDRUS-1D Application Help and user manual for details about heat transport parameters.

- Click on the Next button (Figure 16.11), which will bring up the next dialog window Heat Transport Boundary (Figure 16.12).

16.3.3.10 Heat Transport Boundary (Pre-Processing | Heat Transport | Heat Transport Boundary Conditions)

The Heat Transport Boundary dialog window allows users to specify the upper and lower heat transport boundary conditions (Figure 16.12). HYDRUS-1D includes two types of upper boundary conditions: temperature (Dirichlet-type condition) and heat flux (Cauchy-type condition). Three types of lower boundary conditions are available: temperature (Dirichlet-type condition), heat flux (Cauchy-type condition), and free drainage (a Neumann-type boundary condition with a zero temperature gradient when water flow is directed out of the domain).

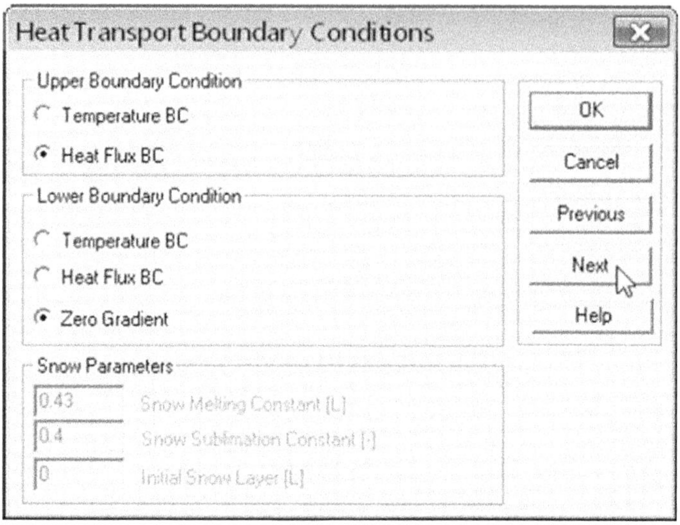

FIGURE 16.12 Heat Transport Boundary Conditions dialog window.

- Check the radio button Heat Flux BC for the Upper Boundary Condition (Figure 16.12).

Note that HYDRUS-1D uses the continuous meteorological data (see Sections 16.3.3.3, 16.3.3.12, and 16.3.3.13) in the surface energy balance equation to estimate the surface heat flux, which is subsequently used as a known Heat Flux BC on the soil surface.

- Check the radio button Zero Gradient for the Lower Boundary Condition.

Note that the heat transfer across the lower boundary was assumed to occur only by convection of liquid water and water vapor.

- Click on the Next button (Figure 16.12), and the next dialog window Time Variable Boundary Conditions will appear (Figure 16.13).

16.3.3.11 Time Variable Boundary Conditions (Pre-Processing | Variable Boundary Conditions)

Note that the number of time-variable boundary records (number records = 1148) is specified in the Time Information dialog window (Section 16.3.3.3).

Time variable input data can be prepared in a Microsoft Excel spreadsheet (Figure 16.14), then:

- Copy the data from the Excel spreadsheet file "Meteorological and variable BC data.xlsx" (Figure 16.14), and then paste it in the spreadsheet cells of the Time Variable Boundary Conditions dialog window (Figure 16.13).
- Use Ctrl V to paste the columns of data (Figure 16.13).

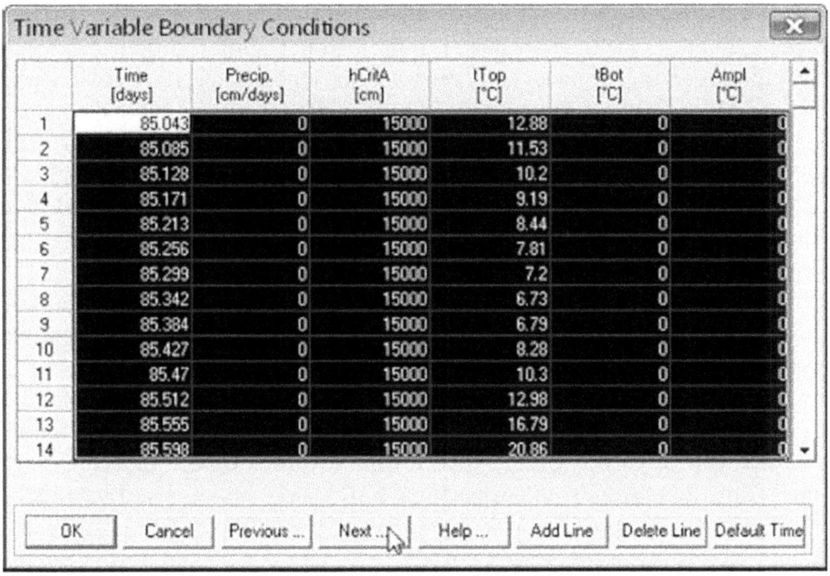

FIGURE 16.13 Time Variable Boundary Conditions dialog window.

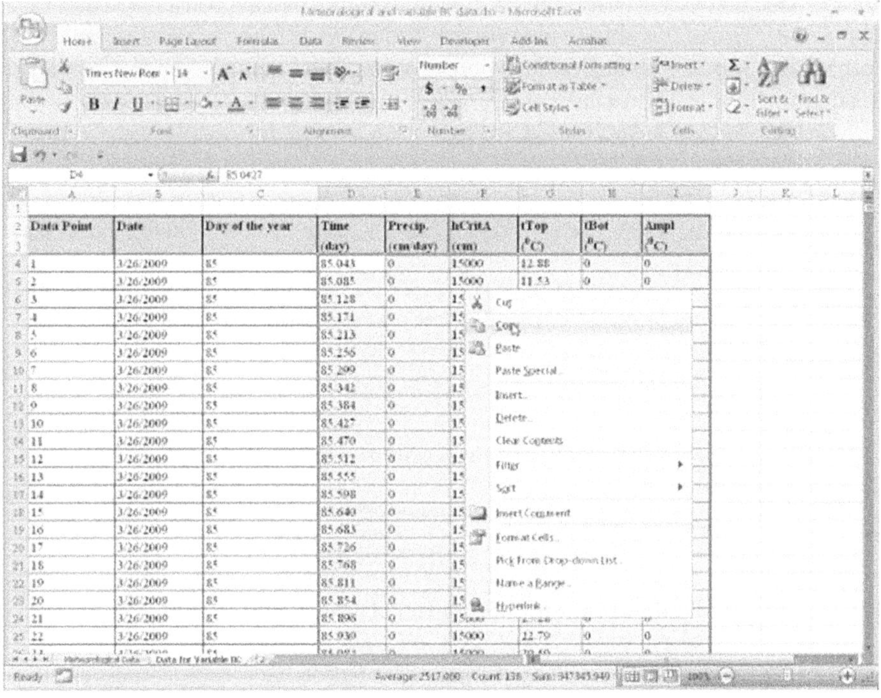

FIGURE 16.14 Time variable boundary records for the tutorial project.

Note that an atmospheric boundary condition with surface layer is imposed at soil surfaces accounting for time variable rainfall and irrigation (Section 16.3.3.8). For the Precip. (cm/days) parameter (Figure 16.13), the amounts of rainfall that occurred during an irrigation event or immediately after irrigation were added to the irrigation rate. For this tutorial example, the experimental plot was irrigated once a week at a rate of 20 cm/day during the period from DOY 85 (March 26) to DOY 134 (May 14) 2009. Hourly or daily rainfall data during this period can be downloaded from the weather station Leyendecker PSRC of the New Mexico Climate Center at http://weather.nmsu.edu.

Note that the tTop and tBot parameters are time-dependent temperatures of the soil surface boundary and the lower boundary, respectively. As the heat boundary flux is specified as the upper boundary condition, values in the tTop are used as the temperature of the incoming water. Hourly or daily air temperature during the period from DOY 85 to 134 2009 can also be downloaded from the weather station Leyendecker PSRC of the New Mexico Climate Center. Data for the "tBot" and "Ampl" columns (Figure 16.13) are not required for this tutorial project.

The hCritA is the absolute value of the minimum allowed pressure head at the soil surface, which was assumed to be −15,000 cm. Note that the absolute value of hCritA must be entered in the Time Variable Boundary Conditions dialog window (Figure 16.13). With the specified atmospheric boundary condition at the soil surface, water evaporates at a potential rate as long as the pressure head at the soil surface is greater than hCritA. Once the hCritA value is reached, the soil is then too dry to deliver the potential rate; the actual evaporation rate is decreased from the potential value.

See HYDRUS-1D Application Help or user manual for details about the time variable boundary conditions.

- Click on the Next button (Figure 16.13), and the next dialog window Meteorological Parameters will appear (Figure 16.15).

16.3.3.12 Meteorological Parameters (Pre-Processing | Meteorological Parameters)

- Check the radio button Solar Radiation under the option labeled Radiation (Figure 16.15).
- Specify "32.2" and "1178" in the Latitude (deg) and Altitude (m) fields, respectively.

HYDRUS-1D estimates the net solar radiation.

- Specify the HYDRUS-1D-recommended Angström values (short wave radiation) values of "0.25" and "0.5" for a and b, respectively.
- Set "0" (zero) for fields a1 and b1 under the option labeled Cloudiness effect on long-wave radiation, and choose the radio button Transmission c. under the option labeled Cloudiness.

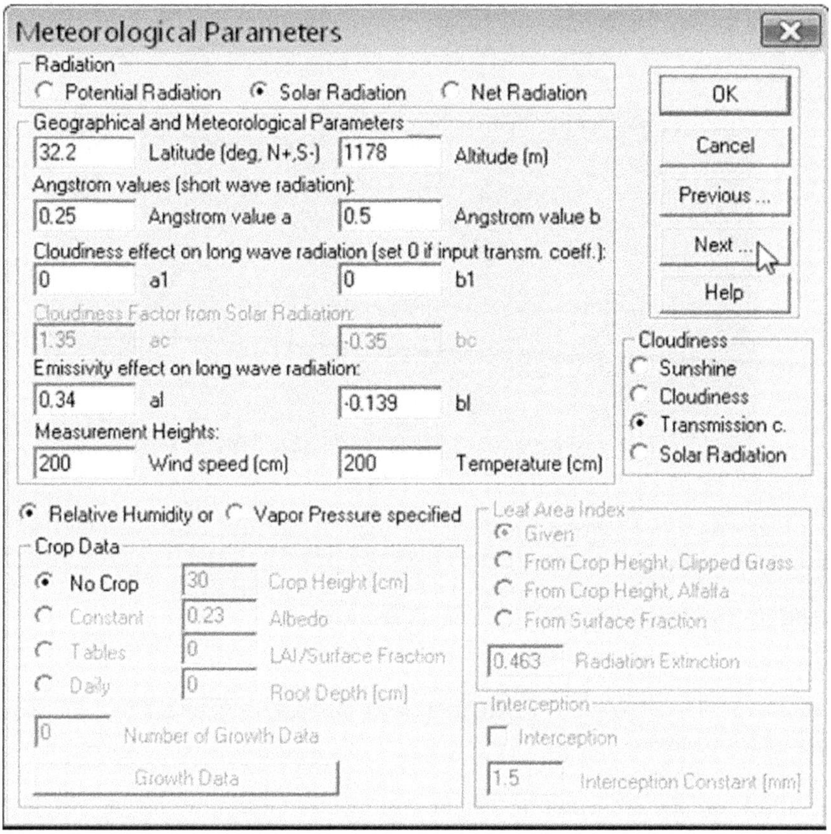

FIGURE 16.15 Meteorological Parameters dialog window.

The cloudiness factor can be estimated from the atmospheric transmission coefficient for solar radiation, which is defined as the ratio of the measured daily global solar radiation and the daily potential global (extraterrestrial) radiation. Note that in HYDRUS-1D, the cloudiness factor can also be calculated from the measured solar radiation data or from the measured sunshine hour data.

Readers are referred to the user/reference manual (Šimůnek et al., 2012) for details on the calculation of the cloudiness factor.

- Specify the HYDRUS-1D-recommended values of "0.34" and "−0.139" for a1 and b2, respectively, for calculating Emissivity effect on long-wave radiation.
- Enter the value "200" cm for heights of wind speed and temperature and humidity measurements, and check the radio button Relative Humidity.
- Check the radio button No Crop under the option labeled Crop Data.

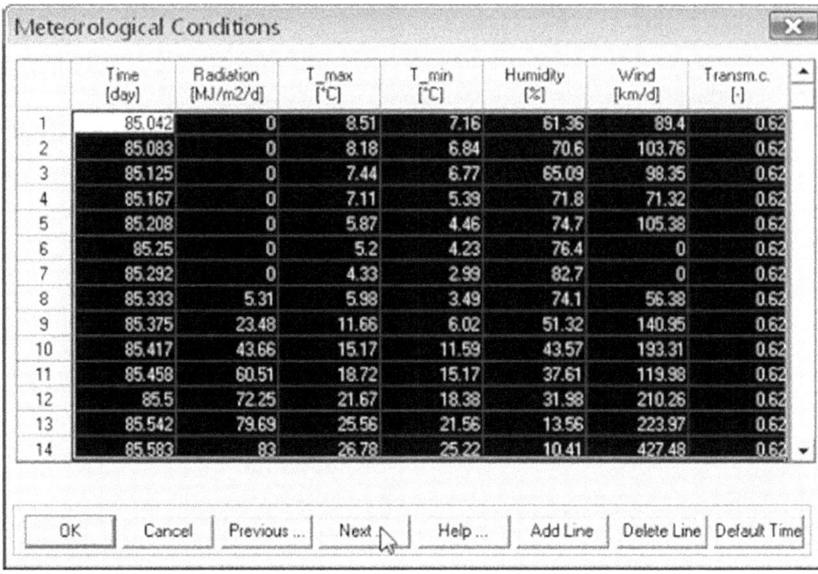

FIGURE 16.16 Meteorological Conditions dialog window.

Readers are referred to the HYDRUS-1D user manual for details about the meteorological parameters.

- Click on the Next button (Figure 16.15), and the next dialog window Meteorological Conditions will appear (Figure 16.16).

16.3.3.13 Meteorological Conditions (Pre-Processing | Meteorological Boundary Conditions)

Note that the number of meteorological records (number records = 1176) is specified in the Time Information dialog window (Section 16.3.3.3). Meteorological input data can be prepared in a Microsoft Excel spreadsheet (Figure 16.17), then:

- Copy the data from the Excel spreadsheet file "Meteorological and variable BC data.xlsx" (Figure 16.17), and then paste (Ctrl V) in the spreadsheet cells of the Meteorological Conditions dialog window (Figure 16.16).

Meteorological data include solar radiation (MJ/m^2/day), T_max (maximum air temperature; °C), T_min (minimum air temperature; °C), relative humidity (%), wind speed (km/day), and the estimated atmospheric transmission coefficient for solar radiation (dimensionless) (Figure 16.16). For this tutorial example, hourly or daily meteorological data during the period from DOY 85 (March 26) to DOY 134 (May 14) 2009 can be downloaded from the weather station Leyendecker PSRC of the New Mexico Climate Center at http://weather.nmsu.edu.

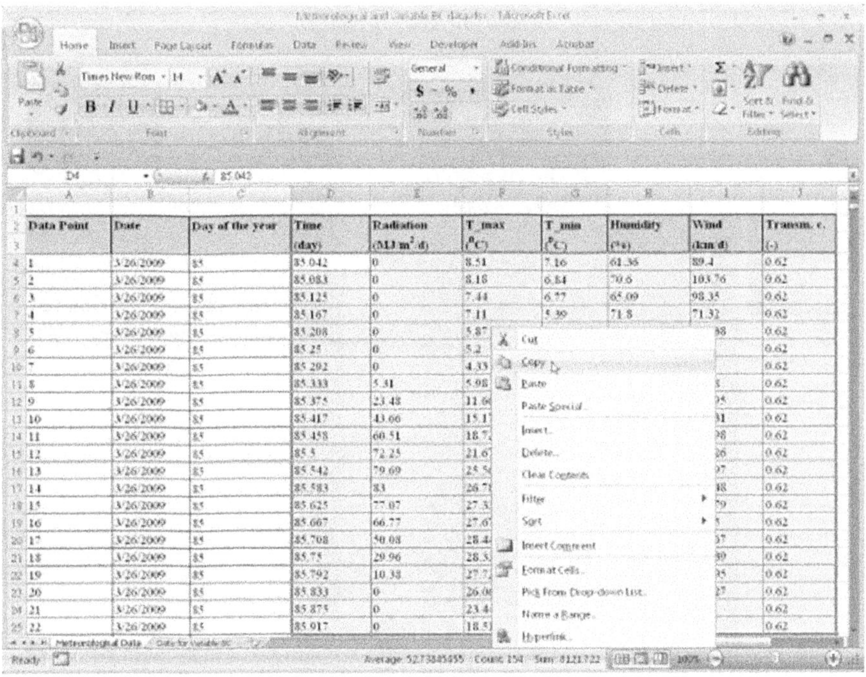

FIGURE 16.17 Meteorological data for the tutorial project.

- Click on the Next button (Figure 16.16), which will call the HYDRUS-1D guide to run the PROFILE module (i.e., Soil Profile-Graphical Editor or Pre-processing | Profile Information) (Figure 16.18).
- Click the OK button (Figure 16.18) to run the PROFILE module (Figure 16.19). Click the Yes button if an alert dialog box appears to save data before running the graphical profile editor.

16.3.3.14 Soil Profile: Graphical Editor (Pre-Processing | Profile Information)

The HYDRUS-1D – Profile Information module window provides both menu commands and toolbar items for the profile discretization, specification of initial

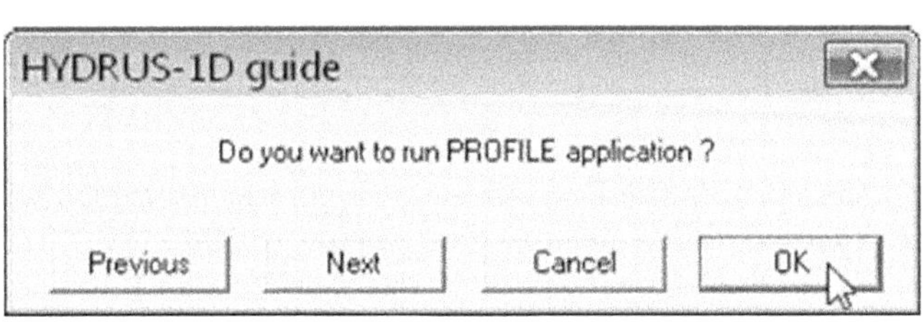

FIGURE 16.18 The HYDRUS-1D guide dialog box for running the PROFILE module (i.e., Soil Profile-Graphical Editor or Profile Information submenu).

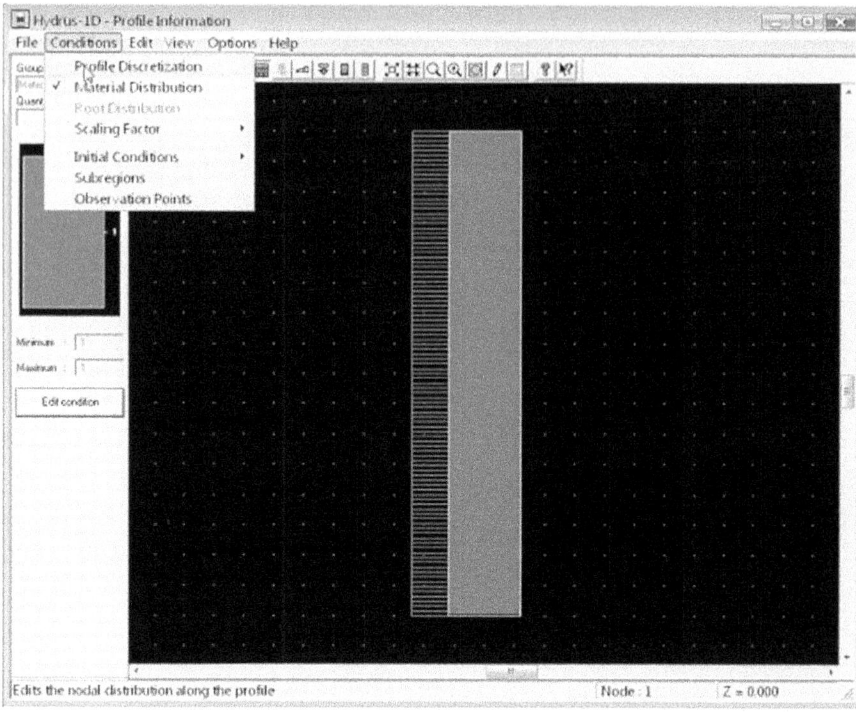

FIGURE 16.19 Profile Information dialog window.

conditions for all processes (e.g., water flow, heat transport, and solute transport), as well as the distribution of soil materials and root water uptake, among other parameters (Figure 16.19). Users can select menu commands or click on toolbars to open dialog windows/screens in the HYDRUS-1D – Profile Information module window, and note that placing the mouse cursor on any toolbar icon displays the corresponding menu/submenu item (Figure 16.19). The dialog window(s) will not appear and submenu(s) or toolbar icon(s) will be disabled (grayed out) for the unchecked processes in the Main Processes dialog window (see Section 16.3.3.1).

16.3.3.14.1 Profile Discretization

- Click Conditions | Profile Discretization (Figure 16.19), which will call up the screen for the profile discretization (Figure 16.20).
- Either click on the Number button in the panel labeled Edit nodes on the left or use the Edit menu command (i.e., click Edit | Number of Nodes). The dialog box Number of nodes will appear (Figure 16.20). Enter the value "101," and then click on the button OK.

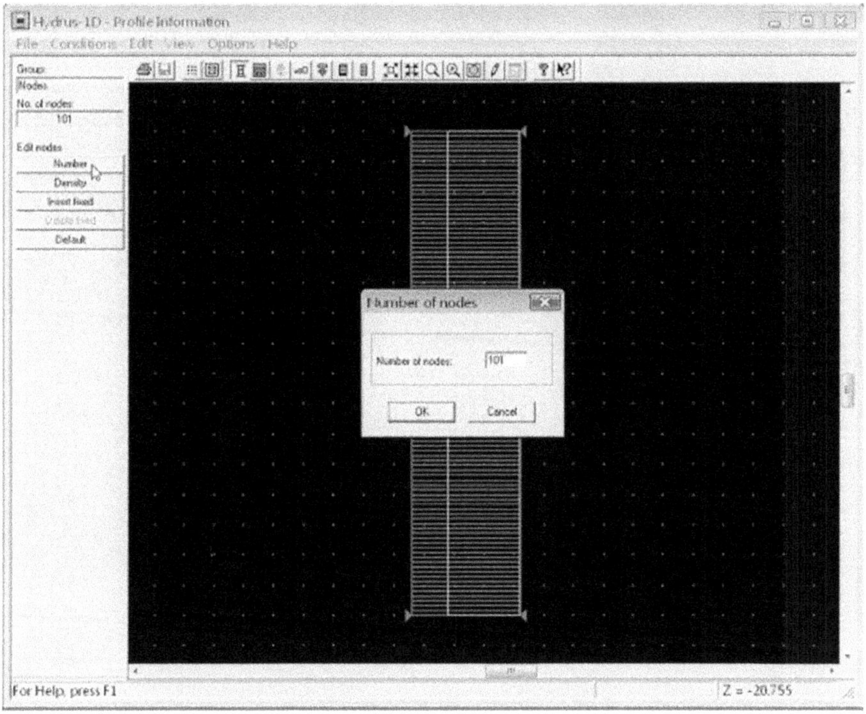

FIGURE 16.20 Profile Information module: Profile discretization.

The 80 cm soil profile is therefore discretized into 101 nodes, and the green horizontal lines across the soil profile indicate the setting of grid spacing of 0.8 cm (Figure 16.20).

As shown in Figure 16.20, users can change/edit the nodal density by using buttons such as Density or Insert fixed in the panel labeled Edit nodes on the left or by using Edit menu commands such as Nodal Density and Insert New Fixed Point. Fixed points are used to adjust the local discretization density of nodal points. The nodal density (a real number between 0.01 and 100) can be specified only at fixed points (nodes) and in terms of an upper and lower density. The default value of the nodal density is one. The default fixed points in the soil profile are placed (marked by purple stars) at the upper and lower boundaries (Figure 16.20). Users can use either the button Insert fixed in the panel Edit nodes on the left or use the Edit menu command (e.g., click Edit | Insert New Fixed Point) to insert at, or delete from, any other point in the soil profile.

16.3.3.14.2 Material Distribution

- Click Conditions | Material Distribution (Figure 16.19).
- As shown in Figure 16.21, either click on the Edit condition button in the pane on the left or click Edit | Edit Condition, which will provide the hand cursor.

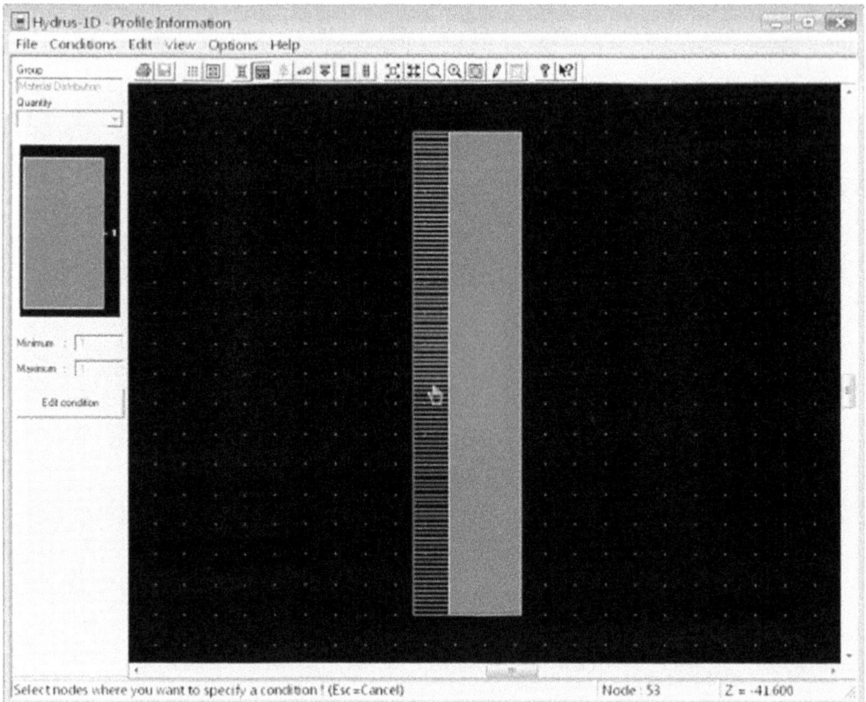

FIGURE 16.21 Profile Information module: Selection of the profile depth(s) using the hand cursor.

Note that the soils 80 cm profile is vertically homogeneous/uniform (Section 16.3.3.2). Use the hand cursor to select all the grid points (i.e., entire soil profile).

- Click first on the topmost green line and then click on the bottom green line (Figure 16.21), which will call the Condition Specification dialog box (Figure 16.22).
- Enter the value "1" (one) in the Material index field, and then click on the OK button in the Condition Specification dialog box (Figure 16.22).

16.3.3.14.3 Scaling Factor

HYDRUS-1D allows users to specify the hydraulic conductivity, pressure head, and water content scaling factors to simplify the description of spatial variability in the unsaturated soil hydraulic properties in the flow domain.

- Click Conditions | Scaling Factors | Hydraulic Conductivity (Figure 16.19).
- As shown in Figure 16.23, either click on the Edit condition button in the pane on the left or click Edit | Edit Condition, which will provide the hand cursor.

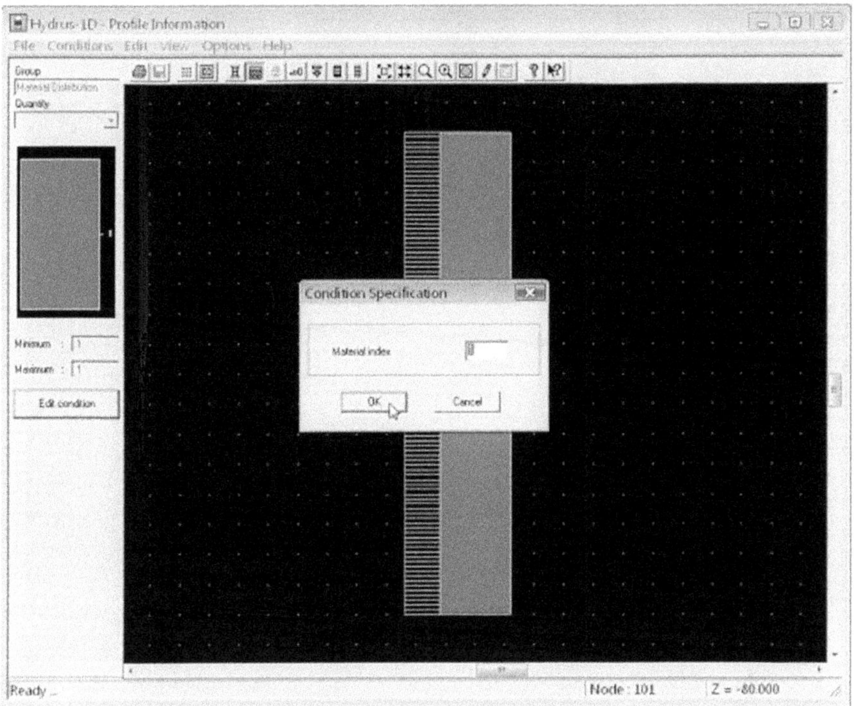

FIGURE 16.22 Profile Information module: Specification of soil material distribution.

- Click the hand cursor first on the topmost green line and then click on the bottom green line (Figure 16.23), which will call the Condition Specification dialog box for the hydraulic conductivity scaling factor (Figure 16.23).
- Set the default values (homogeneous) at the top and bottom fields to “1,” check the box Use top value for both, and click on the button OK in the Condition Specification dialog box.

Note that scaling factors can be used to simulate heterogeneous soil conditions. Users can follow the same procedure (i.e., Conditions | Scaling Factors | Hydraulic Conductivity) to set the water content (click Conditions | Scaling Factors | Water Content) and pressure head (click Conditions | Scaling Factors | Pressure Head) scaling factors. For both the water content and pressure head scaling factors, specify the value “1” (homogeneous) at both the top and bottom fields in the Condition Specification dialog box. See the user manual for details about the scaling procedure implemented in HYDRUS-1D.

16.3.3.14.4 Initial Conditions

- Click Conditions | Initial Conditions | Pressure Head/Water Content (Figure 16.19).

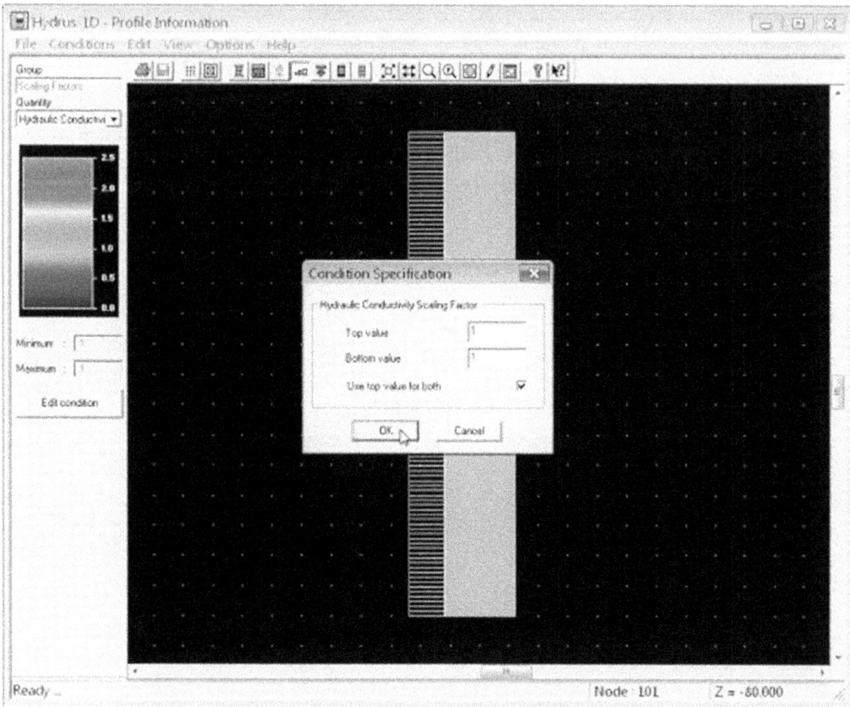

FIGURE 16.23 Profile Information module: Hydraulic conductivity scaling factor window.

- As shown in Figure 16.24, either click on the Edit condition button in the pane on the left or click Edit | Edit Condition, which will provide the hand cursor.
- Click the hand cursor first on the topmost line and then on the bottom line. The Condition Specification/Water content distribution dialog box will appear (Figure 16.24).

Note that the initial condition in terms of the water content is specified in the Water Flow Boundary Conditions dialog window (Section 16.3.3.8).

- Set the same water content value (uniform) at the top and the bottom to "0.17" cm^3/cm^3, check the box Use top value for both, and click on the button OK (Figure 16.24).
- Click Conditions | Initial Conditions | Temperature (Figure 16.24).
- As shown in Figure 16.25, either click on the Edit condition button in the pane on the left or click Edit | Edit Condition, which will provide the hand cursor.

Note that users can also use the Quantity drop-down menu to select Temperature, and then click on the Edit condition button in the panel on the left.

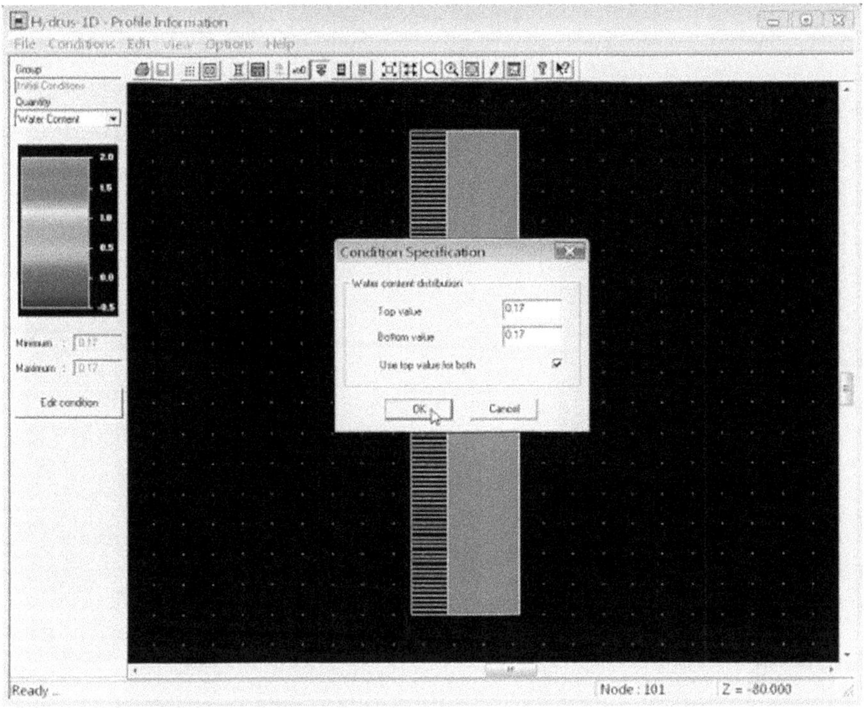

FIGURE 16.24 Profile Information module: Initial condition in terms of the water content.

- Click the hand cursor first on the topmost line and then on the bottom line. In the Condition Specification/Temperature distribution dialog box, set the same temperature value at the top and the bottom to "15"°C, check the box Use top value for both, and click on the button OK in the Condition Specification/Temperature distribution dialog box (Figure 16.25).

16.3.3.14.5 Subregion

- Click Conditions I Subregions (Figure 16.19).
- As shown in Figure 16.26, either click on the Edit condition button in the pane on the left or click Edit I Edit Condition.

Note that the soil profile is vertically homogeneous/uniform (see Section 16.3.3.2).

- Click the hand cursor first on the topmost green line and then on the bottom green line, enter the value "1" (one) in the Subregion index field of the Condition Specification dialog box, and then click on the button OK (Figure 16.26).

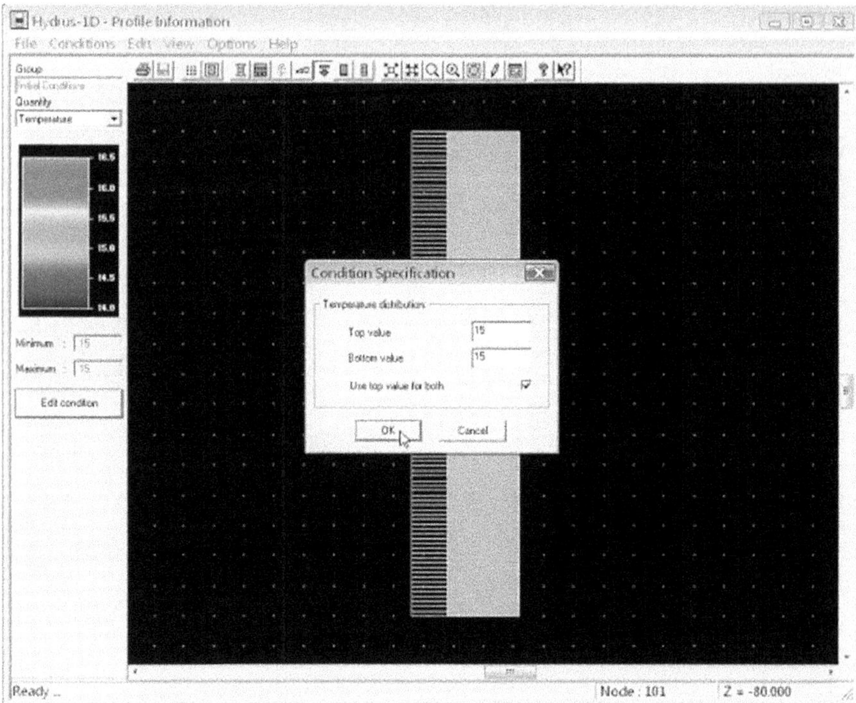

FIGURE 16.25 Profile Information module: Initial condition in terms of the soil temperature.

16.3.3.14.6 Observation Points

- Click Conditions | Observation Points (Figure 16.19).
- As shown in Figure 16.27, either click on the Insert button in the panel labeled Edit observ nodes on the left or click Edit | Insert Observation Point.

When you move the hand cursor over the green line, the node number and depth (Z coordinate) will be displayed in the status bar (if you do not see it, click View | Status Bar at the lower right of the screen) (Figure 16.27). Note that observation points are nodes for which the outputs (e.g., values of the pressure head, water content, and temperature) are saved at each time step.

- To insert an observation point, place the hand cursor over the green horizontal line (i.e., depth), and then click on the depth of interest.
- For the tutorial project, insert four observation points at depths of −4.8 cm (≈5 cm; node 7), −9.6 cm (≈10 cm; node 13), −20 cm (node 26), and −49.6 cm (≈50 cm; node 63) (Figure 16.27). Note that a right-click of the mouse will exit the hand cursor mode.

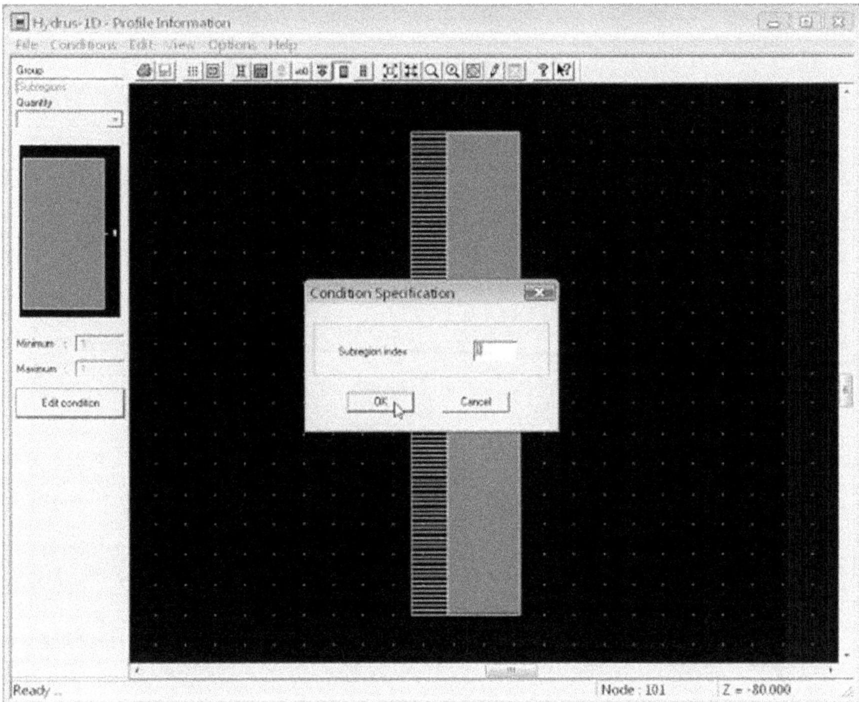

FIGURE 16.26 Profile Information module: Specification of subregion.

Users can delete the observation point(s). For example, either click on the "Delete" button in the panel labeled Edit observ nodes on the left or click Edit | Delete Observation Point, which will provide the hand cursor with a square box. In order to delete, place and click the hand cursor (with the square box) on the observation point of interest (Figure 16.28).

16.3.3.14.7 View and Options Menus

Before you exit the HYDRUS-1D – Profile Information module window (Figure 16.19), it is also necessary to be familiar with the View and Options menus. The View menu allows users to view or hide the toolbar and status bar, to choose the default view of the view window, to enlarge/reduce the content of the view window, to zoom on a certain part of the view window, to view the previous zoom on a certain part of the view window, and to redraw the content of the view window.

- Click Options | Display Options, which will bring up the Display Options dialog box for the flow domain (Figure 16.29).

Options include Display profile contours, Display mesh nodes, Display fixed points, and Numbering of nodes. This window also allows users to select the number of

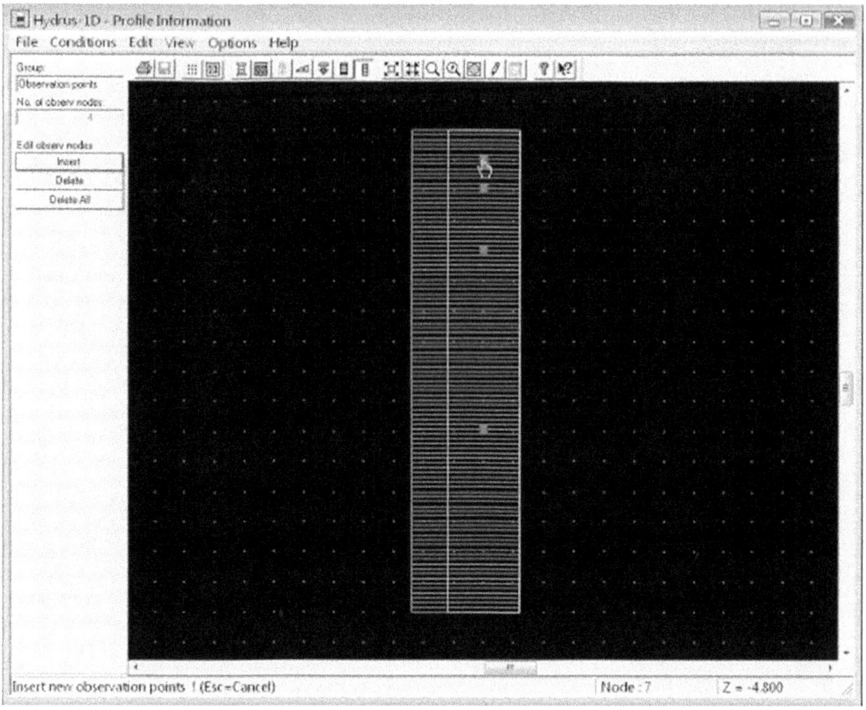

FIGURE 16.27 Profile Information module: Insertion of observation points.

colors to be used, including a monocolor spectrum option. Click on the Default button (Figure 16.29).

- Click Options | Grid, which will bring up the Grid settings dialog box to customize grid nodes of the view window for better orientation in space and for easier identification of the Z coordinate (Figure 16.30).
- Click on the Default button.
- Click Options | Profile Shape; the dialog box Profile width will appear (Figure 16.31).

This window allows users to customize the soil profile view in a view window (Figure 16.31). For example, set the value "80" (an integer between 1 and 100) to the width of the column in which the selected properties are displayed, and the value "40" (an integer between 3 and 100) to the width of the column in which the nodal points are displayed (Figure 16.31).

- Click on the OK button (Figure 16.31) and return to the HYDRUS-1D – Profile Information module window (Figure 16.19).

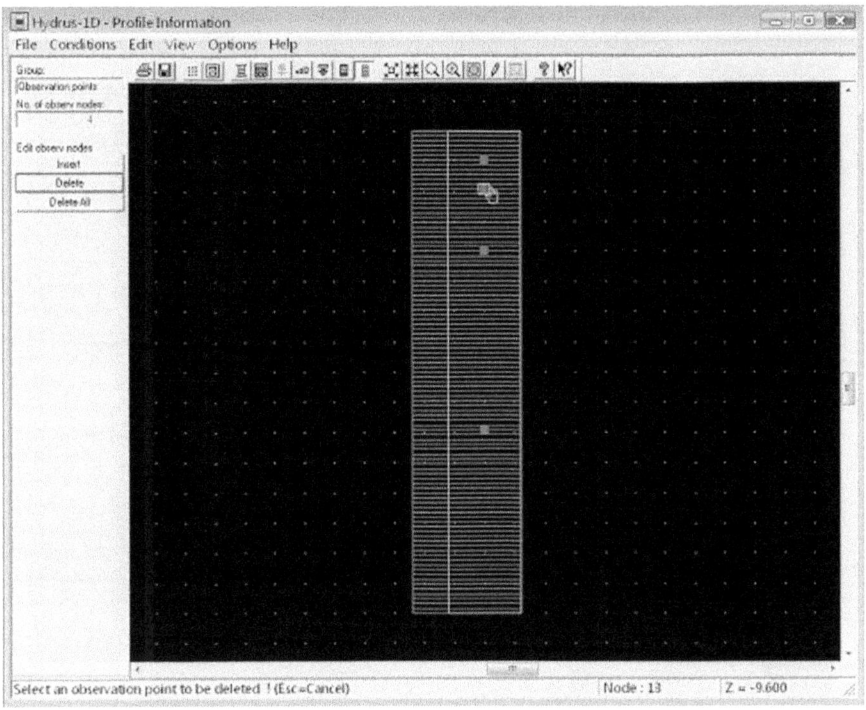

FIGURE 16.28 Profile Information module: Deletion of observation points.

- Click File I Save Data. Click the OK button if an alert message "Old profile will be overwritten and all subsequent data will be deleted. Continue?" appears.
- Click File I Exit and exit the HYDRUS-1D – Profile Information module window.

16.3.3.15 Soil Profile Summary (Preprocessing | Profile Summary)

Once you exit the HYDRUS-1D – Profile Information module window, the Soil Profile Summary dialog window will appear (Figure 16.32). The Soil Profile Summary window summarizes the spatial discretization and distribution of soil properties, initial conditions (water content and soil temperature), and related information in tabular form (Figure 16.32). Note that this dialog window allows users to modify the parameter setup in the external Profile Information module. See HYDRUS-1D Application Help and user manual for the description of parameters in the Soil Profile Summary table.

- Click the button Next (Figure 16.32).
- An alert dialog box (i.e., HYDRUS-1D guide; Figure 16.33) appears, whether or not to run the HYDRUS-1D application. Click the button OK.

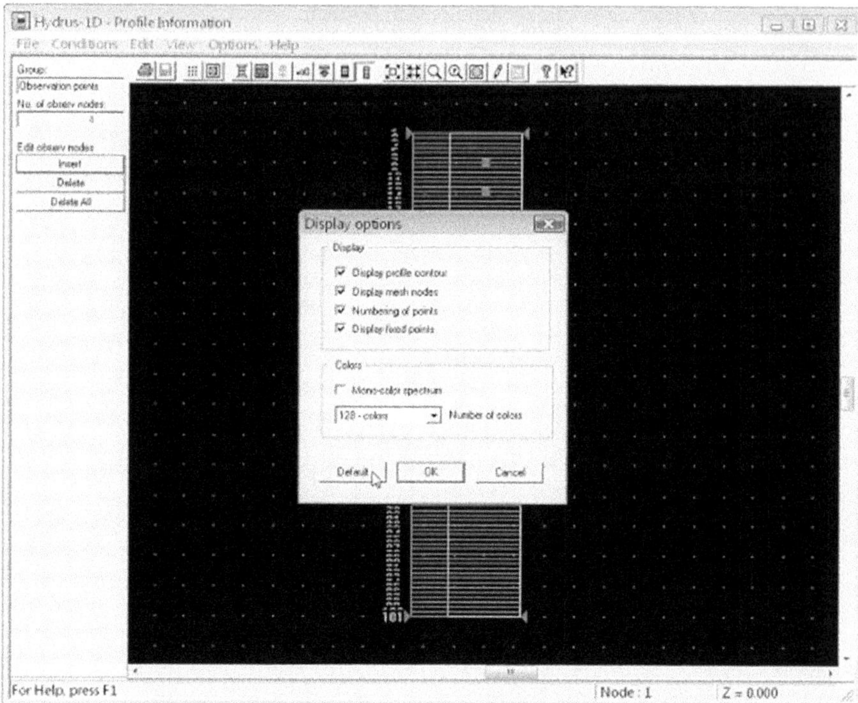

FIGURE 16.29 Profile Information module: Display options for the flow domain.

A disk operating system (DOS) window will appear to show the information (e.g., time, number of iterations, cumulative number of iterations, actual surface flux, and so forth) during the HYDRUS-1D execution. As shown in Figure 16.34, upon the completion of all calculations, the DOS window displays the total time in seconds that completed the calculations.

- Press Enter to close the DOS window (Figure 16.34). The results will be displayed in the Post-processing panel (Figure 16.35).

16.3.4 Working with the HYDRUS-1D Post-Processing Unit

16.3.4.1 Output Files

All output files (*.OUT) for the tutorial project are stored in the project directory or folder (located in C:\Hydrus1D-Tutorial\Hyd1D-Tutorial) (Figure 16.36). Note that all output files are text files, which can be opened with any text editor (e.g., Notepad) or spreadsheet software (e.g., Microsoft Excel). Numbers in the output files are separated by spaces. In order to open an output file using Microsoft Excel's Text Import Wizard (e.g., Microsoft Excel 2007), select Original Data Type: Delimited, and then choose Delimiters: Spaces.

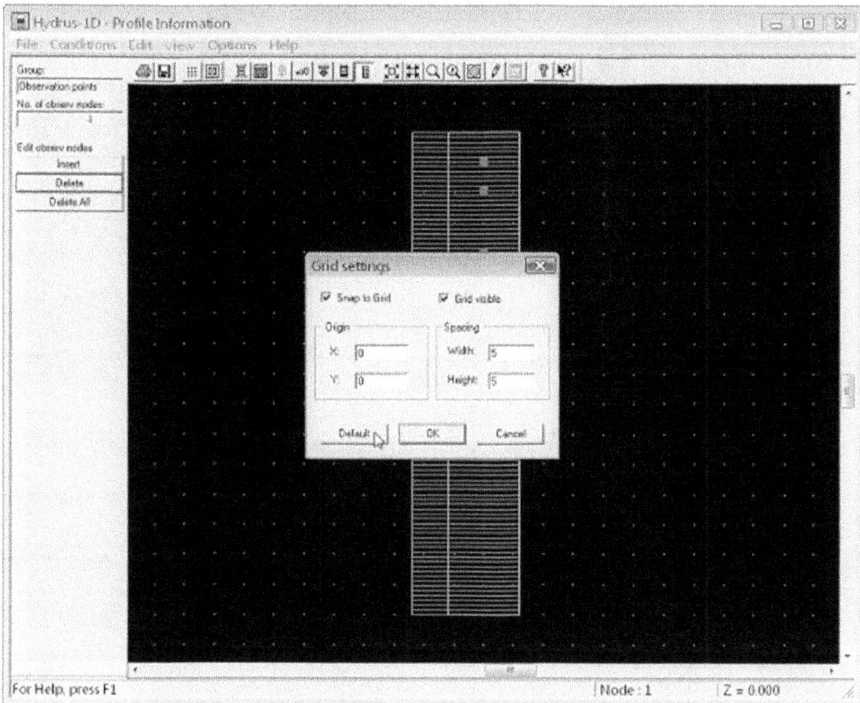

FIGURE 16.30 Profile Information module: Grid setting of the View window.

In HYDRUS-1D, the output is organized into the following three groups (Figure 16.36): (1) T-level information—information in the output files (e.g., "OBS_NODE.OUT," "RUN_INF.OUT," and "T_LEVEL.OUT") is printed at the end of each time step; (2) P-level information—information in the output files (e.g., "BALANCE.OUT," "NOD_INF.OUT," and "NOD_INF_V.OUT") is printed only at selected print times (Section 16.3.3.4); and (3) A-level information—information in the output files (e.g., "A_LEVEL.OUT" and "Meteo.out") is printed each time a time-dependent boundary condition is specified (Sections 16.3.3.3 and 16.3.3.11).

Examples of output text files (Notepad) are depicted in Figure 16.37 ("OBS_NODE.OUT"; transient values of the pressure head, water content, and temperature at specified observation nodes; T-level information), Figure 16.38 ("NOD_INF.OUT"; profile information; P-level information), Figure 16.39 ("NOD_INF_V.OUT"; profile information for liquid water and water vapor fluxes; P-level information), and Figure 16.40 ("Meteo.out"; meteorological information/energy fluxes; A-level information).

Note that some of the input data are printed to the files "I_CHECK.OUT" (i.e., a complete description of the space discretization, the hydraulic characteristic, and the transport properties of each soil material) and "PROFILE.OUT" (i.e., profile information). HYDRUS-1D provides module output files based upon the type of

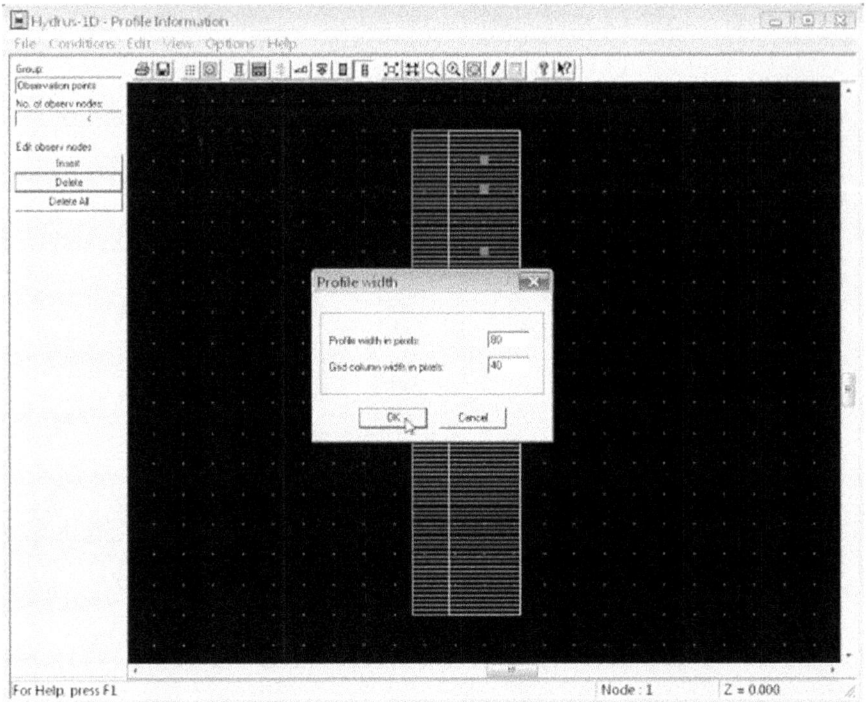

FIGURE 16.31 Profile Information module: Specification of the profile width in the View window.

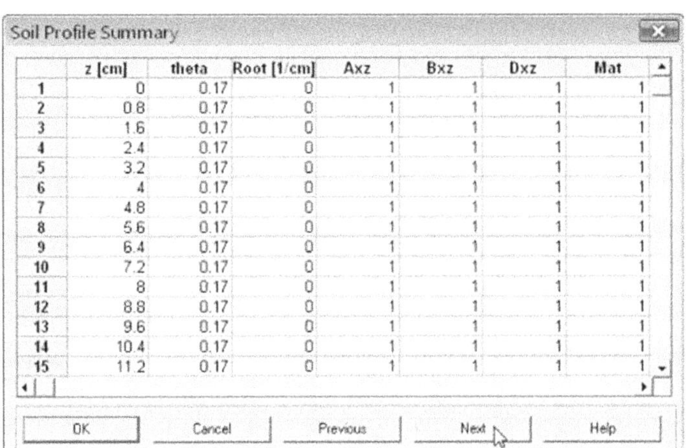

	z [cm]	theta	Root [1/cm]	Axz	Bxz	Dxz	Mat
1	0	0.17	0	1	1	1	1
2	0.8	0.17	0	1	1	1	1
3	1.6	0.17	0	1	1	1	1
4	2.4	0.17	0	1	1	1	1
5	3.2	0.17	0	1	1	1	1
6	4	0.17	0	1	1	1	1
7	4.8	0.17	0	1	1	1	1
8	5.6	0.17	0	1	1	1	1
9	6.4	0.17	0	1	1	1	1
10	7.2	0.17	0	1	1	1	1
11	8	0.17	0	1	1	1	1
12	8.8	0.17	0	1	1	1	1
13	9.6	0.17	0	1	1	1	1
14	10.4	0.17	0	1	1	1	1
15	11.2	0.17	0	1	1	1	1

FIGURE 16.32 Soil Profile Summary window.

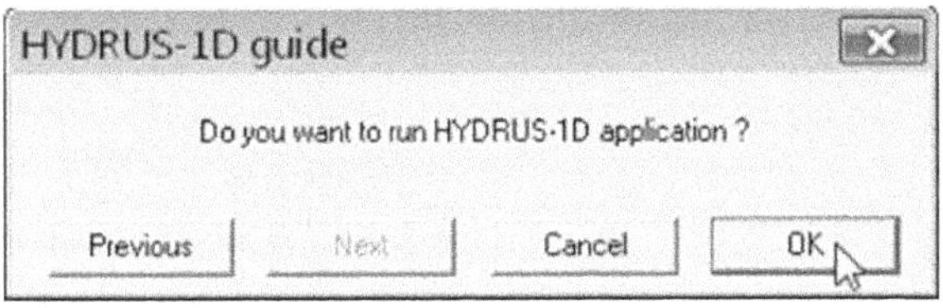

FIGURE 16.33 The HYDRUS-1D guide dialog box for running the HYDRUS-1D application.

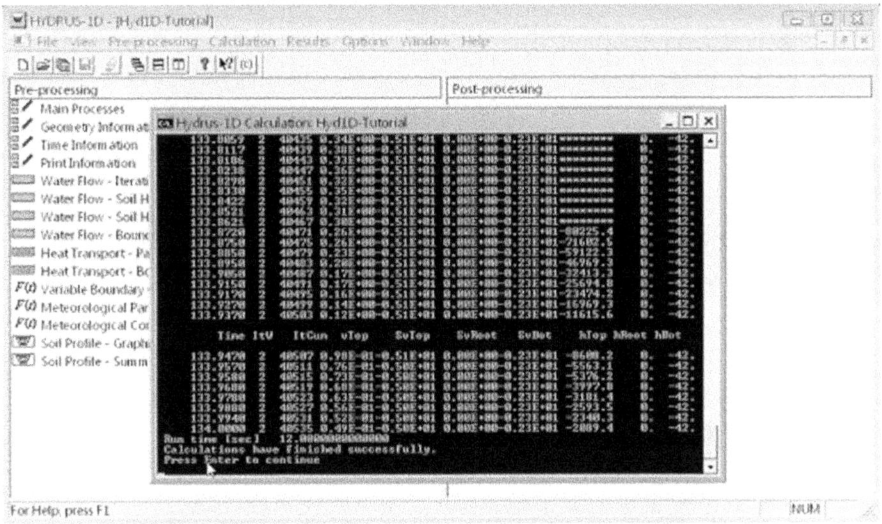

FIGURE 16.34 The DOS window during the HYDRUS-1D execution.

processes specified/selected in the Main Processes dialog window (Section 16.3.3.1). For example, the selection of Solute Transport (Standard Solute Transport option), which has not been selected in Main Processes (Figure 16.3), will create a separate output file "SOLUTE.OUT" (actual and cumulative concentration fluxes; T-level information) for each solute. All output files are described in detail in HYDRUS-1D Application Help and user manual.

16.3.4.2 Results Menu and On-Screen Graphics

As shown in Figure 16.35, the results (Results menu items) are displayed in the Post-processing pane on the right-hand side, which provides users with the on-screen graph options. For example,

- Either double-clicking on Observation Points in the Post-processing pane or clicking Results | Observation Points will produce the Observation Nodes graphical display (Figure 16.41). Note that if an alert message box "File

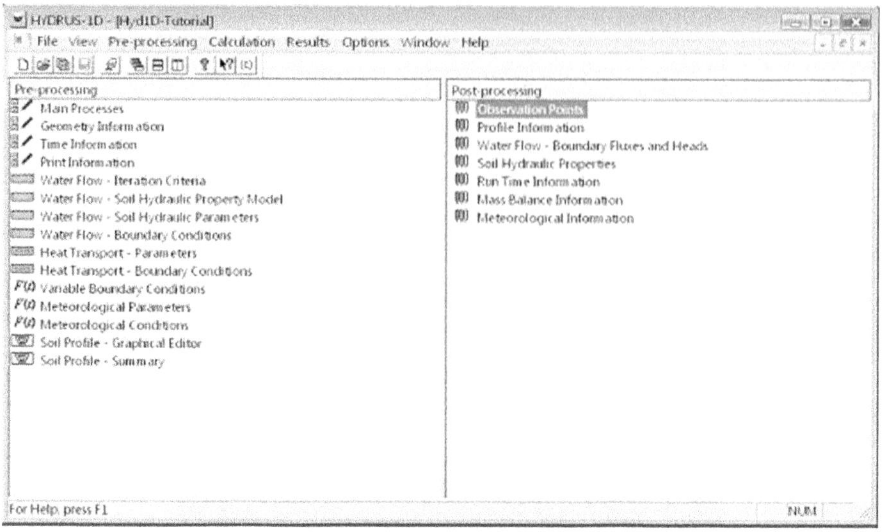

FIGURE 16.35 HYDRUS-1D overall display window (Pre- and Post-processing panels) after the completion of "Hyd1D-Tutorial.hid" project simulation run.

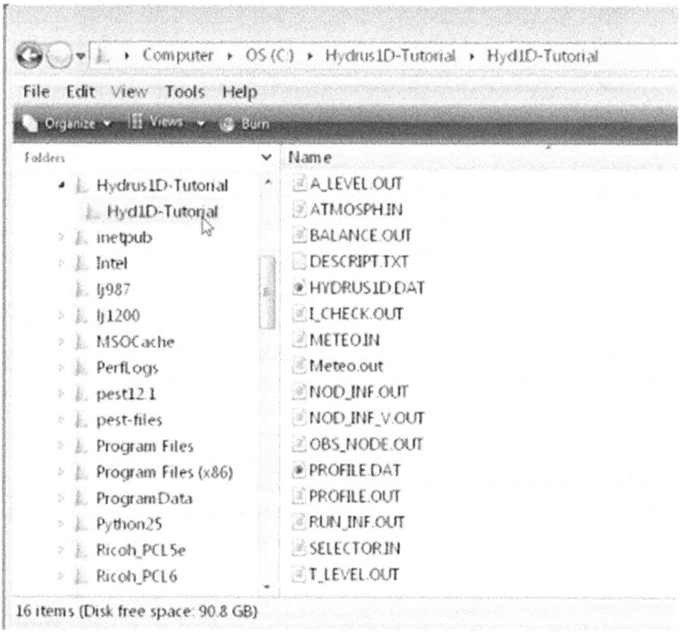

FIGURE 16.36 Input (*.IN) and output (*.OUT) files in the tutorial project folder (C:\Hydrus1D-Tutorial\Hyd1D-Tutorial).

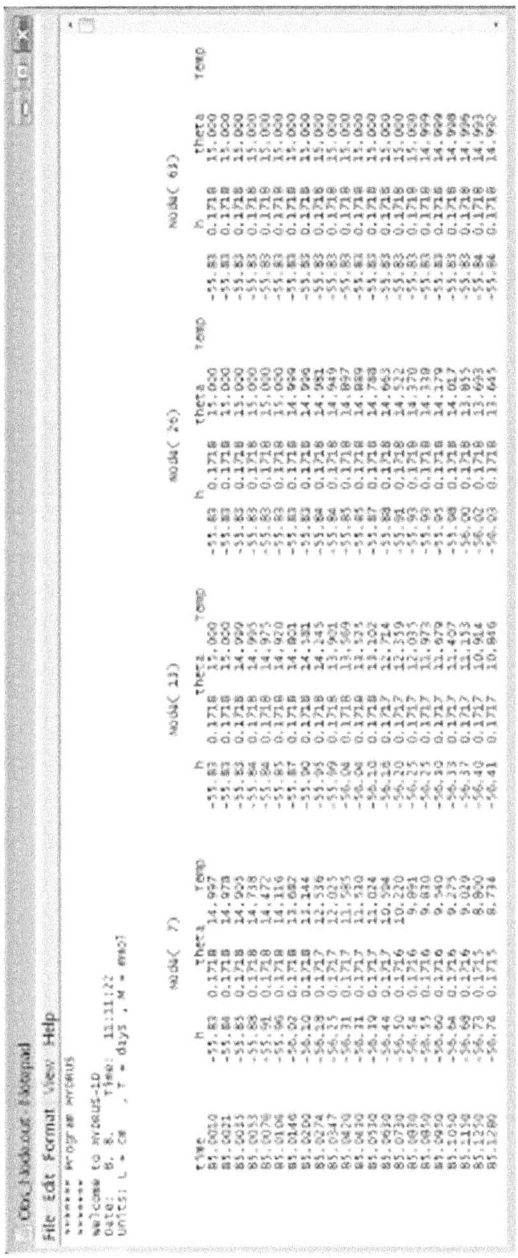

FIGURE 16.37 The output text file "OBS_NODE.OUT": Transient values of the pressure head (h), water content (theta), and temperature (Temp) at specified observation nodes (see Figure 16.27).

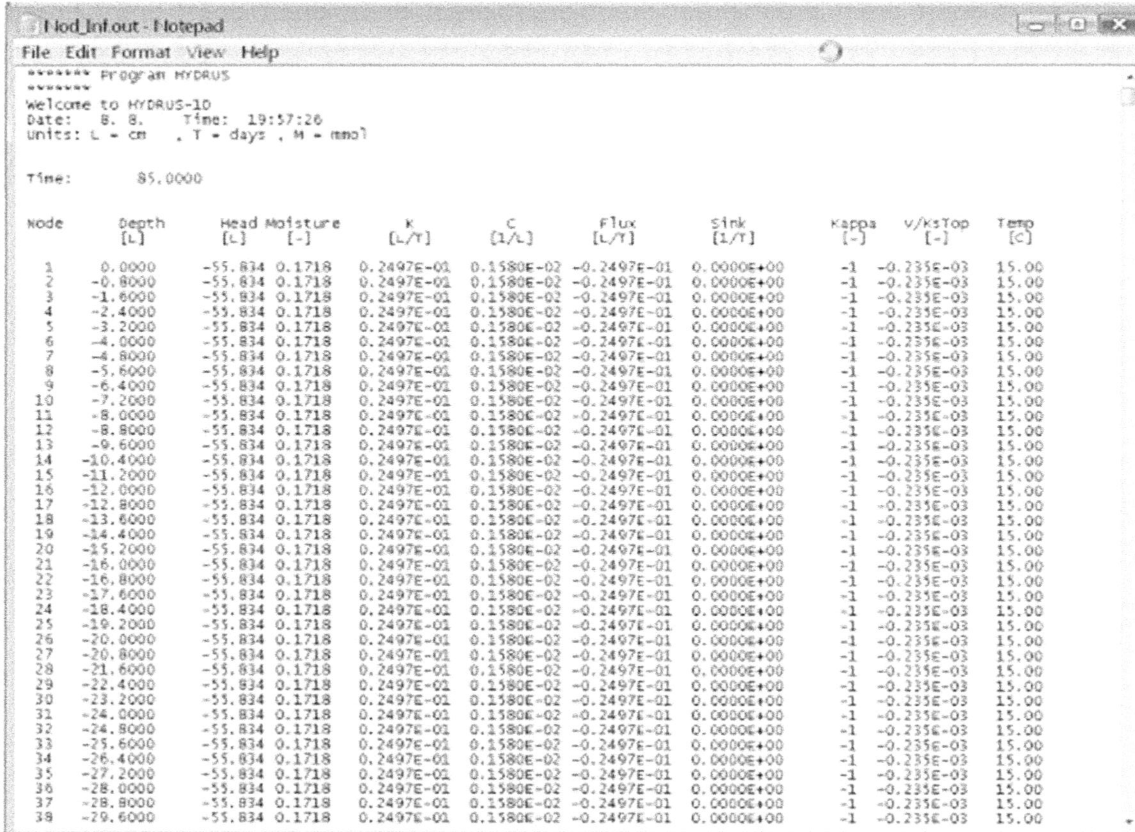

Node	Depth [L]	Head [L]	Moisture [-]	K [L/T]	C [1/L]	Flux [L/T]	Sink [1/T]	Kappa [-]	v/KsTop [-]	Temp [C]
1	0.0000	-55.834	0.1718	0.2497E-01	0.1580E-02	-0.2497E-01	0.0000E+00	-1	-0.235E-03	15.00
2	-0.8000	-55.834	0.1718	0.2497E-01	0.1580E-02	-0.2497E-01	0.0000E+00	-1	-0.235E-03	15.00
3	-1.6000	-55.834	0.1718	0.2497E-01	0.1580E-02	-0.2497E-01	0.0000E+00	-1	-0.235E-03	15.00
4	-2.4000	-55.834	0.1718	0.2497E-01	0.1580E-02	-0.2497E-01	0.0000E+00	-1	-0.235E-03	15.00
5	-3.2000	-55.834	0.1718	0.2497E-01	0.1580E-02	-0.2497E-01	0.0000E+00	-1	-0.235E-03	15.00
6	-4.0000	-55.834	0.1718	0.2497E-01	0.1580E-02	-0.2497E-01	0.0000E+00	-1	-0.235E-03	15.00
7	-4.8000	-55.834	0.1718	0.2497E-01	0.1580E-02	-0.2497E-01	0.0000E+00	-1	-0.235E-03	15.00
8	-5.6000	-55.834	0.1718	0.2497E-01	0.1580E-02	-0.2497E-01	0.0000E+00	-1	-0.235E-03	15.00
9	-6.4000	-55.834	0.1718	0.2497E-01	0.1580E-02	-0.2497E-01	0.0000E+00	-1	-0.235E-03	15.00
10	-7.2000	-55.834	0.1718	0.2497E-01	0.1580E-02	-0.2497E-01	0.0000E+00	-1	-0.235E-03	15.00
11	-8.0000	-55.834	0.1718	0.2497E-01	0.1580E-02	-0.2497E-01	0.0000E+00	-1	-0.235E-03	15.00
12	-8.8000	-55.834	0.1718	0.2497E-01	0.1580E-02	-0.2497E-01	0.0000E+00	-1	-0.235E-03	15.00
13	-9.6000	-55.834	0.1718	0.2497E-01	0.1580E-02	-0.2497E-01	0.0000E+00	-1	-0.235E-03	15.00
14	-10.4000	-55.834	0.1718	0.2497E-01	0.1580E-02	-0.2497E-01	0.0000E+00	-1	-0.235E-03	15.00
15	-11.2000	-55.834	0.1718	0.2497E-01	0.1580E-02	-0.2497E-01	0.0000E+00	-1	-0.235E-03	15.00
16	-12.0000	-55.834	0.1718	0.2497E-01	0.1580E-02	-0.2497E-01	0.0000E+00	-1	-0.235E-03	15.00
17	-12.8000	-55.834	0.1718	0.2497E-01	0.1580E-02	-0.2497E-01	0.0000E+00	-1	-0.235E-03	15.00
18	-13.6000	-55.834	0.1718	0.2497E-01	0.1580E-02	-0.2497E-01	0.0000E+00	-1	-0.235E-03	15.00
19	-14.4000	-55.834	0.1718	0.2497E-01	0.1580E-02	-0.2497E-01	0.0000E+00	-1	-0.235E-03	15.00
20	-15.2000	-55.834	0.1718	0.2497E-01	0.1580E-02	-0.2497E-01	0.0000E+00	-1	-0.235E-03	15.00
21	-16.0000	-55.834	0.1718	0.2497E-01	0.1580E-02	-0.2497E-01	0.0000E+00	-1	-0.235E-03	15.00
22	-16.8000	-55.834	0.1718	0.2497E-01	0.1580E-02	-0.2497E-01	0.0000E+00	-1	-0.235E-03	15.00
23	-17.6000	-55.834	0.1718	0.2497E-01	0.1580E-02	-0.2497E-01	0.0000E+00	-1	-0.235E-03	15.00
24	-18.4000	-55.834	0.1718	0.2497E-01	0.1580E-02	-0.2497E-01	0.0000E+00	-1	-0.235E-03	15.00
25	-19.2000	-55.834	0.1718	0.2497E-01	0.1580E-02	-0.2497E-01	0.0000E+00	-1	-0.235E-03	15.00
26	-20.0000	-55.834	0.1718	0.2497E-01	0.1580E-02	-0.2497E-01	0.0000E+00	-1	-0.235E-03	15.00
27	-20.8000	-55.834	0.1718	0.2497E-01	0.1580E-02	-0.2497E-01	0.0000E+00	-1	-0.235E-03	15.00
28	-21.6000	-55.834	0.1718	0.2497E-01	0.1580E-02	-0.2497E-01	0.0000E+00	-1	-0.235E-03	15.00
29	-22.4000	-55.834	0.1718	0.2497E-01	0.1580E-02	-0.2497E-01	0.0000E+00	-1	-0.235E-03	15.00
30	-23.2000	-55.834	0.1718	0.2497E-01	0.1580E-02	-0.2497E-01	0.0000E+00	-1	-0.235E-03	15.00
31	-24.0000	-55.834	0.1718	0.2497E-01	0.1580E-02	-0.2497E-01	0.0000E+00	-1	-0.235E-03	15.00
32	-24.8000	-55.834	0.1718	0.2497E-01	0.1580E-02	-0.2497E-01	0.0000E+00	-1	-0.235E-03	15.00
33	-25.6000	-55.834	0.1718	0.2497E-01	0.1580E-02	-0.2497E-01	0.0000E+00	-1	-0.235E-03	15.00
34	-26.4000	-55.834	0.1718	0.2497E-01	0.1580E-02	-0.2497E-01	0.0000E+00	-1	-0.235E-03	15.00
35	-27.2000	-55.834	0.1718	0.2497E-01	0.1580E-02	-0.2497E-01	0.0000E+00	-1	-0.235E-03	15.00
36	-28.0000	-55.834	0.1718	0.2497E-01	0.1580E-02	-0.2497E-01	0.0000E+00	-1	-0.235E-03	15.00
37	-28.8000	-55.834	0.1718	0.2497E-01	0.1580E-02	-0.2497E-01	0.0000E+00	-1	-0.235E-03	15.00
38	-29.6000	-55.834	0.1718	0.2497E-01	0.1580E-02	-0.2497E-01	0.0000E+00	-1	-0.235E-03	15.00

FIGURE 16.38 The output text file "NOD_INF.OUT: Profile information for selected print times (Section 16.3.3.4). Information includes nodal values of the pressure head, water content, hydraulic conductivity (K), hydraulic capacity (C), flux, "sink" or root water uptake, ratio between the local hydraulic conductivity and the saturated hydraulic conductivity at the soil surface (kappa), ratio between the local velocity and the saturated hydraulic conductivity at the soil surface (v/KsTop), and soil temperature.

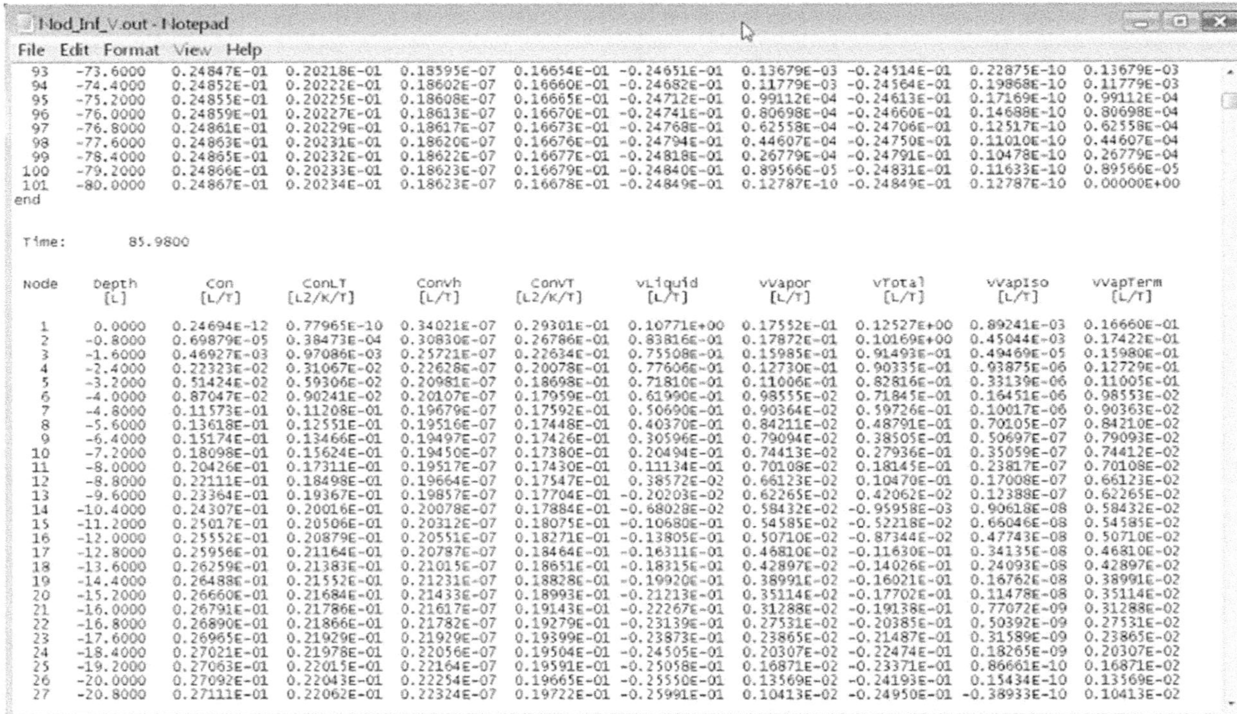

```
Nod_Inf_V.out - Notepad
File  Edit  Format  View  Help
  93   -73.6000  0.24847E-01  0.20218E-01  0.18595E-07  0.16654E-01 -0.24651E-01  0.13679E-03 -0.24514E-01  0.22875E-10  0.13679E-03
  94   -74.4000  0.24852E-01  0.20222E-01  0.18602E-07  0.16660E-01 -0.24682E-01  0.11779E-03 -0.24564E-01  0.19868E-10  0.11779E-03
  95   -75.2000  0.24855E-01  0.20225E-01  0.18608E-07  0.16665E-01 -0.24712E-01  0.99112E-04 -0.24613E-01  0.17169E-10  0.99112E-04
  96   -76.0000  0.24859E-01  0.20227E-01  0.18613E-07  0.16670E-01 -0.24741E-01  0.80698E-04 -0.24660E-01  0.14688E-10  0.80698E-04
  97   -76.8000  0.24861E-01  0.20229E-01  0.18617E-07  0.16673E-01 -0.24768E-01  0.62558E-04 -0.24706E-01  0.12517E-10  0.62558E-04
  98   -77.6000  0.24863E-01  0.20231E-01  0.18620E-07  0.16676E-01 -0.24794E-01  0.44607E-04 -0.24750E-01  0.11010E-10  0.44607E-04
  99   -78.4000  0.24865E-01  0.20232E-01  0.18622E-07  0.16677E-01 -0.24818E-01  0.26779E-04 -0.24791E-01  0.10478E-10  0.26779E-04
 100   -79.2000  0.24866E-01  0.20233E-01  0.18623E-07  0.16679E-01 -0.24840E-01  0.89566E-05 -0.24831E-01  0.11633E-10  0.89566E-05
 101   -80.0000  0.24867E-01  0.20234E-01  0.18623E-07  0.16678E-01 -0.24849E-01  0.12787E-10 -0.24849E-01  0.12787E-10  0.00000E+00
end

 Time:        85.9800

 Node     Depth        Con        ConLT        ConVh        ConVT       vLiquid       vVapor       vTotal      vVapIso      vVapTerm
           [L]        [L/T]     [L2/K/T]       [L/T]      [L2/K/T]      [L/T]        [L/T]        [L/T]        [L/T]        [L/T]

   1     0.0000  0.24694E-12  0.77965E-10  0.34021E-07  0.29301E-01  0.10771E+00  0.17552E-01  0.12527E+00  0.89241E-03  0.16660E-01
   2    -0.8000  0.69879E-05  0.38473E-04  0.30830E-07  0.26786E-01  0.83816E-01  0.17872E-01  0.10169E+00  0.45044E-03  0.17422E-01
   3    -1.6000  0.46927E-03  0.97086E-03  0.25721E-07  0.22634E-01  0.75508E-01  0.15985E-01  0.91493E-01  0.49469E-05  0.15980E-01
   4    -2.4000  0.22323E-02  0.31067E-02  0.22628E-07  0.20078E-01  0.77606E-01  0.12730E-01  0.90335E-01  0.93875E-06  0.12729E-01
   5    -3.2000  0.51424E-02  0.59306E-02  0.20981E-07  0.18698E-01  0.71810E-01  0.11006E-01  0.82816E-01  0.33139E-06  0.11005E-01
   6    -4.0000  0.87047E-02  0.90241E-02  0.20107E-07  0.17959E-01  0.61990E-01  0.98555E-02  0.71845E-01  0.16451E-06  0.98553E-02
   7    -4.8000  0.11573E-01  0.11208E-01  0.19679E-07  0.17592E-01  0.50690E-01  0.90364E-02  0.59726E-01  0.10017E-06  0.90363E-02
   8    -5.6000  0.13618E-01  0.12551E-01  0.19516E-07  0.17448E-01  0.40370E-01  0.84211E-02  0.48791E-01  0.70105E-07  0.84210E-02
   9    -6.4000  0.15174E-01  0.13466E-01  0.19497E-07  0.17426E-01  0.30596E-01  0.79094E-02  0.38505E-01  0.50697E-07  0.79093E-02
  10    -7.2000  0.18098E-01  0.15624E-01  0.19450E-07  0.17380E-01  0.20494E-01  0.74413E-02  0.27936E-01  0.35059E-07  0.74412E-02
  11    -8.0000  0.20426E-01  0.17311E-01  0.19517E-07  0.17430E-01  0.11134E-01  0.70108E-02  0.18145E-01  0.23817E-07  0.70108E-02
  12    -8.8000  0.22111E-01  0.18498E-01  0.19664E-07  0.17547E-01  0.38572E-02  0.66123E-02  0.10470E-01  0.17008E-07  0.66123E-02
  13    -9.6000  0.23364E-01  0.19367E-01  0.19857E-07  0.17704E-01 -0.20203E-02  0.62265E-02  0.42062E-02  0.12388E-07  0.62265E-02
  14   -10.4000  0.24307E-01  0.20016E-01  0.20078E-07  0.17884E-01 -0.68028E-02  0.58432E-02 -0.95958E-03  0.90618E-08  0.58432E-02
  15   -11.2000  0.25017E-01  0.20506E-01  0.20312E-07  0.18075E-01 -0.10680E-01  0.54585E-02 -0.52218E-02  0.66046E-08  0.54585E-02
  16   -12.0000  0.25552E-01  0.20879E-01  0.20551E-07  0.18271E-01 -0.13805E-01  0.50710E-02 -0.87344E-02  0.47743E-08  0.50710E-02
  17   -12.8000  0.25956E-01  0.21164E-01  0.20787E-07  0.18464E-01 -0.16311E-01  0.46810E-02 -0.11630E-01  0.34135E-08  0.46810E-02
  18   -13.6000  0.26259E-01  0.21383E-01  0.21015E-07  0.18651E-01 -0.18315E-01  0.42897E-02 -0.14026E-01  0.24093E-08  0.42897E-02
  19   -14.4000  0.26488E-01  0.21552E-01  0.21231E-07  0.18828E-01 -0.19920E-01  0.38991E-02 -0.16021E-01  0.16762E-08  0.38991E-02
  20   -15.2000  0.26660E-01  0.21684E-01  0.21433E-07  0.18993E-01 -0.21213E-01  0.35114E-02 -0.17702E-01  0.11478E-08  0.35114E-02
  21   -16.0000  0.26791E-01  0.21786E-01  0.21617E-07  0.19143E-01 -0.22267E-01  0.31288E-02 -0.19138E-01  0.77072E-09  0.31288E-02
  22   -16.8000  0.26890E-01  0.21866E-01  0.21782E-07  0.19279E-01 -0.23139E-01  0.27531E-02 -0.20385E-01  0.50392E-09  0.27531E-02
  23   -17.6000  0.26965E-01  0.21929E-01  0.21929E-07  0.19399E-01 -0.23873E-01  0.23865E-02 -0.21487E-01  0.31589E-09  0.23865E-02
  24   -18.4000  0.27021E-01  0.21978E-01  0.22056E-07  0.19504E-01 -0.24505E-01  0.20307E-02 -0.22474E-01  0.18265E-09  0.20307E-02
  25   -19.2000  0.27063E-01  0.22015E-01  0.22164E-07  0.19591E-01 -0.25058E-01  0.16871E-02 -0.23371E-01  0.86661E-10  0.16871E-02
  26   -20.0000  0.27092E-01  0.22043E-01  0.22254E-07  0.19665E-01 -0.25550E-01  0.13569E-02 -0.24193E-01  0.15434E-10  0.13569E-02
  27   -20.8000  0.27111E-01  0.22062E-01  0.22324E-07  0.19722E-01 -0.25991E-01  0.10413E-02 -0.24950E-01 -0.38933E-10  0.10413E-02
```

FIGURE 16.39 The output text file "NOD_INF_V.OUT": Nodal values of the isothermal hydraulic conductivities of liquid water and water vapor (Con and ConVh, respectively), thermal hydraulic conductivities of liquid water and water vapor (ConLT and ConVT, respectively), total liquid and vapor fluxes (vLiquid and vVapor, respectively), total water (liquid and vapor) flux (vTotal), and isothermal (vVapIso) and thermal (vVapTerm) vapor fluxes for selected print times (Section 16.3.3.4). Note that positive fluxes are directed upward.

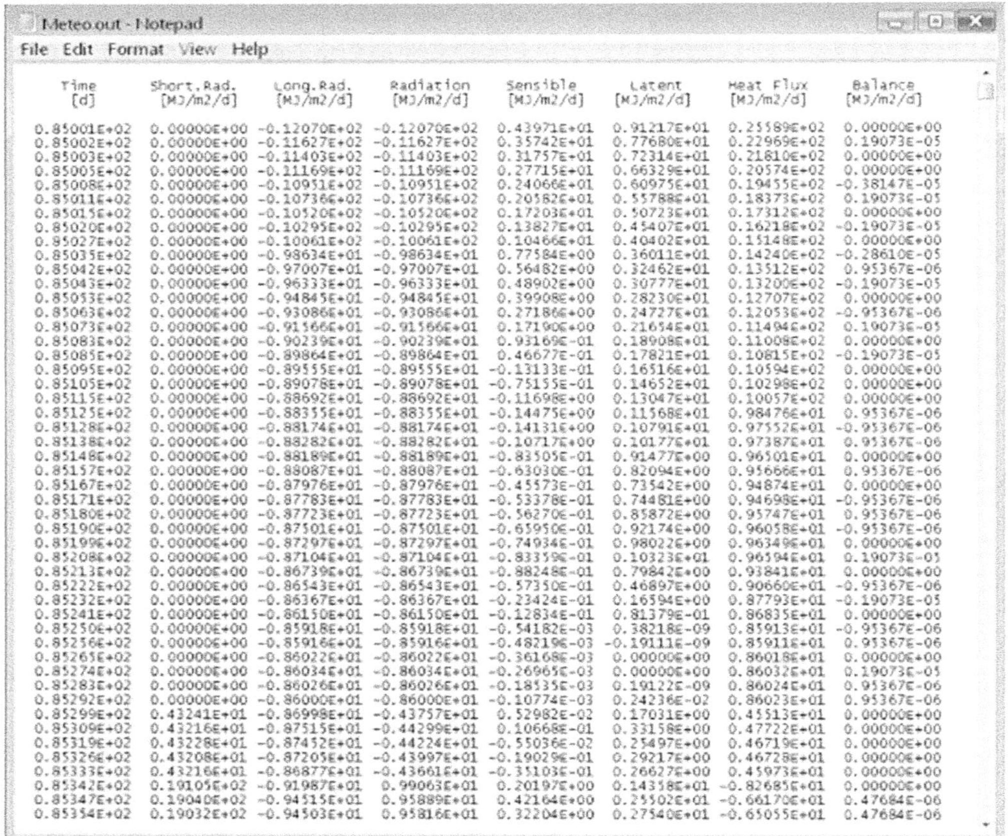

Meteo.out - Notepad

File Edit Format View Help

Time [d]	Short.Rad. [MJ/m2/d]	Long.Rad. [MJ/m2/d]	Radiation [MJ/m2/d]	Sensible [MJ/m2/d]	Latent [MJ/m2/d]	Heat Flux [MJ/m2/d]	Balance [MJ/m2/d]
0.85001E+02	0.00000E+00	-0.12070E+02	-0.12070E+02	0.43971E+01	0.91217E+01	0.25589E+02	0.00000E+00
0.85002E+02	0.00000E+00	-0.11627E+02	-0.11627E+02	0.35742E+01	0.77680E+01	0.22969E+02	0.19073E-05
0.85003E+02	0.00000E+00	-0.11403E+02	-0.11403E+02	0.31757E+01	0.72314E+01	0.21810E+02	0.00000E+00
0.85005E+02	0.00000E+00	-0.11169E+02	-0.11169E+02	0.27715E+01	0.66329E+01	0.20574E+02	0.00000E+00
0.85008E+02	0.00000E+00	-0.10951E+02	-0.10951E+02	0.24066E+01	0.60975E+01	0.19455E+02	-0.38147E-05
0.85011E+02	0.00000E+00	-0.10736E+02	-0.10736E+02	0.20582E+01	0.55788E+01	0.18373E+02	0.19073E-05
0.85015E+02	0.00000E+00	-0.10520E+02	-0.10520E+02	0.17203E+01	0.50723E+01	0.17312E+02	0.00000E+00
0.85020E+02	0.00000E+00	-0.10295E+02	-0.10295E+02	0.13827E+01	0.45407E+01	0.16218E+02	-0.19073E-05
0.85027E+02	0.00000E+00	-0.10061E+02	-0.10061E+02	0.10466E+01	0.40402E+01	0.15148E+02	0.00000E+00
0.85035E+02	0.00000E+00	-0.98634E+01	-0.98634E+01	0.77584E+00	0.36011E+01	0.14240E+02	-0.28610E-05
0.85042E+02	0.00000E+00	-0.97007E+01	-0.97007E+01	0.56482E+00	0.32462E+01	0.13512E+02	0.95367E-06
0.85043E+02	0.00000E+00	-0.96333E+01	-0.96333E+01	0.48902E+00	0.30777E+01	0.13200E+02	-0.19073E-05
0.85053E+02	0.00000E+00	-0.94845E+01	-0.94845E+01	0.39908E+00	0.28230E+01	0.12707E+02	0.00000E+00
0.85063E+02	0.00000E+00	-0.93086E+01	-0.93086E+01	0.27186E+00	0.24727E+01	0.12053E+02	-0.95367E-06
0.85073E+02	0.00000E+00	-0.91566E+01	-0.91566E+01	0.17190E+00	0.21654E+01	0.11494E+02	0.19073E-05
0.85083E+02	0.00000E+00	-0.90239E+01	-0.90239E+01	0.93169E-01	0.18908E+01	0.11008E+02	0.00000E+00
0.85085E+02	0.00000E+00	-0.89864E+01	-0.89864E+01	0.46677E-01	0.17821E+01	0.10815E+02	-0.19073E-05
0.85095E+02	0.00000E+00	-0.89555E+01	-0.89555E+01	-0.13133E-01	0.16516E+01	0.10594E+02	0.00000E+00
0.85105E+02	0.00000E+00	-0.89078E+01	-0.89078E+01	-0.75155E-01	0.14652E+01	0.10298E+02	0.00000E+00
0.85115E+02	0.00000E+00	-0.88692E+01	-0.88692E+01	-0.11698E+00	0.13047E+01	0.10057E+02	0.00000E+00
0.85125E+02	0.00000E+00	-0.88355E+01	-0.88355E+01	-0.14475E+00	0.11568E+01	0.98476E+01	0.95367E-06
0.85128E+02	0.00000E+00	-0.88174E+01	-0.88174E+01	-0.14131E+00	0.10791E+01	0.97552E+01	-0.95367E-06
0.85138E+02	0.00000E+00	-0.88282E+01	-0.88282E+01	-0.10717E+00	0.10177E+01	0.97387E+01	0.95367E-06
0.85148E+02	0.00000E+00	-0.88189E+01	-0.88189E+01	-0.83505E-01	0.91477E+00	0.96501E+01	0.00000E+00
0.85157E+02	0.00000E+00	-0.88087E+01	-0.88087E+01	-0.63030E-01	0.82094E+00	0.95666E+01	0.95367E-06
0.85167E+02	0.00000E+00	-0.87976E+01	-0.87976E+01	-0.45573E-01	0.73542E+00	0.94874E+01	0.00000E+00
0.85171E+02	0.00000E+00	-0.87783E+01	-0.87783E+01	-0.53378E-01	0.74481E+00	0.94698E+01	-0.95367E-06
0.85180E+02	0.00000E+00	-0.87723E+01	-0.87723E+01	-0.56270E-01	0.85872E+00	0.95747E+01	0.95367E-06
0.85190E+02	0.00000E+00	-0.87501E+01	-0.87501E+01	-0.65950E-01	0.92174E+00	0.96058E+01	-0.95367E-06
0.85199E+02	0.00000E+00	-0.87297E+01	-0.87297E+01	-0.74934E-01	0.98022E+00	0.96349E+01	0.00000E+00
0.85208E+02	0.00000E+00	-0.87104E+01	-0.87104E+01	-0.83359E-01	0.10323E+01	0.96594E+01	0.19073E-05
0.85213E+02	0.00000E+00	-0.86739E+01	-0.86739E+01	-0.88248E-01	0.79842E+00	0.93841E+01	0.00000E+00
0.85222E+02	0.00000E+00	-0.86543E+01	-0.86543E+01	-0.57350E-01	0.46897E+00	0.90660E+01	-0.95367E-06
0.85232E+02	0.00000E+00	-0.86367E+01	-0.86367E+01	-0.23424E-01	0.16594E+00	0.87793E+01	-0.19073E-05
0.85241E+02	0.00000E+00	-0.86150E+01	-0.86150E+01	-0.12834E-01	0.81379E-01	0.86835E+01	0.00000E+00
0.85250E+02	0.00000E+00	-0.85918E+01	-0.85918E+01	-0.54182E-03	0.38218E-09	0.85913E+01	-0.95367E-06
0.85256E+02	0.00000E+00	-0.85916E+01	-0.85916E+01	-0.48219E-03	-0.19111E-09	0.85911E+01	0.95367E-06
0.85265E+02	0.00000E+00	-0.86022E+01	-0.86022E+01	-0.36168E-03	0.00000E+00	0.86018E+01	0.00000E+00
0.85274E+02	0.00000E+00	-0.86034E+01	-0.86034E+01	-0.26965E-03	0.00000E+00	0.86032E+01	0.19073E-05
0.85283E+02	0.00000E+00	-0.86026E+01	-0.86026E+01	-0.18535E-03	0.19122E-09	0.86024E+01	0.95367E-06
0.85292E+02	0.00000E+00	-0.86000E+01	-0.86000E+01	-0.10774E-03	0.24236E-02	0.86023E+01	0.95367E-06
0.85299E+02	0.43241E+01	-0.86998E+01	-0.43757E+01	0.52982E-02	0.17031E+00	0.45513E+01	0.00000E+00
0.85309E+02	0.43216E+01	-0.87515E+01	-0.44299E+01	0.10668E-01	0.33158E+00	0.47722E+01	0.00000E+00
0.85319E+02	0.43228E+01	-0.87452E+01	-0.44224E+01	-0.55036E-02	0.25497E+00	0.46719E+01	0.00000E+00
0.85326E+02	0.43208E+01	-0.87205E+01	-0.43997E+01	-0.19029E-01	0.29217E+00	0.46728E+01	0.00000E+00
0.85333E+02	0.43216E+01	-0.86877E+01	-0.43661E+01	-0.35103E-01	0.26627E+00	0.45973E+01	0.00000E+00
0.85342E+02	0.19105E+02	-0.91987E+01	0.99063E+01	0.20197E+00	0.14358E+01	-0.82685E+01	0.00000E+00
0.85347E+02	0.19040E+02	-0.94515E+01	0.95889E+01	0.42164E+00	0.25502E+01	-0.66170E+01	0.47684E-06
0.85354E+02	0.19032E+02	-0.94503E+01	0.95816E+01	0.32204E+00	0.27540E+01	-0.65055E+01	0.47684E-06

FIGURE 16.40 The output text file "Meteo.out": Transient values of energy fluxes such as shortwave and longwave radiation, net solar radiation, sensible heat flux, latent heat flux, and surface heat flux. Note that sensible heat, latent heat, and surface heat fluxes are positive upward. Positive (downward) and negative (upward) values of radiation indicate incoming radiation to the soil and outgoing radiation from the soil surface, respectively.

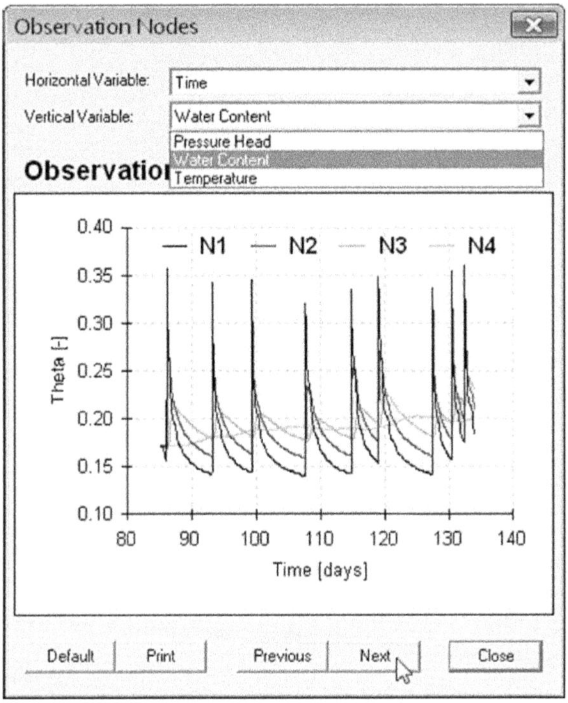

FIGURE 16.41 Observation Points graph option in the Post-processing pane: Water content as a function of time at specified observation nodes.

is too big to be displayed entirely! Automatic selection has been made" appears, click on the OK button.

As shown in Figure 16.41, users can choose a variable (such as water content, pressure head, and temperature) from the drop-down list to display the selected variable as a function of time at each of the four specified observation nodes (Figure 16.27). The graph displayed for Observation Nodes uses the information from the "OBS_NODE.OUT" output file (Figure 16.37). To evaluate the model performance, users can compare HYDRUS-1D-simulated water contents and soil temperatures at different observation points (Figures 16.37 and 16.41) with the measured data at different soil depths. The "selected variable" will appear along the vertical axis, while "time" is used along the horizontal axis.

- In general, the graphical interface also provides different options to customize the on-screen display, including Copy, Paste, Print, and Save As options. Right-click on the graph, which will provide context or pop-up menu options for the graph (Figure 16.42). Moreover, double-clicking on each of the line or curve (series) of the graph, horizontal/vertical axis of the graph, horizontal/vertical axis title of the graph, horizontal/vertical axis

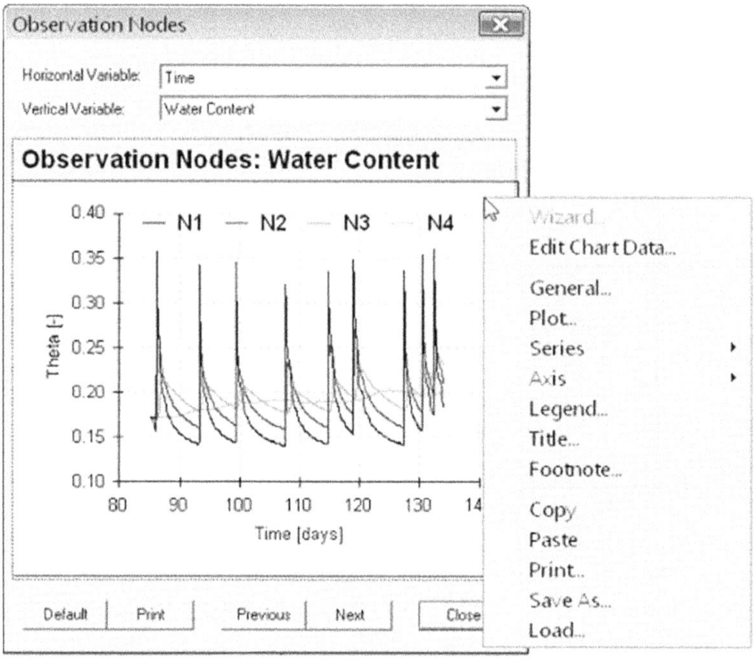

FIGURE 16.42 Observation Points graph option in the Post-processing pane: Context or pop-up menus (right-click menu) for the graph.

label of the graph, and graph title will bring up the respective dialog window with the graph format options.

- Next, either double-click on Profile Informations in the Post-processing pane or click Results | Profile Informations | Basic Information, which will display the graph for basic profile information such as pressure head, water content, hydraulic conductivity, hydraulic capacity, water flux, and temperature (Figure 16.43).
- If an alert message box "Too many curves to display" appears, click on the OK button.

Users can select a variable from the drop-down list to display the selected variable (appears along the horizontal axis) as a function of depth (vertical axis) (Figure 16.43). Information displayed in the Profile Information graph option (Figure 16.43) is read from the "NOD_INF.OUT" output file (Figure 16.38). Note that information in the "NOD_INF_V.OUT" output file for the selected vapor flow option (Figure 16.39) is not visible using the graphical interface of HYDRUS-1D (version 4.17.0140). The "NOD_INF_V.OUT" output file (Figure 16.39) provides the vertical distributions of isothermal and thermal liquid water and water vapor fluxes in the vadose zone at each print time (see Section 16.3.3.4). HYDRUS-1D-simulated vertical profiles of liquid water and water vapor fluxes for both isothermal and thermal components are

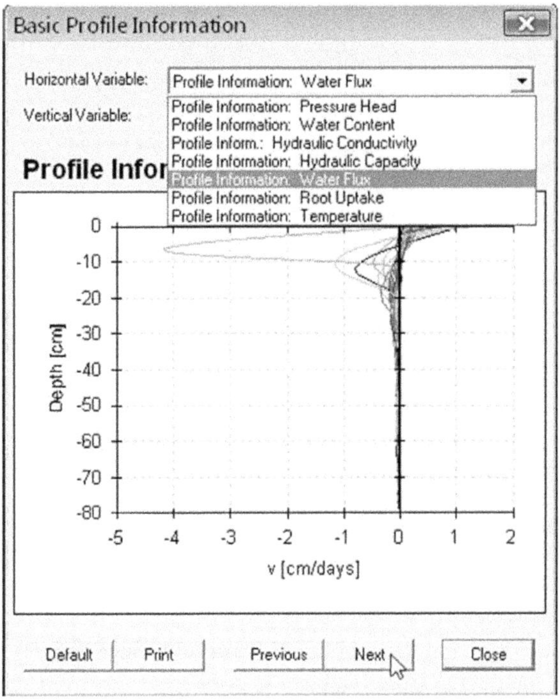

FIGURE 16.43 Profile Informations graph option in the Post-processing pane: Water flux as a function of depth. Note that positive fluxes are directed upward, while the negative sign of the vertical axis (y-axis) indicates depth below the soil surface.

shown in Figure 16.44 to demonstrate the relative importance of various transport mechanisms associated with dry and wet soil conditions that affect total water flux (sum of the total isothermal and thermal liquid water and water vapor fluxes) in the unsaturated zone at a sandy loam field (Deb et al., 2011).

As shown in Figure 16.44, the downward (negative) isothermal liquid water flux dominates the soil water movement during early periods after irrigation. There appears an upward thermal liquid water flux above 15 cm soil depth because of temperature gradients as the surface cools due to higher latent heat fluxes during the early periods after irrigation (discussed later in this Section 16.3.4.2). The contribution of the upward (positive) vapor flux increases with increasing soil drying before and after irrigation. Both isothermal and thermal liquid water and thermal water vapor fluxes are responsible for water vapor near the soil surface. During the progressive soil drying process, the source of upward water vapor flux is also the liquid water transported by the upward isothermal and thermal liquid water fluxes within 15 cm depth (Figure 16.44). Isothermal and thermal liquid water and thermal vapor fluxes decrease from 5 cm depth, indicating that the magnitude of water vapor flux is much higher in the layer near the soil surface. At a soil depth of ~2 cm, the upward water vapor flux starts dominating the upward total water flux (sum of the

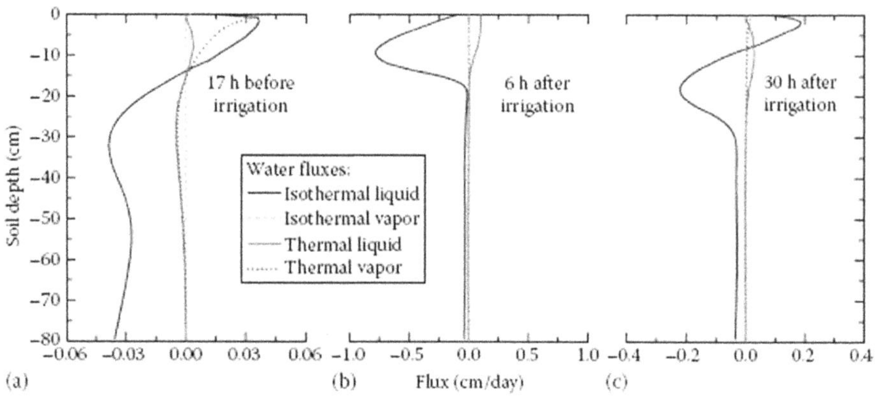

FIGURE 16.44 Simulated vertical distributions of the isothermal and thermal liquid water and water vapor fluxes in the unsaturated zone at a sandy loam field. (a) Water fluxes at 17 h before, and (b) water fluxes at 6 h and (c) 30 h after an irrigation event. Positive and negative values of abscissa (*x*-axis) indicate upward and downward fluxes, respectively, whereas the negative sign of the vertical axis (*y*-axis) indicates depth below the soil surface. (Redrawn from Deb, S. K., et al., *Soil Sci.*, 176, 395, 2011.)

upward total isothermal and thermal liquid water and water vapor fluxes), which can be referred to as the drying front (Deb et al., 2011).

- To display the graphs of temporal changes in actual and cumulative water fluxes and pressure heads across the upper and lower boundaries, either double-click on Water Flow–Boundary Fluxes and Heads in the Post-processing pane or click Results | T-Level Information | Water Flow, which will graphically display Boundary Water Fluxes and Pressure Heads (Figure 16.45).

Information displayed in the graph Boundary Water Fluxes and Pressure Heads (Figure 16.45) is read from the "T_LEVEL.OUT" output file (Figure 16.36). The selected variable will appear along the vertical axis, while time is used along the horizontal axis (Figure 16.45).

Users can view the outputs of the unsaturated soil hydraulic properties (Figure 16.46).

- Either double-click on Soil Hydraulic Properties in the Post-processing pane or click Results | Soil Hydraulic Properties (Figure 16.46).

Users can choose several combinations of dependent (vertical axis) and independent (horizontal axis) variables from the drop-down lists. The pressure head, logarithm of the pressure head, or water content can be selected as the independent variable, while the water content, hydraulic capacity, hydraulic conductivity, logarithm of the hydraulic conductivity, effective water content, pressure head, and logarithm of the pressure head can be selected as the dependent variable.

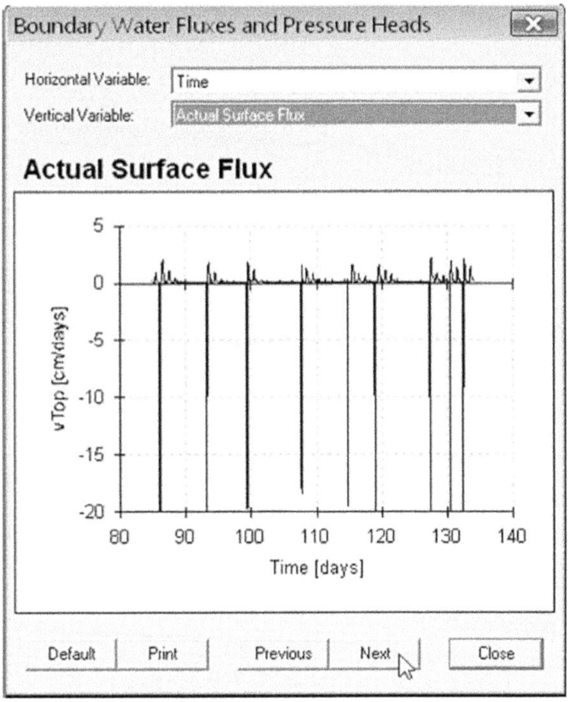

FIGURE 16.45 Boundary Water Fluxes and Pressure Heads graph option in the Post-processing pane: Actual surface flux as a function of time. Note that positive and negative values of actual surface flux indicate evaporation and infiltration, respectively.

Run Time Information (click Results | Run Time Information) in the Post-processing pane displays graphs of temporal changes in the time step, number of iterations necessary to solve the Richards equation at a particular time level, and cumulative number of iterations. Users can plot these variables against either time or time-step number. Note that information displayed in the Run Time Information graph option is read from the "RUN_INF.OUT" output file (Figure 16.36).

- Users can view Mass Balance Information in the Post-processing pane (or click Results | Mass Balance Information), which is an ASCII file containing mass balance information such as the volume of water inflow and outflow rates, mean pressure head and mean temperature in the entire flow domain, actual surface flux (infiltration/evaporation), and actual flux across the bottom of the soil profile (Figure 16.47).

Note that the soil profile is homogeneous/uniform for the tutorial project, and therefore, the mass balance information is presented for the entire flow domain. Absolute and relative errors in the water mass balance of the entire flow domain are also

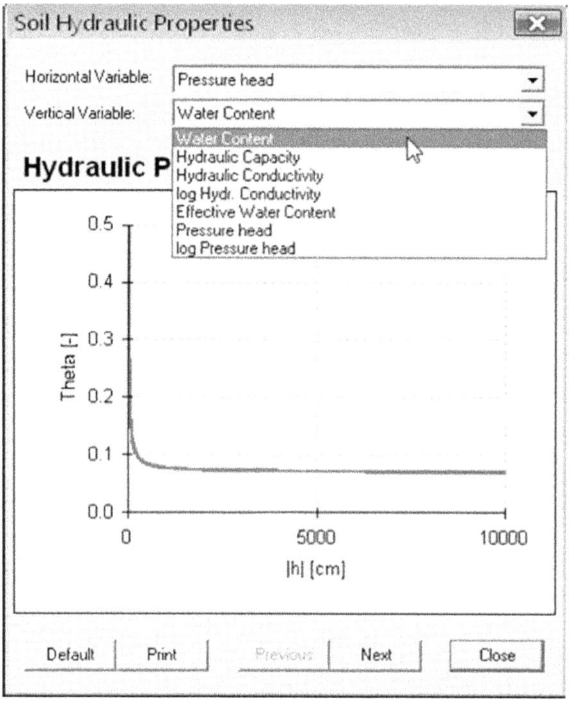

FIGURE 16.46 Soil Hydraulic Properties graph option in the Post-processing pane: Soil water retention curve, i.e., water content as a function of pressure head.

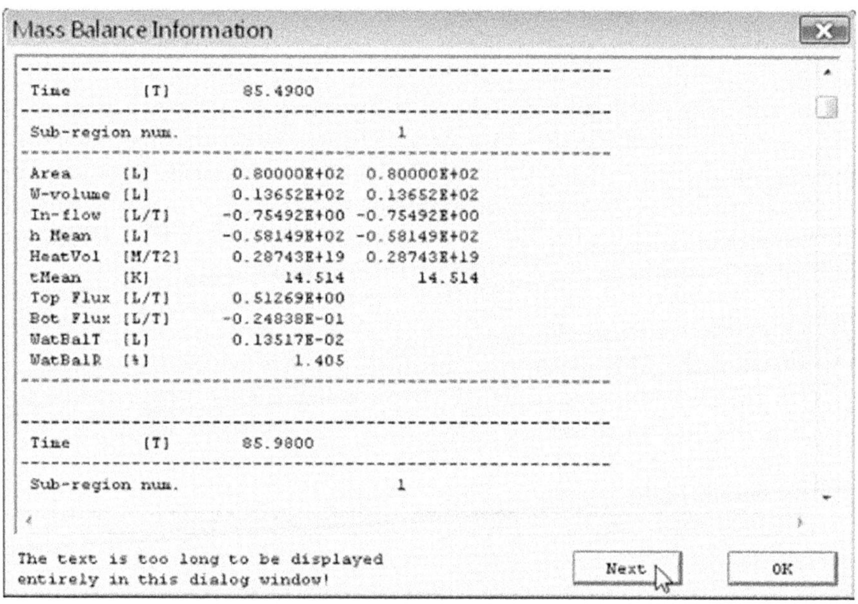

FIGURE 16.47 Mass Balance Information in the Post-processing pane.

given. Information displayed in the Mass Balance Information option in the Post-processing pane is read from the "BALANCE.OUT" output file (Figure 16.36).

- The Meteorological Information option in the Post-processing pane (or click Results | Meteorological Information) (Figure 16.48) uses information from the "Meteo.out" output file (Figure 16.40), which displays graphs of temporal changes in variables in the surface energy balance equation (i.e., energy fluxes at the bare soil surface) such as shortwave and longwave radiation, net solar radiation, sensible heat flux, latent heat flux, and surface heat flux (Figure 16.48).
- Users can select a variable from the drop-down list to display the selected variable (appears along the vertical axis) as a function of time (horizontal axis) (Figure 16.48).

The "Meteo.out" output file (Figure 16.40) provides the dynamics of surface energy components and their contributions due to heat conduction, convection, and vapor transfer at the soil–atmosphere interface for evaluating the surface water and energy balance. As a substantial part of net radiation causing evaporation during early

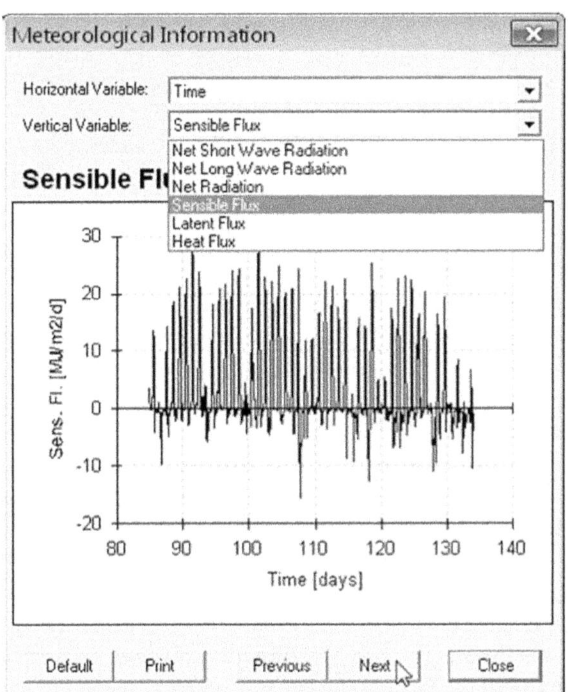

FIGURE 16.48 Meteorological Information graph option in the Post-processing pane: Transient values of energy fluxes at the soil surface, such as sensible heat flux, as a function of time. Note that sensible heat fluxes are positive upward.

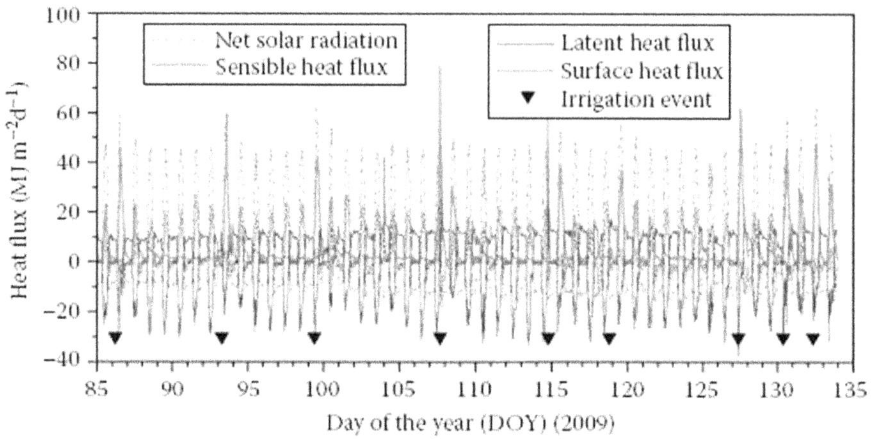

FIGURE 16.49 Temporal variations in the HYDRUS-1D-simulated net radiation, sensible heat flux, latent heat flux, and surface heat flux at a sandy loam field during the calibration and validation periods from DOY 85 (March 26, 2009) to DOY 134 (May 14, 2009). All fluxes are positive upward and negative when energy is transferred downward into the soil. However, positive (downward) and negative (upward) values of net radiation indicate incoming radiation to the soil and outgoing radiation from the soil surface, respectively. (Redrawn from Deb, S. K. et al., *Soil Sci.*, 176, 393, 2011.)

periods after an irrigation event, the latent heat flux becomes a dominant component of the surface energy balance in a sandy loam field (Figure 16.49; Deb et al., 2011). The sensible heat flux is considerably reduced and becomes negative (downward) due to advective effects. The surface heat flux is also reduced because of the increased evaporation from the surface and also because of the increased heat capacity and thermal conductivity of the soil and becomes negative (downward) as the surface is cooler than the underneath soil. When surface heat and sensible heat fluxes are much smaller in magnitude and downward (negative), the latent heat fluxes even exceeds the net radiation received at the surface, particularly during the early periods of several irrigation events (Figure 16.49).

In addition to the response of irrigation, the direction of the surface heat flux, sensible heat flux, and latent heat flux vary depending on the time of day (Figure 16.49). The daytime net radiation is relatively constant during the entire simulation period and slightly larger during the early periods after irrigation because of the reduced albedo as well as the increased absorption of shortwave radiation by the wet surface. The magnitudes of surface heat flux to or from the soil surface differ widely between day and night, and the direction reverses during the morning and evening transition periods. During daytime, especially when the soil has dried, the evaporation rate and the magnitude of upward latent heat flux decrease markedly with soil surface warming and drying. While the daytime surface temperature and upward sensible heat flux increase (positive), the upward latent heat component during the night increases, which can be explained by the lower energy available for evaporation and condensation due to the cooler surface layer increasing the liquid water content at the surface. At night, the surface loses energy by outgoing radiation that can most likely be compensated by gains of heat from the atmosphere and soil as well as from the latent heat of condensation. The surface cooling at night causes a downward (negative) sensible heat flux. Readers are referred to Saito et al. (2006) and Deb et al. (2011) for details about the simulation of coupled liquid water, water vapor, and heat transport in the vadose zone using the HYDRUS-1D model.

PROBLEMS

16.1 Write a two-page summary on the capabilities of the HYDRUS model.
16.2 Write a note on the data required for conducting a water balance using the HYDRUS-1D model.
16.3 Run the HYDRUS-1D model using any available dataset for different soil types and compare infiltration and fluxes at the bottom boundary.
16.4 Collect published literature on the HYDRUS model and list newer developments not covered in this chapter.
16.5 Explain different sink/source terms and boundary conditions.

REFERENCES

Deb, S. K., M. K. Shukla, P. Sharma, and J. G. Mexal. 2011. Coupled liquid water, water vapor, and heat transport simulations in an unsaturated zone of a sandy loam field. *Soil Sci.* 176: 387–398.

Radcliffe, D. E. and J. Šimůnek. 2010. *Soil Physics with HYDRUS Modeling and Applications.* CRC Press, Boca Raton, Fl, p. 373.

Saito, H., J. Šimůnek, and B. P. Mohanty. 2006. Numerical analysis of coupled water, vapor, and heat transport in the vadose zone. *Vadose Zone J.* 5: 784–800.

Šimůnek, J., M. Th. van Genuchten, and M. Šejna. 2008. Development and applications of HYDRUS and STANMOD software packages, and related codes. *Vadose Zone J.* 7: 587–600.

Šimůnek, J., M. Šejna, H. Saito, M. Sakai, and M. Th. van Genuchten. 2012. *The HYDRUS-1D Software Package for Simulating the One-dimensional Movement of Water, Heat, and Multiple Solutes in Variably-saturated Media.* Version 4.15. Department of Environmental Sciences, University of California, Riverside, CA.

van Genuchten, M. Th., F. J. Leij, and S. R. Yates. 1991. *The RETC Code for Quantifying the Hydraulic Functions of Unsaturated Soils.* Version 1.0. EPA/600/2-91/065. U.S. Salinity Lab., Riverside, CA.

17 Flow through the Vadose Zone Using RZWQM

17.1 INTRODUCTION

The purpose of this chapter is to give readers hands-on experience with the Root Zone Water Quality Model (RZWQM2) (Ahuja et al., 2000). This chapter has been written using published data (Deb et al., 2012) and available RZWQM2 documentation. The model application is explained using an example project (or scenario). Information in greater detail is also available in the RZWQM2 User Guide in the model's installation directory (see Section 17.2).

RZWQM2 is a one-dimensional, integrated physical, chemical, and biological process model that simulates crop growth and movement of water, nutrients, and agrochemicals over, within, and below the root zone of a unit area of an agricultural cropping system. RZWQM2 best describes areas where rainfall, soil, and crop conditions are uniform. As the model may need to run for different conditions in heterogeneous situations to characterize the combination, the model operates on a daily time step, whereas hydrological processes and chemical movement in the soil are simulated at subhourly time steps (e.g., 5–60 min).

As a system model, RZWQM2 consists of seven main components or modules aiming to describe a complete agricultural system: water balance, heat and chemical transport, nutrient processes (carbon and nitrogen), plant growth processes, soil chemical processes, evapotranspiration processes, and pesticide processes. A complete description of the RZWQM2 components can be found in Ahuja et al. (2000). A detailed description of the RZWQM2 model parameterization and calibration is given by Ma et al. (2011).

The soil water module of RZWQM2 calculates water balance using the modified form of the Green–Ampt equation during infiltration, while the redistribution of soil water in the soil matrix (between rainfall or irrigation events) is simulated by numerical solution of the Richards equation (see Chapters 9 and 10). RZWQM2 uses a modified version of the Brooks–Corey (BC) relationship to calculate the unsaturated hydraulic conductivity. Plant water uptake is treated as a sink term in the Richards equation. RZWQM2 uses a modified Shuttleworth–Wallace equation to estimate daily or hourly potential evapotranspiration (PET).

The soil nutrient module simulates soil carbon and nitrogen dynamics in conjunction with soil microbial growth. RZWQM2 has a generic plant growth module that can be calibrated for any annual crop, although it has been used mostly for corn, soybean, and wheat. The model also consists of the DSSAT (Decision Support System for Agrotechnology Transfer) version 4.0 crop growth, Quickplant, Quickturf, and Quicktree modules. The pesticide module simulates pesticide movement in the soil,

DOI: 10.1201/9780429264849-17

as well as pesticide uptake by the plant. The RZWQM2 management options include planting and harvesting of crops, crop rotations, tillage, irrigation, fertilization, and pesticide applications.

17.2 INSTALLATION OF THE RZWQM2 MODEL

RZWQM2 (version 4.2 released in 2020) is Windows 7 or later compatible. Although the system can operate in lower resolutions, the display resolution of 1024 × 768 pixels is recommended for all entry screens, plot windows, and output viewer of RZWQM2 to appear as expected. The latest version of the RZWQM2 software can be downloaded (requires registration) from the Agriculture Research Service of the US Department of Agriculture: https://www.ars.usda.gov/research/software/download/?softwareid=412&modecode=30-12-30-25.

- Save the installation file "rzwqmsetup.exe" to your computer.
- Note that the previous version of RZWQM2 software must be uninstalled before installing the latest version of RZWQM2.

To install, run/double-click on the installation file "rzwqmsetup.exe," and follow the instructions of the InstallShield Program. A "What's New," detailing the latest features, will display at the start. Although users may also specify the directory path for the installation, the RZWQM2 default directory (i.e., C:\RZWQM2) is recommended. Follow the installation wizard instructions by clicking on the Next button to advance through the installation. Application installation is complete when Finish is chosen. The user guide for the RZWQM2 "RZWQM2userguide.pdf" is located in the RZWQM2 default directory (i.e., C:\RZWQM2\documentation).

17.3 RZWQM2 TUTORIAL

The tutorial example describes the simulation of deep percolation in a mature pecan orchard using RZWQM2 (version 2.10.2010). For this pecan orchard, deep percolation was defined as the net amount of water percolating below a soil depth of 100 cm because no roots were found below this soil depth (Deb et al., 2012). Soil texture at the orchard was a sandy loam above 40 cm soil depth and sand below it. The orchard was flood-irrigated using both surface water and groundwater Fourteen irrigations were applied to the orchard during the period from March 22 to December 31, 2009.

Meteorological input variables into RZWQM2, including daily minimum and maximum air temperature, wind speed, shortwave radiation, and relative humidity, were measured. The breakpoint rainfall (amount of rainfall where rainfall intensity changes) data were determined from continuous hourly rainfall records. Time domain reflectometry (TDR) sensors were installed at soil depths of 5, 10, 20, 40, 60, 80, and 120 cm to continuously monitor volumetric soil water contents. A 100 cm profile is used that consists of seven soil layers at depth intervals of 0–5, 5–12, 12–20, 20–40, 40–60, 60–80, and 80–99 (i.e., 80–100) cm to coincide with

TDR sensor measurements as well as to consider differences in measured saturated hydraulic conductivity (K_{sat}).

Before we begin the tutorial project, it is essential to be familiar with the RZWQM2 interface. The Windows-based interface of RZWQM2 facilitates menus, submenus, and toolbars; input data management; and user interaction with the simulation model through dialogs, spreadsheets, wizards, help, and output visualization.

17.3.1 Description of the RZWQM2 Workspace

In RZWQM2, the project window opens the default project ("projects") or the last opened project in the Project view (Figure 17.1). A scenario (i.e., New Scenario, New Scenario 2) is a particular treatment (site-specific resources and management) under your project.

The project directory contains all analysis output files for each scenario (Analysis subdirectory), meteorology files (Meteorology subdirectory), and individual scenarios (subdirectory for each named scenario, for example, New Scenario and New Scenario 2 as shown in Figures 17.1 and 17.2). A project may have one or many scenarios (subdirectories). The "Project.rzp" file in the project directory will be recognized by RZWQM2 to open the directory as a project (Figure 17.2). "Project.rzp" should be opened from the RZWQM2 user interface (Figure 17.1) but not from inside

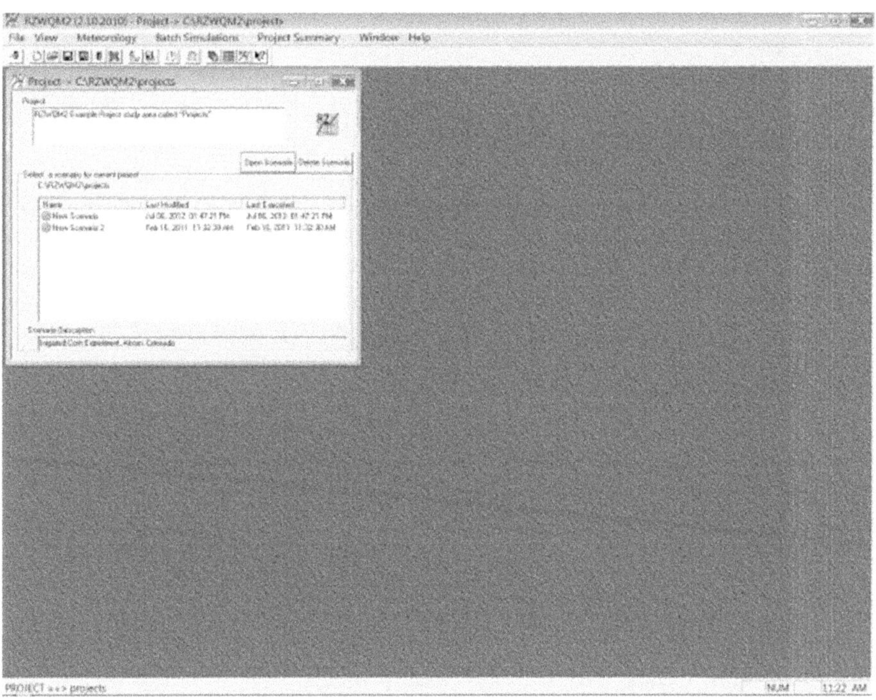

FIGURE 17.1 The RZWQM2 interface: Project view.

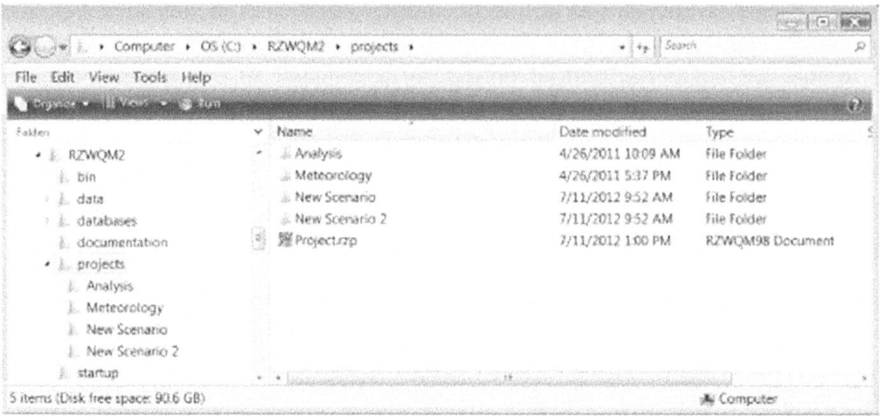

FIGURE 17.2 The RZWQM2 project directory.

a project directory (Figure 17.2). Projects are independent directories of the application and should be located outside the install directory or it may be on any drive.

There are two menus in the RZWQM2 interface: the Project menu and the Scenario menu. The menus provide access to all the commands available when in the Project or Scenario window. The most commonly used menu items are also furnished as toolbar icons in the RZWQM2 interface.

17.3.1.1 Project Menu

The project stores all your simulation information. As shown in Figure 17.3, the File menu contains project file commands: New Project, Open Project, Save Project, Save Project As, Close Project, Delete Project, and Exit the RZWQM2 application. The View menu toggles the status bar. The Meteorology menu contains Generate (using CLIGEN90 weather generator), Edit/Create user-defined data files (*.brk and *.met), and Create A Breakpoint File (*.brk) From Daily/Hourly Rainfall. The Batch Simulations menu allows the execution of multiple scenarios. The Project Summary menu allows analysis and scenario comparisons. There are also menus for Window arrangement and Help.

17.3.1.2 Scenario Menu

Selecting a scenario (e.g., New Scenario or New Scenario 2) and clicking on the Open Scenario button (or selecting New Scenario or New Scenario 2, and opening with the right-click context menu) will result in the scenario window, scenario menus, and toolbars (Figures 17.4 and 17.5). Model input and output files are stored within the Scenario subdirectory.

FIGURE 17.3 Project menus and toolbars.

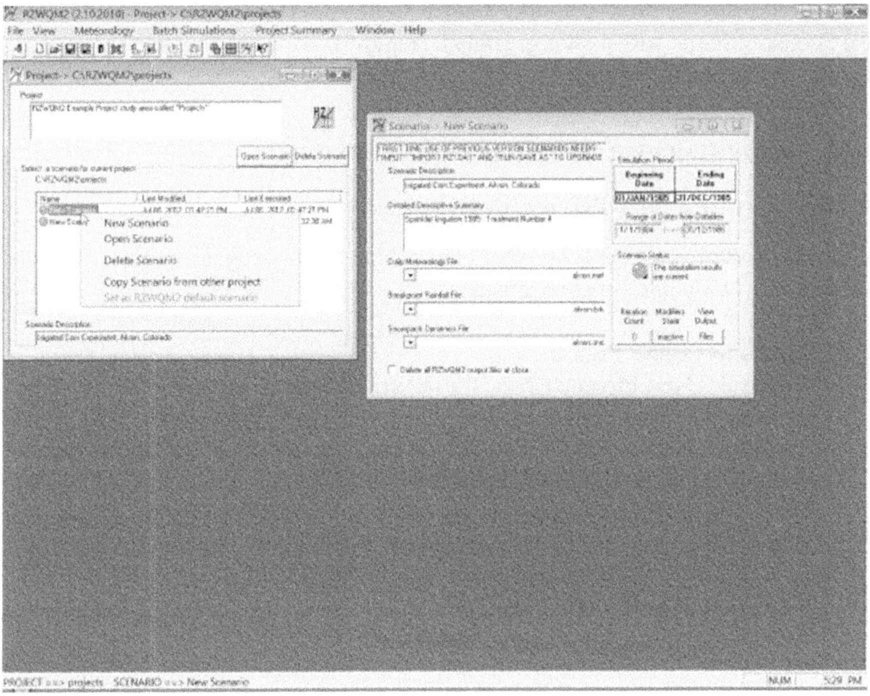

FIGURE 17.4 The RZWQM2 interface: Scenario view.

FIGURE 17.5 Scenario menus and toolbars.

The File menu contains commands specific to Scenario: New, Save, Save As, Exit/Save, and Exit/DO NOT Save. The Input menu includes submenus for user-defined inputs: Site Description, Initial State, Residue State, Management Practices, PRMS Snow Parameters (default file). The Input menu also includes optional entering of experimental data and importing of scenario data (i.e., Import RZWQM*.DAT Files submenu). The Simulation Controls menu accesses simulation output variables, breakthrough curve nodes, debugging output selection, and simulation iteration control. Modifiers are the interface with the capability of modifying weather information and management practices to conduct a what-if analysis of a particular agricultural system. The Run Scenario and Output menu items access the running scenario (simulation), and viewing output graphs and files, respectively. The Scenario Status in Figure 17.4 will be displayed as a red ball to indicate that input data have been changed in the interface and the output files need to be updated, that is, New Scenario or New Scenario 2 needs to be run. Upon completion of the run, the Scenario Status will show a green ball (Figure 17.4), indicating that the output is current to the input provided.

17.4 WORKING WITH THE RZWQM2 PROJECT

17.4.1 Create a New Project and Scenario for the Tutorial Project

From the desktop, execute RZWQM2. The program will open the default "Projects" subdirectory with an example Scenario.

- Click File | Close Project.
- Click File | New Project. Create the project with an example scenario (check the radio button Create project with example scenario) (Figure 17.6). The default scenario is copied from its location in the RZWQM2 directory (generally located in C:\RZWQM2\startup) to your new project. The Create blank project option is not recommended, as it will be completely empty of required data as it creates just the project subdirectory structure. As shown in Figure 17.6, fill in the project Name "RZWQM2-Tutorial" and enter the project Description "Deep Percolation in a Sandy Loam Pecan Orchard," which will always show in the project view.

Note that project file names can include interior space, upper and lowercase letters, and digits 0 to 9. Do not use /\, :, ?, *, ", ', <, >, |, (,), {, }, [,], or any other special characters in a filename. Enter or browse to select the location (i.e., C:\) where you

FIGURE 17.6 Screen for creating a new project.

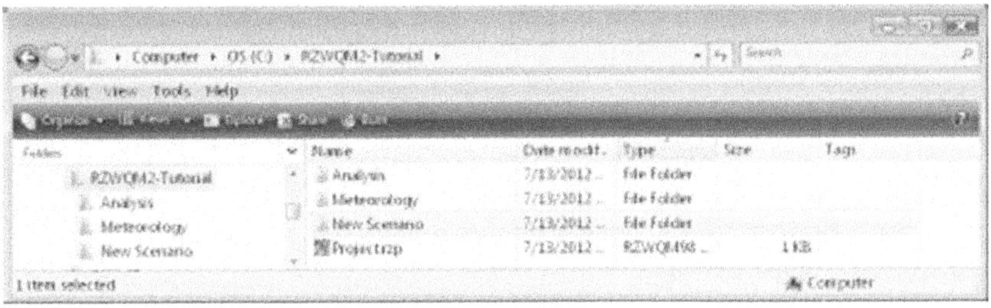

FIGURE 17.7 The directory (C:\RZWQM2-Tutorial) for the RZWQM2 tutorial project.

intend to save the project. Do not create a project under an existing RZWQM2 project directory (i.e., C:\RZWQM2\projects).

- Click OK to complete. It will ask you to create "C:\RZWQM2-Tutorial."
- Click File | Save Project.

The default scenario New Scenario is copied from the RZWQM2 directory (C:\RZWQM2Startup) to C:\RZWQM2-Tutorial (i.e., New Scenario subdirectory in the C:\RZWQM2-Tutorial directory; Figure 17.7). The New Scenario subdirectory can be renamed first by closing the RZWQM2 program, and then by using Windows Explorer (i.e., right-click on the New Scenario subdirectory and then click on the context menu Rename). Note that you must not rename Analysis and Meteorology subdirectories (Figure 17.7).

For this tutorial, we will keep the subdirectory name (New Scenario) unchanged.

- Open "Project.rzp" (C:\RZWQM2-Tutorial\Project.rzp) from the RZWQM2 user interface.
- Select and open New Scenario (Figure 17.8).

New Scenario will show default files in the Daily Meteorology File (*. met), Breakpoint Rainfall File (*.brk), and Snowpack Dynamics File (*.SNO) boxes (Figure 17.8). Accordingly, default beginning/ending dates will be shown under the box labeled Simulation Period. We will create a daily meteorological file ("LamanchaMETmardec09.met"), breakpoint rainfall file ("LamanchaMETmardec09hrly.brk"), and a snow file ("RZWQM2MetData.SNO") for the tutorial project RZWQM2-Tutorial.

- Close the New Scenario window to return to the project (i.e., RZWQM2-Tutorial) window.

17.4.2 Create/Edit Daily Meteorology File for the Tutorial Project

- Click Meteorology | Edit/Create.

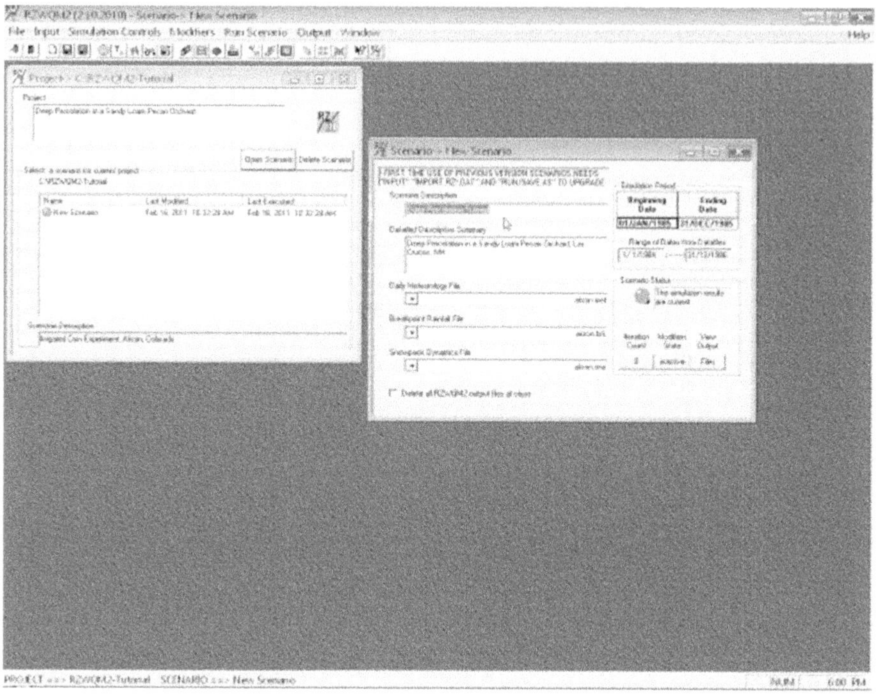

FIGURE 17.8 Default daily meteorology, breakpoint rainfall, and snowpack dynamics files on the New Scenario window.

To create a daily meteorology file (*.met), select "RZWQM2 Daily Meteorology Datafiles (*.met)" from the Files of type drop-down list and enter the new name "LamanchaMETmardec09" in the File name field (Figure 17.9).

- Click on the Open File button (Figure 17.9), which will result in the next screen for manually entering Beginning Date (March 22, 2009) and Ending Date (December 31, 2009) of daily climate (Figure 17.10).
- You can also double-click on the Beginning Date/Ending Date cells to bring up a calendar to set dates. Then, click on the Continue button, which will bring up the next screen to enter daily meteorology data in the fields (Figure 17.11).

For this tutorial example, daily meteorological data (e.g., minimum and maximum air temperatures, wind speed, shortwave radiation, relative humidity, and rainfall) during the period from March 22 to December 31, 2009, can be downloaded from the weather station Fabian Garcia RC of the New Mexico Climate Center at http://weather.nmsu.edu. Daily meteorological data can be prepared in a Microsoft Excel spreadsheet, and then:

- Copy daily meteorological data from the Excel spreadsheet (e.g., "Daily Meteorological Data.xlsx") and paste in the fields (Figure 17.11).

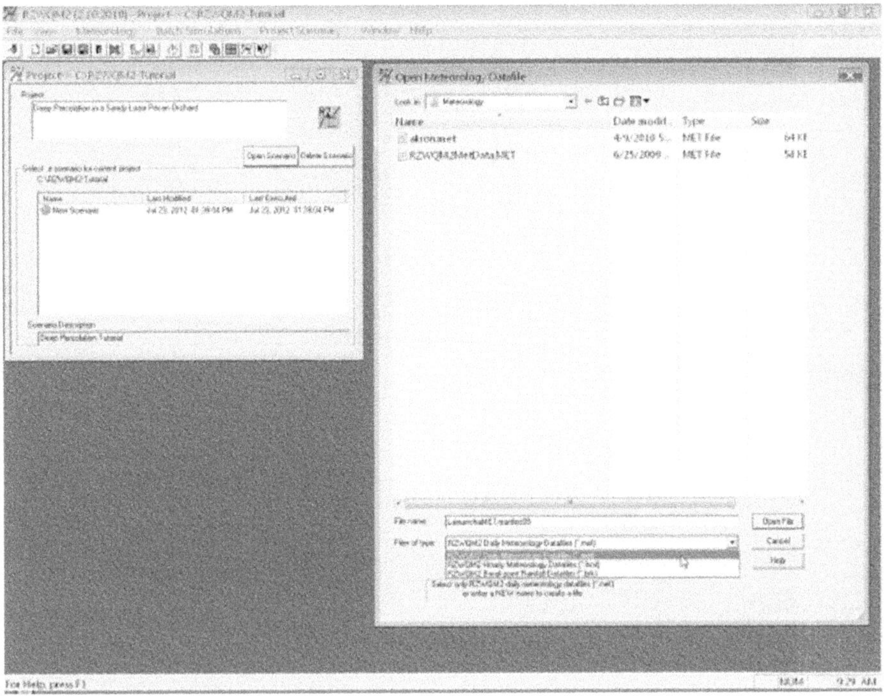

FIGURE 17.9 Edit/create the RZWQM2 daily meteorology datafiles (*.met).

- Click on the Verify Data button, and correct data and repeat until the message reads "Data Successfully Validated" (Figure 17.11).
- Click on Save.

As shown in Figure 17.11, if pan evaporation (not a required input) is entered it will be used as daily potential evaporation; otherwise, PET is calculated from the modified Shuttleworth–Wallace equations. Photosynthetically active radiation (PAR) (not a required input for this tutorial) is optional for the DSSAT crop growth model. The daily meteorology file "LamanchaMETmardec09.met" for the tutorial project is shown in Figure 17.12, which can be edited in the WordPad (or Notepad) text editor.

Note that RZWQM2 also incorporates the CLIGEN90 weather generator so that the user may generate meteorology files for a specific weather station in the United States. Generated weather files use the same root name with different extensions: (*.met) for daily meteorology, (*.hmt) for hourly meteorology, and (*.brk) for breakpoint rainfall. For example, click Meteorology | Generate, which will bring up the CLIGEN90 climate generator window. Readers are referred to the RZWQM2 model user guide (i.e., C:\RZWQM2\documentation\RZWQM2userguide.pdf) for details of the climate generator options.

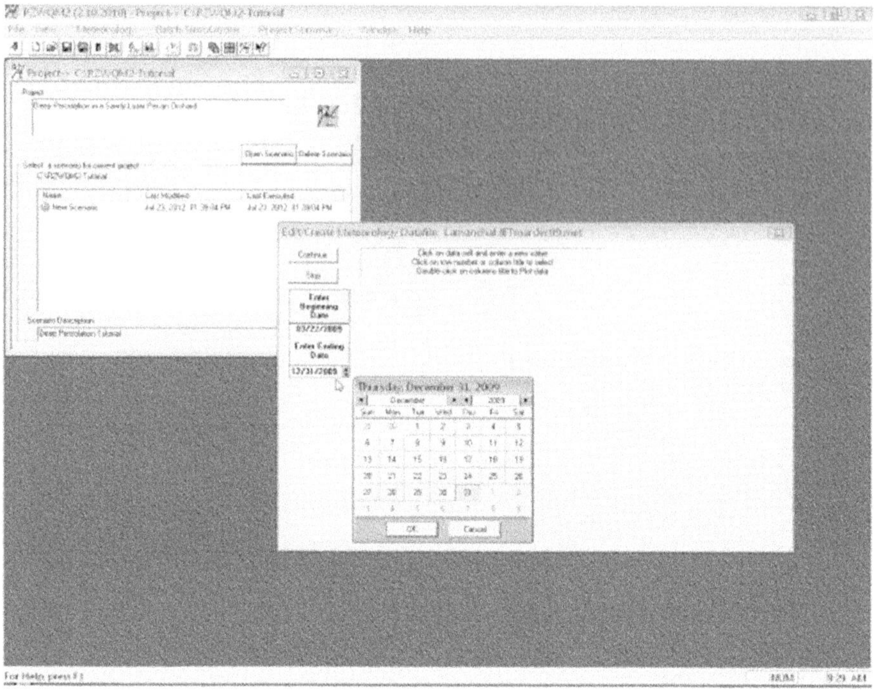

FIGURE 17.10 Beginning date (March 22, 2009) and ending date (December 31, 2009) of the RZWQM2 daily climate for the tutorial project.

17.4.3 Create/Edit Breakpoint Rainfall File for the Tutorial Project

We will use hourly meteorology data (*.hmt) to create a breakpoint file (*.brk) for the tutorial project. This new tool is available in RZWQM2 Version 2.0 and later. Note that daily meteorology (*.met) including rainfall data can also be used to create a breakpoint file (*.brk).

To create the hourly meteorology file (*.hmt):

- Click Meteorology | Edit/Create.
- Select "RZWQM2 Hourly Meteorology Datafiles (*.hmt)" from the Files of type drop-down list and enter the new name "LamanchaMETmardec09hrly" in the File name field (Figure 17.13).
- Click on the Open File button (Figure 17.13), and the next screen will display for the beginning (March 22, 2009) and ending date information (December 31, 2009) (Figure 17.14).
- Click on the Continue button (Figure 17.14) and the next screen will display for hourly meteorology data entries (Figure 17.15).

For this tutorial example, hourly meteorological data (e.g., mean air temperature, wind speed, shortwave radiation, relative humidity, and rainfall) during the period from

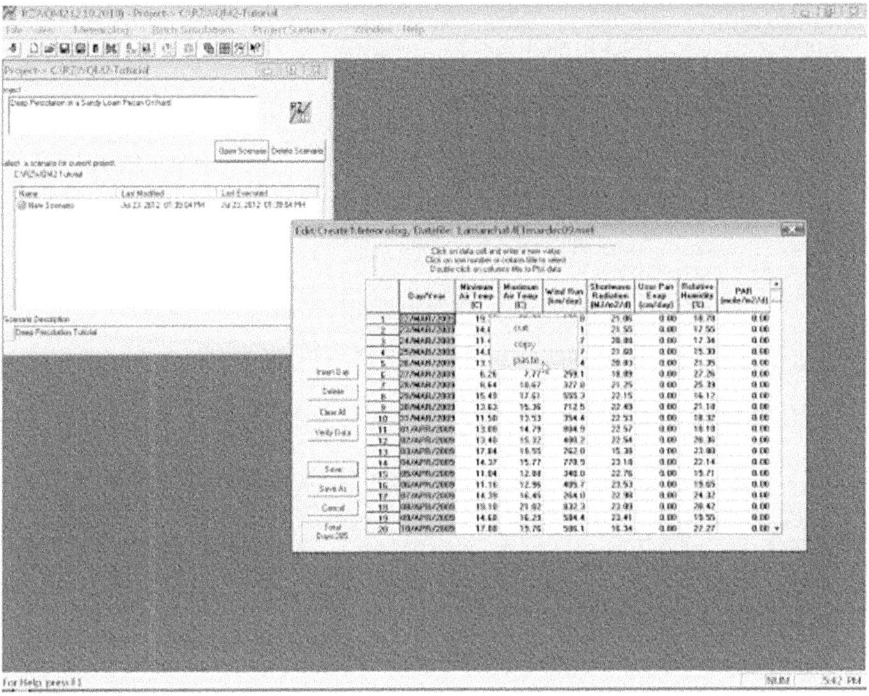

FIGURE 17.11 Edit/create the daily meteorology data input for the tutorial project.

March 22 to December 31, 2009, can be downloaded from the weather station Fabian Garcia RC of the New Mexico Climate Center at http://weather.nmsu.edu. Hourly meteorological data can be prepared in a Microsoft Excel spreadsheet, and then:

- Copy the hourly meteorological data from the Excel spreadsheet (e.g., "Hourly Meteorological Data.xlsx") and paste in the fields (Figure 17.15).
- Click on the Verify Data button, and correct data and repeat until message reads "Data Successfully Validated." Then, click on Save (Figure 17.15).

Note that User Pan Evap (0.0) and PAR data (0.0) are not required inputs for the tutorial project (Figure 17.15). The hourly meteorology file "LamanchaMETmardec09hrly.hmt" for the tutorial project is shown in Figure 17.16, which can be edited in the WordPad (or Notepad) text editor as well.

- Click Meteorology | Create A Breakpoint File From Daily/Hourly Rainfall, which will bring up the Create a *.BRK from Daily or Hourly Rainfall screen (Figure 17.17, left).
- Select the radio button Hourly Data File (*.hmt). In Select Input Meteorology File Name, select "LamanchaMETmardec09hry.hmt" from the drop-down box list (Figure 17.17, left).

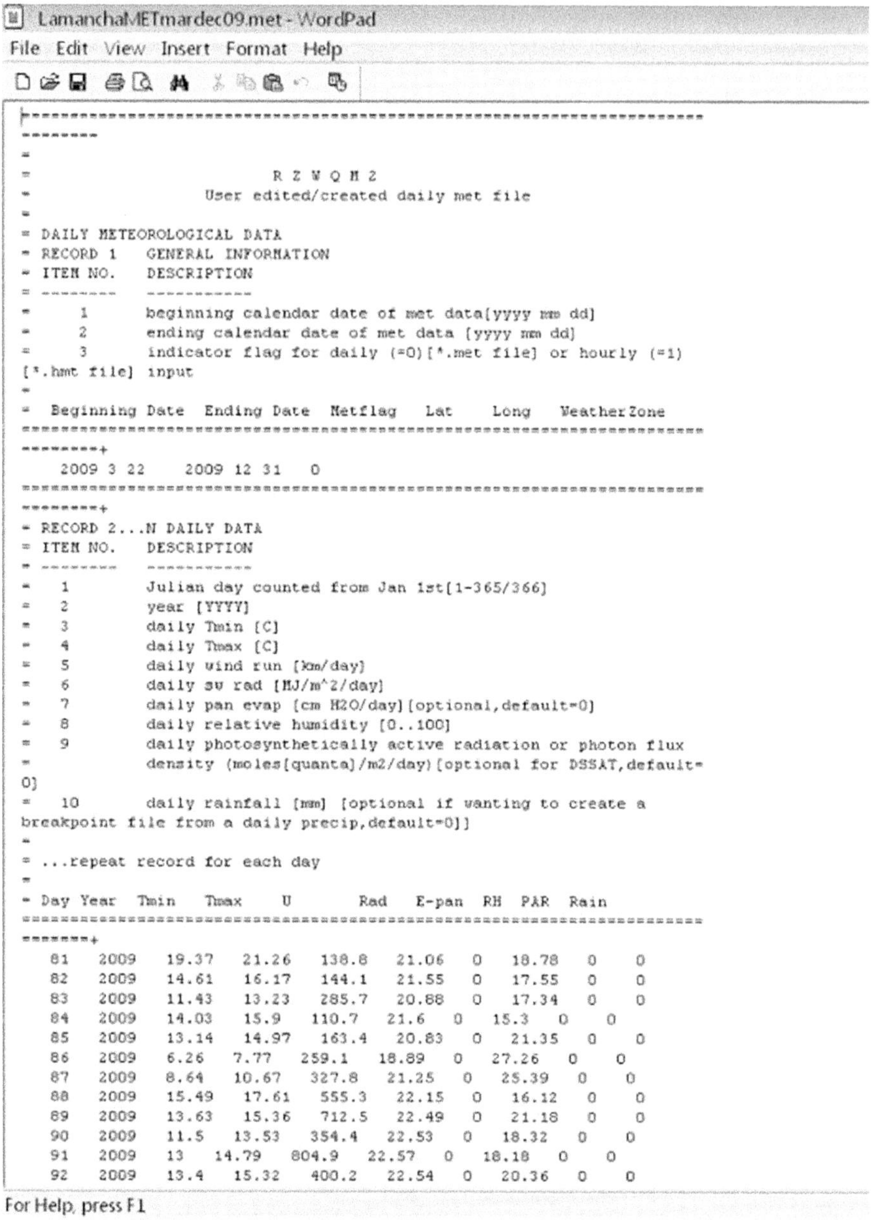
LamanchaMETmardec09.met - WordPad

File Edit View Insert Format Help

```
=======================================================================
========
=
=                          R Z W Q M 2
=                  User edited/created daily met file
=
= DAILY METEOROLOGICAL DATA
= RECORD 1    GENERAL INFORMATION
= ITEM NO.    DESCRIPTION
= --------    -----------
=     1       beginning calendar date of met data[yyyy mm dd]
=     2       ending calendar date of met data [yyyy mm dd]
=     3       indicator flag for daily (=0)[*.met file] or hourly (=1)
[*.hmt file] input
=
=  Beginning Date  Ending Date  Metflag   Lat     Long    WeatherZone
=======================================================================
========+
    2009 3 22     2009 12 31    0
=======================================================================
========+
= RECORD 2...N DAILY DATA
= ITEM NO.    DESCRIPTION
= --------    -----------
=   1         Julian day counted from Jan 1st[1-365/366]
=   2         year [YYYY]
=   3         daily Tmin [C]
=   4         daily Tmax [C]
=   5         daily wind run [Km/day]
=   6         daily sw rad [MJ/m^2/day]
=   7         daily pan evap [cm H2O/day][optional,default=0]
=   8         daily relative humidity [0..100]
=   9         daily photosynthetically active radiation or photon flux
=             density (moles[quanta]/m2/day)[optional for DSSAT,default=
0]
=   10        daily rainfall [mm] [optional if wanting to create a
breakpoint file from a daily precip,default=0]]
=
= ...repeat record for each day
=
= Day Year  Tmin   Tmax    U       Rad   E-pan  RH  PAR  Rain
=======================================================================
=======+
   81   2009   19.37   21.26   138.8   21.06   0   18.78   0    0
   82   2009   14.61   16.17   144.1   21.55   0   17.55   0    0
   83   2009   11.43   13.23   285.7   20.88   0   17.34   0    0
   84   2009   14.03   15.9   110.7   21.6   0   15.3   0    0
   85   2009   13.14   14.97   163.4   20.83   0   21.35   0    0
   86   2009   6.26   7.77   259.1   18.89   0   27.26   0    0
   87   2009   8.64   10.67   327.8   21.25   0   25.39   0    0
   88   2009   15.49   17.61   555.3   22.15   0   16.12   0    0
   89   2009   13.63   15.36   712.5   22.49   0   21.18   0    0
   90   2009   11.5   13.53   354.4   22.53   0   18.32   0    0
   91   2009   13   14.79   804.9   22.57   0   18.18   0    0
   92   2009   13.4   15.32   400.2   22.54   0   20.36   0    0
```

For Help, press F1

FIGURE 17.12 The daily meteorology file "LamanchaMETmardec09.met" for the tutorial project.

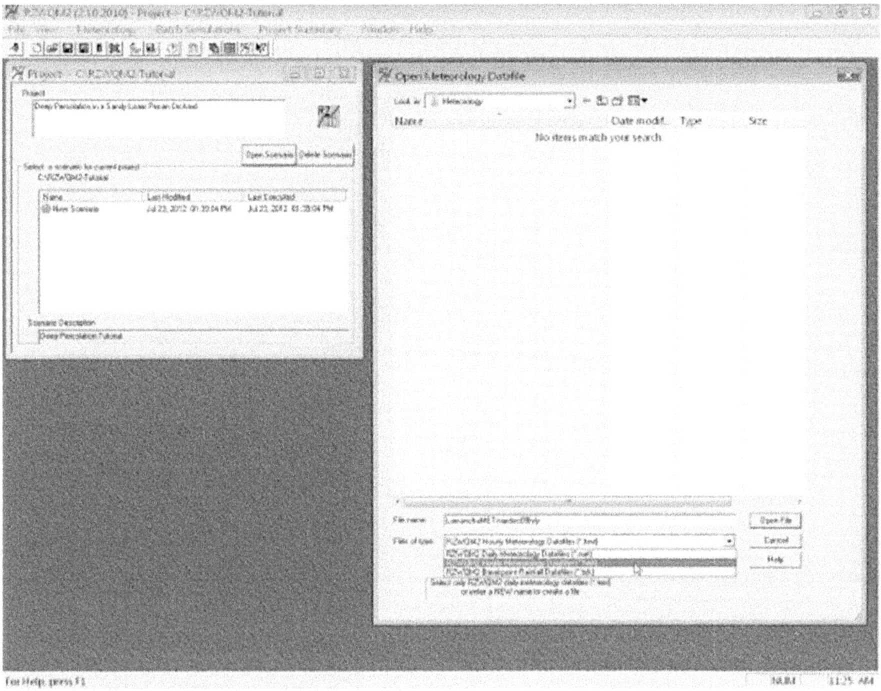

FIGURE 17.13 Edit/create the RZWQM2 hourly meteorology datafiles (*.hmt).

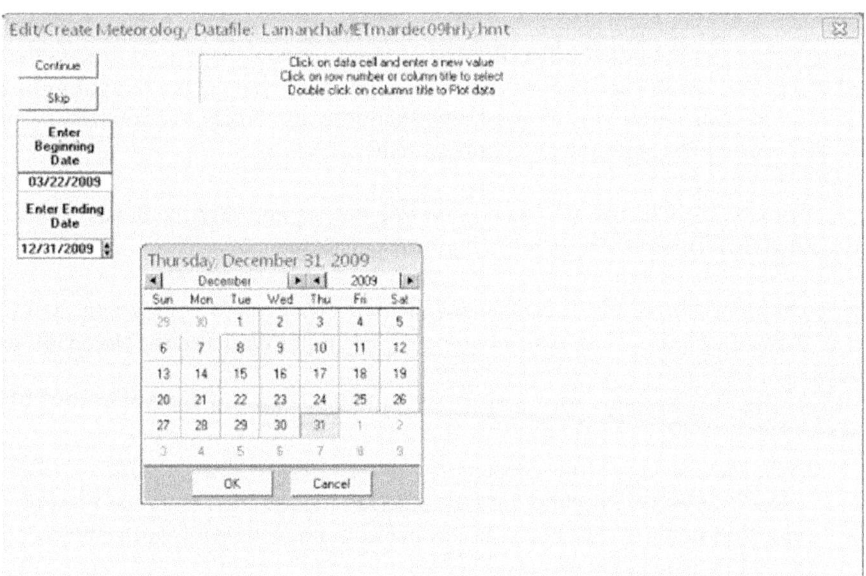

FIGURE 17.14 Beginning date (March 22, 2009) and ending date (December 31, 2009) of the RZWQM2 hourly climate for the tutorial project.

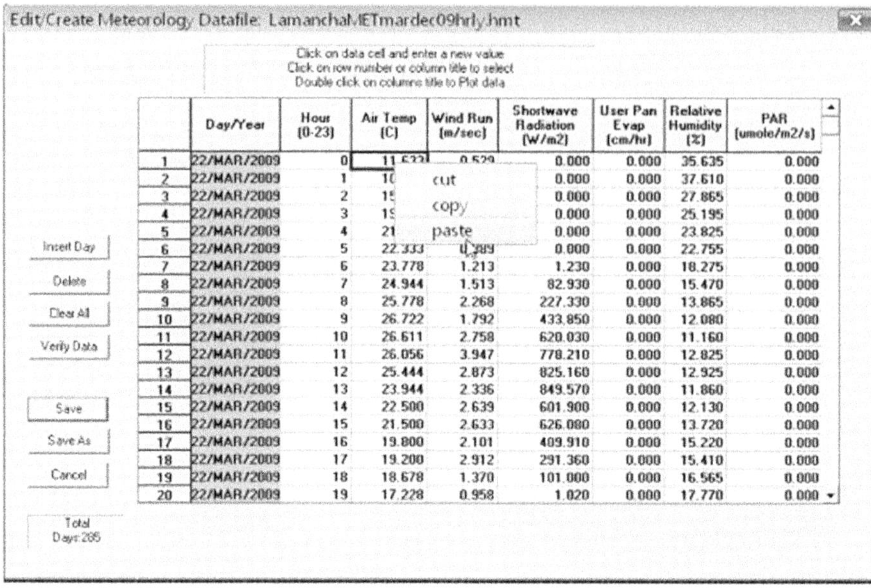

FIGURE 17.15 Edit/create the hourly meteorology data input for the tutorial project.

- Use the default values for Enter Storm Information (i.e., Duration, Minimum Intensity, and Maximum Intensity) (Figure 17.17, left).

Minimum and maximum storm intensity is used to determine the range of intensity acceptable in the breakpoint file. Storm durations are calculated from hourly data, and more than one storm can occur within the day. Note that during rainfall infiltration no evapotranspiration is allowed, and therefore, extremely long rainfall events will affect evapotranspiration and plant growth.

- Click on the Create/Save Breakpoint File button, and save the file as "LamanchaMETmardec09hrly.brk" (Figure 17.17, right).

The breakpoint rainfall file "LamanchaMETmardec09hrly.brk" for the tutorial project is shown in Figure 17.18, which can be edited in the WordPad (or Notepad) text editor.

Users can also directly create a *.brk file. For example,

- Click Meteorology | Edit/Create.
- Select "RZWQM2 Breakpoint Rainfall Datafiles (*.brk)" from the Files of type drop-down list and enter the new name "LamanchaMETmardec09hrly" in the File name field (Figure 17.19).
- Click on the Open File button (Figure 17.19), which will result in the next screen to enter the first storm date (10/APR/2009) and number of

LamanchaMETmardec09hrly.hmt - WordPad

File Edit View Insert Format Help

```
=======================================================================
========
=
=                               R Z W Q M 2
=                    User edited/created hourly met file
=
= HOURLY METEOROLOGICAL DATA
= RECORD 1    GENERAL INFORMATION
= ITEM NO.    DESCRIPTION
= --------    -----------
=     1       beginning calendar date of met data[yyyy mm dd]
=     2       ending calendar date of met data [yyyy mm dd]
=     3       indicator flag for daily (=0)[*.met file] or hourly (=1)
[*.hmt file] input
=
=  Beginning Date  Ending Date   Flag
=======================================================================
=======+
    2009 3 22     2009 12 31    1
=======================================================================
=======+
= RECORD 2...N HOURLY DATA FOR EACH DAY
= ITEM NO.    DESCRIPTION
= --------    -----------
=   1         Julian day counted from Jan 1st[1-365/366]
=   2         year [yyyy]
=   3         hour [0-23]
=   4         hourly air temperature [C]
=   5         hourly wind run [m/s]
=   6         hourly sw rad [W/m2]
=   7         hourly pan evap [cm H2O/hr][optional,default=0]
=   8         hourly relative humidity [0..100]
=   9         hourly photosyntheically active radiation or photo flux
density (umoles[quanta]/m2/sec)[optional for DSSAT,default=0]
=  10         hourly rainfall [mm] [optional if wanting to create a
breakpoint file from hourly precip, defaulty=0 or blank]
=
= ...repeat record for each hour
=
= Day  Year   HOUR  T_AIR   U_HR  Rad_HR  Epan_HR  RH-HR  PAR_HR
Rain_HR
=======================================================================
=======+
   81   2009   0    11.633   0.529   0   0   35.635   0   0
   81   2009   1    10.656   0.492   0   0   37.61   0   0
   81   2009   2    15.944   0.805   0   0   27.865   0   0
   81   2009   3    19.572   0.805   0   0   25.195   0   0
   81   2009   4    21.833   0.847   0   0   23.825   0   0
   81   2009   5    22.333   0.889   0   0   22.755   0   0
   81   2009   6    23.778   1.213   1.23   0   18.275   0   0
   81   2009   7    24.944   1.513   82.93   0   15.47   0   0
   81   2009   8    25.778   2.268   227.33   0   13.865   0   0
   81   2009   9    26.722   1.792   433.85   0   12.08   0   0
   81   2009   10    26.611   2.758   620.03   0   11.16   0   0
   81   2009   11    26.056   3.947   778.21   0   12.625   0   0
```

For Help, press F1

FIGURE 17.16 The hourly meteorology file "LamanchaMETmardec09hrly.hmt" for the tutorial project.

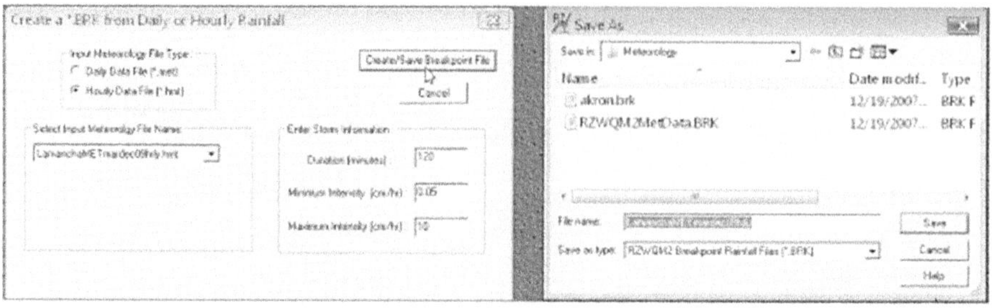

FIGURE 17.17 Creating a breakpoint file from the hourly meteorology file "LamanchaMET mardec09hrly.hmt" (Figure 17.16) containing rainfall.

breakpoints (enter "2") (Figure 17.20a). You can also double-click on the Enter 1st Storm Date field to bring up the calendar to set the date.

- Click on the Continue button (Figure 17.20a), which will bring up the next screen for breakpoint data entry (Figure 17.20b).
- Click first on the Verify Data button and then on the Save button (Figure 17.20b).

The Storm Summary screen will be displayed (Figure 17.21). Users can enter more storms with the Append Storm Below and Insert Strom Above buttons. Users can also delete storm events or edit storm breakpoints. Readers are referred to the RZWQM2 model user guide (i.e., C:\ RZWQM2\documentation\RZWQM2userguide.pdf) for the details of the breakpoint rainfall file description.

Once the files for daily meteorology (i.e., "LamanchaMETmardec09.met") and breakpoint rainfall ("LamanchaMETmardec09hrly.brk") have been created:

- Click File I Save Project.

Note that RZWQM2 requires a snow file (*.SNO), which is described in Section 17.5.1.5.

17.5 WORKING WITH THE RZWQM2 SCENARIO

Open New Scenario to work with the scenario menus (Figure 17.22).

17.5.1 Define the System/lnput Menu

- Click on the Input menu, which gives the submenus: Site Description, Initial State, Residue State, Management Practices, PRMS Snow Parameters, Experiment Data (optional), and Import RZWQM*. DAT Files (Figure 17.22).

Each submenu brings up a set of tabbed dialog screens including Help (or press F1 for help).

LamanchaMETmardec09hrly.brk - WordPad

File Edit View Insert Format Help

```
=======================================================================
========
=
=                             R Z W Q M 2
=              User created breakpoint file from an hourly met file
=
=
==        B R E A K P O I N T   R A I N F A L L   D A T A
==
== Rec 1: calendar code: =1 - date counted as Julian days within year
== Rec 2: control info for a breakpoint event:
==        year, date, no. breakpoints, midnight (day spanning, 1=yes)
==        code, total storm depth [in]
== Rec 3: pairs of cum storm dep [in] & clock time [min],
==        5 pairs per rec
== REPEAT Rec 3 to complete data for an event
==
== REPEAT Rec's 2 & 3 for subsequent events
=
=======================================================================
========
1
     2009       100         2         0       0.0254
        0      1020     0.0254      1080
     2009       141         7         0        0.635
        0       840       0.17       900         0.33       960
0.41     1020         0.57       1080         0.63      1140        0.635
1200
     2009       142         2         0        0.03
        0       240       0.03       300
     2009       142         7         0        0.376
        0       360       0.08       420         0.13       480
0.18      540         0.32        600        0.326       660        0.376
720
     2009       150         6         0        0.381
        0       960       0.16      1020         0.21      1080
0.291     1140        0.351      1200        0.381      1260
     2009       151         3         0         0.3
        0       900       0.15       960          0.3      1020
     2009       174         4         0       0.0508
        0       660       0.01       720         0.02       780
0.0508     840
     2009       179         2         0        0.01
        0        60       0.01       120
     2009       179         7         0        0.168
        0       780     0.0078       840        0.108       900
0.118     960         0.148      1020        0.158      1080        0.168
1140
     2009       183         7         0        0.279
        0      1020        0.1      1080        0.169      1140
0.189     1200        0.209      1260        0.219      1320        0.279
1380
     2009       184         8         0          1
        0       660      0.316       720        0.631       780
0.901      840        0.911       900        0.961       960        0.991
```

For Help, press F1

FIGURE 17.18 The breakpoint rainfall file "LamanchaMETmardec09hrly.brk" for the tutorial project.

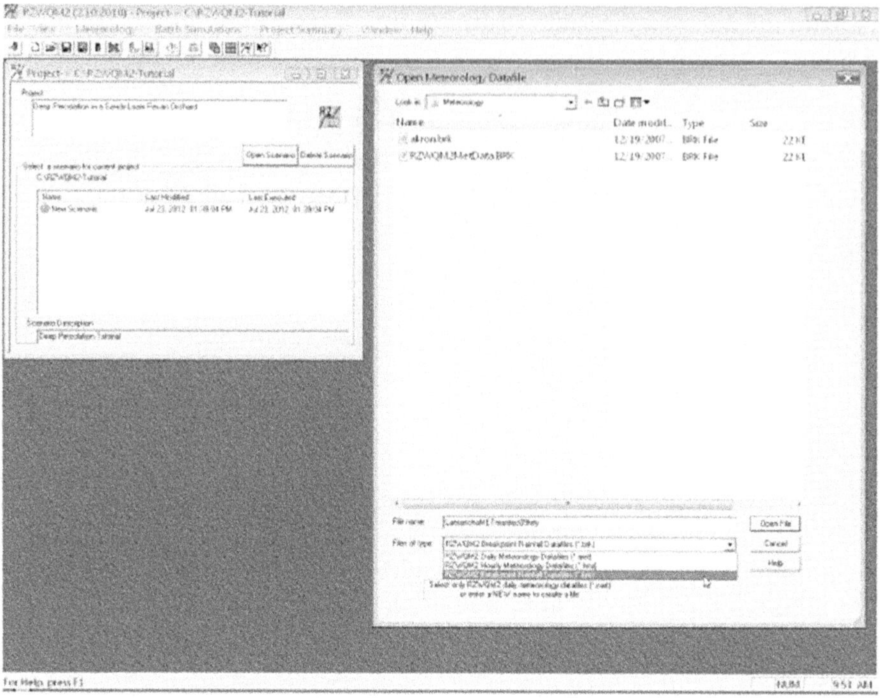

FIGURE 17.19 Edit/create the RZWQM2 breakpoint rainfall datafiles (*.brk).

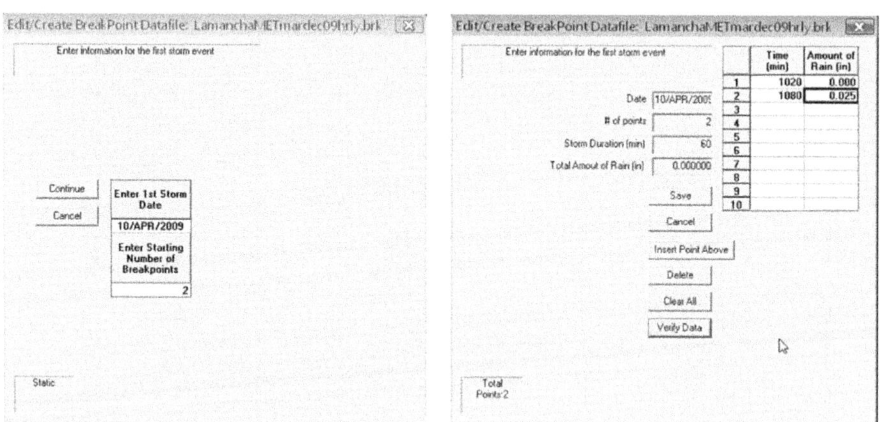

(a) (b)

FIGURE 17.20 (a) Screens for the rainfall date and number of breakpoints. (b) Breakpoint data entry.

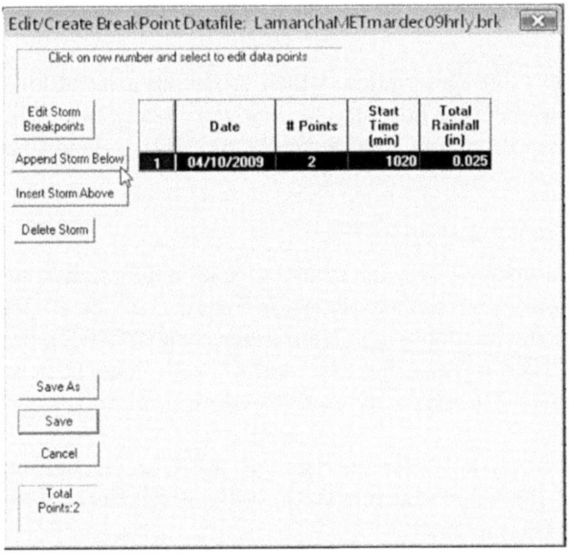

FIGURE 17.21 Summary of the rainfall event after breakpoint data entry.

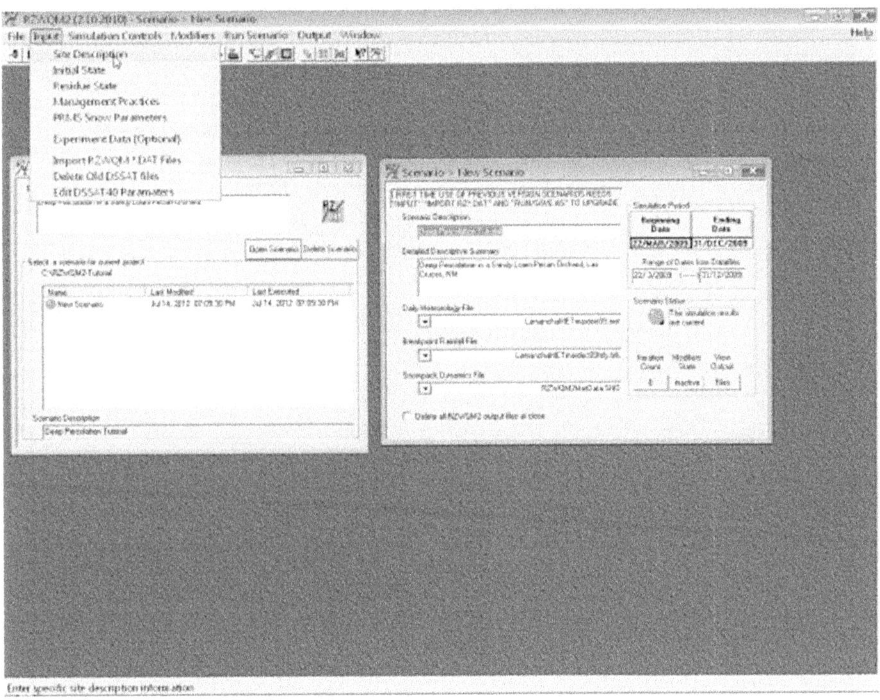

FIGURE 17.22 The RZWQM2 scenario view and input menu items.

17.5.1.1 Site Description

- Click Input | Site Description, which brings up a set of nine tabbed dialog screens: General Information, Horizon Description, Soil Hydraulics, Soil Physical Properties, Hydraulic Control, Background Chemistry, PET, Nutrients, and NH3 (Figure 17.23).

17.5.1.1.1 General Information

The General Information dialog tab includes location information and field specifics. Enter values and/or information as shown in Figure 17.23. Field specifics determine the site's geographic and topographic attributes. Aspect is referenced in degrees to true North. The asterisk (i.e., Longitude* and Climate Zone*) denotes new variables in RZWQM version 2.0., which are used in evapotranspiration calculations.

- Click on Apply to save the data on the General Information screen (Figure 17.23) before selecting the next dialog tab Horizon Description for data entry (Figure 17.24).

Note that clicking on the OK button will also save, but will exit you from the Input submenus entirely.

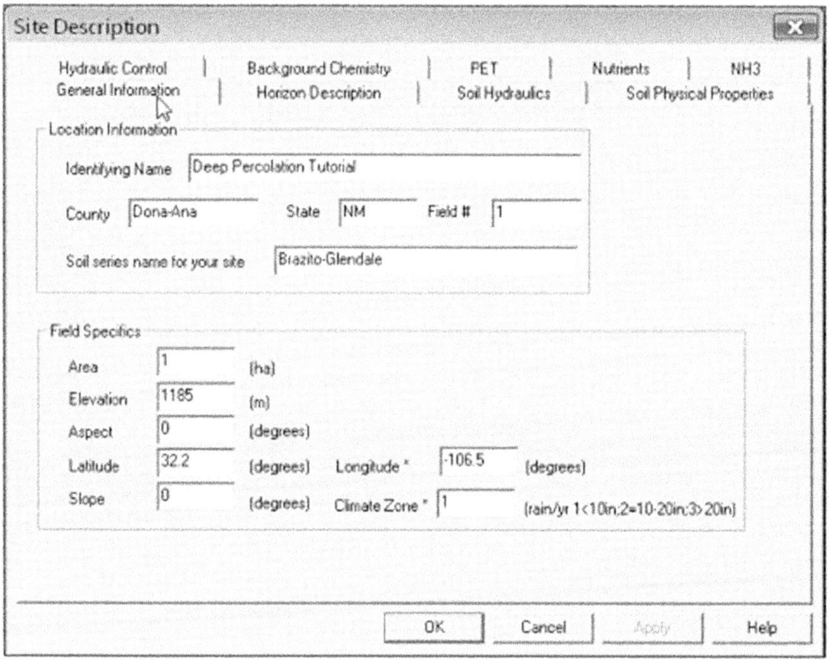

FIGURE 17.23 General information screen.

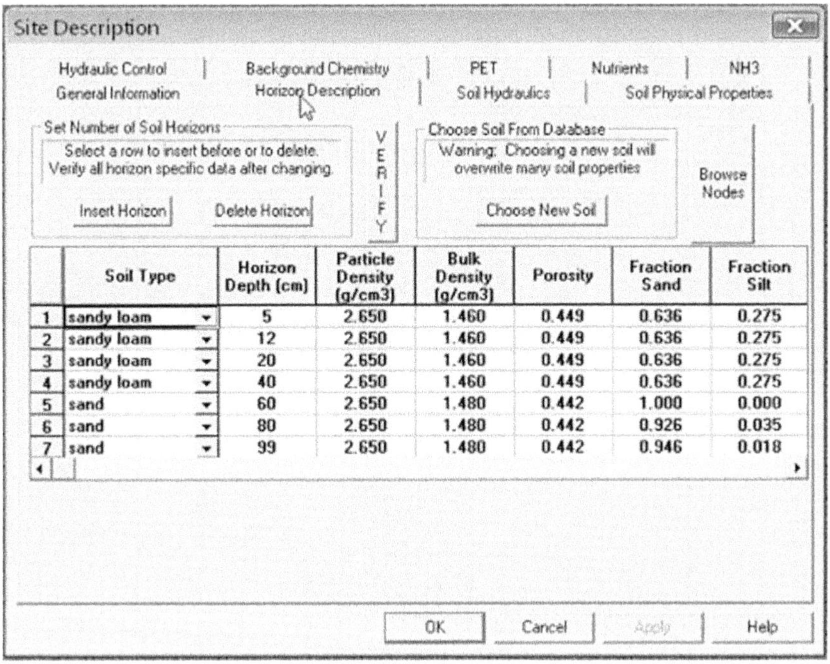

FIGURE 17.24 Soil horizon description screen.

17.5.1.1.2 Horizon Description

The Horizon Description dialog tab allows users to describe the soil horizons and their attributes (Figure 17.24). Note that in RZWQM2 (version 2.10.2010), the maximum soil layers for user input is ten and the maximum profile depth simulated is 30 m.

- As shown in Figure 17.24, set the Soil Type (Texture Class) and depth for each horizon. For the tutorial project, the soil profile depth is 100 cm and seven soil layers at depth intervals of 0–5 cm (horizon depth: 5 cm; sandy loam), 5–12 cm (horizon depth: 12 cm; sandy loam), 12–20 cm (horizon depth: 20 cm; sandy loam), 20–40 cm (horizon depth: 40 cm; sandy loam), 40–60 cm (horizon depth: 60 cm; sand), 60–80 cm (horizon depth: 80 cm; sand), and 80–99 cm (horizon depth: 99 (~100) cm; sand).
- Click VERIFY to verify whether the soil layers you entered can be subdivided into the numerical nodes used in solving the Richards equation.

It may change the last horizon for you or ask you to change one or two horizons to coincide a horizon boundary with a numerical node boundary. Note that all horizons must be not less than 1.0 cm and the first horizon must be greater than 2.0 cm thick. After verifying, values of bulk density and fractions of sand, silt, and clay for each horizon can be entered from Table 17.1.

TABLE 17.1
Soil Properties for the RZWQM2 Tutorial Project

	Particle Size Distribution								
Soil Depth (cm)	**Sand**	**Silt (%)**	**Clay**	**Bulk Density (g/cm^3)**	**K_{sat} (cm/h)**	**FC (cm^3/cm^3)**	**WP (cm^3/cm^3)**	**θ_s (cm^3/cm^3)**	**θ_r (cm^3/cm^3)**
0–5	63.6	27.5	8.9	1.46	1.25	0.3322	0.1576	0.4600	0.085
5–12	63.6	27.5	8.9	1.46	1.25	0.3322	0.1576	0.4600	0.085
12–20	63.6	27.5	8.9	1.46	1.25	0.3322	0.1576	0.4600	0.085
20–40	63.6	27.5	8.9	1.46	1.25	0.2225	0.0700	0.4600	0.051
40–60	100.0	0.0	0.0	1.48	1.40	0.2363	0.0705	0.4420	0.051
60–80	92.6	3.5	3.9	1.48	1.40	0.2363	0.0705	0.4420	0.051
80–99 (≈80–100)	94.6	1.8	3.6	1.48	1.40	0.2363	0.0705	0.4420	0.051

Note: K_{sat}, saturated hydraulic conductivity; FC, soil water content at field capacity (at 1/3 bar); WP, soil water content at wilting point (at 15 bar); and θ_s and θ_s are saturated and residual volumetric water contents, respectively.

- Use RZWQM2 default values for particle density (2.65 for all horizons) and porosity (0.449 for sandy loam and 0.442 for sand horizons).
- Click on the Apply button to save the data on the Horizon Description screen (Figure 17.24) before selecting the next dialog tab Soil Hydraulics (Figure 17.25a).

If measured values of Horizon Description are not available, users can create soil data (using the Choose New Soil button; Figure 17.24) from the RZWQM2 Database (USA). Readers are referred to the RZWQM2 model user guide (i.e., C:\RZWQM2\documentation\RZWQM2userguide.pdf) for the details of the soil database options.

17.5.1.1.3 Soil Hydraulics

RZWQM2 requires the full BC soil hydraulic property parameters, which can be entered directly in the full description (constrained or not constrained; Figure 17.25a). Other methods of estimating the BC parameters are available by choosing one of the following radio buttons: from soil texture (Minimum Input), from 15 (wilting point water content at 15 bar) and 1/3 (field capacity water content at 1/3 bar) bar water content, and from water content correlations with 1/3 bar water content only (one parameter model; Figure 17.25a). All soil hydraulic property estimation methods use 1/3 bar water content. Water content is volumetric in these screen entries.

For estimating BC parameters with minimum input (Figure 17.25a):

- Select the radio button Minimum Input to enter the saturated hydraulic conductivity (K_{sat}) and 1/3 bar water content (FC) value for each soil layer or horizon (Table 17.1).
- Click on the Apply button.

For estimating BC parameters using their correlations with 1/3 bar water content (Figure 17.25a):

- Select the radio button, and then enter horizon K_{sat}, saturation (θ_s) water content, 1/3 bar water content (FC), and residual water content values from Table 17.1.
- Click on the Apply button.

For the tutorial project, we will estimate BC parameters by choosing the radio button Estimate BC Parameters Using 15 and 1/3 bar Water Content (Figure 17.25a).

- Enter horizon K_{sat}, saturation (θ_s), water content K_{sat}, 1/3 bar water content (FC), 15 bar water content (WP), and residual water content (θ_r) values (Table 17.1).
- Click on the button Advanced Richards Equation Parameters. Number of iterations and convergence criteria of the Richards equation can be modified using this screen (Figure 17.25b).

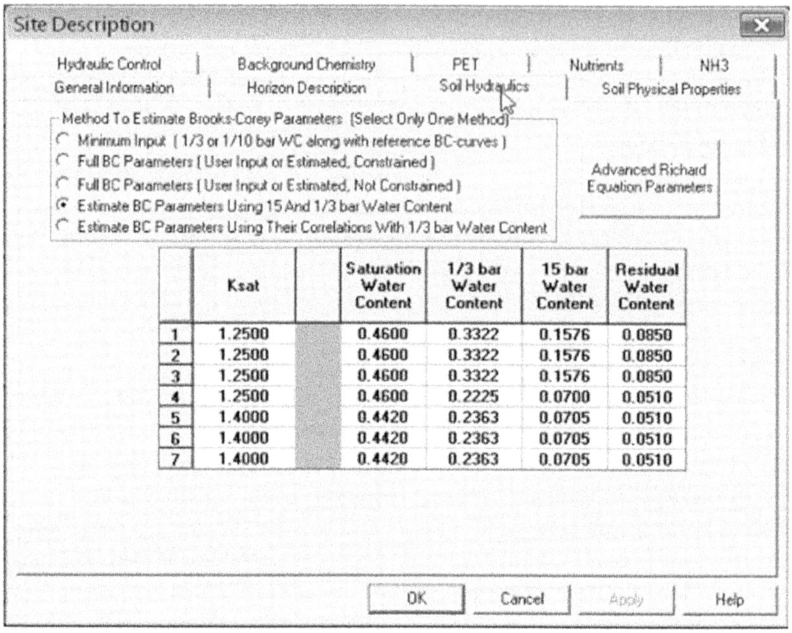

(a)

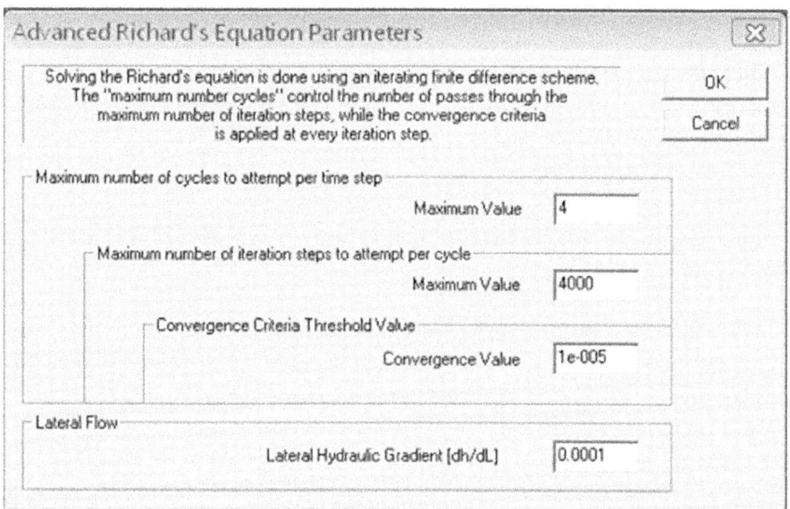

(b)

FIGURE 17.25 Estimate BC Parameters Using 15 and 1/3 bar Water Content option chosen for the tutorial project (a) and advanced Richards equation parameters (b).

For Advanced Richards Equation Parameters, readers are referred to the RZWQM2 model user guide (i.e., C:\RZWQM2\documentation\RZWQM2userguide.pdf). Increase the iterations to 4,000 or greater (usually <10,000) to reach convergence, or an increase in the convergence criterion (1e–005) may be needed. However, setting too high a convergence criterion may cause a mass balance problem.

- Use default values as shown in Figure 17.25b, and then click OK to return to the Estimate BC Parameters Using 15 and 1/3 bar Water Contents screen (Figure 17.25a).
- Click on the Apply button to save the data on the Soil Hydraulics screen (Figure 17.25a) and select the next dialog tab Soil Physical Properties for data entry (Figure 17.26).

17.5.1.1.4 Soil Physical Properties

Th Soil Physical Properties dialog tab allows users to describe input parameters related to heat capacity, and macropore and micropore information (Figure 17.26). Users may further characterize macropores with the Advanced Macropore Parameters button. Readers are referred to the RZWQM2 model user guide (i.e., C:\RZWQM2\documentation\RZWQM2userguide.pdf) for details of the micro-macropore control options. For this tutorial project:

- Unselect both boxes under the option labeled Micro-Macropore Control (Figure 17.26).
- Click on the Apply button to save the data on the Soil Physical Properties screen (Figure 17.26) and select the next dialog tab Hydraulic Control for data entry (Figure 17.27).

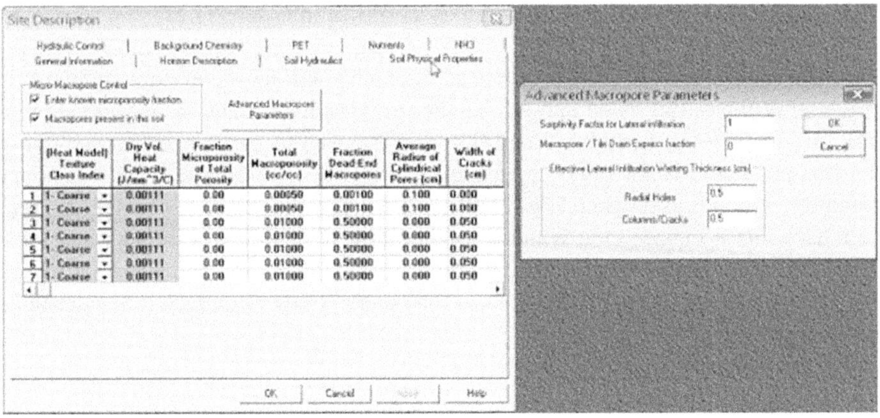

FIGURE 17.26 Soil physical properties screen.

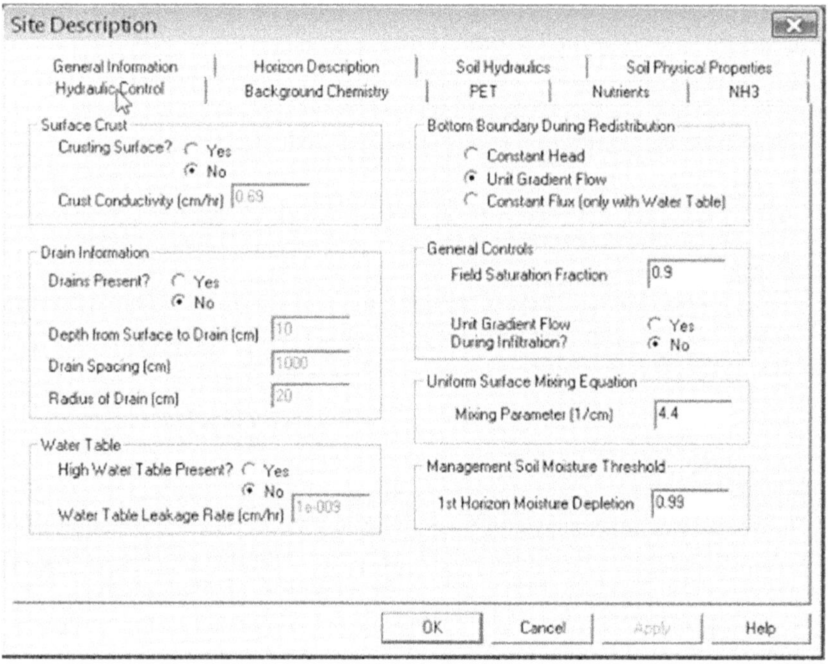

FIGURE 17.27 Hydraulic control of the site for the tutorial project.

17.5.1.1.5 Hydraulic Control

The Hydraulic Control dialog tab allows users to enter input information related to surface crusting, drainage, water table, boundary condition, field saturation, chemical mixing, and threshold water content for agricultural management (Figure 17.27).

- Select radio buttons and enter values as shown in Figure 17.27.

For the bottom boundary, a constant head boundary will fix the bottom node matric potential at the initial value and not let the initial value change during the simulation. A constant flux boundary condition may be used to simulate a leakage out of the bottom of the water table or unsaturated profile. The constant flux boundary condition option is used when High Water Table Present? is set to Yes and the bottom boundary flux is equal to the Water Table Leakage Rate (cm/h). When a water table is present, the constant flux boundary condition should be used and Field Saturation Fraction should be set to 1.0. Do not use the constant flux boundary if a water table is not present.

For this tutorial project:

- Select the radio button Unit Gradient Flow (i.e., free drainage) as the bottom boundary condition because the water table, generally located below

1.5 m from the soil surface at this orchard, did not influence the root zone soil water fluctuations.

With a Unit Gradient Flow boundary, the bottom node is allowed to react to the hydraulic situation above it, and water will move out of the profile at the current hydraulic conductivity rate in the bottom layer.

- Click on Apply to save the data on the Hydraulic Control screen (Figure 17.27) and select the next dialog tab Background Chemistry for data entry (Figure 17.28).

17.5.1.1.6 Background Chemistry

The Background Chemistry dialog tab allows users to specify input information for the chemistry of rainwater or irrigation water (Figure 17.28). The pH is required, while other fields are optional. For the tutorial project, consider all default values as shown in Figure 17.28, including values for pH of rainwater and irrigation water.

Click on the Apply button to save the data on the Background Chemistry screen (Figure 17.28) and select the next dialog tab PET for data entry (Figure 17.29).

17.5.1.1.7 Potential Evapotranspiration

As shown in Figure 17.29, parameters used to calculate PET are entered on the screen.

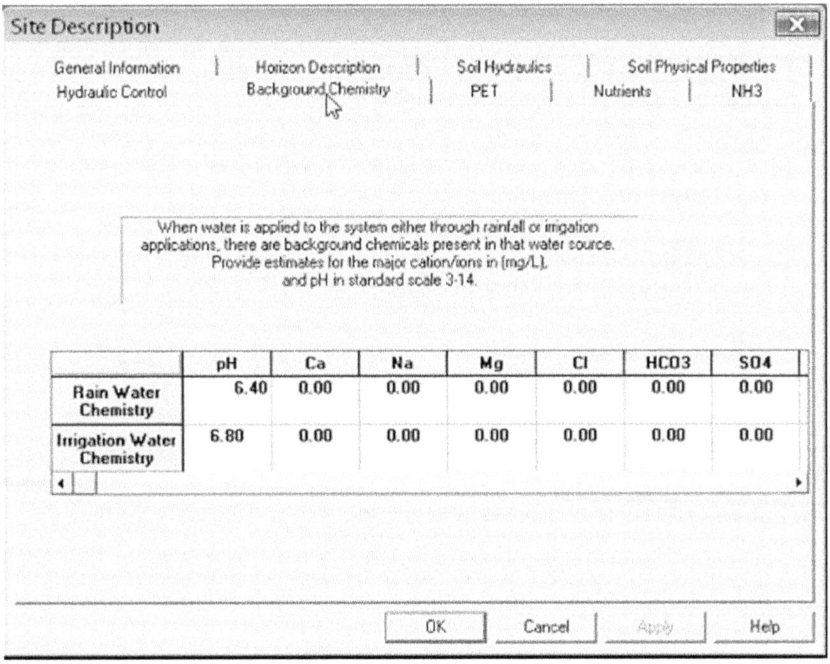

FIGURE 17.28 Background Chemistry screen for rainfall and irrigation water data entry.

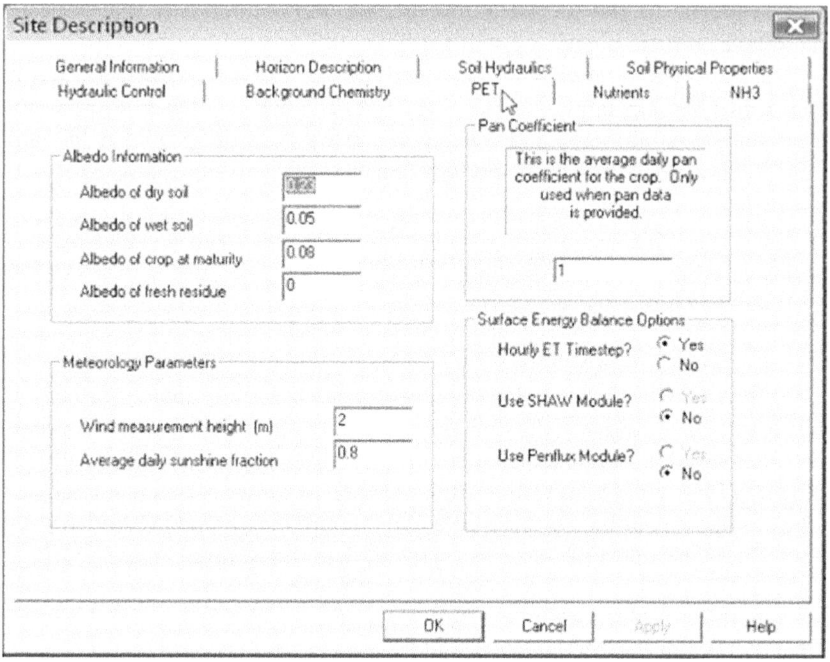

FIGURE 17.29 Information for PET.

- Enter Albedo information, wind measurement height, sunshine fraction, and pan coefficient values as shown in Figure 17.29.

Note that pan data are not provided in the daily meteorology file "LamanchaMETmardec09.met."

- The RZWQM2 version 2.0 simulation option can calculate ET on an hourly time step and check the Yes radio button.

Future options include using the SHAW model (with or without calculation hourly) or use of the Penflux module to calculate energy balance.

- Check the radio buttons No for both SHAW and Penflux modules.
- Then, click on the Apply" button to save the data on the PET screen (Figure 17.29) before selecting the next dialog tab Nutrients for data entry (Figure 17.30).

17.5.1.1.8 Nutrient System Parameters

The Nutrients dialog tab allows users to enter input parameters for the simulation of nutrients (Figure 17.30, left). The default parameters can be retrieved with the Reset C:N Ratios to Default Values or Reset Detailed Parameters to Default Values

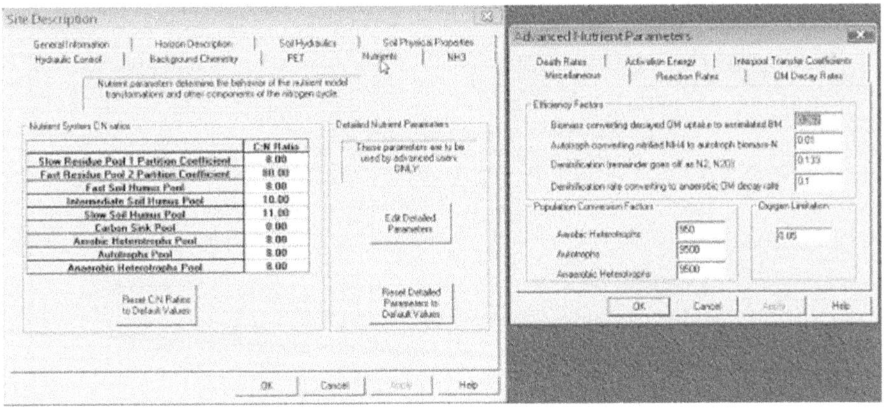

FIGURE 17.30 Nutrient parameters.

buttons. Users may override carbon/nitrogen ratios of each organic pool. Users may modify the nutrient parameters including pool transfer rates, organic matter decay rates, microbial death and activation energies, and rate coefficients of carbon/nitrogen transformations by clicking the Edit Detailed Parameters button (Figure 17.30, right). Readers are referred to the RZWQM2 model user guide (i.e., C:\RZWQM2\documentation\RZWQM2userguide.pdf) for details of the nutrient system parameters.

- For this tutorial project, consider all default values and click on Apply to save the data on the Nutrients screen (Figure 17.30, left) and select the next dialog tab NH3 for data entry (Figure 17.31).

17.5.1.1.9 NH3

The lag time on nitrification due to anhydrous ammonia application and nitrification inhibitors can be defined on the NH3 screen (Figure 17.31).

- Use the default values as shown in Figure 17.31.
- Click on Apply (Figure 17.31), and then the OK button to exit the Site Description submenu, that is, return to New Scenario to work with the other scenario menus (Figure 17.22).
- Click File I Save Scenario, and click OK to the message that appears.

Note that as you are working with model inputs/parameterization or data entry, the Scenario Status in the RZWQM2 scenario interface may be displayed as a red ball.

17.5.1.2 Initialize State/Initial Conditions of the Soil Profile

- In the Scenario window (Figure 17.22), clicking Input I Initial State will bring up the Initial System State screen with three tabbed dialog options: Water and Temperature, Soil Chemistry State, and Pesticide State (Figure 17.32).

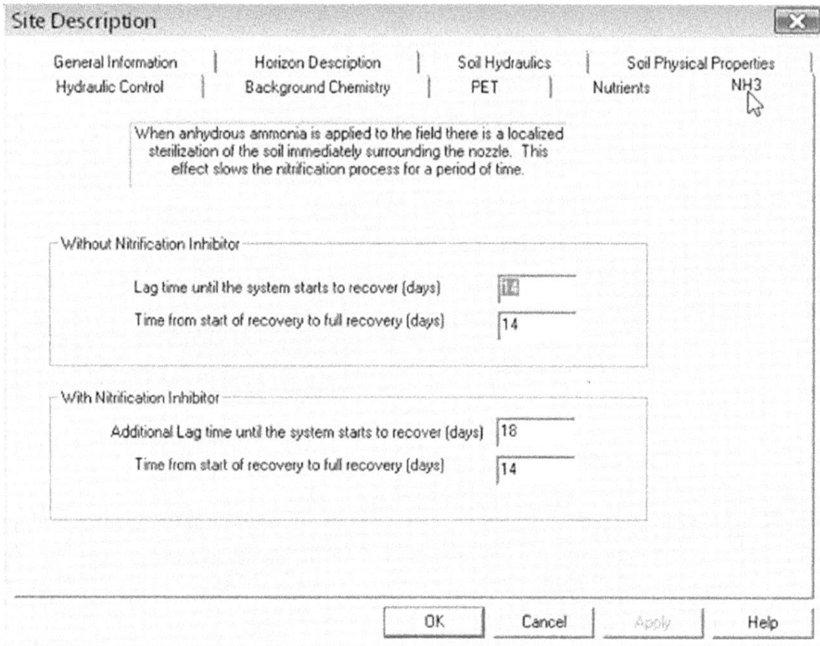

FIGURE 17.31 NH3 data entry screen.

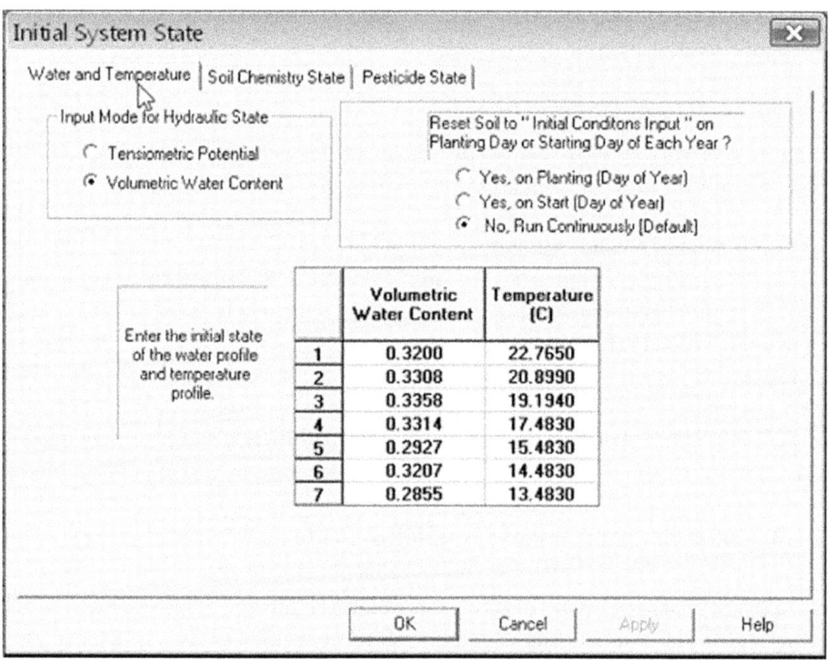

FIGURE 17.32 Initial system state for the soil profile of water content and soil temperature.

17.5.1.2.1 Water and Temperature State

Initial water may be input as volumetric water content or tensiometric potential (Figure 17.32). For this tutorial:

- Select the radio button Volumetric Water Content under Input Mode for Hydraulic State.

Measured volumetric soil water contents (cm^3/cm^3) and soil temperature (°C; extrapolated values below 60 cm) at 5, 10, 20, 40, 60, 80, and 100 cm soil depths on the first day of the simulation were specified as initial conditions (Figure 17.32). Note that a 100 cm profile was used and the modeled seven soil layers were at depth intervals of 0–5, 5–12, 12–20, 20–40, 40–60, 60–80, and 80–99 (i.e., 80–100) cm. Note that the radio button Tensiometric Potential should be selected if a water table exists within the profile, and thus, the location of the initial water table is defined at the soil layer based on a positive potential. RZWQM2 runs continuously through time by default. However, the state of the system at planting or at run start can be reset to the initial state input chosen by the radio buttons Yes, on Planting or Yes, on Start.

- Select the radio button No, Run Continuously for the tutorial project (Figure 17.32).

Note that the Soil Chemistry State dialog tab (Figure 17.32) allows users to initialize the equilibrium chemistry state for each soil horizon with soil pH, CEC, and other cations, including the existence of lime, gypsum, and gibbsite. The Pesticide State dialog tab allows users to specify initial total pesticide concentrations on crop residue and in the soil profile. Only three pesticides may be simulated at one time in the system whether it is a parent or a product of a parent. These initial total amounts are partitioned between absorbed and in solution. Readers are referred to the RZWQM2 model user guide (i.e., C:\RZWQM2\documentation\ RZWQM2userguide.pdf) for the details about Soil Chemistry State and Pesticide State data entries.

For the tutorial project, consider all default values of initial soil chemistry and pesticide states in the Soil Chemistry State and Pesticide State dialog tabs.

- Click on Apply, and then the OK button to exit the Initial System State screen (Figure 17.32) and return to the Scenario window (Figure 17.22).
- Click File | Save Scenario (Figure 17.22), and click OK.

17.5.1.3 Initialize the State of the Soil Residue

Click Input | Residue State, which will bring up the Residue Condition screen with two dialog tabs: Initial Residue/Inorganic N Profile and Surface Residue Properties (Figure 17.33). Users set the amount and type of residue on the surface at the start of the run on the Surface Residue Properties screen. Other properties to be specified are age, height, C:N ratio, surface residue decomposition constant (i.e., fraction of natural incorporation by biologics such as worms), and characteristics of standing

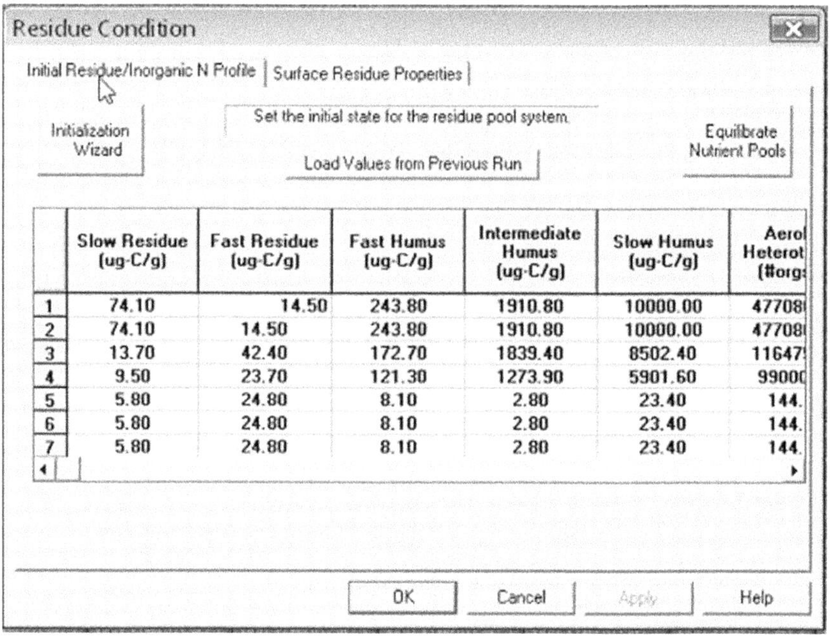

FIGURE 17.33 The Residue Condition screen.

residue. All the simulated carbon/nitrogen pools and soil microbial pools are initialized on the screen Initial Residue/Inorganic N Profile. Users may choose to load the pools from a previous run, follow the step-by-step initialization wizard based on the soil organic carbon measurement, or equilibrate the pools based on general weather information and agricultural practices. Readers are referred to the RZWQM2 model user guide (i.e., C:\RZWQM2\documentation\RZWQM2userguide.pdf) for details of Residue Condition data entries.

For the tutorial project:

- Consider all default values in the Surface Residue Properties and Initial Residue/Inorganic N Profile dialog tabs (Figure 17.33).
- Click on Apply and then the OK button to exit the Residue Condition screen (Figure 17.33) and return to the Scenario window (Figure 17.22).
- Click File | Save Scenario (Figure 17.22), and click OK to the message that appears.

17.5.1.4 Define Agricultural Management Practices

- In the Scenario window (Figure 17.22), click Input | Management Practices.

Seven tabbed dialog screens are available for the parameterization of RZWQM2 management: Crop Selection, Crop Planting, Manure, Irrigation, Fertilization, Pesticides, and Tillage (Figure 17.34, left).

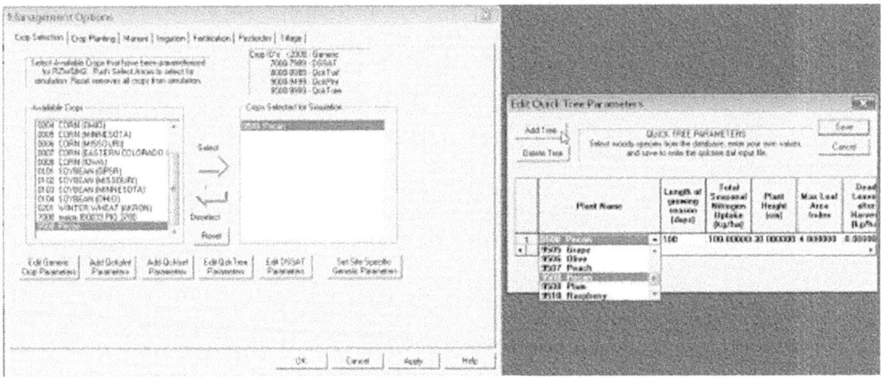

FIGURE 17.34 Screens for crop and crop model selection (left), and tree (Quicktree module) input parameters (right).

17.5.1.4.1 Crop Selection

The user decides whether they will use the RZWQM generic crop growth model, the DSSAT 4.0 crop growth models (CROPGRO and CERES Models), or the Quick Plant, Quick Turf Models, or Quick Tree Models (Figure 17.34, right). The four-digit Crop Identification (Crop ID) numbers indicate which crop model is used: generic crop growth model (numbers less than 2000), DSSAT 4.0 crop growth models (7000–7999), Quick Turf Crops (8000–8999), Quick Plant Crops (9000–9499), and Quick Tree Crops (9500–9999).

For the tutorial project, we will work with the tree module of RZWQM2 (Quick Tree Crops), which is not a crop growth simulation model. Readers are referred to the RZWQM2 model user guide (i.e., C:\RZWQM2\documentation\RZWQM2userguide.pdf) for details about the RZWQM generic crop growth model, the DSSAT 4.0 crop growth models, the Quick Plant, and the Quick Turf Models.

As the tree crop pecan has been added in the right window Crops Selected for Simulation, it shows the Edit QckTree Parameters button (Figure 17.34, left). If the tree crop pecan is not listed in the left window Available Crops (Figure 17.34, left), you will see the Add QckTree Parameters button in Figure 17.34, left.

- Clicking on the Add QckTree Parameters button (Figure 17.34, left) will show the Edit Quick Tree Parameters screen (Figure 17.34, right).
- Select the plant name 9508 Pecan from the drop-down list.
- Enter values for total growing season length (enter 240 days), total seasonal N uptake (120 kg/ha), plant height (800 cm), maximum Leaf Area Index (LAI) (4.0), and maximum rooting depth (80 cm). Use RZWQM2 default values for all other parameters.
- Click first on the Save button and then on Add Tree.

Now, 9508 Pecan will appear in the left window Available Crops (Figure 17.34, left). Select 9508 Pecan and click on Select Arrow to move names from Available Crops (left

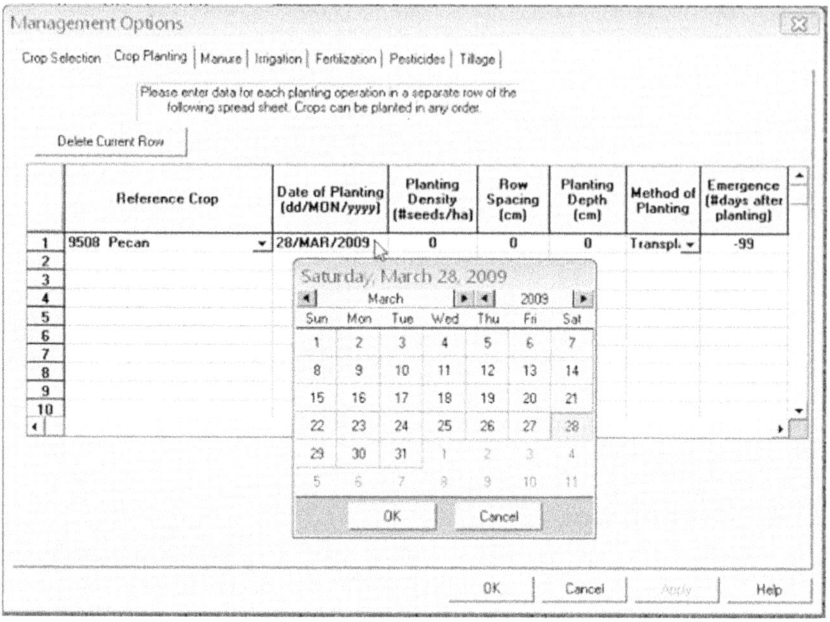

FIGURE 17.35 Crop planting and harvest input.

window) to Crops Selected for Simulation (right window). Note that if you want to remove crop name(s) from right window, select crop name(s) and click on Deselect Arrow.

- Click on the Apply button to save the data on the Crop Selection screen (Figure 17.34, left) and select the next dialog tab Crop Planting for data entry (Figure 17.35).

17.5.1.4.2 Crop Planting

The Crop Planting dialog tab allows users to set up crop management information, such as planting (date, density, row spacing) and harvesting (date, efficiency, type) (Figure 17.35).

- Select the crop 9508 Pecan" from the drop-down crop list.
- Double-click on the Date of Planting field or cell to bring up the calendar to set planting date as 28/MAR/2009.
- Enter Method of Planting (Transplant from the drop-down list), and Harvest Dependency (Harvest Date from drop-down list), and double-click on the Date of Harvest cell to bring up the calendar to set harvest date as 28/DEC/2009.
- Enter Harvest Types (Single/Seeds Only from drop-down list), and enter zero (0.0) for other parameters except for the RZWQM2 default number (–99.0).

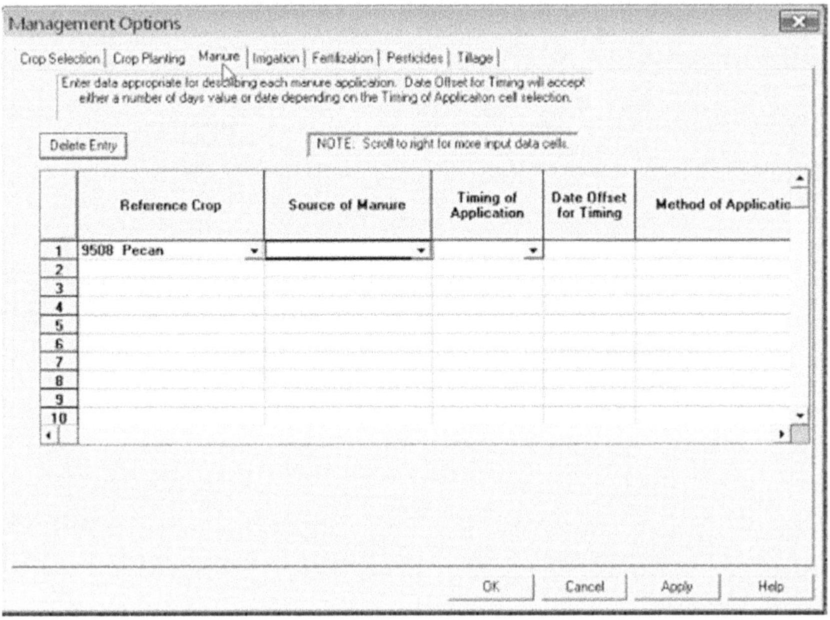

FIGURE 17.36 Manure management input.

- Click on the Apply button to save the data on the Crop Planting screen (Figure 17.35) and select the next dialog tab Manure for data entry (Figure 17.36).

17.5.1.4.3 Manure

The Manure dialog tab allows users to specify manure management inputs, such as manure type, application time, application amount, fraction of carbon, and C/N ratio (Figure 17.36). Readers are referred to the RZWQM2 model user guide (i.e., C:\RZWQM2\documentation\RZWQM2userguide.pdf) for details on manure management data entries. For the tutorial project, manure was not applied to the pecan orchard, so do not enter any values in parameter fields or cells.

- Click on the Apply button to save the data on the Manure screen (Figure 17.36) and select the next dialog tab Irrigation for data entry (Figure 17.37).

17.5.1.4.4 Irrigation

The Irrigation dialog tab allows users to schedule irrigation events and summarize irrigation amounts (Figure 17.37).

- Click on the Add/Edit Irrigation Events button under the box labeled Schedule Irrigation Events to bring up three dialog tabs for the following irrigation timing options: (1) Specific Dates to trigger an irrigation event on

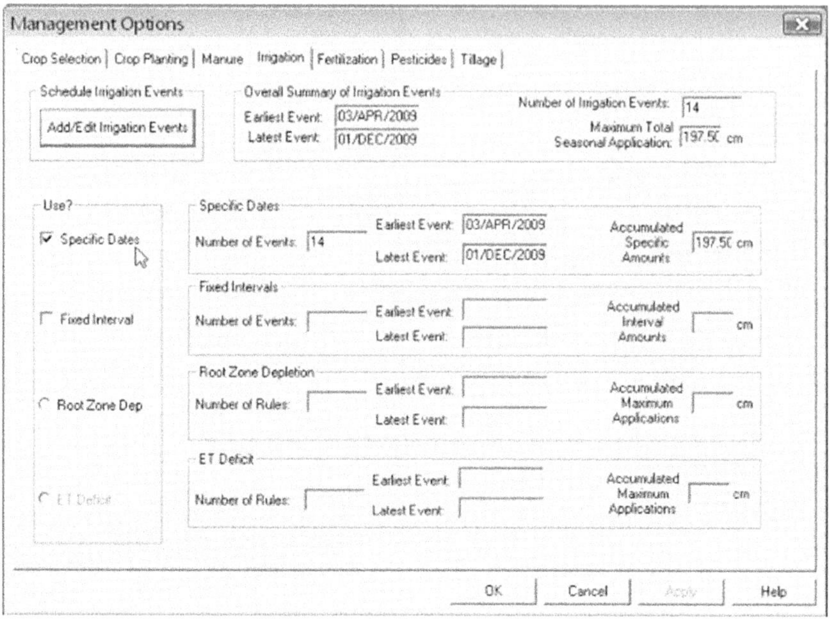

FIGURE 17.37 Summary screen for irrigation events.

a specified date, (2) Fixed Intervals to trigger an irrigation event at specified fixed time intervals within a specified range of dates, and (3) Root Zone Depletion to trigger an irrigation event when the available soil water content falls below a specified threshold within a specified range of dates (Figure 17.38).

- For the tutorial project, click on the Specific Dates dialog tab (Figure 17.38).

Readers are referred to the RZWQM2 model user guide (i.e., C:\RZWQM2\documentation\RZWQM2userguide.pdf) for details about Fixed Intervals and Root Zone Depletion data entries.

- Enter Plant Identification (9508 Pecan from drop-down list) and Type (Flood from drop-down list) (Figure 17.38).
- No entry for Application Rate (only applicable to sprinkler irrigation and no entry in this field for other types).
- Enter Maximum Total Seasonal Application (200 cm).
- From Table 17.2, enter Date and Amount in centimeters. Double-clicking on the Date field or cell will bring up the calendar to set the irrigation date.

For each row, 10 irrigation events (date and amount) can be entered, and continue on another row to complete 14 irrigation event dates and amounts (Table 17.2). Note

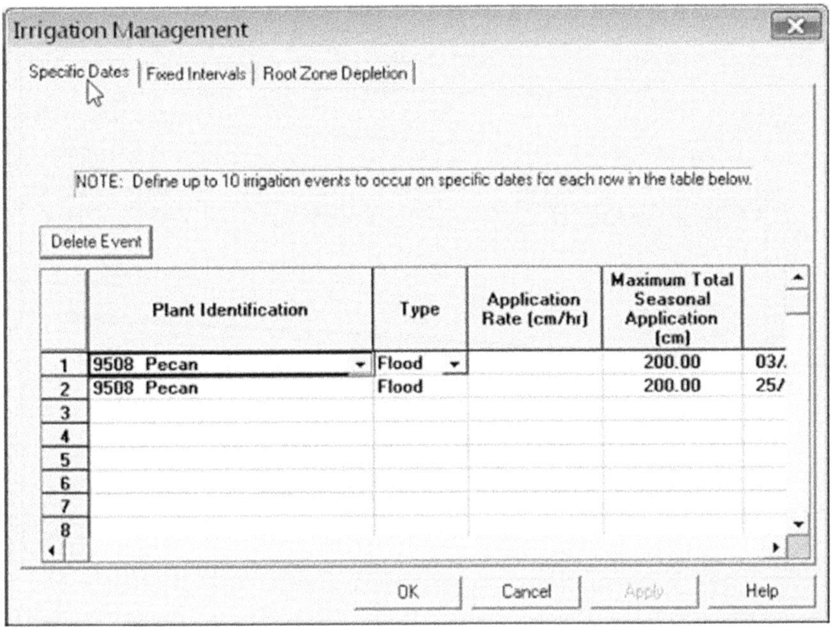

FIGURE 17.38 Irrigation data entry by the Specific Dates option.

TABLE 17.2
Irrigation Event Dates and Amounts for Irrigation by Specific Dates in the RZWQM2 Irrigation Management Screen (Figure 17.38)

Irrigation Event	Date (dd/MON/yyyy)	Amount (cm)
1	03/APR/2009	4.5
2	23/APR/2009	4.5
3	13/MAY/2009	6.0
4	01/JUN/2009	17.4
5	15/JUN/2009	17.7
6	01/JUL/2009	17.4
7	16/JUL/2009	17.9
8	27/JUL/2009	17.8
9	11/AUG/2009	17.6
10	03/SEP/2009	17.6
11	25/SEP/2009	17.6
12	07/OCT/2009	16.5
13	26/OCT/2009	16.5
14	01/DEC/2009	8.5

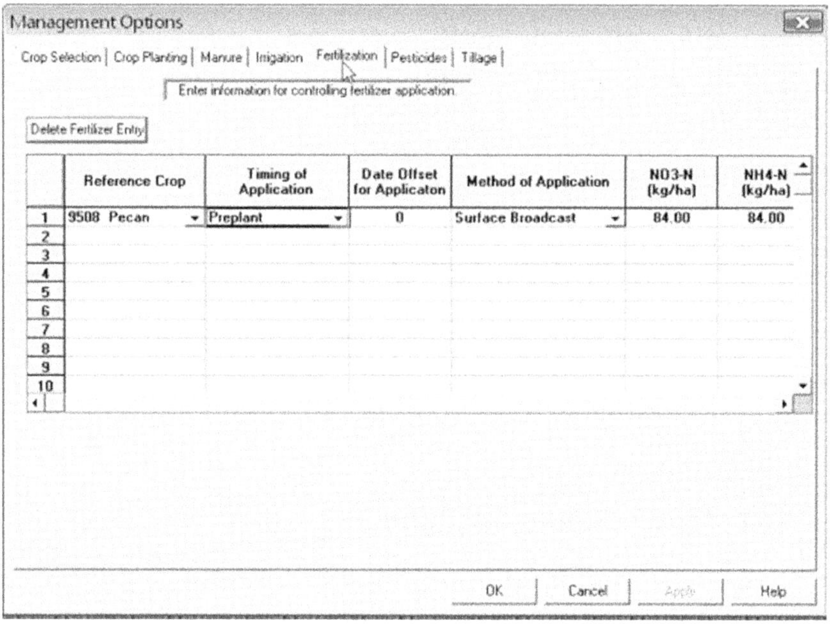

FIGURE 17.39 Fertilizer management input.

that if the value of Maximum Total Seasonal Application is 0.0, irrigation water is unlimited for the irrigation scheduling.

- Click on Apply to save the data on the Irrigation Management screen (Figure 17.38), and then click OK to return to the Irrigation Summary screen (Figure 17.37).
- Select the next dialog tab Fertilization (Figure 17.39).

17.5.1.4.5 Fertilization

The Fertilization dialog tab allows users to schedule inorganic fertilizer applications such as information on fertilizer type, application amount, application method, application time, and a Best Management Practices (BMP) option (Figure 17.39). Note that when BMP is selected, users can select BMP Options from the Method of Application drop-down options, and then users can enter BMP options from drop-down lists in the Chemical Application Option for BMPs and BMP Application Options fields. Readers are referred to the RZWQM2 model user guide (i.e., C:\RZWQM2\documentation\RZWQM2userguide.pdf) for details about Fertilization parameters.

- Select the Reference Crop (9508 Pecan), Timing of Application (Preplant), and Method of Application (Surface Broadcast) from the respective drop-down list. Use RZWQM2 default values for all other parameters (Figure 17.39).

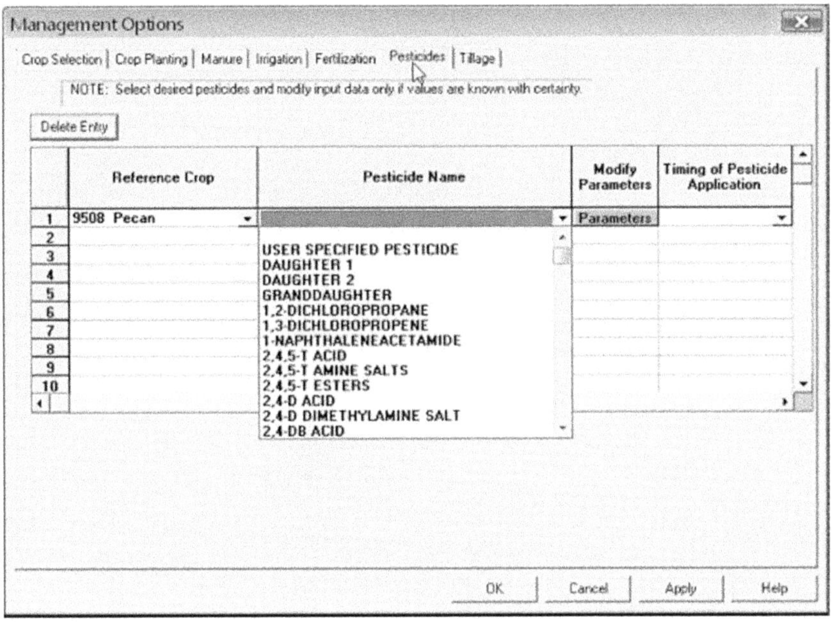

FIGURE 17.40 Pesticide management input.

- Click on Apply to save the data on the Fertilization screen (Figure 17.39), and click on the next dialog tab Pesticides (Figure 17.40).

17.5.1.4.6 Pesticides

The Pesticides dialog tab allows users to specify pesticide management information and pesticide parameters such as information on reference crop, pesticide type, application time, application amount, application method, and daily fraction of pesticide released (for slow-released form) (Figure 17.40). Pesticide parameters may be modified in the Modify Parameters column. Readers are referred to the RZWQM2 model user guide (i.e., C:\RZWQM2\ documentation\RZWQM2userguide.pdf) for details on Pesticides parameters.

For this tutorial project:

- Select the Reference Crop (9508 Pecan). Do not enter any data or keep all cells blank/default values for all other parameters (Figure 17.40).
- Click on the Apply button to save the data on the Pesticides screen (Figure 17.40) and click on the next dialog tab Tillage (Figure 17.41).

17.5.1.4.7 Tillage

The Tillage dialog tab allows users to enter tillage management inputs for the reference crop, including tillage type, tillage time, intensity, and effective depth of disturbance (Figure 17.41). Readers are referred to the RZWQM2 model user guide

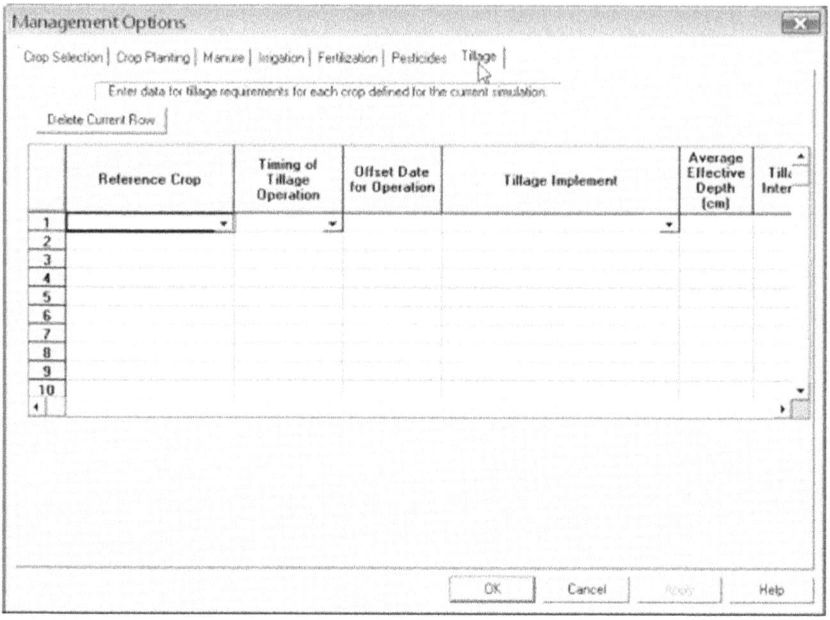

FIGURE 17.41 Tillage management input.

(i.e., C:\RZWQM2\documentation\RZWQM2userguide.pdf) for details on Tillage parameters.

For the tutorial project:

- Leave all parameter fields or cells blank as shown in Figure 17.41.
- Click on Apply and then on the OK button to exit the Management Options screen (Figure 17.41) and return to the Scenario window (Figure 17.22).
- In the Scenario window (Figure 17.22), click File | Save Scenario, and click OK.

17.5.1.5 Define PRMS Snow Parameters

A snow file (*.SNO) for snowpack dynamics is required for RZWQM2. The snow file (*.SNO) can be copied directly from the default scenario. The default snow file can be renamed or left with the same name. For the tutorial project, the default snow file is "RZWQM2MetData.SNO" (located in C:\RZWQM2-Tutorial\Meteorology). Note that the snow file is not daily snow data but an input parameter file for the PRMS snow module. The PRMS snow module is not important for this tutorial project, and therefore, the default snow file "RZWQM2MetData.SNO" may not require editing. However, one can replace all default values with zero (0.0) in text editor.

- Click Input | PRMS Snow Parameters (Figure 17.22). The text file "RZWQM2MetData.SNO" for the tutorial project will be accessed by WordPad from the interface (Figure 17.42).
- Close the WordPad file "RZWQM2MetData.SNO."

RZWQM2MetData.SNO - WordPad

File Edit View Insert Format Help

```
===========================================================================
=
==
==
==          P R M S   S N O W   P A R A M E T E R S
==
==
==
==            WATERSHED:  Lucerne, CO 1994-95
==
==
==
===========================================================================
=
=      GENERAL SNOWPACK DYNAMICS
=
= REC.NUM    DESCRIPTION
= --------   -----------
=   1.1     AVERAGE MAXIMUM SNOWPACK DENSITY [GM/CM3]
=   1.2     INITIAL DENSITY OF NEW-FALLEN SNOW IN [GM/CM3]
=   1.3     FREE-WATER HOLDING CAPACITY OF SNOWPACK
=   1.4     SNOWPACK SETTLEMENT TIME CONSTANT
=   1.5     PRECIP ALL SNOW IF MAX TEMPERATURE BELOW THIS VALUE [F]
===========================================================================
==
0.0  0.0  0.00  0.0  00.0
===========================================================================
=
=      ALBEDO ADJUSTMENTS
=
= REC.NUM    DESCRIPTION
= --------   -----------
=   1.1     ALBEDO RESET - RAIN, MELT STAGE
=   1.2     ALBEDO RESET - RAIN, ACCUMULATION STAGE
=   1.3     ALBEDO RESET - SNOW, MELT STAGE
=   1.4     ALBEDO RESET - SNOW, ACCUMULATION STAGE
===========================================================================
=
0.0  0.0  0.0  0.00
===========================================================================
=
=      SURFACE COVER ADJUSTMENTS
=
= REC.NUM    DESCRIPTION
= --------   -----------
=   1.1     COVER TYPE DESIGNATION FOR HRU
=   1.2     SUMMER VEGETATION COVER DENSITY FOR MAJOR VEGETATION TYPE
=   1.3     WINTER VEGETATION COVER DENSITY FOR MAJOR VEGETATION TYPE
===========================================================================
=
0  0.0  0.0
===========================================================================
=
=      GENERAL ATMOSPHERIC PARAMETERS
=
```

For Help, press F1

FIGURE 17.42 The snowpack dynamics file "RZWQM2MetData.SNO" for the tutorial project.

17.5.1.6 Define Experimental Data Observations

- Click Input | Experiment Data (Optional) (Figure 17.22) and the text file "expdata.dat" can be accessed via WordPad from the RZWQM2 interface (version 2.10.2010).

Users can enter observed data from their experiment and RZWQM2 will perform comparison statistics of observed versus predicted for variables allowed in the format. The format is listed in the Appendix of the RZWQM2 model user guide (i.e., C:\RZWQM2\documentation\RZWQM2userguide.pdf). For the tutorial project, we will not define experimental data in the "expdata.dat" file. Close the WordPad file "expdata.dat."

- Click File | Save Scenario (Figure 17.22), and click OK.

17.5.2 Simulation Controls

- Click Simulation Controls | Select Output variables (Figure 17.22), which will bring up the Program Output Control screen with four dialog tabs: Output Variable Selection, Debugging Flags, Simulation Iteration Control, Misc. Output Items (Figure 17.43).

Each of these four submenus can also be opened from the Simulations Controls menu.

17.5.2.1 Output Variable Selection

The Output Variable Selection dialog tab allows users to select from 98 scalar (value–time) and 51 vector (value–depth–time) output variables (Figure 17.43). Push the downward arrow to selected variables into the lower spreadsheet (Figure 17.43). Upon completion of the simulation run, scalars will be stored in "DAILY.PLT" and vector variables will be stored in "LAYER.PLT" (located in C:\RZWQM2-Tutorial\New Scenario). Note that RZWQM2 will not create these plot (*.PLT) files unless at least one output variable is chosen.

- Click on the Apply button (Figure 17.43) and select the next dialog tab Debugging Flags (Figure 17.44).

17.5.2.2 Debugging Flags

The Debugging Flags dialog tab allows users to select the output files that are most related to their experimental interests by selecting different debug options (Figure 17.44).

- For the tutorial project, select the radio button Water mass balance output generated at daily time step (Figure 17.44).

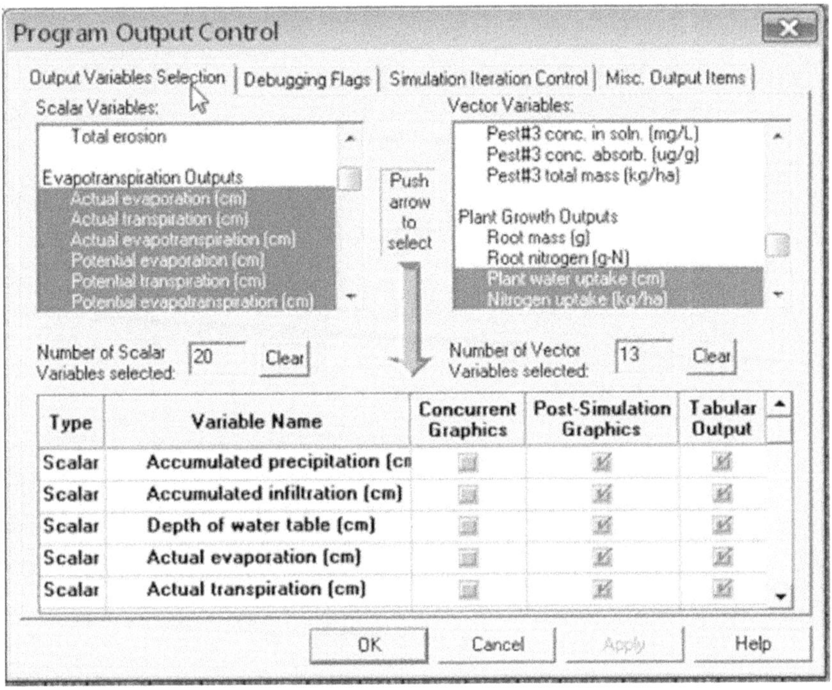

FIGURE 17.43 Output variable selection on the Program Output Control screen.

- Click on the Apply button (Figure 17.44) and select the next dialog tab Simulation Iteration Control (Figure 17.45).

17.5.2.3 Simulation Iteration Control

The Simulation Iteration Control dialog tab allows users to set the number of iterations and define system status for the next simulation (Figure 17.45). Users may set up multiple runs (Multi-Pass Control). The setup screen allows users to select the number of runs of the same scenario, recording system status at the end of each run, and reading in the system status at the beginning of a subsequent run.

- For the tutorial project, check boxes as shown in the Simulation Iteration Control screen (Figure 17.45) and enter zero (0) for Number of Passes.
- Click on the Apply button (Figure 17.45) and select the next dialog tab Misc. Output Items (Figure 17.46).

17.5.2.4 Misc. Output Items

The Misc. Output Items dialog tab allows users to select node numbers where chemical breakthrough values are monitored (Figure 17.46).

- For the tutorial project, shut down the inorganic chemical equilibrium model (check radio button No) or the microbial growth simulation (check radio button No).

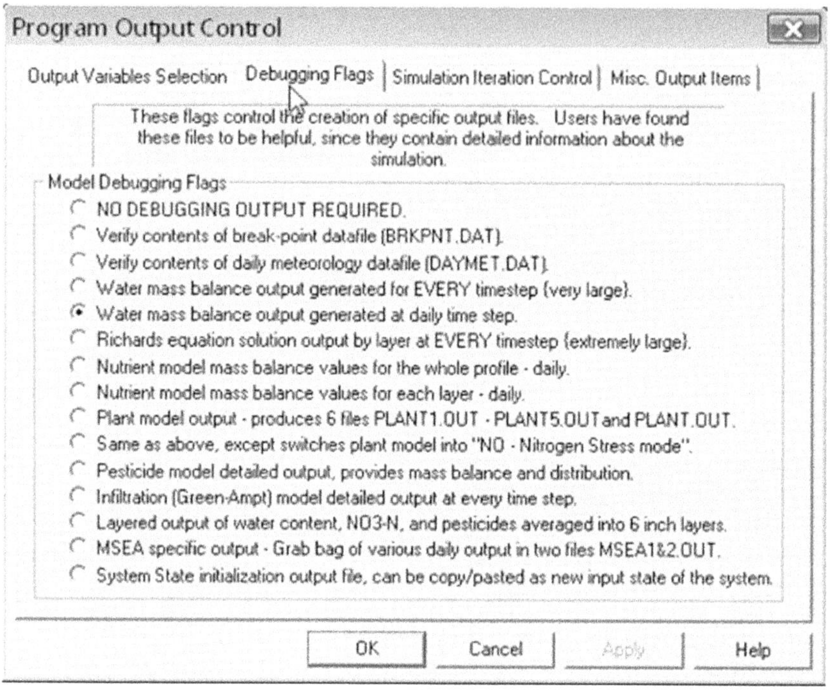

FIGURE 17.44 Debugging flags under Simulation Controls.

- Click on Apply and then on OK to exit the Program Output Controls screen (Figure 17.46) and return to the Scenario window (Figure 17.22).
- Click File | Save Scenario (Figure 17.22), and then click OK.

17.5.3 Modifiers

Clicking on the Modifiers menu (Figure 17.22) will bring up the Simulation Adjustments screen (Figure 17.47), which allows users to modify weather information (Driving Variable Adjustments) and management practices (Management Practice Adjustments) to conduct a what-if analysis of a particular agricultural system. Clicking on the Adjustments button under the box labeled Driving Variable Adjustments (Figure 17.47) will allow users to modify weather information such as air temperature, rainfall, wind speed, solar radiation, relative humidity, pan evaporation, and CO_2. For the tutorial project, skip the Driving Variable Adjustments box.

Clicking on the Adjustments button under the box labeled Management Practice Adjustments (Figure 17.47) will allow users to modify fertilizer amount and timing, manure application amount and timing, irrigation amount and timing, pesticide application amount and timing, and timing of tillage. For this tutorial project, skip the Management Practice Adjustments box. Readers are referred to the RZWQM2

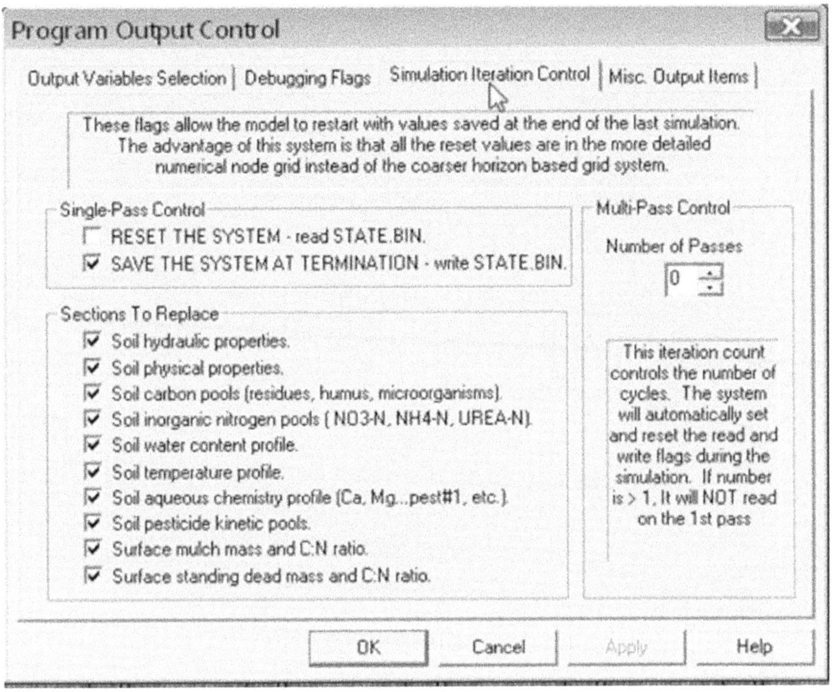

FIGURE 17.45 Simulation iteration control options.

model user guide (i.e., C:\RZWQM2\documentation\RZWQM2userguide.pdf) for Driving Variable Adjustments and Management Practice Adjustments dialog tab descriptions.

- Click OK to exit the Simulation Adjustments screen (Figure 17.47) and return to the Scenario window (Figure 17.22).
- Click File | Save Scenario (Figure 17.22).

17.5.4 Run the Scenario

Before running the scenario for the tutorial project (Figure 17.22), check that input data files/file names are correctly selected in each of the boxes labeled Daily Meteorology File ("LamanchaMETmardec09.met"), Breakpoint Rainfall File ("LamanchaMET mardec09hrly.brk"), and Snowpack Dynamics File ("RZWQM2MetData.SNO"). If these files do not appear in the boxes, click on the arrow at the right of the field for options. Also, make sure that New Scenario is saved by clicking File | Save Scenario.

- Click on the Run Scenario menu to execute New Scenario. You will see a running progress window at the center of your monitor, notifying the status of the execution (Figure 17.48).

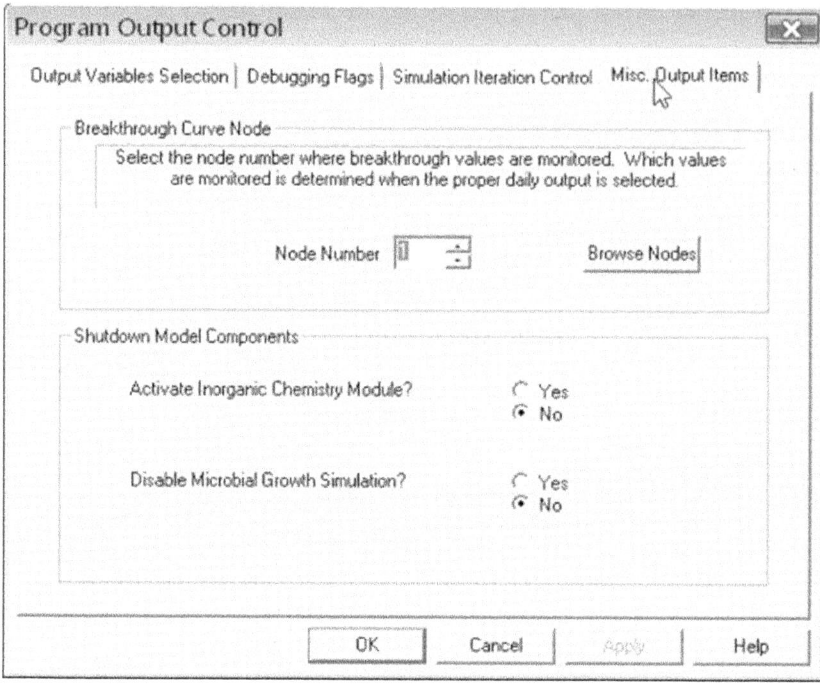

FIGURE 17.46 Options for monitoring a specific soil numerical node and shutting down model components.

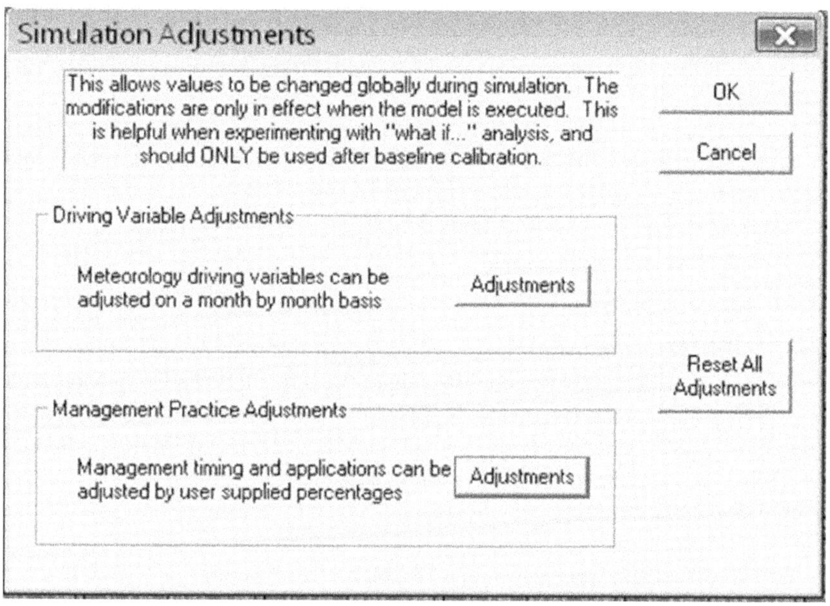

FIGURE 17.47 Simulation adjustments for meteorological variables or management practices.

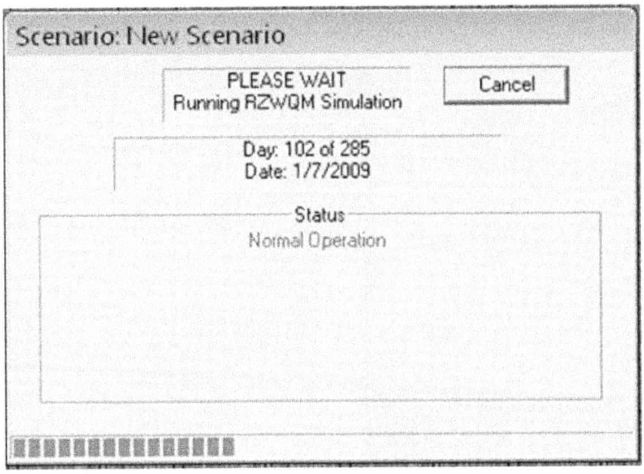

FIGURE 17.48 The progress screen for the simulation run.

17.5.5 Scenario Output

17.5.5.1 View Scenario Output Files

- Click the menu item Output | View Output Files or click on the View Output Files button on the New Scenario window (Figure 17.49).

The four types of files from the RZWQM2 interface for a simulation run are (1) *.OUT files (output results), (2) *.DAT (input parameter files), (3) *.PLT (for graphic plotting), and (4) *.LOG (daily recording of the simulation run). It is recommended to view the RZWQM.LOG file to check the completion of the simulation run.

The output files are opened with the WordPad text editor for viewing purposes. Examples of output files are depicted in Figures 17.50 through 17.52. Information in the output file "ACCWAT.out" (Figure 17.50) can be used to evaluate temporal variations in different water balance components such as infiltration, evapotranspiration, total soil water storage in the profile, deep percolation, etc. during the simulation period. Information in the output file "MBLWAT. out" (Figure 17.51) can be used to evaluate daily water balance components, including daily deep percolation (i.e., seepage out of drain) below the root zone, particularly for each irrigation event date as presented in Table 17.2. Information in the output file "LAYER.PLT" (Figure 17.52), for example, temporal variations in simulated soil water content at different soil depths, can be compared with the measured soil water contents at corresponding soil depths. Readers are referred to Deb et al. (2012) for details about the simulation of deep percolation in a flood-irrigated, mature pecan orchard using both RZWQM2 (version 2.10.2010) and the daily water balance estimation method.

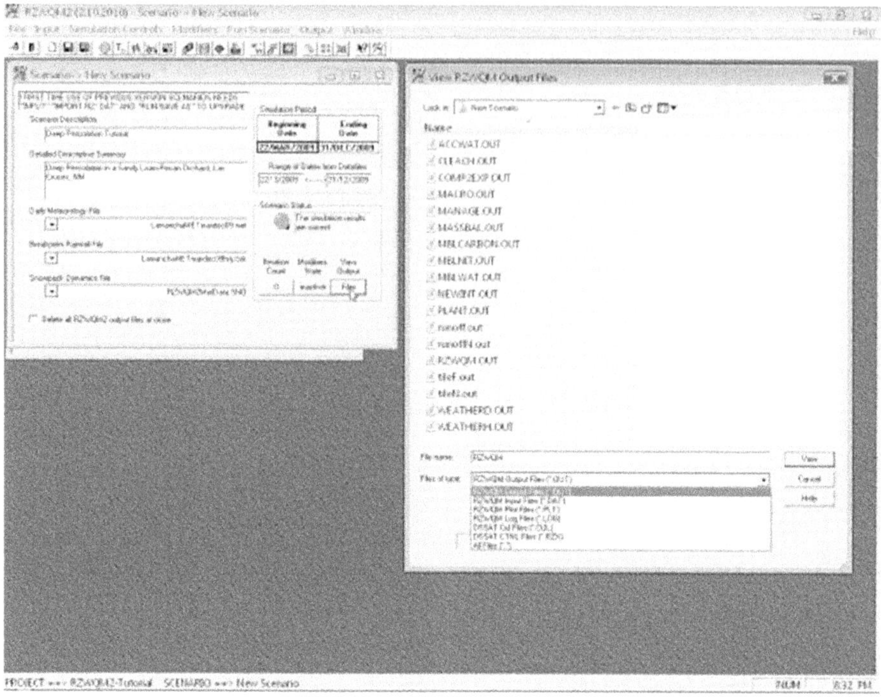

FIGURE 17.49 View output files.

All output files are stored in the New Scenario subdirectory (located in C:\RZWQM2-Tutorial\New Scenario). Output files in this New Scenario subdirectory can also be opened using Microsoft Excel. Description of RZWQM2 output files (*.OUT, *.PLT, and *.LOG) can be found in the Appendix of the RZWQM2 User Guide (i.e., C:\RZWQM2\documentation\RZWQM2userguide.pdf).

17.5.5.2 Plot Model Outputs

- Click Output | Autoplot (Figure 17.22), which will bring up the AutoPlot Selection screen (Figure 17.53).
- Check the box LAYER.PLT and click OK to view the 3D plots (value–depth–time) of the selected variables (Figure 17.54).

Similarly, users can also check the box DAILY.PLT to view the 2D plots (value–time) of the selected variables.

17.6 SCENARIO COMPARISONS

Scenario comparisons can be made at the project window (Figure 17.1). Each successful RZWQM2 Scenario run produces a standard output file for analysis in the

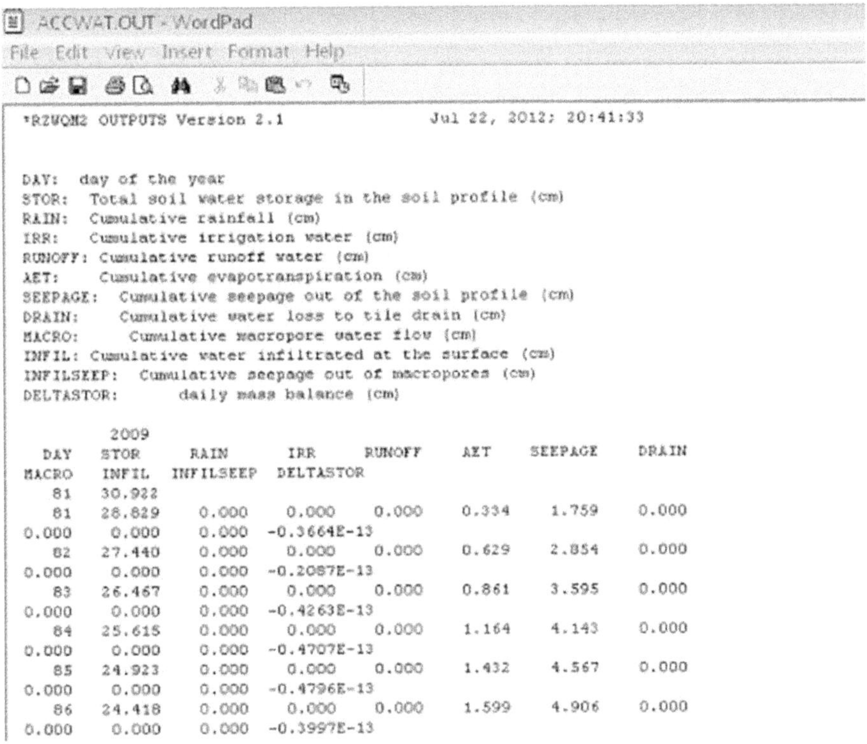

```
ACCWAT.OUT - WordPad
File Edit View Insert Format Help

*RZWQM2 OUTPUTS Version 2.1                    Jul 22, 2012; 20:41:33


DAY:   day of the year
STOR:   Total soil water storage in the soil profile (cm)
RAIN:   Cumulative rainfall (cm)
IRR:    Cumulative irrigation water (cm)
RUNOFF: Cumulative runoff water (cm)
AET:     Cumulative evapotranspiration (cm)
SEEPAGE:   Cumulative seepage out of the soil profile (cm)
DRAIN:     Cumulative water loss to tile drain (cm)
MACRO:       Cumulative macropore water flow (cm)
INFIL: Cumulative water infiltrated at the surface (cm)
INFILSEEP:  Cumulative seepage out of macropores (cm)
DELTASTOR:        daily mass balance (cm)

          2009
   DAY   STOR       RAIN       IRR       RUNOFF      AET     SEEPAGE      DRAIN
MACRO    INFIL   INFILSEEP  DELTASTOR
    81   30.922
    81   28.829     0.000     0.000     0.000      0.334      1.759      0.000
0.000     0.000     0.000  -0.3664E-13
    82   27.440     0.000     0.000     0.000      0.629      2.854      0.000
0.000     0.000     0.000  -0.2087E-13
    83   26.467     0.000     0.000     0.000      0.861      3.595      0.000
0.000     0.000     0.000  -0.4263E-13
    84   25.615     0.000     0.000     0.000      1.164      4.143      0.000
0.000     0.000     0.000  -0.4707E-13
    85   24.923     0.000     0.000     0.000      1.432      4.567      0.000
0.000     0.000     0.000  -0.4796E-13
    86   24.418     0.000     0.000     0.000      1.599      4.906      0.000
0.000     0.000     0.000  -0.3997E-13
```

FIGURE 17.50 The output file "ACCWAT.out": Accumulated daily water storage, runoff, actual evapotranspiration, seepage (deep percolation), drainage, macropore flow, and infiltration.

project's Analysis subdirectory (located in C:\RZWQM2-Tutorial\Analysis), for example, for the tutorial project "New Scenario.ana." Deleting a scenario on the project window will not remove "New Scenario.ana" from the Analysis subdirectory, but it can be deleted using Windows Explorer.

Since only one scenario, New Scenario, is considered in this tutorial project, we can view the project summary for New Scenario.

- If the New Scenario window or menu items are open, click File | Exit and Save Scenario (Figure 17.1).
- Click the Project Summary menu on the Project window (Figure 17.1), which will bring up the Project Summaries screen (Figure 17.55).
- Select "New Scenario.ana" under the box labeled Choose Which Scenarios to Summarize, and select a variable under the box labeled Choose Multiple Variables to Summarize. Multiple variables can be selected by using the control key with the selection (Figure 17.55).

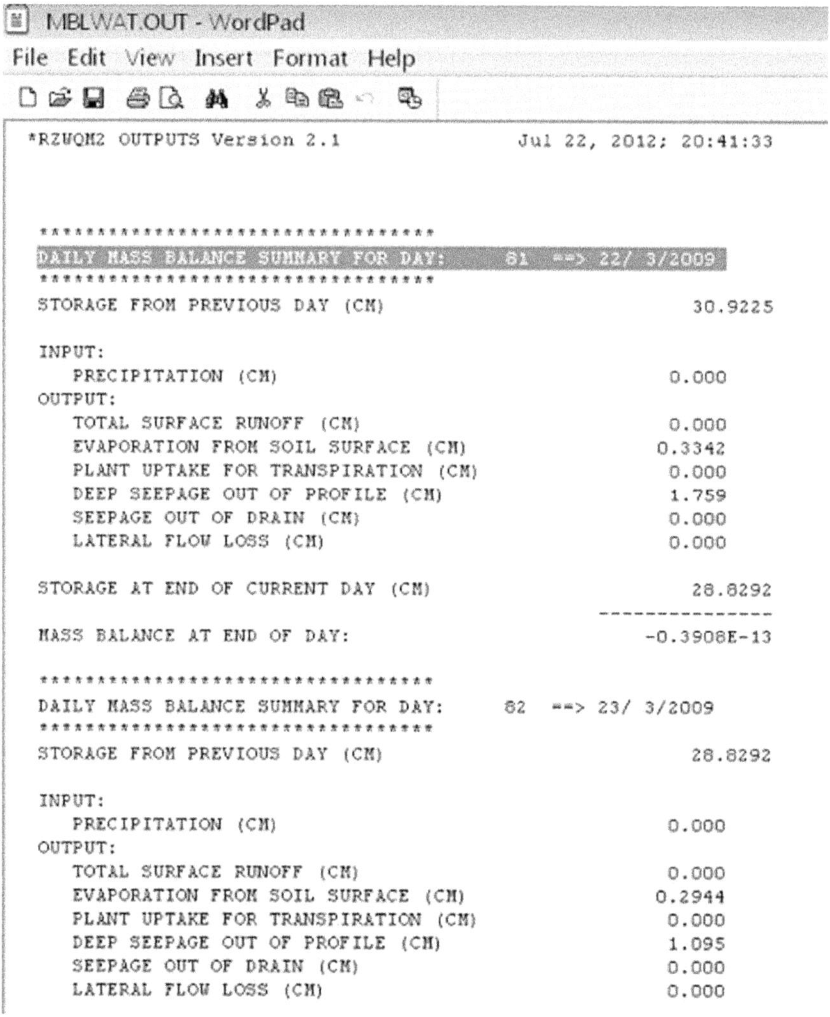


```
*RZWQM2 OUTPUTS Version 2.1                         Jul 22, 2012; 20:41:33



 ***********************************
 DAILY MASS BALANCE SUMMARY FOR DAY:      81  ==> 22/ 3/2009
 ***********************************
 STORAGE FROM PREVIOUS DAY (CM)                              30.9225

 INPUT:
    PRECIPITATION (CM)                                    0.000
 OUTPUT:
    TOTAL SURFACE RUNOFF (CM)                             0.000
    EVAPORATION FROM SOIL SURFACE (CM)                   0.3342
    PLANT UPTAKE FOR TRANSPIRATION (CM)                   0.000
    DEEP SEEPAGE OUT OF PROFILE (CM)                      1.759
    SEEPAGE OUT OF DRAIN (CM)                             0.000
    LATERAL FLOW LOSS (CM)                                0.000

 STORAGE AT END OF CURRENT DAY (CM)                          28.8292
                                                        ---------------
 MASS BALANCE AT END OF DAY:                             -0.3908E-13

 ***********************************
 DAILY MASS BALANCE SUMMARY FOR DAY:      82  ==> 23/ 3/2009
 ***********************************
 STORAGE FROM PREVIOUS DAY (CM)                              28.8292

 INPUT:
    PRECIPITATION (CM)                                    0.000
 OUTPUT:
    TOTAL SURFACE RUNOFF (CM)                             0.000
    EVAPORATION FROM SOIL SURFACE (CM)                   0.2944
    PLANT UPTAKE FOR TRANSPIRATION (CM)                   0.000
    DEEP SEEPAGE OUT OF PROFILE (CM)                      1.095
    SEEPAGE OUT OF DRAIN (CM)                             0.000
    LATERAL FLOW LOSS (CM)                                0.000
```

FIGURE 17.51 The output file “MBLWAT.out”: Detailed daily water balance information.

The asterisk (*) next to a variable in the interface indicates that the variable was added or its definition was changed in RZWQM2 version 2.0. However, only one variable can be compared at one time when multiple scenarios are selected.

- Select the summary’s time resolution from the drop-down list (i.e., whole simulation period, yearly, month-by-month, and month-by-year) and the radio button for the summary statistics (i.e., Minimum, Maximum, Minimum and Maximum, Average, Average & Standard Deviation, and Summation) (Figure 17.55).
- Click on the Generate Summary button (Figure 17.55), which will result in a summary sheet.

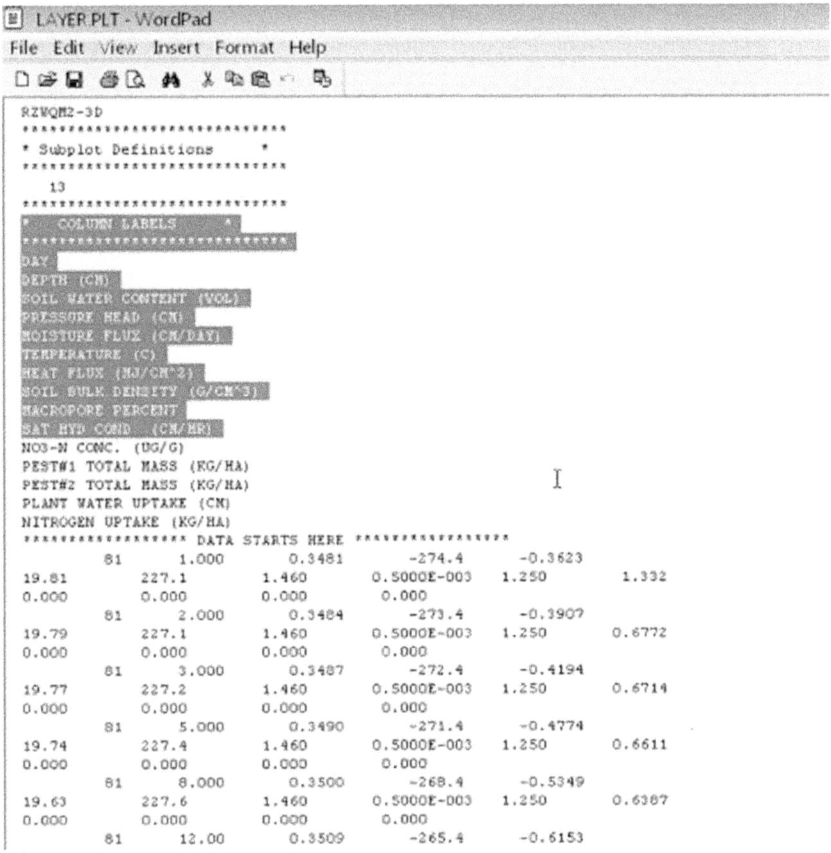


```
RZWQM2-3D
*****************************
* Subplot Definitions       *
*****************************
   13
*****************************
*    COLUMN LABELS          *
*****************************
DAY
DEPTH (CM)
SOIL WATER CONTENT (VOL)
PRESSURE HEAD (CM)
MOISTURE FLUX (CM/DAY)
TEMPERATURE (C)
HEAT FLUX (MJ/CM^2)
SOIL BULK DENSITY (G/CM^3)
MACROPORE PERCENT
SAT HYD COND  (CM/HR)
NO3-N CONC. (UG/G)
PEST#1 TOTAL MASS (KG/HA)
PEST#2 TOTAL MASS (KG/HA)
PLANT WATER UPTAKE (CM)
NITROGEN UPTAKE (KG/HA)
****************** DATA STARTS HERE *****************
             81        1.000         0.3481         -274.4        -0.3623
19.81           227.1         1.460         0.5000E-003    1.250          1.332
0.000           0.000         0.000         0.000
             81        2.000         0.3484         -273.4        -0.3907
19.79           227.1         1.460         0.5000E-003    1.250         0.6772
0.000           0.000         0.000         0.000
             81        3.000         0.3487         -272.4        -0.4194
19.77           227.2         1.460         0.5000E-003    1.250         0.6714
0.000           0.000         0.000         0.000
             81        5.000         0.3490         -271.4        -0.4774
19.74           227.4         1.460         0.5000E-003    1.250         0.6611
0.000           0.000         0.000         0.000
             81        8.000         0.3500         -268.4        -0.5349
19.63           227.6         1.460         0.5000E-003    1.250         0.6387
0.000           0.000         0.000         0.000
             81        12.00         0.3509         -265.4        -0.6153
```

FIGURE 17.52 The output file "LAYER.PLT": All the information to generate 3D (value–depth–time) plots of user-selected variables in RZWQM2.

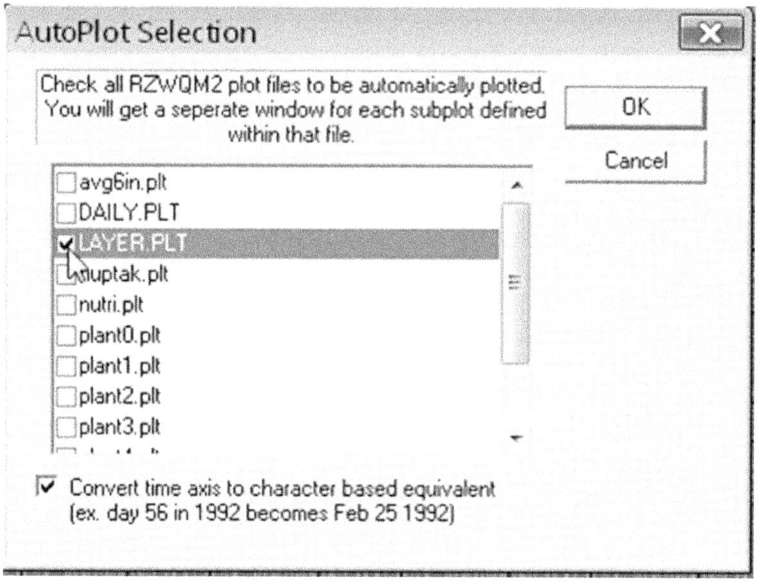

FIGURE 17.53 The file selection for on-screen graphics.

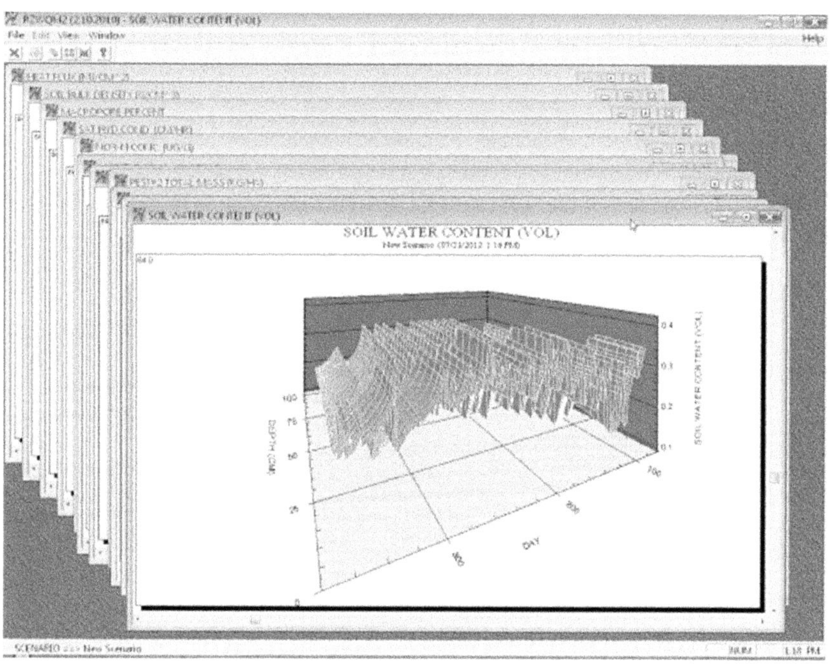

FIGURE 17.54 On-screen graphics of selected variables using the file "LAYER.PLT" (Figure 17.52).

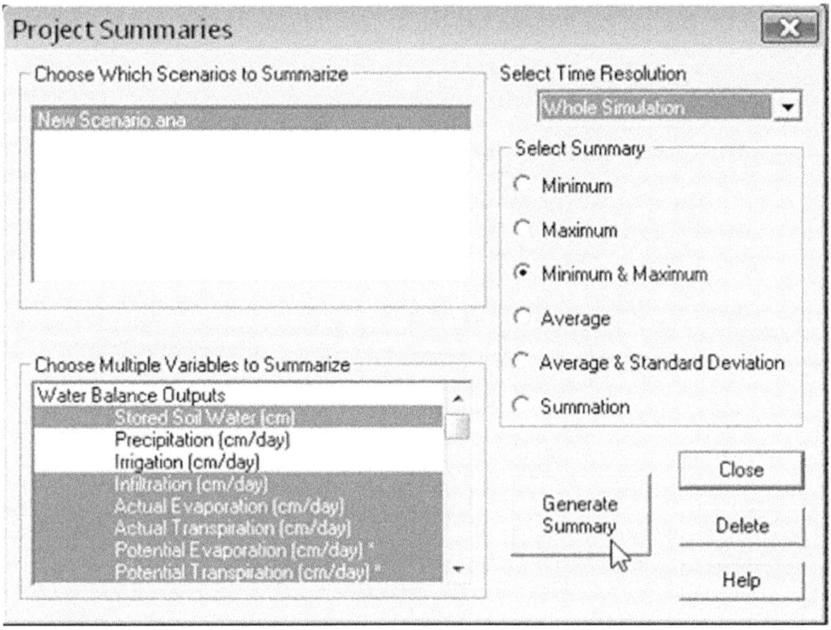

FIGURE 17.55 Project Summary dialog screen.

PROBLEMS

17.1 Write a summary on RZWQM2 capabilities.

17.2 Write a note on the data required for conducting a water balance using RZWQM.

17.3 Run the RZWQM using any available dataset for different soil types and compare infiltration and fluxes at the bottom boundary.

17.4 Collect published literature on RZWQM and list newer developments not covered in this chapter.

17.5 Explain different sink/source terms and boundary conditions.

REFERENCES

Ahuja, L. R., K. W. Rojas, J. D. Hanson, M. J. Shaffer, and L. Ma. 2000. *The Root Zone Water Quality Model: Modeling Management Effects on Water Quality and Crop Production.* Water Resources Publications, LLC, Highlands Ranch, CO.

Deb, S. K., M. K. Shukla, and J. G. Mexal. 2012. Simulating deep percolation in flood-irrigated mature orchards with RZWQM2. *Trans. ASABE* 55: 2089–2100.

Ma, L., L. R. Ahuja, S. A. Saseendran, R. W. Malone, T. R. Green, B. T. Nolan, P. N. S. Bartling, G. N. Flerchinger, K. J. Boote, and G. Hoogenboom. 2011. Chapter 1: A protocol for parameterization and calibration of RZWQM2 in field research. In: L. R. Ahuja and L. Ma (eds.), *Methods of Introducing System Models into Agricultural Research ASA*, CSSA and SSSA Book Series, Madison, WI, pp. 1–64.

18 Special Topics
Soil Erosion, Contaminant Remediation, and Salt Removal

18.1 INTRODUCTION

The aim of this chapter is to provide more knowledge about different solutes that could also be contaminants. Undergraduate and graduate students of environmental science and environmental engineering study various types of contaminants that are found in the soil naturally or due to anthropogenic activities. Their focus is simultaneously on soil or environmental remediation. To do that, students learn about various types of solutes and contaminants; properties of the contaminants; soil–contaminant–plant interactions that determine how a contaminant moves through the soil, from soil into the plant, or is retained by the soil; and how a contaminant is subsequently slowly released by the soil. Contaminant sorption and movement are discussed in Chapter 15. Contaminants (or solutes) are present naturally in soil. These could include inorganic (salts, heavy metals) and organic (oil, gas) contaminants. Some contaminants can be added along with applications of pesticides or fertilizers to the soil or due to a spill. Inorganic contaminants can increase soil salinity and sodicity, and render a field uncultivable. Contaminants can also move into an area due to flooding or soil erosion. Therefore, a proper understanding of types of soil erosion and various erosion processes are needed. Both wind and water cause erosion and together a lot of soil as well as attached chemicals and contaminants can be transported to agricultural, urban, rural areas or to water bodies. It is very hard to measure actual erosion taking place from the soil as well as from a hill or a mountain. Several empirical and physically based models are available for erosion prediction. In this chapter, soil erosion types, processes, contaminants and contaminant remediation techniques as well as soil salinity and sodicity are discussed.

18.2 SOIL EROSION

Soil erosion is the removal of the top soil layer due to water, wind, snow, ice, plants, and animals. Soil erosion is both a naturally occurring process as well as a manmade or anthropogenic process. Human activities can result in accelerated erosion as well as erosion control. Natural processes, including weathering, dissolution, abrasion, corrosion, and transportation, can cause material to be worn away from the earth's surface. In agriculture, the top soil is the most fertile soil

DOI: 10.1201/9780429264849-18

and its removal due to erosion can result in loss of productivity and it has the potential to disrupt the food chain. In spite of this, the impact of soil erosion on crops is neglected, likely because it a very small slice of soil over a large area and it is not visible to eyes.

The main agents of erosion that wear away the earth's surface are gravity, wind (eolian erosion) or water (eluvian erosion), which also includes ice (mainly glaciers). Erosion does not include mass wasting (movement down slope; slope failure), unless erosion is being discussed in a tectonic context, in which case exhumation may be the better term to describe erosion. Erosion is an extensive and chronic problem, and human activities, particularly agriculture and overgrazing, accelerate erosion.

Eastern Asia, South Eastern Asia, Southern North America, and sub-Saharan Africa are some of the areas where agricultural lands are most vulnerable to erosion (Borrelli et al., 2017; Pimentel et al., 1995). The southwestern United States has a very high rate of soil erosion. For example, some estimates by the National Resources Conservation Service of the US Department of Agriculture (USDA NRCS, 2018) show that, on average, the annual soil erosion rate for cultivated lands in New Mexico is 53 Mg ha^{-1} (23.6 tons/acre), more than 97% of which was due to high winds in the area. Some estimates from USDA NRCS (2018) indicated that the rate of wind erosion in New Mexico has increased from about 16 tons/acre in 1982 to 23 tons/acre in 2015. Soil is generally considered nonrenewable because it takes 200 to 1,000 years for one inch of soil to form.

The Dust Bowl of the 1930s in the US was precipitated by drought and bad land-use practices. The Dust Bowl led to the creation of the Soil Erosion Service, which developed into the Soil Conservation Service (SCS) in 1935 and was critical to the future of federal soil conservation activities. The SCS was later renamed the Natural Resources Conservation Service (NRCS).

18.3 COST TO SOCIETY

Soil erosion puts a huge financial burden on societies. The costs to society include both direct costs (costs that can be estimated) and indirect costs (difficult to estimate). Some of the direct costs for farmers and society are the costs associated with loss of productivity, loss or replacement of nutrients, and loss of organic matter. A horizon (eroded first) is richest in nutrients because it is the zone of fertilizer placement, nutrient cycling, and highest organic matter content. The erosion can create problems for field operations, for example, operating equipment and irrigation. Generally, direct costs are much smaller than indirect costs.

Some examples of indirect costs include soil compaction, loss of soil porosity, loss of water-holding capacity of the soil, soil flooding, and depression storage. Lack of water storage capacity in eroded soils is often erosion's most damaging effect, especially in drier regions where water is the main limiting factor, for example, in the southwestern US. The indirect costs could also be associated with sediment migration from the field or area to subsequent deposition in surface water sources. Soil leaving the field because of erosion can have as much as two to three times the nutrients and one to five times the organic matter than the soil left behind. Along with soil particles, nutrients, pesticides, and other contaminants can be deposited in the surface water sources. These can cause eutrophication and adversely affect aquatic and wildlife. Cleanup of sediments moved via wind or water can be very expensive.

The most costly effect of erosion is sedimentation, usually an off-site cost. According to some reported estimates, more than 12,300 hectare-meters (1 million acre-feet) of sediment settles annually in reservoirs, reducing water storage by the same volume. Sediment reduces the carrying capacity of waterways, reduces storage capacity of reservoirs and lakes, produces turbidity that may harm aquatic plants and fish, and increases the cost of water purification.

One other indirect cost of erosion is associated with the impacts on the ecology. Erosion can cause deleterious impacts for species of the ecosystem due to the alterations in seasonal flow, water temperature, nutrient loads, and trace metal accumulation, and could result in reductions in species diversity with increases in the non-native species population.

Many indirect consequences of erosion involve human health and values. The clogging of streams and reservoirs by erosion and the formation of swampy areas is one of the principal obstacles to effective control of malaria, encephalitis, and other mosquito-transmitted diseases. Non-native animal and vegetation species often become invasive when introduced into environments lacking natural predators or other mechanisms to keep the dominating species in check. Lack of ownership, values, and understanding of long-term devotion to land stewardship results in disregard for erosion.

The government can improve or encourage more degradation. The Federal Water Pollution Control Act (FWPCA) of 1972 focused attention on eroded soil as an extensive water pollutant; the Clean Water Act of 1977 amended section 208 and requires states to develop plans to control "non-point" sources of pollution such as eroded soil.

18.4 PHYSICAL AND CHEMICAL PROCESSES OF EROSION

The physical processes of erosion are called corrasion or mechanical erosion. Corrasion can be in the lateral or vertical direction and can widen a valley or deepen it, respectively. The examples of corrasion include weathering and movement of a rock in the floodplain of a river or in an ocean. The continuous flow of water causes the abrasion or disintegration of rock. Similarly, blowing winds also cause corrasion. The chemical processes of erosion are known as corrosion or chemical erosion. Corrosion is erosion by chemical factors including solutions or water with suspended chemical sediments. Corrosion results in a substance or rust formation. However, most examples of erosion include both corrasion and corrosion.

18.5 COASTAL EROSION

The United States has more than 95,000 miles of coastline and 3.4 million square miles of ocean within territorial waters. Most erosion events at the coasts are the result of storms. Since slope in these areas is so gentle, a small rise in the sea level produces a large inland shift of the shoreline. This increases the threats to coastal development, transportation infrastructure, tourism, freshwater aquifers, and fisheries, which are already stressed by human activities. The global sea level has already risen by 4 to 8 inches (10–20 cm) in the past century and models suggest this rise is very likely to accelerate. The best estimate is that the sea level will rise by an additional 19 inches (48 cm) by 2100, with an uncertainty range of 5 to 37 inches (13–95 cm).

18.6 TYPES OF EROSION

There are two main agents of erosion, namely, water and wind. Based on the mechanism, water erosion can be by sheet, rill, or gully. Wind erosion mechanisms are classified as creep, saltation, and suspension.

18.6.1 Water Erosion

Sheet erosion is the removal of soil in thin but uniform layers, mainly with surface runoff. Water flows on the surface uniformly like a sheet. Sheet erosion occurs when rainfall intensity is more than the infiltration rate of the soil. Sheet erosion takes place extensively in coastal areas, beaches, floodplains, and hillslopes. Erosion can start with the impact of raindrops on the soil surface, also known as "splash erosion," that detaches the soil particle to be removed with runoff.

Rill erosion occurs when runoff water creates small channels on the soil surface with water moving down the slope. Rill erosion causes water to cut through the soil because runoff shear stress is greater than a soil's shear strength. The rills can be up to 30 cm deep. Rill erosion removes fertile topsoil and soil organic matter, exposes root systems, and adversely affects root development. If erosion is not controlled, these channels become deeper and the erosion is known as gully erosion. Gullies are formed when water flows through these narrow channels, cutting the soil laterally and deeper. These gullies cannot be repaired simply by tillage.

18.6.2 Wind Erosion

Creep is the process of soil particle movement while in contact with the surface. Motion involves rolling and sliding, and predominantly involves coarse to medium grains (sand), and occurs during lower wind speeds. Particle sizes are usually larger than 500 microns in diameter.

Saltation is the bouncing of soil particles often initiated by ballistic impact. The particle becomes airborne and attains a height of up to 25 cm. Particle sizes in the process of saltation range from 70 to 500 microns in diameter. Short jump lengths and low energy processes resulting in a slow movement between creep and saltation modes is also known as "reptation." Particles smaller than 70 microns in diameter can stay airborne for a longer duration and the process is known as "suspension."

Particles smaller than 20 microns in diameter can stay in suspension for a very long time and travel long distances.

18.7 EFFECTS OF VEGETATION ON WIND EROSION

The roots of vegetation, such as grass, bind soil particles together to resist erosion. Vegetation can also help lessen or deflect wind, intercept windborne sediment, and keep soils moist, making them less susceptible to wind-related erosion. Crop or plant residue dissipates the energy available to cause erosion. Vegetation absorbs the energy of falling raindrops, lessens detachment, reduces sealing of pores in the soil surface, and promotes good water infiltration.

18.8 EROSION CONTROL

For controlling water and wind erosion, several similar measures are adopted. The common measures include keeping the soil covered, making breaks to slow the speed of runoff water or wind, conservation or no-tillage, and strip cropping. Kinetic energy is dissipated if soil is covered. Wind erosion can also be controlled by keeping the soil moist or applying soil binders. Contour farming, leaving crop residue on the surface, applying mulch, and planting trees along the edge of the farm are other measures to control water and wind erosion.

18.9 MAJOR FACTORS AFFECTING EROSION

Four factors that affect erosion are climate, soils, topography, and land use.

Climate. The most important climatic variable is rainfall erosivity, which is related to rainfall amount (how much it rains) and rainfall intensity (how hard it rains). Temperature is another important climate variable because temperature and precipitation together determine the longevity of biological materials like crop residue and applied mulch used to control erosion.

Soils. Soils vary in their inherent erodibility as measured in a standard test involving a unit plot. A unit plot is 72.6 ft (22.1 m) long and 3.3 ft (1 m) wide on a 9% slope and is maintained in continuous tilled fallow (no vegetation) using periodic tillage up and down slope to leave a "seedbed-like" soil condition. The USDA NRCS has assigned soil erodibility values for most cropland and similar soils across the US. The soil erodibility index (EI) is the measure selected to determine whether a soil map unit is highly erodible.

Topography. Slope length, steepness, and shape are the topographic characteristics that most affect rill and interrill erosion. Site-specific values are required for these variables to determine erosion.

Land use. Land use is the single most important factor affecting rill and interrill erosion because type of land use and land use condition are features that can be most easily changed to reduce excessive erosion.

18.10 EROSION MODELING

Modeling provides a quantitative estimate of soil erosion across different land uses, soils, topography, and rainfall conditions, and it can provide a broad guideline for choosing effective measures to control erosion. Over the past several decades, several empirical and physically based erosion prediction models have been developed. Some of the water erosion prediction models include the Universal Soil Loss Equation (USLE), USLE-2D, Modified USLE, Revised USLE (RUSLE), RUSLE 2, Agricultural Non-Point Source Pollution Model (AGNPS), Environmental Policy Integrated Climate (EPIC), Soil Erosion Risk Assessment in Europe (SERAE), Soil and Water Assessment Tool (SWAT), and Water Erosion Prediction Project (WEPP). In this chapter, we will only describe some of the models; for an overview of several of these models, readers are referred to Shukla (2011).

18.10.1 Universal Soil Loss Equation (USLE)

The USLE was developed to predict the long-term average rate of erosion on a watershed using rainfall, soil type, topography, and land use information. USLE predicts the total soil erosion amount due to sheet and rill erosion on a single slope. It does not take into account the erosion from gullies. There are five factors associated with USLE (note: L and S are combined) and each factor is assigned a numerical value. Since the values of each factor can have a large temporal variability, erosion predicted by the USLE is a long-term average.

$$A = R\ K\ LS\ C\ P$$

A is the erosion loss (tons/acre; or tons/acre × 2.24 = tons/ha). R is the rainfall and runoff factor, and is the product of the kinetic energy. The greater the rainfall intensity and duration, the higher the soil erosion. K is the soil erodibility factor and is the average soil loss in tons/ha (tons/acre) for a soil. It is estimated from soil texture, soil organic matter, soil structure, and soil permeability. L is the length of slope (less important than S and other factors), and S is the slope gradient. Usually L and S are combined to give the length and steepness of the slope as a ratio to a reference condition. The steeper and longer the slope, the higher the risk for erosion. C is the cover and management factor. It includes type and density of crop covers, stage of crop growth, and management of crop. It determines the relative effectiveness of soil and crop management systems to preventing soil loss. P is the practices for erosion control (tillage contour, strip cropping, terracing, etc.). Five factors are determined using the USDA NRCS database (for example, https://efotg.sc.egov.usda.gov/references/public/IA/Universal_Soil_Loss_Equation1.pdf) and then multiplied to obtain total soil loss per unit area.

Example 18.1

Consider a watershed shown below. Part of the watershed is in woods and the rest of the area is part of a small arms-training area. The watershed is divided into three areas: A, B, and C.

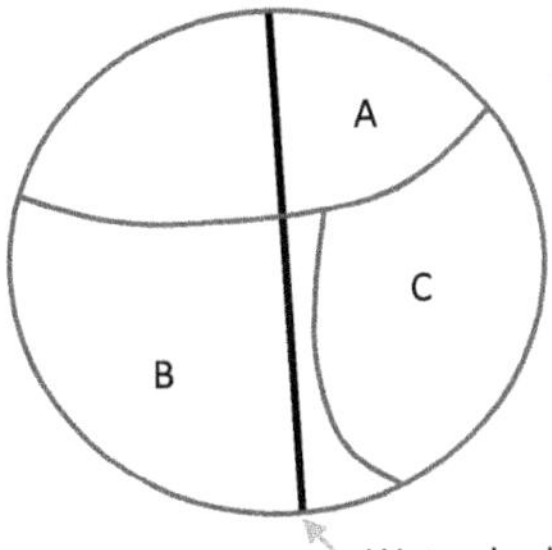

A: Mature forest with underbrush and undisturbed litter. Area 13 acres, slope length 540 feet, slope steepness 8%, loamy sand soil.
B: Open ranch with 40% surface coverage, grassy vegetation. Area 25 acres, slope length 740 feet, slope steepness 8%, sandy loam soil.
C: Open ranch with 40% surface coverage, grassy vegetation. Area 2 acres, slope length 610 feet, slope steepness 8%, loamy coarse sand.

Calculate the total erosion for each of the subwatersheds.
Use the USDA NRCS tables to obtain the values of each factor.

USLE Factors	**A**	**B**	**C**
Rainfall and runoff erosivity factor, R	350	350	350
Soil erodibility factor, K	0.10	0.24	0.15
Topographic factor, LS	2.3	2.6	2.4
Cover management factor, C	0.003	0.10	0.10
Conservation practice factor, P	1.0	1.0	1.0
Average annual erosion, A (ton/acre/year)	0.24	21.8	12.6
Area of section, acres	13	25	12
Total gross erosion, tons	3	545	151

18.10.2 Revised Soil Loss Equation, Version 2 (RUSLE2)

RUSLE2 is the computerized version of USLE, but it estimates daily soil loss from sheet and rill erosion caused by rainfall and runoff. RUSLE2 uses six factors for climatic erosivity, soil erodibility, slope length, slope steepness, cover management, and the supporting practices factor for soil loss estimation. The functional form of the equation is the same as the USLE; however, the values of each factor are daily values. The key element in RUSLE2 equation is the product of RK, which produces a daily sediment production estimate for unit-plot conditions. RUSLE2 uses a subfactor approach to calculate C (cover and management), and also gives more credit for surface residue and residue partially incorporated near the surface for slowing erosion. The equation is applied to a much broader range of conditions including forest roads, reclaimed sites, and construction sites. Details of the RUSLE2 can be referred elsewhere on the NRCS website.

18.10.3 Watershed Erosion Prediction Project (WEPP)

The WEPP model development was done by the USDA Agriculture Research Service (ARS) National Soil Erosion Research Laboratory (NSERL) in West Lafayette, Indiana (Flanagan et al., 2007). The WEPP is a physically based soil erosion prediction model that combines hydrology, soils, plants, and erosion mechanisms to predict sheet and rill erosion from hillslopes and small watersheds. WEPP does not account for gully erosion or mass wasting. WEPP has several customized interfaces for easy use with rangelands and forests. Further information about the model can be assessed at the USDA website (https://www.fs.usda.gov/ccrc/tool/watershed-erosion-prediction-project-wepp).

18.10.4 Soil and Water Assessment Tool (SWAT)

The SWAT model also predicts soil erosion from small watersheds to basin scale and is used widely for soil erosion prevention (Arnold et al., 1998). SWAT was developed at the USDA ARS Grassland, Soil and Water Research Laboratory in Temple, Texas (Arnold et al., 1998). The area of the basin is divided into subbasins, which are further divided into hydrologic response units (Deb and Shukla, 2011). SWAT utilizes the Modified USLE for water erosion prediction. The SWAT model was developed over time and it can simulate the hydrology of several watersheds simultaneously. SWAT is a comprehensive model and further details can be read at https://swat.tamu.edu/.

18.10.5 Wind Erosion Model

Soil loss due to wind erosion can be calculated using the following equation:

$$E = I\,C\,K\,L\,V$$

E is the erosion loss (tons/acre; or tons/acre × 2.24 = tons/ha), I is the soil erodibility index based on texture and aggregation (0 for stony to over 300 for nonaggregated sands), C is the climate based on wind speed and effective soil moisture, K is the surface roughness (1.0 for smooth to 0.5 for rough surfaces), L is the width of field (0 for small protected areas to 1 for wide open areas; usually 0.8–1.0), and V is the vegetative cover (amount and type of cover left in the field). The complexity of the tables and determination of values of L and V have resulted in the development of computer models like that developed by the USDA ARS Wind Erosion Prediction System (WEPS). WEPS is a process-based, daily time-step model that can predict soil erosion by simulating the fundamental processes controlling wind erosion. For more information about the WEPS model, please refer to https://data.nal.usda.gov/dataset/wind-erosion-prediction-system-weps.

18.11 EROSION AND CONTAMINATION

Soil erosion removes fertile soil from the area. However, along with the soil, several contaminants are also removed from the soil. These contaminants could be sorbed in the soil or dissolved in the runoff water. These move to the rivers and other surface water sources, and decrease the carrying capacity of these waterbodies, as well as influence the water quality and adversely affect aquatic, wild, and other lives supported through those water sources. The soil contaminants include fertilizers, pesticides, and herbicides from agricultural areas. However, heavy metals and carcinogens could also be released when erosion is taking place from industrial areas, landfills, or waste disposal sites. These could cause severe environmental problems for the community as well as require remediation and removal of these contaminants from the contaminated areas.

18.12 CONTAMINANTS AND REMEDIATION

Any physical, chemical, microbial, biological, or radiological material added to or present in soil or water can be a contaminant. Fertilizers applied to soil as nutrients can also be called contaminants when they leach into groundwater. Common examples of contaminants are heavy metals, salt, bacteria, pesticides, nitrate, antibiotics, bleach, and other organic material.

18.13 SOILBORNE CONTAMINANTS

Soil contaminants can be substances that exceed naturally occurring levels and pose risks to human health. However, in urban areas, contamination is predominantly caused by human activities. The most common soil contaminants are petroleum-based. Hydrocarbons from diesel fuel and gasoline are widespread problems, as are polyaromatic hydrocarbons (PAHs). Common contaminants in soil are heavy metals, radon, agrochemicals (pesticides, herbicides, fertilizers), lead, arsenate, asbestos, and other metals. Contaminated soils can leach toxic elements into groundwater or as runoff reaches surface water (ponds, reservoirs, rivers) sources.

There are many different ways people can be exposed to soilborne contaminants. It could be due to ingesting the contaminated soil, breathing airborne dust, absorption through the skin, or eating food grown in contaminated soils. There are two types of contaminated sites: a brownfield and a superfund site. A brownfield is an area of land that was used for commercial or industrial uses and then abandoned. Some brownfields are actually perfectly usable and have not been targeted for redevelopment, but a majority of them have some form of contamination that needs attention before the land can be used. Managing brownfields is complex because of liability issues. But by developing brownfields, people are ensuring that land is fully utilized, rather than spreading out and utilizing previously undisturbed land that might be providing habitat for animals or be an environmental buffer.

A superfund site is an abandoned site that is known to be polluted and requires long-term cleanup prior to development. In the US, there are about 14 million acres of abandoned and classified former industrial sites and dumps. The conditions at different sites can vary widely. Sites can range from individual abandoned factories with little or no hazardous materials involved to part of a toxic stew that blights an entire region.

18.14 CONTAMINANT REMEDIATION

There are two distinct classes of soil remediation. They are in situ or on-site, and ex situ or off-side. On-site cleanups are often preferred because they are cheaper. On the other hand, excavating a contaminated area and transporting the contaminated soil to a remote site before cleaning it can often be more complete. There are two main types of contaminant remediation. One of them is physical and chemical remediation, while the other is bioremediation.

18.14.1 Physical and Chemical Remediation

Physical remediation utilizes the physical properties of contaminant for remediation. In the chemical remediation method, various chemical reagents are used to alter the properties of a contaminant. The purpose of these methods is to make the contaminant immobile or to chemically convert it to a less harmful form.

Sealing. A contaminant can be sealed by pouring concrete over a small area. This method is frequently used because it ensures that the contaminants are no longer exposed to the environment. However, it does not remove contaminants from the area. This method is used for brownfield reclamation.

Physical barriers. Physical barriers are installed to prevent the movement of contaminants. This allows time for the contaminant to undergo natural decay. For example, if the half-life of an organic contaminant is 6 months, then the concentration of the contaminant will be half at the end of the 6 months. One-fourth of the contaminant will remain at the end of the year due to the natural process of degradation. However, this method does not immediately remove contaminants from the area. Some simple examples are placing a plastic cover between the clean soil and the contaminated soil, or covering the contaminated soil with plastic sheets to prevent any precipitation to enter the soil and winds to blow the soil away.

Vacuuming. Contaminants from the surface can be removed via vacuuming. These can include plastics, metals, dust, and other contaminated materials. This process physically removes the contaminants from the area.

Dig and bury. In the dig and bury method, contaminated soil is excavated from the area and is then removed to a location and buried in the ground. This method is workable when contamination has occurred in a small area, otherwise costs associated with digging and transport could make this method very expensive. This method physically removes a contaminant from the area. However, it only stores the contaminants at another location, does not treat or remove them, and is not a permanent solution. Burial grounds for contaminants have expanded over the years, primarily to accommodate nuclear waste from the Rocky Flats, Colorado, factory that made plutonium triggers for nuclear weapons. The 88-acre plot at the Idaho National Laboratory now includes 16 pits and 54 trenches. All of them are unlined and about 600 feet (~200 m) above the Snake River Aquifer, North America's second-largest unified aquifer.

Example 18.2

A contaminant spill took place in California. It is estimated that over many years, approximately 60 tons (1 ton = 2,000 pounds) of the contaminant were deposited in a major city over a 1.2-mile stretch of freeway median that is 0.5 miles wide. Since the contaminant does not readily leach, the city assumes all of it is in the top 1 inch of soil and that it is evenly distributed over the median in this 1.2-mile stretch. The city is considering scraping the top 1 inch off the median of the entire 1.2-mile stretch and trucking it to a nearby remediation facility. (We are not considering a vapor form of the contaminant.)

- The bulk density of soil is D_b = 1.20 g/cm^3. Soil water content of the contaminated soil to be trucked is 0.20 g/g.
- Each truck holds a volume of 14.0 yd^3 of soil, and the indigenous soil had contaminant concentration of 0.15 mg/kg soil.

Answer the following questions:

a. Covert bulk density in g/cm^3 to kg/m^3 and soil water content in g/g to soil water content in cm^3/cm^3? (Density of water can be assumed to be 1 g/cm^3.)
b. How many lb of contaminant fell on each yd^2? How many kg of contaminant fell on each m^2 of soil?
c. How many m^3 and yd^3 will be removed? How many trucks will it take to haul away the contaminated soil?
d. What is the wet weight and dry weight of the contaminated soil in each truck in kg and lb?
e. How many kg of contaminant is in the top 1 inch of soil? (Hint: You know the soil weight for a given volume of soil, now figure the weight of lb added.)

Solutions

a.

Bulk density = 1.20 g/cm^3 = 1.20 g × 1 kg/1000 g × (1/cm^3) × (100 × 100 × 100 cm^3)/1 m^3
= 1.20 /1000 × 100 × 100 × 100 kg/m^3 = 1200 kg/m^3
Soil water content = 0.20 (g/g) × BD/Dw = 0.20 × 1.20 (g/cm^3)/1 (g/cm^3) = 0.24 (cm^3/cm^3)

b.

Area = 1.2 × 0.5 mile^2 = 0.6 mile^2 × (1760^2) yd^2/1 mile^2 = 1858560 yd^2
Area = 1858560 yd^2 × (0.9144)^2 m^2/yd^2 = 1539074 m^2
How many lb of contaminant fell on each yd^2? = 60 ton/1858560 yd^2
= 60 ton × 2000 lb/ton/1858560 yd^2 = 0.07 lb/yd^2
How many kg of contaminant fell on each m^2 of soil? = 60 ton/1539074 m^2
= 60 ton × 1000 kg/ton/1539074 m^2 = 0.04 kg/m^2

c.

How many m^3 and yd^3 will be removed? (1 m = 1.0936 yd)
1539074 m^2 × 1 inch = 1539074 m^2 × 1 inch × 0.0254 m/inch
= 39093 m^3 or (39093 × 1.3 = 50821 yd^3)
How many trucks does it take to haul away the contaminated soil?
50821 yd^3/14 yd^3 = 3631

d. What is the wet weight and dry weight of the contaminated soil in each truck in kg and lb?
Dry weight of soil = Bulk density × Volume of soil = 1200 kg/m^3
× 39093 = 46911600 kg
Dry weight of soil = 46911600 kg x 2.205 lbs/kg = 103422374.59 lbs
Wet weight of soil = Dry wt × (1 + w) = 46911600 kg × (1 + 0.2) = 56293920 kg
Wet weight of soil = 56293920 kg x 2.205 lbs/kg = 124106849.5 lbs

e. How many kg of contaminant is in the top 1 inch of soil?

0.15 mg/kg = 0.15 mg/kg × (1 kg/1000000 mg) × 46911600 kg = 7.04 kg

Leaching. Some contaminants can be leached to deeper soil depths or to the shallow groundwater tables. Alternately, these contaminants can move into the soil and are collected by surface or subsurface (tile) drains usually on the edge of a field. The collected water can then be reclaimed using appropriate methods. Leaching is a common practice to remove salts from the root zone in arid and semi arid regions of the world.

Hydraulic barriers. A hydraulic barrier isolates the contaminated area and protects the adjoining area from contaminant exposure. Some examples of hydraulic barriers are drains (subsurface or tile drains), and injection and capture wells. Aquifer remediation wells (ARWs) are widely used in the US for benefits associated with the control of groundwater contamination. These wells can be used for different specific purposes, for example, for introducing remediation agents (i.e., chemicals or microorganisms) into contaminated aquifers to neutralize the contamination. These wells can also be used to increase groundwater flow through the contaminant zone in an aquifer to aid in contaminant removal, form hydraulic barriers to contain contaminant plumes; and reinject treated groundwater for aquifer recharge after an on-site pump-and-treat system.

Solidification. The solidification method turns the contaminated soil into a solid mass. Contaminant remains in the soil but is immobile. Various mixtures of materials, including cement, can be mixed and pumped into the soil for immobilizing the contaminated soil. This method does not work when contamination is deep into the soil nor for organic contaminants, unless extra steps are taken and gases are trapped. Solidification can be achieved by in situ thermal vitrification that melts soil into glass and permanently traps waste. The process uses large amounts of electricity to raise temperatures up to 2000°C, and contaminants, for example, radionuclides and heavy metals, are retained within the vitrified product. The off-gases emitted from the molten material need to be correctly treated, cooled, scrubbed, and filtered to remove any released materials before discharge.

Incineration. When it is too expensive or too difficult to haul away contaminated soil from the area, the contaminants are burned in situ. Generally, the technique is used for organics spread within small areas. This method removes contaminants but residues are left behind.

Electroosmosis. Electroosmosis is the movement of water, and whatever is contained in the water, through porous media by applying a direct current (DC) field. When DC is applied to the soil volume, pore water and most contaminants move from the anode (the positive electrode) to the cathode (the negative electrode). Under the influence of a DC field, the rows of cations on the soil particle surface start sliding along toward the cathode by electrical attraction. This ion movement, called electromigration, is very important when treating soils contaminated with metals, nitrates, sulfates, or other inorganic compounds. Remediation of clay and silt-laden soils is generally a big challenge in the soil remediation arena, and electroosmosis is a useful technique because, unlike hydraulic conductivity, the electroosmotic flow rate is not sensitive to pore size. The boundary layer movement toward the cathode provides the motive force for the bulk pore water; the size of the pore is not important. Use of electroosmosis for removal of organics reduces the need for specialized electrodes. The second major benefit is the electricity applied to the soil directly results in heating of the soil. The soil warming only increases the mobilization of volatile organics.

Stabilization. Stabilization is the chemical change in the contaminant that makes it less mobile, toxic, or soluble. Stabilization involves binding reagents that are mixed with the contaminated soil. The stabilization and solidification processes are also known as encapsulation. These two processes are sometimes done prior to landfilling.

Pump and treat. Pump and treat is a common method of groundwater contamination remediation. In this method, groundwater is pumped and kept in a holding tank before undergoing treatment. The pump is located downstream of the contaminated plume of groundwater. The treated water can be put back in the aquifer or used for suitable purposes. A series of these types of wells can also be used to inject chemicals for remediation and subsequent removal.

Venting. Venting is particularly used for remediating organic contaminants. Venting can be active in which air is pumped into and out of the contaminated area. It can also be passive in that case vents stay open. The fumes are released into the atmosphere if the gaseous form of contaminants is not toxic or hazardous. Otherwise, gases must be trapped for further treatment prior to release.

18.14.2 Bioremediation

Bioremediation is the removal or cleanup of pollutants through biological processes. It usually involves the degradation of an organic pollutant to basic and "natural" constituents. It can also involve the transformation, mobilization, or immobilization of organic and inorganic pollutants. The method utilizes the ability of microorganisms to carry out some type of biological transformation of a problem pollutant, usually in soil, sediments, or water.

There are several advantages to bioremediation. Some of these are the method is less expensive, causes less disturbance, and instead of moving or sequestering the contaminant into another environments, the method removes the contaminant. The disadvantages of the method are that it is mainly useful only for organics, takes a long time for remediation to complete, and relies on the uncertainty of biology. Similar to physical and chemical remediation, there are two types of bioremediation: in situ and ex situ. Examples of in situ bioremediation are treating an oil, pesticide, or solvent spill; and treatment of nitrate contamination in groundwater or high biological oxygen demand (BOD) waters. BOD is the amount of dissolved oxygen required by aerobic microbes to break down organic contaminants in water. Examples of ex situ treatment include pump and treat, and dig and treat.

Strategies of bioremediation involve knowledge of the waste types, microorganisms, and pollutant characteristics. Most organics are handled by degradation or "mineralization" of pollutants down to natural products (CO_2, water, etc.). If the organic is degradable, it may be "transformed" to a nontoxic form or "immobilized" to prevent its movement. Some inorganics, for example, nitrate, can be transformed and removed from the system by denitrification. Some inorganics may be mineralized to make them soluble and easily removed, or immobilized so that they remain in place. For example, Cr^{+6}, which is toxic, poorly sorbed, and mobile, can be reduced to Cr^{+3}, which is less toxic, strongly sorbed, and immobile.

To increase bioremediation efficiency, the environment needs to be more desirable for degradation and encourage microbial growth. Temperature, moisture, aeration, pH, microbial types, nutrients, etc. are required in certain amounts for optimum biological activity. Otherwise, organisms are present to carry out the job, albeit in low numbers, and the environment is limiting their activity. Bioremediation of groundwater entails adding nutrients (organic and inorganic) and electron acceptors (O_2 or alternatives).

A pollutant can have a variety of metabolic roles: carbon source, electron source (energy), electron acceptor, source of inorganic nutrients (nitrogen and phosphorus). Microorganisms use the pollutant as a carbon and energy source; problems occur if the pollutant concentration is too low (not enough energy present) or too high (becomes a toxicant). The addition of other carbon and energy sources encourages degradation; and increases population, diversity, and cometabolism (simultaneous degradation of two compounds).

Microbes can use the pollutant as an alternative electron acceptor; the change of the pollutant to a different form may make it nonmobile and nonhazardous. For example, changing uranium peroxide (or chromate) to uranium oxide (or chromium oxide), that is, UO_4^{-2} to UO_2 or CrO_4^{-2} to Cr_2O_3.

Alternately, a contaminant can be made more mobile so that it can leave the system, for example, transforming selenite into methyl selenium or nitrate into nitrogen gas. Organic matter stimulates microbial activity that encourages degradation of the pollutant. Incorporation of green manures into soil is a good example of enhancing microbial activity.

Some of the limitations to bioremediation are related to the physical environment. A groundwater temperature between 11°C and 15°C, good aeration, and the presence of electron acceptors are good for microbial activity. Inorganic nutrients are added (nitrogen, phosphorus, micronutrients, etc.) to enhance microbial activities, particularly for oil/fuel spills, and pH is kept near 6.5. However, competition between microorganisms for pollutants or added nutrients/electron acceptors may slow degradation. Competition between indigenous and introduced microbes limits the effectiveness of the introduced microorganisms. With a mixture of contaminants, if some portions of the mixture are toxic, microbial growth may not occur.

18.15 INORGANIC WASTE DISPOSAL

Inorganic wastes are the chemical substances from minerals. Some common examples are sand, silt, silica, calcium, magnesium, iron, sodium, chloride, glass, and aluminum cans. Community recycling centers help dispose of most of them. However, due care must be taken because some inorganics usually accompany organics.

18.16 LAND FARMING

Land farming refers to controlled application of wastes to soil whereby soil becomes the ultimate site of disposal; it is both a treatment and disposal method. The objectives of land farming are to decompose organic waste and immobilize it. Soil is a large composting system utilizing basic biology, chemistry, and physics in a semi-controlled environment for contaminant remediation. There are several advantages to land farming. It is benign to the environment, uses natural processes, eliminates wastes (not sequesters them), uses less energy, and permits reuse of the site. The disadvantages of land farming include that the method is not appropriate for inorganics and certain organics; requires good management, monitoring, and record keeping; and can be easily misused.

18.17 PLANNING A DISPOSAL SITE

For planning a disposal site, prior knowledge of site characteristics, waste characteristics, monitoring, reuse of land requirements, and governing laws are required.

Site characteristics. The site should preferably be close to the contaminant sources or contaminated area but away from residential areas, such that the possibility of trespassing as well as contaminant-laden dust emissions or runoff do not reach people. The site should be level or with little slope to prevent runoff. Soil at the disposal site should have moderate soil texture and pH, and the water table at the site should be deep.

Waste characteristics. These are regulated by various agencies including the Environmental Protection Agency (EPA). Degradability of waste should be known. Wastes containing high levels of metals, nondegradable or slowly degradable organics (PCBs), radionucleotides, and very volatile substances are not suitable for land farming.

Monitoring. The site should be continuously monitored. Periodic samples should be collected at the surface soil, in the unsaturated zone, and from the water table to determine contaminant concentrations. Air and dust samples should also be collected to make sure contaminants are not becoming airborne.

Reuse of land. If the land was managed and monitored correctly, it can be used for many different purposes, with the exception of growing direct food chain crops, such as vegetables.

Governing laws. The EPA is required by the Resource Conservation and Recovery Act (RCRA) to issue standards. Because of the concern for landfill problems, land farming regulations (which were established with landfill regulations) are much more restrictive than they need to be and actually discourage land farming. Other requirements are mandatory public hearings and registration with the EPA. There are certain restrictions about earthen perimeters and liners, 24-hour security, and the company must demonstrate that the waste will be degraded.

Example 18.3

The total area of a golf course is 70 acres. During the summer, water was applied at the rate of 2 acre-inch per acre once every week for 12 weeks. During spring and fall, water was applied at the rate of 1 acre-inch per acre once every week for 16 weeks. The water contains 25 mg/L of NO_3–N.

a. How many hectare-meters and cubic meters are applied per year to the fairways and roughs of the golf course (28 weeks)?
b. How many kg of NO_3–N are applied with the irrigation water/year to the fairway and rough?
c. Did the rate of N application (plant available) exceed the recommended rate of fertilization? The recommended rate was 220 lb/acre.

Solutions

a. $(2 \times 12 \times 70) + (1 \times 16 \times 70) = 2800$ acre-inch
$= 2800$ acre $\times$ (1 ha/2.25 acre) $\times$ inch $\times$ (1 m/39.37 inch)
$= 28.62$ ha-m $= 28.62$ ha $\times$ (10000 m^2/ha) m $= 286200$ m^3

b. 25 mg/L $\times$ 286200 m^3 $\times$ 1000L/m^3 $= 7154$ kg
c. 220 lb/acre $= 220$ lb $\times$ (1 kg/2.204 lb) $\times$ 70 $= 6987$ kg.

The total NO_3-N applied was 7154 kg that was higher than recommended. So yes, the applied fertilizer exceeded the recommended rate.

18.18 SOIL SALINITY AND SODICITY

Soil salinity generally denotes amounts of dissolved salts or ions in soil solution, also known as aqueous phase. The process of increases in salt content of the soil is known as soil salinization. Salts occur naturally within soils, surface water and groundwater, and in oceans. They can also be incorporated in the soil as anthropogenic inputs of fertilizer application, irrigation, seawater intrusion, and salt applications to deice roads. The common ions causing soil salinization include sodium, potassium, calcium, magnesium, and chloride. The soil salinity is represented as the electrical conductivity (EC) of the saturated soil paste extract. The EC is generally expressed as dS/m (old popular unit is mmhos/cm) and the parts per million (ppm) salts is calculated using meq/l multiplied by the equivalent weight of each ion. Generally, the following equations are used to convert EC to TDS (total dissolved solids) and ppm salt:

$$\text{TDS (mg/l)} = 640 \times \text{EC (dS/m)}, \text{ for water assuming a typical mixture of salts and in the range of 0.1–5 dS/m}$$

$$\text{TDS (mg/l)} = 800 \times \text{EC (dS/m)}, \text{ for water assuming a typical mixture of salts and in the range of >5 dS/m}$$

Soils in arid regions usually are classified as alkaline soils and essentially contain Na, K, Ca, Mg, Cl, SO4, and HCO_3. About 95% of the alkalinity is from HCO_3.

When a large proportion of sodium ion as opposed to other ions is present in the soil, then that soil is known as sodic soil. The process of increases in soil sodium content is known as sodicity. Sodium is strongly bound to clay particles with negative charge itself or by displacing other cations. Sodium weakens the bond between soil particles and causes swelling and dispersion with attendant loss of soil structure. Soil sodicity is expressed as the sodium adsorption ratio (SAR) and exchangeable sodium percentage (ESP). The individual cations, including Na, Ca, K, and Mg, are measured, and SAR and ESP are measured as follows:

$$SAR = \frac{\left[Na^{+}\right]}{0.5*\left[Ca^{2+} + Mg^{2+}\right]}$$

$$ESP = \frac{\text{Exchangeable } Na}{\text{Cation exchange capacity}}$$

The concentration of ions is expressed in meq/l or cmol_c. Exchangeable Na and cation exchange capacity (CEC) are in cmol_c/kg. Determination of SAR is easier than the determination of CEC and exchangeable cations.

18.19 ENVIRONMENTAL CONSEQUENCES AND CAUSES

Soil salinity and sodicity have important environmental consequences for both agriculture and nonagriculture sectors. As far as agriculture is concerned, soil salinization or sodification causes yield reduction. This can lead to changes in the cropping pattern of an area. However, as soil salinity or sodicity further increases, it can cause the complete failure of crops and abandonment of the land. The shallow groundwater also gets increasingly salty and can no longer be used as a source of drinking water. Other consequences are that soil salinization increases desertification. The drainage as well as runoff water causes downstream water contamination, which is detrimental to aquatic life, wildlife, and other natural waterbodies. In ancient times, many civilizations collapsed due to soil salinity or sodicity, including Mesopotamia, the kingdoms of Sumer and Akkad, and the Indus valley civilization in Pakistan.

Generally, soil salinization is naturally controlled by the weathering and leaching of salts due to sufficient precipitation. Evaporation does bring salty water to the soil surface and can cause salt deposition on the surface or near surface after water is evaporated. This is a problem in arid and semiarid regions, where sufficient water is not available to push the salts into deeper soil depths. Some other causes of soil salinization include faulty or poor drainage systems, a shallow water table, the low hydraulic conductivity of soil, irrigation with low-quality water, and poor irrigation management.

18.20 SALT EFFECTS ON PLANTS

Plants have evolved mechanisms to sense abiotic stresses associated with drought and salinity (salt stress) via a complex physiological and biometabolic processes (Cao et al., 2023). These mechanisms protect plant at cell, tissue, organ, and the whole plant levels. Salts effect plants in two ways: one of them is due to osmotic effects and the other is specific ion effects. Salty soils have lower osmotic potential, which normally does not affect total soil water potential in nonsaline soils. Reduced osmotic potential decreases the water availability to plants, thus plants appear to suffer from drought. By the time visual symptoms occur, yield depression has already happened. Plant leaves are smaller with a darker shade of green or even blue-green, slightly thickened and waxy, and could have tip or marginal leaf burn. A sand and a clay could have the same soluble-salt content expressed as percent dry-weight basis, but the soil-solution concentration when near the wilting point (%) could be ten times higher for sand than clay soil.

Plant response to combined abiotic stresses (water and salt) cannot be adequately predicted by studying one stress at a time. Although in the beginning both water and salt stresses cause physiological water deficit in plants, however, prolonged stresses affect growth and production (Chaves et al., 2009). Plants are affected differently by different ions including Cl, SO_4, HCO_3, Na, Ca, Mg, Li, and B. Direct toxicity is caused by chloride and sodium, while indirect toxicity is due to nutrient imbalances (Ca or Mg replaced by Na). Indirect toxicity can also occur with destruction of soil properties. Plants absorbing salts directly through leaves through spray or sprinkler irrigation are much more susceptible to specific ion effects. Often plants do not translocate salts to leaves from roots but resist by adjusting (opening or closing) stomata. Under prolonged salt stress, plants experience hyperionic and hyperosmotic stresses due to excessive uptake of specific ions (Hussain et al., 2016; Cao et al., 2023). However, impact of salt stress depends on several factors including stress duration, intensity, recovery effectiveness, and plant species (Kapoor et al., 2020).

Generally, water stress decreases photosysnthetic rate of a plant due to the direct influence on photosysnthetic apparatus, decrease in the activity of photosysnthtic enzyme as well as other physiological and biochemical activities. These can be grouped into stomatal and non-stomatal limitation. Stomatal closure decreases water uptake but protects from salt stress. However, with prolonged stress stomatal limitation changes to non-stomatal limitation because photosysnthetic apparatus gets damaged (Grassi and Magnani,

2005). However, the effects of irrigation with saline waters as well as effects due to soil salinity and soil sodicity are more pronounced in glycophytes (salt sensitive plants) than halophytes (salt tolerant plants) (Bedirhanoglu et al., 2022; Yang et al., 2020; Ben Ali et al., 2020; 2022; Flores et al., 2016). Further information on draught stress, salt stress, and combined stresses in plants and their effect on plant growth and production, plant tolerance mechanisms, and strategies are given in Cao et al. (2023).

18.21 REMEDIATION AND RECLAMATION OF SALINE SOILS

There are three general steps in reclamation of soils: provide internal drainage (if necessary), exchange sodium, and leach. Generally, leaching (applying enough water to penetrate through the root zone of soil) soils with good quality water low in sodium is enough for reclaiming saline and sometimes sodic soils. The amount of water applied is more than the evapotranspiration requirement of the crop. In limited water (scarce irrigation and low precipitation) areas, a mulch can be placed on the surface to hasten leaching and slow evaporation. The salt content of the mulch must be low. The amount of water to apply depends on other factors; however, as a rule of thumb, about 30 cm of water is required to remove 70%–80% of the salt for each 30 cm depth of soil to be leached of salt. Unsaturated water flow through intermittent water additions are more efficient than continuous leaching, as the former requires about 70% of the water needed to complete leaching.

Sodic soils are usually dispersed and water flow through the soil (percolation) is slow. It is difficult to leach salts out by applying water, although sodic soils do have some drainage. A better option is to exchange Na with Ca (Mg, H, or other possibilities), and then leaching with good quality water or water that contains Ca, Mg, or H ions. Gypsum is often used as a source of Ca to precipitate exchangeable Na. Sulfuric acid can also be used, however, care must be taken to ensure that the water for leaching has cations in it to prevent dispersion. Initially, a gypsum requirement (GR) is obtained, which is the amount of gypsum required to reclaim a sodic soil. The bulk density of the soil and depth of soil to reclaim are needed to determine the amount of total soil mass. The ESP of the soil before reclamation, the target ESP of the soil after reclamation, and the CEC are needed to calculate the amount of Na^{+} to exchange with Ca^{+2}. Because gypsum is not pure, an efficiency of 75% is usually used. The disadvantage of using gypsum is that the process is very slow and it may take years to reclaim the soil. Tillage, including deep plowing, may improve soil drainage and improve leaching efficiency. The other option of reclamation is to grow crops such as barley that can tolerate poorly aerated soil conditions.

18.22 LEACHING REQUIREMENT

The leaching requirement (LR) is the amount of water needed to remove excess salts from soils, over and above consumptive use or the fraction of the water applied that leaches. If the LR is followed, the correct amount of water is applied to a crop that maintains a desirable soil salinity without using excess water for leaching. The water requirement (WR) of the crop becomes the consumptive use of the crop plus the LR. The correct use of the LR to determine the WR of a crop relies on data for the consumptive use of a crop, the salt tolerance of a crop, and the salinity levels of the soil (within or at the bottom of the root zone) and irrigation water. The LR thus indicates how much water is needed over and above consumptive use and is one tool in managing salinity for agriculture.

Simply put, the LR is the depth (D) of drainage water (dw) collected below the root zone divided by the depth of applied irrigation water (iw):

$$LR = \frac{D_{dw}}{D_{iw}}$$

Amount of salt in a volume of water = Water quantity × [Salt]. The goal is to maintain a salt balance, therefore:

$$D_{dw} * EC_{dw} = D_{iw} * EC_{iw}$$

$$LR = \frac{D_{dw}}{D_{iw}} = \frac{EC_{iw}}{EC_{dw}}$$

where EC_{iw} and EC_{dw} are the electrical conductivity of irrigation and drainage waters, respectively.

However, the EC_{dw} is hard to obtain; therefore, the USDA Salinity Laboratory has determined that a good estimate of the EC_{dw} is $(5 \times EC_{soil}) - EC_{iw}$, where EC_{soil} is the average EC of a soil saturated paste extract taken over the entire rooting depth. Thus, the LR can be easily determined by taking soil samples and irrigation water samples, and calculated as follows:

$$LR = \frac{EC_{iw}}{\left[5 * \left(EC_{soil}\right) - EC_{iw}\right]}$$

EC_{dw} or $5(EC_{soil}) - EC_{iw}$ is a chosen level of salinity and is crop dependent; it may be at 100% production or it may be <100% production; it could be 2 dS/m for pecans or 6 dS/m for cotton. The crop-specific values can be obtained from the USDA National Salinity Laboratory website or other resources.

Example 18.4

If the EC of irrigation water is 1.0 dS/m and saturated paste soil EC for desirable pecan yield is 2.0 dS/m, calculate LR. If 120 cm of water is needed to grow pecans annually, how much water must be applied to account for the LR?

$$LR = \frac{EC_{iw}}{\left[5 * \left(EC_{soil}\right) - EC_{iw}\right]}$$

$$LR = \frac{1}{\left[5 * (2) - 1\right]} = \frac{1}{9} = 0.11$$

Total water to be applied = 120 + (120 × 0.11)= 133.20 cm

Example 18.5

The saturated paste extract soil ECe is 4.9 dS/m and irrigation water EC is 0.7 dS/m. The field is going to be planted to peppers that requires 90 cm of water during the entire growth period. Using following Table, calculate the leaching requirement and total water needed for 75% and 90% yield.

Crop	Yield potential, ECe (Ayers and Westcot, 1985)				Max ECe
	100%	90%	75%	50%	
Barley	8.0	10.0	13.0	18.0	28
Cotton	7.7	9.6	13.0	17.0	27
lettuce	1.3	2.1	3.2	5.2	9
Pepper	1.5	2.2	3.3	5.1	9
Tomato	2.5	3.5	5.0	7.6	13
Onion	1.2	1.8	2.8	4.3	8

For the Table above, the desirable soil EC is 3.3 dS/m for 75% yield of pepper, and 2.2 dS/m for 90% yield. Use LR equation and procedure in example 18.5 to calculate total water needed for the desired yields.

For desirable yield of 75% at EC= 3.3 dS/m; LR = 0.044 and Total water needed = 93.99 cm

For desirable yield of 90% at EC= 2.2 dS/m; LR = 0.068 and Total water needed = 96.12 cm

For higher yields, more leaching needs to be done.

PROBLEMS

18.1 What are different types of erosion? How does vegetation affect total erosion loss?

18.2 As an environmental consultant, you come across a site where an arsenic spill has occurred. Write a detailed plan of action for the removal of arsenic from the site.

18.3 List some of the physically based soil erosion prediction models.

18.4 If in Example 18.1, the areas under A, B, and C are increased by 40%, 30%, and 25%, respectively, calculate total annual soil loss.

18.5 Using data in Example 18.2 as needed, if a contaminant is up to a depth of 2 inches and concentration is 0.4 mg/kg, calculate how many kg of contaminant are in the top 2 inches of soil?

18.6 Using data in Example 18.3 as needed, calculate how many kg of NO_3–N are applied with the irrigation water/year to the fairway and rough. The water contained 45 mg/L of NO_3–N.

18.7 Using data in Example 18.4 as needed, calculate the total water that must be applied if the irrigation requirement is 100 cm.

REFERENCES

Arnold, J. G., R. Srinivasan, R. S. Muttiah, and J. R. Williams. 1998. Large area hydrologic modeling and assessment, Part I: Model development. *J. Am. Water Resour. Assoc.* 34: 73–89.

Ayers, R. S, and D. W. Westcot. 1985. Water quality for agriculture. *FAO Irrigation and Drainage Paper 29*, Rev.1, Rome, Italy.

Bedirhanoglu V., H. Yang, and M. K. Shukla. 2022. Reducing water salinity at flowering stage decreases days to flowering and promotes plant growth and yield in chile pepper. *Hort Science 575*(9): 1128–1134.

Ben Ali A., M.K. Shukla, B. Schutte, and C. Gard. 2020. Irrigation with RO concentrate and brackish groundwater impacts pecan tree growth and physiology. *Ag Water management.* https://doi.org/10.1016/j.agwat.2020.106328.

Ben Ali A., M.K. Shukla, M. Marsalis and N. Khan. 2022. Irrigation with desalinated and raw produced waters: Effects on soil properties, and germination and growth of five forages. *Ag Water Management.* https://doi.org/10.1016/j.agwat.2022.107966.

Borrelli, P., D. A. Robinson, L. R. Fleischer, E. Lugato, C. Ballabio, C. Alewell, K. Meusburger, S. Modugno, B. Schütt, V. Ferro, V. Bagarello, K. Oost Van, L. Montanarella, and K. Panagos. 2017. An assessment of the global impact of 21st century land use change on soil erosion. *Nat. Commun.* 8: 1–13. https://doi.org/10.1038/s41467-017-02142-7

Cao, H., R. Ding, S. Kang, T. Du, L. Tong, Y. Zhang, J. Chen, and M. K. Shukla. 2023. Drought, salt, and combined stresses in plants: effects, tolerance mechanism, and strategies. *Advances in Agronomy.* https://doi.org/10.1016/bs.agron.2022.11.004.

Deb, S. and M. K. Shukla. 2011. An overview of some soil hydrological watersheds models. *@CAB International: Soil Hydrology: Land Use and Agriculture* (ed. M. Shukla), CABI, Massachusetts, USA pp. 75–116.

Flanagan, D. C., J. E. Gilley, and T. G. Franti. 2007. Water erosion prediction project: Development history, model capabilities, and future enhancements. *Trans. Am. Soc. Agric. Biol. Eng.* 50: 1603–1612.

Flores A., M.K. Shukla, D. Daniel, A. Ulery, B. Schutte, G. Pichionni and S. Fernald. 2016. Evapotranspiration changes with irrigation using saline groundwater and RO concentrate. *J. Arid Environments* 131: 35–45.

Grassi, G., Magnani, F., 2005. Stomatal, mesophyll conductance and biochemical limitations to photosynthesis as affected by drought and leaf ontogeny in ash and oak trees. *Plant Cell Environ.* 28: 834–849.

Hussain, M.I., Lyra, D.-A., Farooq, M., Nikoloudakis, N., Khalid, N., 2016. Salt and drought stresses in safflower: A review. *Agron. Sustain. Dev.* 36: 1–31.

Kapoor, D., Bhardwaj, S., Landi, M., Sharma, A., Ramakrishnan, M., Sharma, A., 2020. The impact of drought in plant metabolism: How to exploit tolerance mechanisms to increase crop production. *Appl. Sci.* 10: 5692.

Pimentel, D., C. Harvey, P. Resosudarmo, K. Sinclair, D. Kurz, M. McNair, S. Crist, L. Shpritz, L. Fitton, R. Saffouri, and R. Blair. 1995. Environmental and economic costs of soil erosion and conservation benefits, *Science* 267: 1117–1123. https://doi.org/10.1126/science.267.5201.1117

Shukla, M. 2011. *Soil Hydrology: Land Use and Agriculture. @CAB International: Soil Hydrology: Land Use and Agriculture* (ed. M. Shukla),CABI, Massachusetts, USA p. 434.

Xianbo Z., H. Yang, M. K. Shukla, and T. Du. 2023. Proposing a crop-water-salt production function based on plant response to stem water potential. *Ag Water Management.* 278–https://doi.org/10.1016/j.agwat.2023.108162.

Yang H., T. Du, X. Mao, and M.K. Shukla. 2020. Modeling tomato evapotranspiration and yield responses to salinity using different macroscopic reduction functions. *Vadose Zone Journal.* https://doi.org/10.1002/vzj2.20074.

Index

F

X

Z

For Product Safety Concerns and Information please contact our EU
representative GPSR@taylorandfrancis.com
Taylor & Francis Verlag GmbH, Kaufingerstraße 24, 80331 München, Germany

www.ingramcontent.com/pod-product-compliance
Ingram Content Group UK Ltd.
Pitfield, Milton Keynes, MK11 3LW, UK
UKHW022212190926
475702UK00003B/323
9780367210168